Creative Evolution

First published in French in 1907, Henri Bergson's *L'évolution créatrice* is a scintillating and radical work by one of the great French philosophers of the nineteenth and twentieth centuries. This outstanding new translation, the first for over a hundred years, brings one of Bergson's most important and ambitious works to a new generation of readers.

A sympathetic though critical reader of Darwin, Bergson argues in *Creative Evolution* against a mechanistic, reductionist view of evolution. For Bergson, all life emerges from a creative, shared impulse, which he famously terms *élan vital* and which passes like a current through different organisms and generations over time. Whilst this impulse remains as forms of life diverge and multiply, human life is characterized by a distinctive form of consciousness or intellect that is modeled upon how objects or parts of objects are juxtaposed in space. Yet as Bergson brilliantly shows, the intellect's fragmentary and action-oriented nature, which he likens to the cinematograph, means it alone cannot grasp nature's creativity and invention over time. A major task of *Creative Evolution* is to reconcile these two elements. For Bergson, the answer famously lies in intuition, which brings instinct and intellect together and takes us "into the very interior of life."

A work of great rigor and imaginative richness that contributed to Bergson winning the Nobel Prize for Literature in 1927, *Creative Evolution* played an important and controversial role in the trajectory of twentieth-century philosophy and continues to create significant discussion and debate. The philosopher and psychologist William James, who admired Bergson's work, was writing an introduction to the first English translation of the book before his death in 1910.

This new translation includes a foreword by Elizabeth Grosz and a helpful translator's introduction by Donald Landes. Also translated for the first time are additional notes, articles, reviews, and letters on the reception of *Creative Evolution* in biology, mathematics, and theology. This edition includes fascinating commentaries by philosophers Maurice Merleau-Ponty, Georges Canguilhem, and Gilles Deleuze.

Henri Bergson (1859–1941) was born in Paris, the year Darwin's *On the Origin of Species* was published. Initially drawn equally by the sciences and philosophy, at the age of eighteen Bergson won a prestigious prize for solving a mathematical problem. Choosing philosophy, he attended the École Normale Supérieure and the University of Paris before working as a school teacher in Angers and Clermont-Ferrand while completing his doctorate at the University of Paris in 1889. He worked for eight years at the Lycée Henri-IV before taking a position as Chair of Greek and Roman Philosophy at the Collège de France in Paris 1900. His weekly lectures soon attracted beyond-capacity crowds, and his visits abroad to England and the United States filled venues and reportedly caused the first-ever traffic jam on Broadway in New York City. Bergson engaged with some of the leading contemporary thinkers, including a famous debate with Einstein in 1922 over the nature of time. He influenced Marcel Proust, Thomas Mann, and the philosopher William James, and was a pioneering figure in the Modernist intellectual movement of the early twentieth century.

Henri Bergson

Creative Evolution

Translated by Donald A. Landes,
with a foreword by Elizabeth Grosz

 Routledge
Taylor & Francis Group

LONDON AND NEW YORK

Praise for this new edition

"This new translation by Donald Landes captures the mesmerizing work that turned Henri Bergson into one of the century's most provocative thinkers—with expert annotations, correspondence and additional material by influential thinkers from William James to Gilles Deleuze."
- Jimena Canales, University of Illinois, Urbana-Champaign, USA

"Henri Bergson, who personally oversaw the translation of all his books into English, would be delighted by this new edition of his greatest work. Donald Landes's translation is exquisite and the extensive editorial notes are indispensable for the serious study of *Creative Evolution*."
- Alexandre Lefebvre, University of Sydney, Australia

"This splendid new translation provides an exceptional, scholarly tool for serious specialists as well as all readers interested in Bergson's work. It will swiftly become the definitive reference text for all Anglophone Bergson scholarship."
- Christina Howells, University of Oxford, UK

"A major event in post-Kantian philosophy. Featuring a lucid introduction, helpful translator's notes and a judicious selection from Arnaud François's illuminating critical dossier, this fine translation of *Creative Evolution* means that English-language Bergson scholarship has begun to gain the serious editions of his texts that it deserves."
- Mark Sinclair, Roehampton University, UK

"This superb translation will introduce a new generation to Bergson. Landes's cogent introduction and editorial notes and the accompanying dossier of correspondence, reception and commentaries not only situate *Creative Evolution* in relation to Bergson's oeuvre, but also to the myriad scientific and philosophical sources informing his thought. An outstanding achievement."
- Mark Antliff, Duke University, USA

"This wonderful new translation of Bergson's classic *Creative Evolution* is warmly welcomed, as are the rich introduction, comprehensive editorial notes and thoughtful selection of commentaries. There are many improvements to the original translation published over a century ago."
- Emily Thomas, Durham University, UK

"*Creative Evolution* is essential reading today. To translate it well requires a serious engagement with Bergson's entire body of work, sustained philosophical attention, a feel for context (including discoveries in thermodynamics) and, most importantly, enormous care. Happily, this is what Donald Landes offers us here."
- Suzanne Guerlac, University of California, Berkeley, USA

Cover image: Georges Braque, *Paysage de l'Estaque (L'Estaque Landscape)*, 1906.
Oil on canvas, Centre Pompidou, Musée national d'art moderne, Paris.
© 2014 Artists Rights Society (ARS), New York / ADAGP, Paris

Portrait of Henri Bergson (1859–1941), 1919. France Excelsior collection. © Alamy.

This edition published in 2023
by Routledge
4 Park Square, Milton Park, Abingdon, Oxon OX14 4RN

and by Routledge
605 Third Avenue, New York, NY 10158

Routledge is an imprint of the Taylor & Francis Group, an informa business

Creative Evolution, by Henri Bergson, originally published as *L'évolution créatrice*, 1907.

English Translation © 2023, Routledge.

Foreword © 2023, Elizabeth Grosz.

Translator's Introduction and Translator's notes, selection and editorial matter
© 2023, Donald A. Landes.

"Correspondence, Reception and Commentaries," compiled and edited by Frédéric Worms and
Arnaud François, © 2013, Presses Universitaires de France (*L'évolution créatrice*, "Quadrige"
series, 12th edition 2013). English Translation © 2023, Routledge.

"The Ideas of Bergson" by Maurice Merleau-Ponty, from *Nature: Course Notes from the Collège de
France*, translated by Robert Vallier. Evanston: Northwestern University Press, 2003, pp. 51–70.
Originally published in French under the title *La nature: Notes, cours du Collège de France*.
Course notes and compilation copyright © 1995 by Éditions du Seuil. English translation
copyright © 2003 by Northwestern University Press. Published 2003. All rights reserved.

British Library Cataloguing-in-Publication Data
A catalogue record for this book is available from the British Library

Library of Congress Cataloging-in-Publication Data
Names: Bergson, Henri, 1859–1941, author. | Landes, Donald A, translator.
Title: Creative evolution / Henri Bergson ; translated by Donald A Landes,
with a foreword by Elizabeth Grosz.
Description: Abingdon, Oxon ; New York, NY : Routledge, [2023] |
Includes bibliographical references and index.
Identifiers: LCCN 2022008139 (print) | LCCN 2022008140 (ebook) |
ISBN 9781138689251 (hbk) | ISBN 9781032319216 (pbk) | ISBN 9781315537818 (ebk)
Subjects: LCSH: Life. | Evolution. | Metaphysics.
Classification: LCC B2430.B4 E713 2023 (print) | LCC B2430.B4 (ebook) |
DDC 113/.8–dc23/eng/20220614
LC record available at https://lccn.loc.gov/2022008139
LC ebook record available at https://lccn.loc.gov/2022008140

ISBN: 978-1-138-68925-1 (hbk)
ISBN: 978-1-032-31921-6 (pbk)
ISBN: 978-1-315-53781-8 (ebk)

DOI: 10.4324/9781315537818

Typeset in Joanna
by Newgen Publishing UK
Printed and bound by CPI Group (UK) Ltd, Croydon, CR0 4YY

Contents

Correspondence, Reception, and Commentaries[a]

a Selections translated by Donald A. Landes unless otherwise indicated.

Abbreviations of Bergson's Works Cited in This Translation

AC   "L'âme et le corps" ["The Soul and the Body"] in
L'énergie spirituelle

CV   "La conscience et la vie" ["Life and Consciousness"] in
L'énergie spirituelle

DPP   "Introduction (deuxième partie): De la position des
problèmes" ["Introduction II: Stating the Problems"] in
La pensée et le mouvant

DS   *Les deux sources de la morale et de la religion* [*The Two Sources of Morality and Religion*]

E   *Essai sur les données immédiates de la conscience* [*Time and Free Will: An Essay on the Immediate Data of Consciousness*]

IM   "Introduction à la métaphysique" ["Introduction to
Metaphysics"] in *La pensée et le mouvant*

MM   *Matière et mémoire* [*Matter and Memory*]

MRV   "Introduction (première partie): Croissance de la vérité:
Mouvement rétrograde du vrai" ["Introduction (Part I):
Growth of Truth: Retrograde Movement of the True"] in
La pensée et le mouvant

PC   "La perception du changement" ["The Perception of Change"]
in *La pensée et le mouvant*

PR   "Le possible et le réel" ["The Possible and the Real"] in *La pensée et le mouvant*

VOR   "La vie et l'œuvre de Ravaisson" ["The Life and Work of Ravaisson"] in *La pensée et le mouvant*

FOREWORD

Elizabeth Grosz

Henri Bergson (1859–1941) developed a new kind of philosophy, indeed a new way of understanding not only ourselves but the world, and even the cosmos, a philosophy adequate to all of life in its multiplicity, even those forms we cannot have yet encountered. He aimed to create a philosophy that comprehends the movements of evolution and, by adhering to the methods by which evolution itself works, developing an understanding of life that accounts not only for the evolution of bodies but also for the evolution of consciousness. Such a philosophy, as he recognizes in his earlier writings, *Time and Free Will* (1889) and *Matter and Memory* (1896), must take *durée*, duration, not as mathematical or numerical time, time measurable by counting, a time linked by moments, but as lived time, the continuous flow of acts, lives, and species as its object of philosophical reflection. This new English translation of *Creative Evolution* by Donald A. Landes comes over a century after the first English translation by Arthur Mitchell in 1911. It opens Bergson's entirely original writings to new generations, exploring interpretations and arguments that have developed and elaborated Bergson's work further—as we see in this volume, in the careful readings of this text provided by philosophers, such as Ruyer, Canguilhem, Merleau-Ponty and Deleuze, as well as interpretations of his work provided by mathematicians, theologians and biologists, and the extensive work undertaken by physicists and cosmologists since its original publication. While ignored and passed over for a time, Bergson's work, and especially as developed in this great text on the creativity of evolution itself, has seen a resurgence in interest in the last few decades, to a large extent because of Deleuze's strong reliance on his concepts throughout his writings. Bergson's insights remain significant for physicists, biologists, and philosophers, even in the present, because of his insistence on both the internality of life—its inaccessibility to direct observation—and its connectedness to all forms of existence, even those that may occupy the most remote planets. This book seems even more relevant in the present than it was at the time of its initial publication. It addresses how we are to best conceive life in all its modalities, including the human, without reduction, that is, how we are to understand life from within and develop a philosophy adequate to its unique particularity. It is truly a remarkable text, one that remains both untimely and prophetic.

Creative Evolution explores a philosophy of the development of life on earth, a philosophy in the wake of Darwin. Bergson elaborates what he understands as the common force of life, a vital impulse or *élan vital* that all life shares and that connects it to a common history, and a continuous line of evolutionary elaboration. This original impulse remains, even as the forms of life that constitute species diverge more and more from each other and from their progenitors, constituting the (arbitrary) divisions between fungal, plant, and animal life, and between humans and other orders of animal existence. If there is a common force that life shares, there is also a shared tendency to differentiate and diverge as much as possible within given and changing environments. These diverging tendencies lead to differentiating kinds of engagement with their worlds, instinct on the one hand, and intellect on the other. While many species exist and thrive according to the veracity and power of their instincts, which themselves evolve along with bodily form, intellect also accompanies the life of higher organisms, and, according to Bergson, finds its highest and most complete form in human life. Unlike instinct, intellect is unable to live or think *durée*. Its function is to enable us to deal with, fabricate or reproduce objects—the practical concerns of survival and everyday life. Its orientation is mechanical, causal, mathematical. In dividing the continuity of duration into distinct moments, phases and periods, and objects into geometrically decomposable forms, which can be brought together or separated according to the task at hand, the intellect not only decomposes objects and the space they occupy; it also decomposes durée in isolatable instants, providing snapshots of objects and reconstructing these partial views through what he calls "the cinematographic image," still images stripped of movement, or to which movement can be added artificially. The intellect can do no more than capture the outlines of an object and decompose it into its perceptible parts. Its function is analytic, dividing rather than creating. This makes it incapable of understanding durée, the continuous flow of time, the continuously expanding movements of growth, development, and divergence, the inherence of continuous change.

Bergson has argued that matter can be divided; indeed, it gives itself up to division, into isolated systems according to instinct or intellect. Instinct provides an internal, direct, and unreflective relation to matter, a means of mobilizing action on or with the objects or properties that one needs. Intellect requires reflection and is capable of utilizing abstract or geometrical methods for dividing objects and knowing them from the outside. Instinct and intellect are two means by which we address and engage with objects, things laid out in space. Instinct is spontaneous, species-based, providing a knowledge from within of its objects, while intellect decomposes, divides, simplifies, and recontextualizes objects and qualities. Instinct and intellect tend to diverge over evolutionary time, differentiating themselves as modes of dealing with matter while retaining an inward if unreflective connection to each other.

The center of Bergson's innovative evolutionary philosophy, its most creative concept and method, is that of intuition, which partakes in certain characteristics of both instinct and intellect, while creating new ways of addressing what each must leave out—the inner connections and continuities between things and living beings that must be ignored or put aside in order to measure and divide and thus calculate and predict their modes of combinations. Bergson argues that intuition may be how we, humans, can self-consciously return to the creative impulse that underlies all of life, unifying and connecting what instinct and intellect divide. Intuition is the capacity to connect, from within oneself, through a kind of sympathy, with the inner orientations of things, of species, of life itself, and of the whole from which life constitutes itself. Intuition, connected to both instinct and intellect, goes beyond each to the extent that it seeks out the whole, the entirety of life undivided from its necessary connections with matter. Hard to sustain, difficult to articulate, intuition is the coming-to-awareness of our enmeshment with all of existence, our place as a small bud in an ever-expanding tree of life that connects all living beings, and the material universe that surrounds them, to the continuous undivided fanning out or elaboration of materiality under the impact of time and becoming. Intuition is how we live and comprehend uninterrupted continuity.

Intuition is the elaboration and intensification of the sympathy that directs instinct, and to this extent, intuition is a development of instinct. Instinct knows its objects not reflectively but sympathetically, from a kind of resonance between the object and the very interior of the living being. It knows from within; its knowledge is not directly articulable, but generates action aimed at a goal, at satisfaction. Intuition elaborates and intensifies this instinctive access to objects, but it acquires, Bergson suggests, a disinterested rather than an invested relation to this object. If instinct directs us through an attention to life, especially one's own, and intellect directs us to inert matter (even forms of life) spread out in space, then intuition is a disinterested instinct, an expansive kind of intellect that aims at the very interior of life. Taking elements from both instinct and intellect, and insinuating itself into their procedures in order to connect them and work through them, intuition can highlight the limitations of intellect by showing what it lacks—an understanding of the whole, a knowledge that is not only practical but faithful to the contours, complexity and inner operations of its objects; and the limitations of instinct, which functions for its own benefit without necessarily developing itself further. Instinct is a contracted intuition directed not to the whole but to its own needs and satisfaction; intellect is a reduced and practically oriented intuition.

Bergson aims to create a philosophy not based on scientific models, which strive for repeatability, but one in many ways closer to art and its practices, though no less rigorous than the sciences. The artist finds a way to extract from the interior of an object, through an intuitive act, some of the forces and energies it contains and to represent them in a framed object, to partake in something of the object directly. But art aims at particular objects and specific representations. Philosophical intuition may find in itself the capacity, like art, to discern the inside forces and qualities not only of particular objects but of life itself. Using the resources of intellect to show the limitations of the intellect, using instincts and their fringes of indetermination, intuition creates a way of connecting with what is interior to life, following life's contours and intensities, dilating consciousness itself so that it is more and more in touch with life, that is, with the forces of "reciprocal interpenetration and indefinitely continuous creation."[a] This internal knowledge has no parts, is fundamentally indivisible even as it operates in various streams and directions, making clear that our place as humans is not apart from the world, certainly not above it, but is concentrated in its (and our own) inner operations. Such intuition, when it can be sustained, is properly philosophy, not science or art, but a language and forms of knowledge that aim at understanding ourselves, the world in which we live, and the forms of life with which we share the world, expansively, without reduction or simplification.

Intuition places us in durée, just as perception and intellect position us in space. Intuition places me in my durée, in the historical flow of memory that constitutes my individual life. It also places me in durée in general, a more dilated and broader form, one where my durée is directly connected to those that come before me, to a genealogy of human, animal, and even plant and microbial ancestors, as well as those living beings that associate with me now, and their lineages and predecessors, connecting us directly to all the filaments of life by different degrees of intensity and connectedness. I cannot have a direct knowledge of this whole, but I live it in my bodily form and in the inherited languages, knowledges, concepts, and practices that I may acquire. I can understand, through intuition, a place—or many—in which I participate in the great wave of life that shares my duration and all those that preceded me, that connects us directly to the élan that carries all and from which materiality can be cut out into separable objects or systems but in which it fully inheres. Intuition connects me, by degrees of dilation or contraction, to the whole of materiality and the whole of ideality.

Understanding the intertwined relations between instinct, intellect and intuition enables us to see not only how life accompanies and is made possible by particular configurations of matter, and how it is not separate from the material world, but recognizes its heterogeneity, its connections, similarities, and differences. Life acts in concert with

a Henri Bergson, L'évolution créatrice [1907], critical ed. Arnaud François, series ed. Frédéric Worms (Paris: Presses Universitaires de France, 2013), 179. The French pagination of L'évolution créatrice is included in the margins of this translation.

and through materiality to enhance and complicate each. What intellect requires—the infinite possibilities for the division of objects and of space—can be restored to the whole, the reality, from which they have been cut out in precisely those moments when the exigencies of action are temporarily relieved. We can then undertake the philosophical project of understanding both connections and differences, both tendencies for change and modes of resistance to change, both instinct and intellect, as complementary "parts" of a single, ever-changing continuity that is comprised of differences, variations, things and processes, a universe of time as well as space.

In *Creative Evolution*, Bergson offers us a way of understanding ourselves and the world, not through mastery or knowledge, but as connected but diverging forces, as forms of difference that contain the traces of their common origins and history in their present (and futures). This understanding remains fundamentally unexplored even today, although it provides an inventive, even urgently relevant, way of understanding our place as humans, as historically and geographically located, as sexed, raced, gendered, and enculturated subjects, as beings with differences in identity, interests, orientations who nonetheless share a collective history and a wide range of bodily forms, forces, and capacities. In creating a philosophy that places *durée* at the center of life, Bergson develops a future philosophy, one that more adequately understands the human in a world that connects it to countless other living beings and forces. He produces a new humility for the human, showing us a way of opening ourselves to the forces, living and non-living, that surround and compose us, and a new openness for the human, which can seek a future uncontained by but prepared for in this shared past.

Translator's Introduction

Donald A. Landes

It is in vain that we force the living being into one or another of our frameworks. All of the frameworks crack. They are too narrow and, above all, too rigid...

— Henri Bergson[a]

If Gilles Deleuze is correct that "Duration, Memory, [and] *Élan Vital* mark the three major stages of Bergson's philosophy," then *L'évolution créatrice*—Bergson's third major book, the one in which he develops the notion of *élan vital*—might be seen as the culmination of Bergsonism.[b] Published in 1907, it was quickly translated into several languages and sparked interest the world over. Not only did *L'évolution créatrice* represent an important contribution within philosophy, it also struck a chord in the wider debates of the day and solidified Bergson's place in culture as a famous intellectual. His weekly lectures at the Collège de France in Paris regularly attracted beyond-capacity crowds, and his visits abroad filled venues and even reportedly caused the first-ever traffic

a Henri Bergson, *L'évolution créatrice*, below, vi. All citations to *L'évolution créatrice* are to the French pagination, which is included in the margins of this edition.
b Gilles Deleuze, *Bergsonism*, trans. Hugh Tomlinson and Barbara Habberjam (New York: Zone Books, 1991), 13.

jam on Broadway in New York City.[a] Throughout the twentieth century, *L'évolution créatrice* continued to be recognized in France as central to Bergson's philosophy. In the years that followed Bergson's death in 1941, key thinkers returned to this book to offer commentaries, particularly on the remarkable and remarkably difficult metaphysical reflections found in Chapter III: Georges Canguilhem in 1943, Maurice Merleau-Ponty in 1956–1957, and Deleuze in 1960.[b] And, arguably the two most important book-length commentaries on Bergson's work published in French—Vladimir Jankélévitch's *Henri Bergson* (1931/ 1959) and Deleuze's *Bergsonism* (1966)—both highlighted the importance of *L'évolution créatrice* in Bergson's corpus.[c]

And yet, whether it is because Bergson's concept of *élan vital* (originally translated into English as "the vital impetus," but which will be left in French in this translation for reasons explained below) seemed to associate this book with a somewhat outdated vitalism, or whether because his focus on evolutionary theory makes parts of it read as an "applied" Bergsonism (like *Durée et simultanéité*, his later book on Einstein) rather than as a foundational text, *L'évolution créatrice* has seemed to many English-language scholars as an "optional" work in his corpus.[d] Nevertheless, it is far more than a Bergsonian reading of the major theories in contemporary evolutionary thought, a task primarily restricted to Chapter I. True, Bergson applies his earlier concepts to the problem of life and its evolution, and he extends his criticisms of other philosophical theories to unmask the metaphysical assumptions

---

a See Emily Herring, "Henri Bergson, celebrity," *aeon*, May 6, 2019, https://aeon.co/essays/ henri-bergson-the-philosopher-damned-for-his-female-fans

b Selections from these commentaries are included below in the section *Notable commentaries.*

c Deleuze, *Bergsonism*; Vladimir Jankélévitch, *Henri Bergson*, ed. Alexandre Lefebvre and Nils F. Schott, trans. Nils F. Schott (Durham and London: Duke University Press, 2015).

d Keith Ansell-Pearson, *Thinking Beyond the Human Condition* (London: Bloomsbury, 2018), 92. Ansell-Pearson is correct to note the exception of Elizabeth Grosz, whose work has regularly engaged with the key notions of *Creative Evolution*. For more, see her foreword to this translation. See also: Elizabeth Grosz, *Time Travels: Feminism, Nature, Power* (Durham and London: Duke University Press, 2005); Elizabeth Grosz, "Bergson, Deleuze and the Becoming of Unbecoming," *parallax* 11, no. 2 (2005): 4–13; Elizabeth Grosz, *Becoming Undone: Darwinian Reflections on Life, Politics, and Art* (Durham and London: Duke University Press, 2011). For Bergson's book on Einstein, see Henri Bergson, *Durée et simultanéité* [1922], critical ed. Élie During, series ed. Frédéric Worms (Paris: Presses Universitaires de France, 2009).

shaping evolutionary theory. But *L'évolution créatrice* is also a radical expansion of Bergson's philosophy beyond the individual experience of time (in the lived and flowing account of time that he names *durée*, another term that will be left in French for reasons explained below) and memory (by which we experience the presence of the virtual and thus the real status of the past). The book offers, rather, a genuine philosophy of becoming that encompasses reality as a whole, from the smallest particles of matter to the universe itself. To accomplish this, Bergson sketches the very confrontation between matter and life as it evolves across millions of years, with quite radical and fascinating metaphysical, epistemological, and even ethical consequences. As such, *L'évolution créatrice* takes Bergson's thinking beyond individual *durée* and offers an intuition of *durée* (and hence the virtual) as already at work in and *as* reality itself.[a]

Given the complex place of this book in Bergson's philosophy, the arguably spotty history of its reception in English, and the fact that it assumes at least some familiarity with the key concepts from Bergson's first two books, I begin this Translator's Introduction with a brief discussion of four of Bergson's earlier concepts relevant for *L'évolution créatrice*: *durée*, sensorimotor systems, the virtual, and intuition. I will then offer an overview of the book's argument. Finally, I will explain some of my translation decisions and the scholarly apparatus included in this edition. By bringing this book into English for the second time, I hope to invite a new look at Bergson's most developed metaphysical position and create new interest in what his philosophy has to tell us about life and the interdependency of the matter, individuals, species, and ecosystems that make it up. This is the Bergson that has influenced Canguilhem, Jankélévitch, Merleau-Ponty, Gilbert Simondon, Deleuze, and Elizabeth Grosz. This is, I believe, the Bergson that we need today in conversation with phenomenology, philosophy of science, contemporary biology, environmentalism, and beyond.

---

a For more on these metaphysical stakes of *L'évolution créatrice*, see Jankélévitch, *Henri Bergson*, 41; Deleuze, *Bergsonism*, 91–94; and Ansell-Pearson, *Thinking Beyond the Human Condition*, Chapter 5.

DURÉE, SENSORIMOTOR SYSTEMS, THE VIRTUAL, AND INTUITION

Bergson's first book, *Essai sur les données immédiates de la conscience* (1889), was published in English with an added main title as *Time and Free Will: An Essay on the Immediate Data of Consciousness*.[a] The English main title, "Time and Free Will," is justified by the content of the book, but the English subtitle, which is a literal translation of the original French title ("*Essai sur…*"), might be somewhat misleading for an English reader familiar with either phenomenology or analytic philosophy of mind. Bergson does not conceive of consciousness as a substantial entity to whom data may or may not be presented. Consciousness exists *in* and *as* the unfolding of its immediate givens; consciousness is inseparable from its givens, and they are immediately known from within. By contrast, external perception, language, the intellect, and science are all forms of *mediated* knowledge and hence relative. Nevertheless, the power of these mediating instruments renders the task of coinciding with our own immediate experience very difficult. Language and concepts translate our inner experience of time into the structures of space, namely *discontinuity* among distinct objects that can be described through measurable quantities or external relationships. As much as perception, language, and concepts provide practical advantages for our action and social being, they do not adequately capture our more profound experience of time, which is *continuity* among interpenetrating moments that can only be described through quality and inner relationships. In other words, Bergson's philosophy is, from the beginning, a "critique" of the usual or natural habits of human intellect—when they are illegitimately applied to time, irresolvable philosophical dilemmas are set in motion. Much of Bergson's work aims to show that we simply cannot *think* our inner experience of time through the intellect and its logic of discontinuity, immobility, and space.

Given its importance across Bergson's work, as well as the potentially misleading connotations of the term "duration" in English, I have

a Henri Bergson, *Essai sur les données immédiates de la conscience* [1889], critical ed. Arnaud Bouaniche, series ed. Frédéric Worms (Paris: Presses Universitaires de France, 2007); *Time and Free Will: An Essay on the Immediate Data of Consciousness*, trans. F. L. Pogson, 3rd ed. (Mineola, NY: Dover, 2001). Henceforth cited as *E*, with pages referring to the French edition.

chosen to preserve the French word *durée*. For instance, "duration" might lead the English-speaking reader to think of a measurable length of time, but *durée* is precisely not quantifiable in this way. *Durée* is qualitative time, psychological time that is continuous succession and continuous change. Language and the intellect artificially divide *durée* up into quantifiable, immobile, or spatialized parts, and thus can never fully reconstruct concrete or original time itself.[a] It is worth looking at how the concept of *durée* emerges in Bergson's thinking.

At the beginning of *Time and Free Will*, Bergson focuses on the non-quantifiable nature of psychic states. Despite how language and the intellect might characterize it, a psychic or mental state is not a self-contained block, nor is it made up of specific parts. For example, Bergson argues that an emotion is more intense insofar as it colors and influences other psychic states, so long as we keep in mind that even these supposedly distinct psychic states themselves are ultimately interpenetrating and indefinite, not juxtaposed and delimited like external objects.[b] He thus distinguishes between external phenomena and inner phenomena. To complete the analysis, Bergson puts this somewhat static picture of conscious states into motion. After showing how external phenomena are indeed adequately captured by the *homogeneous* fields of space that we impose upon them, he plunges into the depths of consciousness and finds a different kind of multiplicity, an indistinct or confused multiplicity that is characterized as a heterogeneous "succession without distinction."[c] Nevertheless, it takes an immense effort not to fall back into thinking time as space, parts outside of parts or moments juxtaposed along a line. If we succeed, then duration will be "restored to its original purity, will appear as a wholly qualitative multiplicity, an absolute heterogeneity of elements which pass over into one another."[d] This is *durée*, and the deeper we go into our unmediated experience, the more we connect with this qualitative reality.

Bergson's first book thus posits something like a dualism between space and *durée*, matter and consciousness. But his goal was not to take a strong position on the existence of the external world, but rather to find

a See the "Translation Decisions" section below for more on my decision to maintain *durée* in French in this translation.
b *E*, 6.
c *E*, 75 (*Time and Free Will*, 101).
d *E*, 172 (*Time and Free Will*, 229).

a place for freedom, which he discovers not in space, but in durée. As to the existence of the world, his strongest claim is that if things themselves seem to participate in something like durée, then this is the result of some "inexpressible reason."[a]

*

Deleuze argues that Bergson increasingly presents our personal or *psychological durée* as a first step toward grasping *ontological durée*.[b] As a result, *durée* becomes something like the fabric of reality, drawing Bergson toward a kind of monism. But how might our *durée* be in contact with others? And can we really speak of a single universal *durée*? These questions in part animate Bergson's second book, *Matière et mémoire* (1896), where the question of dualism is indeed present from the very first line: "This book affirms the reality of spirit [*esprit*] and the reality of matter, and tries to determine the relation of the one to the other by the study of a definite example, that of memory."[c] To explore this relationship, Bergson somewhat leaves behind the internal perspective of consciousness adopted in Time and Free Will to consider the role of the body of living beings, particularly humans, and the relationship between the body (action, perception, and matter) and consciousness (memory). With the argument of *L'évolution créatrice* in mind, I will here only discuss two theoretical contributions in this second book: "sensorimotor systems" and the logic of the "virtual."

Bergson begins with a somewhat tricky metaphysics of "images" as a way of finding a middle ground between realism and idealism. Whereas realism and idealism share a common misconception about perception (and by extension, knowledge)—that it is essentially disinterested and "speculative"—Bergson's observations show perception (and human intellect) to be oriented toward an entirely different purpose: *action*. This emerges from his account of living bodies, which are able to introduce indetermination and novelty into an otherwise mechanistic

---

a *E*, 171 (*Time and Free Will*, 227). This is discussed by Deleuze, *Bergsonism*, 48.

b Deleuze, *Bergsonism*, 48–49.

c Henri Bergson, *Matière et mémoire* [1896], critical ed. Camille Riquier, series ed. Frédéric Worms (Paris: Presses Universitaires de France, 2008), 1; *Matter and Memory*, trans. Nancy Margaret Paul and W. Scott Palmer (New York: Zone Books, 1991), 9. Henceforth cited as *MM*, with pagination referring to the French edition.

world. These bodies are designed to hesitate and choose their reactions, and the history of life reveals an increasing ability to choose as we move up the scale of living organisms. The nervous system is like a "telephonic exchange" and the sensorimotor system is "entirely directed toward action," not toward "pure knowledge."[a] An animal's consciousness will be richer and more intense as a function of its possible action and the variety of its available choices.

In the course of his analysis, Bergson considers a theoretical form of perception that would be free of all memory, a "pure perception" by which we touch upon the reality of things.[b] This surely gives us hints about the nature of matter as being very much like space, but Bergson will insist that matter is only a tendency toward the ideal of pure space. Moreover, he demonstrates how spirit is not itself a product or epiphenomenon of matter, but must in fact be different in kind from matter. This leads Bergson to argue that there must be some independent way in which the past preserves itself in what we might call the virtual. Since the "virtual" is different in kind from matter, it seems that mind has "an independent reality."[c] As such, memory cannot be reduced to traces held in the physical brain. Whereas matter tends toward homogeneity and juxtaposition in space, memory tends toward heterogeneity and interpenetration, much like the structure of *durée* discussed above.

Now, the "virtual" should not be mistaken for what is merely "possible," since the virtual is indeed *real*. The virtual can be *actualized*, but since it is heterogeneous and interpenetrating in itself, its actualization into matter and the present, given their structures of homogeneity and juxtaposition, the virtual is forced to crystallize into matter, without exhausting itself—just like each one of my gestures expresses my personality without exhausting it, since my personality is the virtual that will continue along and crystallize anew in new situations. Or again, consciousness is coextensive with memory, not merely with the brain, even if the brain is the "gesture," so to speak, that allows consciousness to actualize itself at any particular moment. Bergson thus conceives of the body (or, more specifically, the sensorimotor system) as being the most contracted point of the past cutting into the future. The virtual is like a cone, touching material reality at the point of the acting body

a MM, 26–27 (*Matter and Memory*, 30–31).
b MM, 33–37.
c MM, 76 (*Matter and Memory*, 72).

and yet preserving the past in its entirety at various levels of contraction all the way up to pure dream.[a] Near the end of *Matière et mémoire*, Bergson even suggests a kind of monism of *durée*, despite the dualism of its modes (matter and memory). He writes: "[T]he material universe itself, defined as the totality of images, is a kind of consciousness."[b]

The logic of the virtual and the role of the sensorimotor systems will be essential to keep in mind for Bergson's theory of life in general and for the oriented, if not predetermined, directions of evolution. In fact, *Matière et mémoire* concludes with precisely a call to think living bodies in terms of evolution, insofar as we see living bodies "even in their simplest forms" as capable "of spontaneous and unforeseen movements."[c] Evolution must be a progressive development of living bodies endowed with increasingly complex nervous systems. But as much as we see the complexification of the organism in space, we do not "see" the "growing and accompanying tension of consciousness" in time.[d] This snowballing of the past available to certain bodies makes them more and more "capable of creating acts of which the inner indetermination, spread over as large a multiplicity of the moments of matter as you please, will pass the more easily through the meshes of necessity."[e] Developing this image of life overcoming necessity by way of living bodies is precisely the goal of *L'évolution créatrice*.

*

But what philosophical methodology allows us to think the virtual and the forms of heterogeneous and interpenetrating multiplicities at work in *durée*, life, and consciousness? Not only is the distinction between the forms of knowledge (instinct, intellect, and intuition) a key part of *L'évolution créatrice*, but the book as a whole might be seen as an initiation into a philosophy of becoming in general. It is worth sketching Bergson's philosophical notion of "intuition" as distinguished from intellect, presented most explicitly in his 1903 article "Introduction à la métaphysique."[f]

a See *MM*, 167–92.
b *MM*, 264 (*Matter and Memory*, 235).
c *MM*, 280 (*Matter and Memory*, 248).
d *MM*, 280 (*Matter and Memory*, 248).
e *MM*, 280 (*Matter and Memory*, 249).
f Henri Bergson, "Introduction à la métaphysique [1903]," in *La pensée et le mouvant*, critical ed. Arnaud Bouaniche, Arnaud François, Frédéric Fruteau de Laclos, Stéphane Madelrieux, Claire Marin, and Ghislain Waterlot, series ed. Frédéric Worms (Paris: Presses

According to Bergson, there are two ways of knowing an object. First, there is external knowledge. We analyze an object and break it up into parts or isolated moments, and then we deploy symbols or concepts to reconstruct its unity. This provides us at best with a "relative" knowledge of the object, and human intellect has, according to Bergson, a natural inclination to this form of knowledge. Second, there is an internal or inner knowledge by which we "enter into" the object without any symbolic or linguistic mediation.[a] Bergson describes this as an effort of "sympathy," a gearing-into the heterogeneous and qualitative unfolding of a durational reality as it is being made, and the result is a sort of immediate, perfect, and *absolute* knowledge.[b] This is "intuition." And yet, given that language follows the first model, it appears that *what* we sympathize with in intuition is essentially "inexpressible."[c] Any intellectual reconstruction of the feeling of intuition will be partial and imperfect. If we hope to communicate our intuitions to others, rather than offer a direct, phenomenological description, we must triangulate that intuitive experience through multiple images or design concepts tailored to the singularity of the experience in question. This is not to say that intellect must be rejected in favor of intuition; rather, Bergson wants to identify the legitimate use or proper domain of each. Intellect, with its capacity for showing predictable repeatability, is apt for the positive sciences, social life, and action generally. Intuition, in a technical sense, is useless for any of these practical activities, and is better suited for art and metaphysics.

How does intuition relate to *durée* and the virtual? At this stage, Bergson posits that the one thing that we know via intuition is "our own person in its flowing through time, the self which endures."[d] Because the person is thus a durational reality, each moment is a "partial expression" of a virtual, not a "part" of a whole.[e] As Bergson writes: "There is no feeling, no matter how simple, which does not virtually contain the past

Universitaires de France, 2013), 177–237; "Introduction to Metaphysics," in *The Creative Mind*, trans. Mabelle L. Andison (New York: Philosophical Library, 1946), 133–69. This chapter is henceforth cited as IM, with pagination referring to the French edition.

a   For more on this point, see Elizabeth Grosz's foreword to this translation.
b   IM, 181 ("Introduction to Metaphysics," 135).
c   IM, 182 ("Introduction to Metaphysics," 135).
d   IM, 182 ("Introduction to Metaphysics, 136).
e   IM, 192.

and present of the being which experiences it, which can be separated from it and constitute a 'state,' other than by an effort of abstraction and analysis."[a] In other words, intuition grasps the whole that is expressed, as virtual, in each of its partial expressions. Intuition—and not intellectual judgment—is how we grasp an entire person in a single one of their gestures, and this is because intuition immerses us in durée, and durée is always psychological in nature, that is, always like a consciousness or a virtual.[b] By extension, intuition will have to be the way we grasp the movement of life in general through life's partial expressions in species and individuals. As a result, intuition is not introspection, insofar as it is not limited to merely looking within. Intuition can gear into durée, wherever it may appear. Bergson writes:

> [T]he intuition of our own durée, far from leaving us suspended in the void as pure analysis would do, places us into contact with a whole continuity of durées that we must try to follow either in a downward or in an upward direction: in both cases we can dilate ourselves indefinitely through a more and more vigorous effort; in both cases we can transcend ourselves.[c]

As such, L'évolution créatrice attempts to provide an intuition of other durées, and perhaps even a universal durée. Thus, the book is an invitation beyond the limitations of the human condition. As much as the intellect is appropriate as a foundation for the physical sciences of matter, it will be intuition that is needed for a philosophy of life.

SUMMARY OF *CREATIVE EVOLUTION*

Introduction

From the very first lines of the introduction, Bergson makes it clear that L'évolution créatrice is shaped by the various concepts I have just outlined above, and particularly his critique of human intellect. Human intellect is not a timeless and pure transcendental perspective; it emerges from within the concrete evolution of life so as to ensure "the perfect

---

a IM, 190 ("Introduction to Metaphysics, 142).
b See, for instance, Deleuze, *Bergsonism*, 34, 37, and 76–77.
c IM, 210 ("Introduction to Metaphysics," 158, translation modified).

insertion of our body into its surroundings."[a] Since the intellect thereby carves up our experience into static and juxtaposed parts, it turns out to be "incapable of conceiving of the true nature of life," which is durée.[b] The intellect, when used properly, has a significant value; but when it attempts to explain life, it leads into endless contradictions and debates, such as the competing positions within evolutionary theory.

According to Bergson, we might still find a way to grasp the nature of life and reality. Not only does human intellect at least touch upon something of the absolute (i.e., matter), it also is just one of several forms of consciousness that have emerged. The question becomes: what is the principle or source of these various forms of individual consciousness? Bergson suggests that, by gearing together with the other forms of consciousness, we might catch a glimpse of their source. But for this to be possible for us, given our primarily intellectual nature, human intellect must remain surrounded by a "vague nebulosity" of consciousness from which it crystallized and by which it allows us to connect with those other durées. Of course, left on its own, the intellect has at best "a confused sense of these powers."[c] This is where the empirical study of evolution becomes methodologically crucial for metaphysics. We must "catch sight" and become aware of these powers that we feel within ourselves by finding them at work in nature along other lines of evolution.[d] The study of nature will thus, in a sense, jog our organic memories of instinct and torpor, and initiate us into a more complete grasp of the nature of life and reality as becoming.

This is the context of Bergson's important claim that the theory of knowledge and the theory of life must be pursued together. The theory of life has too often accepted the static and spatializing categories of human intellect, resulting in an account of life that works for action and science, but that misses the deeper moving and flowing reality of life in general. On the other hand, a theory of knowledge that forgets how the intellect itself is but one product of life will take the intellect for granted, failing to account for its genesis and failing to understand how we might get beyond it. The study of evolution is not merely an applied

---

a Bergson, *L'évolution créatrice*, v. Again, citations to *L'évolution créatrice* are to the French pagination, which is included in the margins of this translation.
b *Ibid.*, vi.
c *Ibid.*, ix.
d *Ibid.*

Bergsonism—it is the call for an expanded consciousness, touching upon that principle or source that engendered the intellect (and perhaps even matter itself). This philosophy of becoming would be a "true evolutionism," and because it would not be a self-enclosed theoretical block to be encompassed by a single intellect, but rather an open and creatively evolving movement, Bergson's metaphysics here requires a radically new kind of collaborative and intuitive philosophy.

Chapter I: On the Evolution of Life. Mechanism and Finality

In the first chapter, Bergson offers an initial account of the evolution of life by exploring an extended analogy between psychological *durée* and life as an organic movement in general. He thus demonstrates the inadequacy of human intellect, limited as it is to the categories of mechanism and finalism, for the project of grasping life, and shows how the uncritiqued categories of the intellect ultimately limit the explanatory scope of evolutionary theories to merely the details of evolution. The chapter culminates with an important section defining the notion of *élan vital*.

As already discussed, *durée* characterizes things that exist as becoming. For a consciousness, the past hangs over the present, virtually; it crystallizes into images when they are useful for action and shapes the various dispositions we call personality or character. Continuously taking up the past toward the future, each new moment is a sort of creation in the overall trajectory of a life. As Bergson writes, "for a conscious being, to exist is to change, to change is to mature, and to mature is to go on endlessly creating oneself."[a] But now Bergson explicitly holds that *durée* cannot be limited to some subjective feature of a personal consciousness, for even inorganic bodies participate in a kind of *durée* "analogous to our own" (as he shows with his famous example of the sweetened water).[b] Matter itself cannot be identified with abstract space, but rather must be defined as a tendency toward spatialization and the creation of only *relatively* closed systems. As a whole, the universe itself must *endure*, like a consciousness, imparting *durée* even to mechanical systems.[c] This has quite striking consequences: since *durée* is psychical in nature, any

a *Ibid.*, 7.
b *Ibid.*, 9.
c *Ibid.*, 11.

time we encounter "invention, the creation of forms, and the continuous development of the absolutely new," we are dealing with something like consciousness.[a] And surely even the most rudimentary individual living bodies meet this definition. As Bergson writes: "*Wherever something is living, there is a register, open somewhere, in which time is being inscribed.*"[b]

But can we say that life itself is "invention and incessant creation"?[c] Does evolutionary theory (Bergson uses the term "transformism") help us to answer this question about life in general? The core part of Chapter I is devoted to exploring how transformism is ultimately limited for having accepted the intellectual categories that structure either mechanism or finalism. But throughout, we begin to catch sight of Bergson's own positive position. He argues that life itself is a kind of *becoming*: "[T]his current of life, passing through the bodies that it had organized [*organisé*] one after the other, passing from generation to generation, was divided among species and distributed to individuals without thereby losing anything of its force, and rather intensifying as it advanced."[d] This might serve as a first definition of the *élan vital*, as a "continuity of progression" that is oriented but unpredictable, and, like consciousness, "creating something at every instant."[e]

After a detailed exploration of the inadequacies of the main positions of radical mechanism and radical finalism, Bergson explores the various scientific attempts to explain evolution. He demonstrates how these theories, relying upon the categories of the intellect, are unable to grasp the whole from which species and individuals emerge. Bergson proposes a kind of inverted finalism in which the harmony is at the source and not at the "end" projected according to a predetermined plan. In short, the unity of life as a movement would be unintelligible if we were to accept mechanism and overly determined if we were to accept finalism. Darwin and De Vries both fail to explain why life itself evolves through variation; Eimer's account of external causes cannot explain the presence of identical organs along different lines of evolution; and neo-Lamarckian theories fail to grasp the kind of connection at issue between generations. In short, the empirical study of evolution reveals a *deeper* psychological

---

a *Ibid.*
b *Ibid.*, 16.
c *Ibid.*, 23.
d *Ibid.*, 26.
e *Ibid.*, 28–29.

cause at work, something that must have the nature of a consciousness without being itself an individual consciousness.

These considerations lead Bergson to explicitly posit this original source, which he names *élan vital*, an impulse or unified movement that passes from generation to generation. This common *élan* can explain why the same eye structure, for instance, appears along seemingly unrelated lines of evolution. The eye is only complex when analyzed by the intellect; for life, the creation of an eye is a *simple* gesture, an actualization of a virtual. Life does not build the eye on the model of human fabrication, but rather through dissociation, canalizing a general vision (or sensitivity at a distance) into a specific circuit. Given the importance of visual perception for action, the *élan vital* would include a certain "march toward vision," not as a preconceived end, but rather as the tendency toward increasing the organism's awareness of options.[a] The eye's structure then is the same between different branches of evolution because it is a negotiation between this oriented *élan* and the similar obstacles it encounters.

Chapter II: The Diverging Directions of Life's Evolution: Torpor, Intellect, and Instinct

Bergson here turns his attention to the diverging directions of life's evolution, focusing on how there are competing and complementary tendencies within the movement of life in general. It is in this chapter that he most explicitly explores the genesis of (and relationships between) instinct, intellect, and, to some extent, intuition. Overall, Chapter II aims to establish the psychical nature of life and to demonstrate the particular place human beings occupy in this movement.

The chapter begins with an important image for conceiving of this movement. Evolution is not a single trajectory; it is like a shell that explodes into fragments, but the fragments in turn explode into further fragments, and so on. The details of each explosion will be a function of the force of the original explosive and the resistances that this force encounters. That original force is Life, but it does not have the power to simply enter into the material world; it must first make itself as small as possible, inserting itself into matter by adopting its very movements,

a *Ibid.*, 97.

which is why it is difficult to distinguish between the most basic organisms and mere physico-chemical processes. But having gained this foothold, life grows and begins to face new obstacles.

For Bergson, life is psychical in nature, and thus evolves much like a personality throughout a person's life. But whereas each individual consciousness evolves only by leaving aside the roads not taken, life is able to explore many roads at a time. Although it strikes various dead ends or even U-turns, life in general continues to evolve in a particular direction: toward free and creative action. For life, even the "roads not taken," insofar as they remain virtual, are not forever lost. As much as a species might emphasize a certain tendency (such as the tendency toward a closed society of ants, or the open society of humans), the other tendencies remain present because they were, prior to the bifurcation, *vaguely mutually complementary* expressions of a common *élan*.[a] Thus, rather than focusing on the details of evolution (variation, adaptation, etc.), Bergson proposes a broader look at the directions of evolution along the main lines, that is, the various ways that life expresses itself in matter. As such, it will not be the enacting of a predetermined plan, but rather an open and ongoing creative evolution.

Given both the shift in focus to life in general and the idea of a common *élan* expressing itself across the diversity of this movement, Bergson suggests that *"a group,"* such as animals or plants, should *"no longer be defined by the possession of certain characteristics, but rather by its tendency to accentuate them."*[b] What distinguishes plants from animals is neither the details of nutrition nor the varieties of reproduction, for in the details there are always exceptions and hybrids. Rather, plants tend toward a certain mode of nil consciousness called "torpor," whereas animals tend toward an awakened and active consciousness. Consciousness and unconsciousness are thus tendencies, not substantial properties or entities. Each remains virtually present to the other, and so may even reappear in extreme situations. As a result, we can say the two main kingdoms are united in their source but divided in their expression of competing tendencies. Bergson thus defines this source as an effort to introduce as much indetermination into the material world as possible. Life cannot create energy; it must devise ways of collecting energy

a *Ibid.*, 102.
b *Ibid.*, 107. Italics in the original.

and storing it for use in increasingly free and unpredictable action. This defines the relationship of mutual dependency between plants and animals, each expressing tendencies that were "contained" in the virtual and hence already *virtually* compatible, even if *actually* in competition with each other.[a] So, if plants tend toward torpor and the accumulation of energy, animals tend toward action and the use of energy. Now we can see the *empirically* privileged place of the sensorimotor system, which, when complex and supported by other systems for maintenance and repair, can be "a genuine *reservoir of indetermination*."[b]

Of course, the immense and diversified effort of life in general to overcome the obstacles of matter has mostly led to dead ends or failures. After exploring some of these dead ends or some of the threats matter posed (such as the immobility resulting from protective carapaces), Bergson turns to the two most successful lines. Life, insofar as one of its tendencies was "looking for mobility, for suppleness, and [...] for a variety of movements," was led to vertebrates and arthropods.[c] And again, these groups are not defined by certain properties, but rather by their accentuation of certain tendencies. In insects (particularly Hymenoptera), we find the perfection of *instinct*; in vertebrates (culminating in humans), we find *intellect*. And, because of the common source, each tendency remains to some extent available to the other through a kind of "fringe" or virtual from which they both emerged. In practice, they do in fact encroach upon each other, especially in humans, where instincts remain quite obviously present.[d]

Developing the idea of intellect, Bergson characterizes the tendency of intellect as oriented toward *fabrication*, in the sense of constructing artificial instruments according to a plan and out of pre-existing parts. By contrast, an instinct is a lived knowledge of the use of a natural instrument. Instincts have certain advantages (such as their immediate reactions) but are also inflexible. Intellect has other advantages (such as an indefinite flexibility) but is hesitant and requires effort. The intellect only fully takes possession of itself in humans. Nevertheless, both of these tendencies are expressions of consciousness more generally. They are both forms of "knowledge," but instinct is *enacted* knowledge,

a *Ibid.,* 119.
b *Ibid.,* 127.
c *Ibid.,* 133.
d For an excellent discussion of these terms, see Grosz's foreword to this translation.

whereas intellect is thought or conscious knowledge. In the end, however, this is a difference of degree, so Bergson turns his attention to the objects of these knowledges to distinguish them—and here we find a difference in nature between their respective objects. Instinct is oriented toward the things of life, whereas intellect is oriented toward form and the relationships among what is inert. The intellect does not emerge to do metaphysics; it emerges to guide action, and as such it is adapted to making use of inert matter, objects, and closed systems. This is where it "feels at home," with space as its natural inclination. "We are only at ease in the discontinuous, in the immobile, and among the dead. The intellect is characterized by a natural incomprehension of life."[a] On the other hand, instinct touches upon life but cannot think it.

And yet, even science finds itself called to look for continuity beyond the categories of the intellect. Modern biology rejects the idea of a linear hierarchy in favor of diverging lines of evolution, and the purest forms of intellect and instinct are the endpoints of different lines. But the lines are not thereby wholly unconnected, for they are two tendencies within the domain of the mind, each haunted by the whole that contains them both in virtuality. The instinct does not "know" what it is doing, but it feels itself drawn into the lines of force, as if communicating internally. To truly understand instinct, then, we must not translate it into tacit intellectual knowledge; we must grasp it as sympathy, which is precisely the definition that Bergson uses to create a bridge to his concept of intuition. Intuition is "instinct that has become disinterested, self-conscious, and capable of reflecting upon and indefinitely enlarging its object."[b] Rather than remaining outside life, intuition takes us "into the very interior of life."[c] Ultimately, the fact that consciousness splits into instinct and intellect shows the double aspect of the real—matter and life—and shows how the theory of knowledge must involve both a study of evolution and a metaphysics able to account for the genesis of these faculties and their objects.

If we look at evolution through this lens, it becomes possible to say that a "vast current of consciousness" insinuates itself in nature, drawing matter into its organic development. By entering into matter, it was slowed and divided up, but some branches of life falling into torpor

a *L'évolution créatrice*, below, 166.
b *Ibid.*, 178. See also Grosz's foreword to this translation.
c *L'évolution créatrice*, below, 178.

provided the fertile soil for other branches to awaken to increasing action. Life is thus "consciousness launched through matter."[a] Intellect, with its ability to circulate disinterestedly, will provide the space needed for intuition to flourish. As such, "consciousness" is, as a whole, the "motor principle" of evolution, and it provides insight into the place of humans as the "*raison d'être*" of the evolution of life on earth, but only in a manner of speaking—only insofar as humans express the freeing of that consciousness that is the source of all of the evolution of life, and all of life in a sense expresses itself in humans as merely its most advanced partial expression.[b]

Chapter III: On the Meaning of Life, the Order of Nature, and the Form of the Intellect

In Chapter III, Bergson offers a deeply metaphysical reflection on the relationship between life, the intellect, and matter. He begins by reminding us of the reciprocal determination between matter and intellect that has been suggested thus far, as well as the fact that the intellect, for its part, is a condensation or crystallization within a larger reality that he names "Consciousness in general." The striking goal of this chapter is to illustrate the simultaneous "genesis" of intellectuality and materiality, with the clue being that instinct (along with intuition) is tuned to *durée*, life, and consciousness, while intellect is modeled upon "brute matter."[c] Here Bergson proposes a philosophical method of working back up to the higher form of existence as well as a subtle logic of inversion or interruption to explain a metaphysical (and in this case, ideal) genesis. This chapter is as difficult as it is fascinating, and it has elicited philosophical comment from Merleau-Ponty, Canguilhem, and Deleuze, included below.

According to Bergson, as audacious as his metaphysical project for this chapter might seem, it is not as audacious as classical metaphysics, whose fundamental error is to have assumed that reality is fixed and entirely intelligible to human intellect. As we have seen, intellect evolves for action, not speculative knowledge.[d] In an argument that is

a *Ibid.*, 182–83.
b *Ibid.*, 183; 186.
c *Ibid.*, 187.
d *Ibid.*, 192.

foundational for his later critique of the history of philosophical systems (Chapter IV), Bergson shows how taking the intellect for granted leaves philosophy caught in a vicious circle. Avoiding this circle requires finding a logic that will allow us to see the genesis of the intellect by catching sight of something larger from which it emerges.

To set up this argument, Bergson turns inward, where we find one tendency that moves toward intensity and "pure *durée*," and another that goes in the direction of relaxation into space, that is, toward "intellectuality" and "materiality." The idea is that although our consciousness only takes steps toward materiality by relaxing into spatialized perception and language, perhaps matter itself is precisely this same movement, "just pushed even farther, and that physical reality is simply an inverted psychical reality."[a] In short, "matter *extends* itself into space without being absolutely *extended* in space."[b] Bergson's earlier dualism between matter and consciousness and his implicit "monism" of *durée* both prove inadequate, for matter now appears as precisely the relaxation or interruption of spirit, a difference in nature that is now between tendencies rather than substances.

The intellect, conceiving of matter as a positive reality rather than as an interruption of a more fundamental positive reality (Consciousness), resists this logic. To illustrate, Bergson discusses how the intellect reifies the practical notion of disorder.[c] According to Bergson, there are two kinds of order: an "automatic" order, which is purely causal, and a "willed" order, which is the work of consciousness. In both cases, the mind finds itself in things: in the former, as intellect; in the latter, as invention and creativity. The mistake is to think that the absence of one is anything other than the presence of the other. That is, disorder is not the absence of all order, but rather the presence of the opposite order from the one desired or expected by a consciousness. Moreover, the mere interruption of one kind of order does not give disorder, but rather the *other* order. Since there is always one of the two orders, disorder is simply the practical concept we use in social life when we encounter the "wrong" order. In other words, disorder is a pseudo-idea that causes pseudo-problems. This logic of inversion is key to grasping what Bergson means in the section "The Ideal Genesis of Matter," namely that

a *Ibid.*, 203.
b *Ibid.*, 204.
c *Ibid.*, 221–36.

the origin of matter is Consciousness or the virtual, and that the genesis of matter is an ongoing and creative process.

These pages are extremely difficult, but the reader should keep in mind the logic of interruption and the notion of the virtual. For here Bergson posits a virtual or "ideal" Consciousness that, when interrupted, falls into (or falls *as*) matter, and yet also continues along in the direction of the creation of the new. With this in mind, it is clear why Bergson explores the metaphysical significance of the second law of thermodynamics, that is, entropy. Matter is characterized as a certain movement toward the even distribution of heat among bodies, a movement toward homogeneity. The law of entropy leads Bergson to describe the tendency of matter as that of "a thing *unmaking itself*."[a] Now, this thing unmaking itself must have first been *made*. In other words, the material universe must be the vestige of the opposite movement: will, consciousness, or *durée*. And so the origin of the material world must itself be immaterial, virtual, or "ideal." Consciousness is a movement upward; the material world interrupts this, and falls into (and falls as) matter.

Life too seems to be *durée*, but now *within* the material universe. As such, Bergson argues, life must be an attempt to go back up the incline that matter is falling down. Life is Consciousness coursing through matter. As a mere flicker of that original energy fallen into the world, it has lost the power to create matter, but it continues to work in that original direction. In short, "life is like an effort to lift up the weight that falls. True, it only manages to slow the fall. But at least it can give us an idea of what the lifting up of the weight would be."[b] Life is what remains of that original creative gesture of making; life is a "*reality that is making itself through the reality that is unmaking itself*."[c]

Given that this creating is not the activity of a thing but rather a pure activity in itself, Bergson does not hesitate here to define God as unceasing life, action, and freedom, hence as the principle or source of those very durational realities we can catch sight of in ourselves.[d] The intellect breaks up these simple realities into an infinitely complex set of parts outside of parts. Thus we "close our eyes to the unity of the *élan* that, passing through the generations, connects individuals to

a *Ibid.*, 246.
b *Ibid.*, 247.
c *Ibid.*, 248. Italics in the original.
d *Ibid.*, 249–51.

individuals, or species to species, and makes the series of living beings as a whole into a single immense wave coursing through matter."[a] To do philosophy, we can no longer merely try to see "through the eyes of the intellect alone"; we must join with the mind as a whole by twisting back toward the origin.[b]

Given this metaphysics, Bergson is now able to evaluate the meaning of evolution and the place of humans in the world. Here he redefines the *élan vital* as "a need for creation," hence the movement within our world that strives to continue in the direction of the movement that was interrupted so as to create our world. But since the *élan vital* is finite and unable to create out of nothing, it is forced to work with and against matter. Thus, "life as a whole—animal life and plant life together— appears in essence to be an effort for accumulating energy and releasing that energy into flexible and deformable canals at the end of which it will accomplish an infinite variety of works."[c] The details as to how life on any particular planet reaches these goals are contingent, unpredict- able, and open, but evolution will always be oriented according to these needs. In other words, life "is of the psychological order."[d] Life must be a perpetual movement from this virtual to the actual, never fully fitting into the categories of materiality or intellectuality, such as unity and multiplicity. As Bergson concludes:

> If life is comparable to an impulse or an *élan* when it is in contact with matter, then when conceived of in itself, life is an immensity of virtu- ality, a mutual encroachment of thousands and thousands of tenden- cies that will, nevertheless, only be "thousands and thousands" once they have been externalized in relation to each other, that is, spatialized. [...] Matter actually divides up what was only virtually multiple.[e]

If we now wish to grasp the movement of evolution as a whole and the place of humans therein, we must say that Life is a "march toward thought."[f] Given the brain and the flexibility of human freedom thanks to language and society, humans are in a privileged position insofar as

a *Ibid.*, 251.
b *Ibid.*
c *Ibid.*, 254.
d *Ibid.*, 258.
e *Ibid.*, 259.
f *Ibid.*, 261.

they break the bonds of determinism. Humans represent a "unique and exceptional success" of Life at a given moment of its evolution, but are not thereby distinct or independent from nature. Humans are but a partial expression of something that was "*indecisive and vague*" in the virtual.[a] Moreover, life moves as a whole. The disharmonies and conflicts are reabsorbed into a unified march toward thought and freedom, and all the animals and plants are our traveling companions.[b]

Philosophy, then, must "seize upon these fleeting intuitions" to grasp that whole from which we have emerged as intellect, and thus to go beyond the intuition of our own *durée* to an intuition of life as an *élan* and as an immense *virtuality* expressing itself as it grows and evolves. Such a philosophy will not be the work of one person, nor even of humanity as a whole; it is the work of the *élan vital* in and through all of its expressions. This is what motivates the rousing yet enigmatic flourish at the end of this chapter, in which Bergson describes species and individuals as an immense army galloping shoulder to shoulder, across time and space, in a forward charge "capable of overpowering all resistances and of clearing so many obstacles, perhaps even death," that is, as capable of overcoming matter and reconnecting with universal *durée*.[c]

Chapter IV: The Cinematographic Mechanism of Thought and the Mechanistic Illusion. A Glance at the History of Systems. Real Becoming and False Evolutionism

In the final chapter, Bergson aims to clarify his philosophical position both in terms of how it addresses important illusions and in terms of how it stands in contrast to the classical "systems" of philosophy. The chapter begins with a discussion of two illusions. First, as we have already seen, despite the fact that consciousness senses the reality of movement and becoming, it nonetheless models itself in humans according to the structures of the intellect. Here Bergson will focus on how this takes place, namely because of what he will name the "cinematographic mechanism of thought." The second illusion has the same origin, and

a *Ibid.*, 266.
b *Ibid.*, 267.
c *Ibid.*, 271.

consists in believing that a void or nothingness must exist prior to being. Unmasking these two illusions will provide the foundations for Bergson's critique of the various philosophical "systems" discussed in the second part of the chapter.

Tackling the second illusion first, Bergson observes that philosophers are haunted by the question of nothingness: "why do I exist," and "why does something exist rather than nothing?" These questions indicate an implicit belief that existence is a victory over nothingness. In other words, we have a deep belief that there "is *less* in the idea of 'nothing' than there is in the idea of 'something.'"[a] And yet, perhaps this idea of "nothing" is a pseudo-idea in the manner of "disorder," as we saw above. If we try to imagine or think of "nothingness," we might do so by abolishing (in imagination or thought) one thing after another in the belief that ultimately we would reach a pure void. But according to Bergson, when we say "there is no longer anything there," what we tacitly mean is that the wrong thing is found where I was looking for the thing I wanted. After all, he posits, "there is no absolute void in nature."[b] Nothingness is a merely practical pseudo-concept, and thus it is theoretically inadmissible when we wish to speak of the whole. As a result, we must accept the seemingly paradoxical idea that there is *more* in nothingness or negation than there is in existence, not less.[c] Unlike affirmation, negation is not a self-sufficient act; it is only half of an act, the other half being an implicit affirmation of existence. To say "this table is not white" is to assert the existence of a white table and then to negate it. It is thus a second-order judgment on a judgment, whereas "this table is black" is a direct judgment applied to a thing. Negation is practical insofar as it has a pedagogical or social role to play—I might not say (or care) what color the table actually is, but since you are shopping for a white table, I want to alert you that your possible judgment "this table is white" is wrong. Bergson thus defines something like a systematic or structural error of intellect, which brings us back to the first illusion: the cinematographic mechanism of thought.

The intellect, designed for action, fixes itself on the immobile goal and only retains of the world around it the aspects that allow it to reach that goal. The qualities of matter, which we sense are in fact ever

a *Ibid.*, 276.
b *Ibid.*, 281.
c *Ibid.*, 285–86.

changing, are condensed and solidified into states or blocks. But these are just so many "views or snapshots" that we take of what is in fact continuously changing.[a] Thus, "whether it is a question of a qualitative movement, an evolutionary movement, or a movement in extension, the mind is organized so as to take stable views or snapshots of instability."[b] Bergson thus likens the intellect to the cinematograph, an early film-based motion picture device that could serve as both a camera and a projector, insofar as it reconstructs movement by capturing a series of views or snapshots and projecting these immobile moments in succession. He writes: "Whether it is a question of thinking, expressing, or even simply perceiving becoming, we hardly do anything other than set into motion a sort of inner cinematograph. We might thus summarize all of this by saying that the *mechanism of our ordinary knowledge is cinematographic in nature*."[c] As Bergson quickly shows, this already sets us on the path toward Zeno's paradoxes. Trying to find out what happens between two snapshots will only lead to further divisions, and so on to infinity. We can never get to real becoming; we are stuck with an artificial reconstruction. To get to real becoming, "we would have to place ourselves back within that reality," and this is precisely what philosophy has failed to do across the history of systems.[d]

The remainder of Chapter IV focuses on the history of philosophy as a function of cinematographic thinking. The obvious connections to Zeno's paradoxes are explored first, which Bergson suggests can only be avoided if we renounce "the intellect's cinematographic habits" and sympathize with movement rather than translating it into juxtaposed stopping points along a line.[e] Bergson goes so far as to argue that the Greek philosophy of Ideas, which posits a universal and stable being that is above the contingent realities of becoming, is the natural outcome of applying "the cinematographic mechanism of the intellect to the analysis of the real."[f] As such, ancient Greek thought still captures the "natural inclination" of human intellect—we are all, so to speak, born Platonists.

a *Ibid.*, 301.
b *Ibid.*, 303.
c *Ibid.*, 305. Italics in the original.
d *Ibid.*, 307.
e *Ibid.*, 312.
f *Ibid.*, 314.

In terms of modern science, the goal is to facilitate our action upon matter. Thus, it is again naturally at home in the cinematographic method. After exploring some superficial differences between ancient and modern science, Bergson proposes that the essential difference is that ancient science has a static understanding of time, whereas modern science takes time into account. But modern science adopts the intellect's natural conception of time, that is, spatialized time.[a] Real time (and real becoming) thus escape the cinematographic method of modern science. This "time-as-length" will never capture "time-as-invention."[b]

In modern metaphysics, we again find adherence to this natural cinematographic method, and so a blindness to invention and growth. This translates into the shared assumption that, rather than a universe that *endures* (and thus grows and evolves), we are in a universe where, in principle, "*everything is given*."[cd] Bergson carefully works through the similarities and differences between Descartes, Spinoza, and Leibniz on this question. Initially, Descartes fails to return to real time after having discovered universal mechanism in space. To overcome the vast difficulties faced with a universal mechanical science, modern metaphysics was forced to become radically dualist between body and soul, quantity and quality, ultimately sweeping the latter terms away as mere appearances or epiphenomena. "Having cut all ties" between soul and body, modern philosophy had to establish some form of parallelism (such as in Spinoza) or pre-established harmony (as in Leibniz). Ultimately, given their assumptions from the natural structure of the intellect, all of these approaches "are repulsed by the idea of a reality that would create itself as it advances, or in other words, the idea of an absolute *durée*."[e]

In the final two sections, Bergson turns to his two key interlocutors in this book: Kant and Herbert Spencer. Although he applauds Kant's ability to stop the slide back into ancient metaphysics and his basic intuition that the intellect must be "critiqued," Bergson argues that Kant too accepts the basic premise of a universal science or knowledge. Unable to embrace intuition, Kant did not see the need to question the origin of the intellectual frameworks themselves and was unaware of anything other

a *Ibid.*, 336.
b *Ibid.*, 341–42.
c *Ibid.*, 344.
d *Ibid.*, 349.
e *Ibid.*, 353.

than intellectual experience. In terms of Spencer, whose evolutionism is a constant (if at times implicit) target throughout *L'évolution créatrice*, Bergson recognizes his attempt to go beyond the abstract metaphysics of modern philosophy and to engage with concrete *durée*. But according to Bergson, Spencer took a shortcut. Rather than following the genesis of the intellect, he reconstructed evolution "from the fragments of the evolved," and so he too fell back into the cinematographic method.[a] Spencer's "false" evolutionism fails to embrace *creative* evolution. Only a "true evolutionism" could grasp how the intellect and materiality emerge as mutually complementary. Perhaps, suggests Bergson, this is the direction we see emerging in quantum physics and other contemporary sciences.

This long journey allows Bergson to leave behind the limited perspective of the mind's contact with itself and the individual experience of the virtual in memory so as to open up a philosophy that provides a "profound examination of becoming in general."[b]

NOTES ON THIS TRANSLATION

For readers familiar with the French original or with the previous English version, one of the more visually striking changes in this new translation will be the insertion of Bergson's section titles directly into the body of the text. This follows my decision to insert Merleau-Ponty's paragraph titles into my translation of his book *Phenomenology of Perception*, and I hope that here too the inclusion of these markers serves to give some guidance to the reader and some air to an otherwise dense text. In addition, if the reader is interested in following or comparing Bergson's original French, this edition provides two components. First, there is a bilingual presentation of the full Table of Contents. Since these titles include many of Bergson's key concepts, this provides something of a running glossary of my translation decisions. Second, the French pagination included throughout this translation is from the now-standard French version of *L'évolution créatrice* (2013), edited and annotated by Arnaud François, which is part of the series of critical editions of

a *Ibid.*, 363.
b *Ibid.*, 369.

Bergson's works edited by Frédéric Worms and published by Presses Universitaires de France.

It is also worth providing a brief explanation regarding the footnote and endnote structure and its relation to the French scholarly edition just mentioned. There are two kinds of footnotes that appear in the body of the text. First, all of the original notes by Bergson appear as footnotes, preceded by "Bergson's note." Bergson's references have been updated where needed. In addition, I have included a number of translator's footnotes, preceded by "TN" so as to immediately differentiate them from Bergson's notes. Both Bergson's notes and my translator's notes are indicated in the body of the text by superscript letters. In addition, this volume also includes an extensive set of editorial endnotes. These endnotes are a selection and translation of the editorial notes by Arnaud François from the French edition mentioned above. These endnotes are indicated in the body of the text by numbers. In the few cases in which I have added editorial material to the endnotes, this is again indicated by "TN." The editorial endnotes have been edited with the English reader in mind, which has occasionally resulted in the paraphrasing of some of the longer original endnotes or in the updating of references. For the few instances where my changes have been significant, I have placed an "*" at the end of the note, indicating that additional information or resources may be available in the corresponding endnotes in the French edition. Finally, to help the reader navigate the endnotes, I have indented the notes that refer to Bergson's footnotes rather than to the body of the text.

As the reader will quickly come to appreciate, François's extensive editorial endnotes have the effect of creating an incredibly useful network of references through his citations to other moments in L'évolution créatrice itself and to relevant moments in Bergson's other works. Indeed, one of my main goals with this English version was to provide this remarkable research tool to the English reader. This did, however, present some difficult decisions in terms of what paginations to prioritize. After all, there are various English editions of Bergson's books, and in the future there will surely be new English translations of the new French critical editions. Given that Worms has established a standard and stable pagination for the French editions of Bergson's works, I have opted to prioritize the French paginations when citing any of Bergson's works. Indeed, the most likely user of this network of citations will be the advanced student or scholar, who will also appreciate the references to the French editions. As such, all of the internal references in this translation refer to the French pages, which are found in the margins. In terms of the references to Bergson's other works—for example, "MM, 33"—the reader will need to look to the French critical edition of *Matière et mémoire*. For readers who do not have access to these editions, a document providing a rough correspondence between the French standard pagination and the most common English versions of Bergson's main works can be found on my personal website (www.donaldlandes.com). The English pagination, however, has been provided in the rare cases where I have included a quotation from those versions.

The reader will also find a selection of additional materials related to the reception of and debates surrounding *L'évolution créatrice*. These materials are largely based on the selected readings included in the French edition, but they have been reorganized and some additional materials have been added. The section includes relevant private correspondence, several debates between Bergson and his interlocutors from biology, mathematics, and theology, and notable philosophical commentaries.

TRANSLATION DECISIONS

One of the main translation decisions that will stand out for the reader is my decision to leave *durée* and *élan vital* in French. First, as mentioned above, Bergson's notion of *durée* is difficult to capture in English, given that the most obvious literal translation, "duration," is potentially misleading. The relationship between "time" (*temps*) and "duration" (*durée*) is not the same in French and English. Bergson opts for *durée* to refer to a non-quantifiable, continuous, heterogeneous, and qualitative time, such as the time lived by a consciousness, whereas the English term "duration" invokes precisely a quantitative length of time, the exact opposite of Bergson's *durée*. Less literal translations, such as "lived time" or "temporality," however, would risk over-translating in the direction of the phenomenological tradition. For these reasons, and also to mark this technical term in relation to Bergson's other works, I have used *durée* throughout this translation, and opted for "endure" where Bergson uses the verb *durer*.

If *durée* serves as a bridge to Bergson's other works, *élan vital* is nearly exclusively found in *L'évolution créatrice*. Given the subtlety of what Bergson is trying to convey with this image, this concept is already difficult to grasp in French. The formulation itself is interesting because—like "creative evolution," which seems to bring together the normally competing positions of evolutionism and creationism—*élan* is a term usually associated with mechanism and *vital* with its opposite, classical vitalism, two positions that Bergson explicitly criticizes. *Vital* of course refers to the domain of life and vitality; *élan*, for its part, has many connotations, including "momentum," "surge," "thrust," "fervor," "impulse," "impulsion," and "impetus," with this last option having been used by previous translators of Bergson's works. For Bergson, the *élan vital* is not a technical term but rather an image, and as such, his argument relies upon all of these various connotations remaining active. This is why the choice of "impetus" risks, I believe, overemphasizing the "physical force" aspects of this diverse set of connotations and obscuring an important "psychological" sense: what Bergson calls the *exigence de création* ("need for creation"). To preserve the richness of his image and the connection between the *élan vital* and the various other occurrences of *élan* by itself in the text, I have left these terms in French. But what exactly does Bergson intend by this enigmatic idea of an *élan* of life? And how can we understand the psychical aspects of this term without falling into vitalism or finalism? To help the reader grasp this French term, it is worth exploring Bergson's image more closely here.

Although *élan vital* rarely appears elsewhere in his writings, Bergson feels the need to offer an extended explanation of it in his final book, *Les deux sources de la morale et de la religion* (1932). There he explains that the concept first emerged because life, as it appears in organisms, does not admit of merely physico-chemical explanation, nor can accidental and external causality account for how life evolves from species to species. For after all, it is obvious that "the evolution of life occurred in certain definite directions."[a] But these directions are not merely imposed from the milieu, nor handed down by individuals through acquired traits. As such, there must be some kind of *poussée interne* ("internal thrust or impulse") that courses through all of life's manifestations and that orients life's evolution in its march toward higher complexity. In addition, theories of variation and adaptation, which have their use in explaining the details of specific evolutionary moments, fail to account

a *Ibid.*, 113.

for the general fact that life is like a "faculty of resolving problems."[a] In other words, life is not merely physical; it is in some ways psychical in nature, like a consciousness. Moreover, by recognizing life itself as a kind of unified (though diverse) movement, life becomes the kind of durational object that we can *sympathize* with, that is, it becomes available to intuition rather than merely the external view of intellect. As a result, by observing this *élan vital* expressing itself along the main lines of evolution, we can grasp it as not merely a force, but as a "virtual" reservoir that contains these individuated characteristics in a pre-individual state of "reciprocal implication" or interpenetration. Thus, Bergson's paradoxical formulation allows us to grasp life as being like a consciousness or virtual in the process of actualization through its contact with matter.[b] The human being's privileged place along the line of intellect thus remains connected to instinct (and life in general) within a virtual that predates their separation in matter. The *élan vital* is, in this sense, that virtual and "simple reality" that interrupts itself into species and individuals, and that falls into matter while nonetheless continuing along and growing. *Élan vital* thus names an oriented yet creative evolution, beyond mechanical causality and any notion of a predetermined plan or finalism. Its status is real but virtual, and in continuous negotiation with matter as both its instrument and its obstacle.

This overview connects with the various occurrences of the term in *L'évolution créatrice*, where Bergson writes "of an *original élan* of life that passes from one generation of germs to the next through the intermediary of the mature organisms that form the link or hyphen between the germs."[c] This first definition captures the negotiation of this virtual with matter, allowing us to grasp the more physical aspects of the image. Later in the text, Bergson insists on the psychical aspects, writing: "The *élan of life* that I am discussing consists, in short, in a need for creation."[d] This use of the image shows that the *élan* is less a substantial force, and more something like a will or a desire, language that evokes Schopenhauer and Nietzsche. Here the *élan vital* is a non-localized effort to introduce "the largest possible amount of indetermination and

a *Ibid.*
b *Ibid.*, 115.
c *Ibid.*, 88.
d *Ibid.*, 252.

freedom" into the world.[a] This is the kind of effort, will, or desire we associate with consciousness, so long as we can think of a non-personal type of consciousness.[b] And finally, the *élan vital* is not an artifact of a lingering classical vitalism, because it does not aim at a projected goal. It is oriented not like a pre-existing plan, but rather in the manner of a virtual. As we already saw above, "when conceived of in itself, life is an immensity of virtuality, a mutual encroachment of thousands and thousands of tendencies that will, nevertheless, only be 'thousands and thousands' once they have been externalized in relation to each other, that is, spatialized."[c]

In addition to using *élan* by itself, Bergson also relies upon a series of related terms to capture the concept. One such term, already mentioned above, is *poussée* ("thrust," "burst," "impulse," as well as "sprout") and the related verb *pousser*. This term is particularly useful in connecting both mechanical and biological concerns, and I have thus at times highlighted the presence of these terms in the text. In addition, Bergson often uses the terms *virtuel* or *virtualités* in relation to the *élan*, and it is important to note that the virtual should not be identified with the "possible." On Bergson's logic, quite marvelously identified by Deleuze, the movement of the "possible" is toward the real (*réalisation*), whereas the virtual is already real, its movement is toward the "actual" (*actualisation*). This terminology is not always consistent in Bergson, but it is essential for the underlying logic of the virtual explained above.[d]

It is also worth noting some of the peculiarities surrounding Bergson's use of the terms *intelligence* and *esprit*. The French word *intelligence* refers to both the faculty of knowing (the "intellect") and the quality of reason ("intelligence"). Throughout the text, Bergson relies upon these multiple meanings, making it necessary to choose each time between "the intellect" and "intelligence." The reader should keep this ambiguity in mind: the use of "the intellect" in this translation is not intended to evoke a substantial entity, but rather a certain activity or capacity. In addition, the French word *esprit* can be translated as either "mind"

a *Ibid.*
b *Ibid.*, 262.
c *Ibid.*, 259.
d For more, see Henri Bergson, "Le possible et le réel [1920/1930]," in *La pensée et le mouvant*, 99–116; "The Possible and the Real," in *The Creative Mind*, 73–86. See also Deleuze, *Bergsonism*, 94–98.

or "spirit." Bergson generally means something more along the lines of "mind," as in a psychical or mental state, yet "mind" risks obscuring the important aspect of "spiritualism" that resonates in the French text. On the other hand, "spirit" and spiritualism have too strong of a religious connotation in English. I have generally opted to translate *esprit* as "mind" or "psychical," occasionally adding "spirit" or "[*esprit*]" wherever the link to the French spiritualist tradition is important. In this same direction, it should be mentioned that Bergson uses the French word *science* in an older or more literary sense that refers not only to the discipline of science or the positive sciences, but also more loosely to "knowledge" in general. In these cases, I have translated *science* as "knowledge." He also occasionally capitalizes the term, intending us to understand the supposed intelligible structure of reality in itself, following its use in the German idealist tradition, specifically Kant, Hegel, and Fichte.

Bergson often employs the verb *se représenter* for the activity of thinking, which can be translated as "to imagine," "to picture," "to think about," or "to represent to oneself." The last option, although at times necessary, might be heard in English as more connected to early analytical philosophy than in fact it is, since Bergson rarely thinks of images or sense data being presented *to* a separable mind. Thus, I have often used the terms "to imagine" or "to picture" in a less technical sense. In a related formulation, Bergson regularly uses *vues* (translated as "views or snapshots") that the intellect takes of an object. Given the relatively non-technical meaning of "view" in English, I have added "snapshots" to ensure that the idea of a captured and immobilized image (such as on the film of the cinematograph) is not lost in the translation.

Throughout the text, Bergson deploys a complex network of terms that invoke traveling images of pathways, railways, roadways, trajectories, journeys, and so on. These images are essential for his various arguments about the diverging lines of the evolutionary movement and his criticisms of Zeno's worldview and the cinematographic method of thought. I have tried to maintain these images as much as possible in the English. Another directional image that becomes increasingly central is the distinction between a downward movement or tendency (*descente*) and an upward movement or tendency (ascent or *montée*). This family of terms has been preserved in the English, and it should be noted that the intellect follows a natural inclination (*pente*) downward, like matter, whereas intuition and consciousness are regularly presented as ascending. In a related set of terms, Bergson uses *se dépasser*, or more rarely *se transcender*, to mean "to transcend" or "to go beyond" oneself, a particularly important notion for his general call to go beyond the human condition insofar as the human condition is limited to the intellect.[a] The term "to transcend" is reserved in this translation for when Bergson explicitly uses *se transcender*. In addition, Bergson uses *Principe* ("Principle") in its metaphysical sense of "source" or primary metaphysical reality. To avoid a misreading of "principle" as merely a "fundamental proposition, premise, or law," I have often translated it as "principle or source."[b] Another important differentiation is among Bergson's various locutions derived from the verb *faire* ("to make" or "to do"): *fait* (as in "made") and *tout fait* (as in "ready-made") primarily refer to things as fixed and spatialized by the intellect; whereas *se faire* ("to make oneself") and *se faisant* ("being made" or "in the making") refer to *durée*, consciousness, and becoming in general.

In terms of vocabulary related to biology, it is important to first note Bergson's use of the adjective *organisé* ("organized" or "organic") to refer to living beings. The more literal translation as "organized" is potentially misleading in English, since this might imply that there was an organizer, whereas for Bergson organic beings are not organized from the outside, but are rather an internal process of maturation and organic development (*organisation*). This leads to another key distinction between *organisation* ("organic development") and *fabrication*. The former is the kind of development or creation that takes place in living beings or systems. *Fabrication* and the adjective *fabriqué* could be translated as "manufacturing" and "manufactured," yet this might imply a specialized, industrial kind of making, for instance one that involves large-scale machinery. For Bergson, *fabrication* characterizes any process of making insofar as it is guided by intellect, that is, a making that is guided by a predetermined plan and that involves assembling parts, regardless of scale. I have thus opted for "fabrication" and "fabricated," asking the reader to suspend any further associations with more colloquial uses of fabrication to mean "concocted" or "invented" with deceitful intent. So the key distinction then is between organic development (with living

a See Ansell-Pearson, *Bergson: Thinking Beyond the Human Condition*, particularly the introduction.
b *L'évolution créatrice*, 275.

bodies that develop organically) and fabrication (with fabricated objects made according to a plan).[a] On a related point, Bergson occasionally uses *évolution* in the more specific sense of the "development" of a particular individual rather than the overall movement of life. In these cases, I have often used "development" to avoid confusion. Finally, Bergson's use of the term *progrès* (literally, "progress") does not usually refer to a process of improvement, but rather to a simple "progression" or "continuing along."

Given that this is a historical document, I have only rarely updated Bergson's gendered formulations, in which he follows the French linguistic customs predominant at the time. For example, *il* ("he") for the subject, *l'homme* ("man") for human beings, and the linguistic implication that *un philosophe* ("a philosopher") will always be a man.

Finally, in terms of the various technical vocabularies that Bergson borrows from the biological sciences (germen, germ-plasma, acquired characters or characteristics, orthogenesis, sudden variations, etc.) and the numerous organisms named either in Latin or by their common names, I have made every effort to adopt standard contemporaneous usage as found in the English versions of the texts Bergson cites, leaving questions of modernization or corrections to the editorial endnotes provided in this volume. For instance, I have preserved Bergson's style of capitalizing the names of animals.

TRANSLATOR'S ACKNOWLEDGMENTS

I would first like to acknowledge a sincere debt of gratitude to Arnaud François, whose incredible philosophical and scientific research, as well as his significant editorial work on internal and external citations, indexing, and the compiling the dossier of readings, created a truly remarkable French scholarly edition that serves as the basis for this translation. It has been a genuine pleasure bringing this material into English. I particularly want to thank him and the series editor of the critical editions of Bergson's works in French, Frédéric Worms, for their permission to adjust this scholarly apparatus as needed for the English reader. I appreciate their trust and flexibility on all of these points.

---

a *Ibid.*, 93, 140–43.

I also want to acknowledge the work of Arthur Mitchell, the first translator of *Creative Evolution*, published in English in 1911, only four years after its original French publication. Although Mitchell's translation was officially authorized by Bergson (who was nearly fluent in English), it is unclear to what degree Bergson himself was involved in the translation decisions. For example, Mitchell's "Translator's Note" includes a much more effusive acknowledgment of contributions by William James, H. Wildon Carr, and editorial assistant Millicent Murby. Despite some of its shortcomings (which I will not list here), Mitchell's translation helped to bring *L'évolution créatrice* to English readers. Although I only consulted Mitchell's version during the final revisions of my own translation, I want to acknowledge that his decisions often helped to shape or reinforce my own.

This project would not have been possible without the expert guidance, patience, and support of Tony Bruce (Senior Publisher, Philosophy) and his assistant at Routledge, Adam Johnson. It has been a pleasure working with them again on a major translation. Thank you sincerely to Elizabeth Grosz for her generosity toward this project and for providing the excellent Foreword. I would also like to acknowledge the three anonymous reviewers for their careful reading, helpful suggestions, and enthusiasm for this project.

This translation would also not have been possible without the support of many members of my local university community. In particular, I would like to acknowledge Sophie-Jan Arrien, co-director of the *Laboratoire de philosophie continentale*; Pierre-Olivier Méthot, for his expert advice and willingness to share resources on Canguilhem; and Victor Thibaudeau, who served as Dean of the *Faculté de philosophie* at the beginning of the long *durée* of this project. Finally, thank you to the many undergraduate and graduate students who have worked with me on Bergson during these years. Beyond Université Laval, I would like to note that this project would never have gotten off the ground without the advice and support of Alia Al-Saji at McGill University and Leonard Lawlor at Penn State University. And a particular thank you to Rawb Leon-Carlyle for his various contributions. Finally, part of the research for this project was made possible by a grant from the *Fonds de recherche du Québec – Société et culture*.

But it is without doubt that this translation only exists thanks to the real and virtual contributions made by my partner, Kathleen Hulley.

Her direct contributions include not only her incredibly careful work researching and compiling the scholarly bibliographies included here, her indefatigable editing of the massive number of footnotes and endnotes that complete this translation, and her multiple readings of drafts, but also her expert translation of two of the articles included in the reading section below. But none of this is to mention her virtual contributions, which permeate every line of this translation. From her unwavering moral support and patience, her linguistic and philosophical intelligence in discussions about difficult translation decisions, to her deep understanding of the vicissitudes of the academic and philosophical labor that goes into a translation of this magnitude, I thank her for all she has done to make this an inestimably better translation and a joyful experience.

PUBLICATION ACKNOWLEDGMENTS

The majority of the material in this book is either in the public domain or the translation rights have been granted by Presses Universitaires de France. We would like to acknowledge as well the special permissions granted as follows. "Bergson and the Ammophila Sphex" (1959), by Raymond Ruyer, translated by Tano S. Posteraro, published with kind permission of Taylor and Francis Group; "The Ideas of Bergson," from *Course on Nature at the Collège de France* (1956–1957), by Maurice Merleau-Ponty, translated by Robert Vallier, published with the permission of Éditions Seuil and Northwestern University Press; and "Lecture Course on Chapter Three of Bergson's 'Creative Evolution,'" by Gilles Deleuze, translated by Bryn Loban, published with the permission of Johns Hopkins University Press. I would also like to thank Arnaud François and Frédéric Worms for their permission to select and edit the critical apparatus as needed for the English reader.

Table des matières bilingue

Bilingual Table of Contents

a In his Table of Contents, Bergson divides his sections into groups as indicated. I have followed his grouping of sections here and also added an asterisk in the body of the text to indicate the breaks between these groups of sections.

b This subsection title was not originally included in Bergson's Table of Contents, but
   appears nonetheless as a distinct running header.

Chapter IV
The Cinematographic Mechanism of Thought and the
Mechanistic Illusion. A Glance at the History of Systems.
Real Becoming and False Evolutionism   **238**

---

c This section title does not appear in the original Table of Contents, but rather as a running header from French pages 238–51. Since this constitutes a relatively coherent section, I have included it here.

d This section title does not appear in the original Table of Contents, but rather as a running header from French pages 252–71. Since this constitutes a relatively coherent section, I have included it here.

e This section is a slightly revised version of material that Bergson published earlier: Henri Bergson, "L'idée de néant," *Revue philosophique de la France et de l'étranger* 62, no. 11 (1906): 449–66. I have added the article's title as a section title to help mark off this previously published material.

---

f   TN: Bergson's original Table of Contents indicates that this section begins on page 323.
    The discussion of modern science and the comparison with ancient science, however,
    clearly begin on page 328 in the French pagination.

INTRODUCTION

The history of the evolution of life, incomplete as it still may be, already allows us to catch sight of how the intellect has been constituted by an uninterrupted progression along a line that ascends through the series of Vertebrates up to man.[a] This history shows, in the faculty of understanding, which is itself an extension of the faculty of action, an increasingly precise, complex, and supple adaptation of the consciousness of living beings to the conditions of existence in which they find themselves. From this it follows that our intellect, in the strict sense of the word, is meant to ensure the perfect insertion of our body into its surroundings and to imagine the relations among external things—in short, to think matter.[b] This will indeed be one of the conclusions of the

---

a TN: For more on the translation of the various terms that Bergson uses in relation to *l'intelligence humaine* ("human intellect"), see the Translator's Introduction. It is also worth noting that Bergson uses the word *homme* for both "man" (as in mankind) and "man" as a singular noun. For more on Bergson's use of gendered language, see the Translator's Introduction.

b TN: For more on Bergson's variable use of the pronominal verb *se représenter* (which I have translated here as "to imagine," but which could also be translated as "to conceive of," "to picture," or "to represent to oneself"), see the Translator's Introduction. For the interchangeability of these terms, see below, 285.

DOI: 10.4324/9781315537818-1

present essay.[1] We will see that human intellect feels at home so long as we allow it to remain among inert objects, particularly among solids, where our action finds its footing and where our industriousness finds its tools.[2] We will see that our concepts have been formed in the image of solids and that our logic is, above all else, the logic of solids. And we will see that, for the same reasons, our intellect excels in geometry, where the kinship between logical thought and inert matter is revealed and where the intellect, after the lightest possible contact with experience, need only follow its natural movement to go from one discovery to the next with the certainty that experience marches along behind it and will invariably prove it right.

But it also follows from this that our thought, in its purely logical form, is incapable of conceiving of the true nature of life and the deep meaning of the evolutionary movement.[a] Our thought was created by life, in specific circumstances, so as to act upon specific things. How could it encompass life, of which it is merely an emanation or an aspect? Deposited along the way by the evolutionary movement, how could it be applied to the entire evolutionary movement itself? One might as well claim that the part is equal to the whole, that the effect can reabsorb its own cause, or that the pebble left behind on the beach sketches out the form of the wave that carried it there.[3] In fact, we clearly sense that none of the categories of our thought (unity, multiplicity, mechanical causality, intelligent finality, etc.) can be unequivocally applied to the things of life—who can say where individuality begins or ends, whether the living being is one or many, or whether the cells come together to form the organism or the organism divides up into cells? It is in vain that we force the living being into one or another of our frameworks. All of the frameworks crack. They are too narrow and, above all, too rigid for what we would like to fit into them. Moreover, our reasoning, so sure of itself when it circulates among inert things, feels ill at ease on this new terrain. It would be very difficult to cite a single discovery in biology owing to pure reason alone. And, more often than not, when experience finally shows us how life goes about obtaining a certain result, we find that its way of proceeding is precisely the one that we would never have thought of.

And yet, evolutionary philosophy immediately extends the explanatory procedures that were successful for brute matter to the things of

a TN: *...se représenter la vraie nature de la vie, la signification profonde du mouvement évolutif.*

life. It had begun by showing the intellect to be a local effect of evolu-   vii
tion, a perhaps accidental glimmer of light that illuminates the comings
and goings of living beings within the narrow passage open to their
action. Then, all of a sudden, and forgetting what it had just told us, this
philosophy transforms this lantern, used in the depths of a cavern, into
a Sun capable of illuminating the world. Boldly and with the powers of
conceptual thought alone, it begins the ideal reconstruction of all things,
even life. True, this philosophy runs into such formidable difficulties
along the way, and sees its logic leading to such strange contradictions,
that it quickly renounces its initial ambition. It no longer claims to
recompose reality itself, but only to provide an imitation of the real, or,
rather, a symbolic image of it. The essence of things escapes us and will
always escape us; we move among relations and the absolute is beyond
our competence, so let us come to a halt before the Unknowable.[4] But
having initially shown so much pride in human intellect, this is surely
an excess of humility. If the living being's intellectual form has been
molded, little by little, upon the reciprocal actions and reactions of certain bodies and their material surroundings, how could this form not
provide us with something of the very essence of which those bodies
are made? Action cannot move in the unreal.[5] Now, I could accept
that a mind [esprit] born to speculate or to dream remains outside of
reality, that it deforms and transforms reality, perhaps even that it creates
reality, such as when we create the shapes of men or animals that our
imagination recognizes in a passing cloud.[a] But an intellect oriented
toward the action that will be accomplished and toward the reaction that
will follow, palpating its object so as to continuously receive a moving
impression of it, such an intellect is one that touches something of the
absolute.[6] Had philosophy never shown us the contradictions that our
speculation runs up against and the impasses that it leads to, would the   viii
idea of casting into doubt the absolute value of our knowledge have ever
even arisen? But these difficulties and contradictions are born when we
apply our habitual forms of thought to objects upon which our industriousness has no purchase and for which, consequently, our frameworks
have not been made. By contrast, intellectual knowledge, insofar as it

---

a TN: For more on the translation of *esprit* by "mind," and the occasional use of "spirit," as
   well as on related questions to Bergson's place within the French "spiritualist" tradition,
   see the Translator's Introduction.

relates to a certain aspect of inert matter, must surely present to us a faithful impression of it, being itself an exact stereotype of the particular object. This knowledge only becomes relative if it then claims to show us life itself, that is, the stereotypist who took the impression.[a]

Must we thus give up trying to gain a deeper understanding of the nature of life? Must we be content with the mechanistic representation of life that the understanding will always give us, a representation that is necessarily artificial and symbolic since it reduces the overall activity of life to the form of a particular human activity, an activity that is itself but a partial and local manifestation of life, an effect or a residue of the vital process?

This would indeed be necessary if life had employed all that it contained of psychical virtualities to create pure understandings, that is, to produce geometers.[b] But the evolutionary line that ends in man is not the only one. Along other diverging pathways, other forms of consciousness have developed that have neither been able to free themselves from external constraints nor been able to gain control over themselves in the way that human intellect has, but that are no less expressions of something immanent in, and essential to, the evolutionary movement. By bringing these forms together, and then by making them fuse with the intellect, would we not thereby obtain a consciousness that is coextensive with life itself and capable of obtaining an integral, though perhaps fleeting, vision of life by suddenly turning back toward the vital thrust that it senses behind itself?[c/7]

ix

---

a TN: Bergson here uses the term *clichée* in the technical sense, for a stereotype relief printing plate, and the *clicheur* is the stereotypist, the technician who creates the plate.

b TN: ...*virtualités psychiques*... Bergson uses various forms of the word *virtuel* in this work, which can be translated as either "virtual" or "potential." To distinguish these uses from his equally frequent use of *potentiel* and *puissance* ["power" or "potentiality"], as well as to make apparent his important influence on later uses of "virtual" in French philosophy, I have chosen to translate *virtuel* and *virtualités* with forms of the word virtual in English. For more on the translation of *psychique* as "psychical" or "mental," and the related terms *psychologique* and *esprit*, see the Translator's Introduction and the Index.

c TN: ...*la poussée vitale*... The word *poussée* in fact belongs to two families of metaphors at work throughout the book and appears even in the final sentence. First, it can be translated as it is here as "thrust" ("impulse" or "momentum"), and as such invokes images from mechanics. Second, the related verb *pousser* ("to sprout" or "to germinate") connects to images from biology. Although Bergson does not use the word *élan* here, this is a first allusion to what he will call below the *élan vital*. See the Translator's Introduction.

Even so, it will be said that we do not go beyond our intellect since we observe these other forms of consciousness with or through our intellect. And this objection would be correct if we were pure intellects; i.e., if surrounding our conceptual and logical thought there did not remain a vague nebulosity made of the very same substance from which the luminous core we call the intellect was formed.[8] Certain complementary powers [*puissances*] of the understanding reside in this vague nebulosity. We have but a confused sense of these powers so long as we remain locked up within ourselves, but they will become clear and distinct when they catch sight of themselves at work, so to speak, in the evolution of nature.[a/9] In this way, they will learn what efforts they have to make so as to intensify and to dilate in the very direction of life.

This is to say that the *theory of knowledge* and the *theory of life* appear to me to be inseparable. A theory of life not accompanied by a critique of knowledge is forced to accept at face value the concepts that the understanding makes available to it; it can do no more than enclose the facts, be they willing or not, within pre-existing frameworks that it takes to be definitive. As such, it establishes a symbolism that is convenient, and perhaps even necessary, for the positive sciences, but it does not achieve a direct vision of its object. On the other hand, a theory of knowledge that does not place the intellect back into the general evolution of life will teach us neither how the frameworks of knowledge were constituted nor how we can expand them or go beyond them. These two areas of research—the theory of knowledge and the theory of life—must come together and, through a circular process, propel each other forward indefinitely.

Together they will be able to resolve philosophy's most difficult   x
questions through a method that is more reliable and that remains closer to experience. For, were they to succeed in their common project, they would allow us to witness the formation of the intellect and, thereby, the genesis of that matter whose general configuration is sketched out by our intellect.[10] They would unearth the very roots of nature and of the mind. They would substitute for Spencer's false evolutionism—which

---

a TN: Bergson's use of the verbs s'*éclaircir* ("to become clear") and se *distinguer* ("to become distinct") alludes to the Cartesian natural light of the intellect with its clear and distinct ideas, as reflected in the translation I have chosen here and throughout.

consists first in cutting up an already evolved reality into smaller bits, themselves no less evolved, then recomposing that reality out of these fragments and thereby taking for granted everything that it was supposed to explain—a true evolutionism that would follow reality in its genesis and growth.[a]

But a philosophy of this type is not made in a day. Unlike philosophical systems, properly so-called, where each was the work of a man of genius and appeared as a block, to take or to leave, this new philosophy can only be constituted through the collective and gradual efforts of many thinkers—and indeed of many observers as well—completing, correcting, and improving upon each other.[11] As such, the present essay does not aim to resolve the most difficult problems in a single stroke. Rather, it simply aims to define the method and to give a first glimpse of its possible application with regard to certain key points.

The plan for accomplishing this was in fact marked out by the topic itself. In the first chapter, I try out on the evolutionary movement the two ready-to-wear garments that our understanding has at its disposal: mechanism and finality.[b/12] I show that neither of these two garments fit properly, but that one of them could be re-cut and re-sewn, and in this new form fits somewhat better than the other. In Chapter II,

xi

---

a TN: Bergson regularly uses the verb *se donner* ("to assume," "to take for granted," or "to take as given") when criticizing other philosophers for the fallacy of begging the question or of making unjustified (and unacknowledged) metaphysical assumptions. Bergson's choice of the somewhat colloquial phrase *se donner par avance* for this technical argument helps to tie together his various criticisms of philosophical "systems." The formulation returns particularly often in his criticisms of Spencer. See below, 367.

b Bergson's note: Moreover, it is hardly a new idea to consider life as transcendent to both finality and mechanism. In particular, this idea is discussed with profound insight in three articles by C. Dunan on the problem of life [Charles Dunan, "Le problème de la vie," *Revue philosophique de la France et de l'étranger* 33, no. 1 (January 1892): 1–35; no. 2 (February 1892): 136–63; no. 5 (May 1892): 519–48]. I have met with Monsieur Dunan more than once during the development of this idea. Nevertheless, the perspectives that I offer on this point, as well as on related questions, are the very same ones that I had previously formulated, now already several years ago, in my *Essai sur les données immédiates de la conscience* (1889). Indeed, one of the principal goals of that essay was to show that psychological life is neither unity nor multiplicity, that it transcends both the *mechanical* and the *intelligent* since mechanism and finalism only make sense when there are "distinct multiplicities," "spatiality," and consequently an assemblage of pre-existing parts. "Real *durée*," by contrast, signifies both undivided continuity and creation. In the present book, I apply these same ideas to life in general, itself conceptualized, thereby, from the psychological point of view.

to go beyond the perspective of the understanding, I attempt to reconstitute the main lines of evolution that life has followed alongside the one that led to human intellect. In this way, the intellect is placed back within its generative cause; it becomes a question of grasping this cause in itself and of following its movement.[13] This is more or less the project I attempt, however incompletely, in the third chapter. The fourth and final chapter is meant to show how our understanding itself, by disciplining itself in a certain manner, can prepare the way for a philosophy that goes beyond it. For this, a glance back at the history of systems became necessary, in addition to an analysis of the two main illusions that threaten human understanding from the moment it begins to speculate about reality in general.

1

ON THE EVOLUTION OF LIFE. MECHANISM AND FINALITY[a]

[ON *DURÉE* IN GENERAL][b]

1   The existence of which we are most certain and that we know the best is unquestionably our own, for when it comes to all other objects, the notions we have of them can be judged to be external and superficial, whereas we perceive ourselves inwardly and profoundly.[1] But what exactly do we observe in this way? In this privileged case, what

a   TN: Bergson uses *finalité* ("finality") and *finalisme* ("finalism") interchangeably to refer to theories in which natural processes or evolutionary change are in some way directed toward a predetermined end or goal. He only rarely uses "teleological," perhaps because in *L'évolution créatrice* he attacks the idea of finalism for its role in biology rather than in philosophy.

b   TN: I have chosen to preserve the French word *durée* in this translation to help mark its role as a technical term in Bergson's thought and to avoid potential misleading connotations in the literal equivalent "duration," which often implies a measurable length of time. For Bergson, *durée* is specifically not a quantifiable length, as he makes clear already in the second paragraph of this chapter. *Durée* is more of a qualitative time, a continuous succession and continuous change that can only artificially be divided into quantifiable, immobile, or spatialized parts. For more, see the Translator's Introduction.

DOI: 10.4324/9781315537818-2

is the exact meaning of the word "exist"? Let me briefly return to the conclusions of an earlier work.[2]

The first thing I observe is that I pass from one state to another. I am hot or cold, I am happy or sad, I work or do nothing, I look at what is around me or think about something else. Sensations, feelings, volitions, ideas—my existence is divided up among these modifications and they color it in turn. As such, I change continuously. But even this does not say enough. The change in question is far more radical than one might at first have believed.

In fact, I speak about each of my states as if it formed a block. I say that I change, of course, but to me the change seems to reside in the passage from one state to the next. As for each state, taken separately, I like to believe that it remains what it is during the entire time that it is occurring. And yet, the slightest effort to pay attention to these states would reveal to me that there is no emotion, no idea, and no volition that is not itself being modified at each moment. If a mental state ceased to vary, its durée would cease to flow. Consider the most stable of internal states: the visual perception of an immobile external object. Even though the object remains the same, and despite the fact that I look at it from the same side, from the same angle, and on the same day, the vision that I now have of it nevertheless differs from the one I just had, even if only because my vision has aged by an instant.[3] My memory is there, and it pushes something from that past into this present. The state of my soul, advancing along the course of time, continuously swells with the durée that it gathers—it, so to speak, snowballs upon itself. When it comes to more profoundly inner states (sensations, emotions, desires, etc.), this will be even more the case since they, unlike a simple visual perception, do not correspond to an invariable external object. But it is convenient to ignore this uninterrupted change, and to only take note of it when it becomes large enough to transmit a new attitude to the body or to give a new direction to the attention. This is the precise moment we discover that we have changed states. The truth is that we change constantly and that the state itself is already a state of change.

This is to say that there is no essential difference between passing from one state to another and persisting in the same state. If the state that "remains the same" is more varied than we generally believe, then inversely the passage from one state to another resembles a single extended state much more than we usually imagine—the transition is continuous. And yet, precisely because we turn a blind eye to the incessant variation of each psychological state, when that variation becomes considerable enough to force itself upon our attention, we are obliged to speak as if a new state had been juxtaposed to the previous one. We assume that this new state in turn remains invariable, and so on indefinitely. The apparent discontinuity of psychological life thus comes from the fact that we focus our attention upon it through a series of discontinuous acts. Where there is in fact nothing but a gentle slope, we believe that we see the steps of a staircase because we follow the broken line of our acts of attention. True, our psychological life is full of unforeseen events [pleine d'imprévu]. Thousands of sudden events arise that seem to contrast sharply with those that came before or to have no connection with those that follow. But the discontinuity in their appearances stands out against the continuity of a background from which they emerge and to which they owe the very intervals that separate them. These events are the timpani strikes that ring out every now and again in the symphony. Our attention focuses on them because it finds them more interesting, but each one is carried along by the flowing mass of our whole psychological existence. Each one is simply the best illuminated point of a moving zone that includes all that we feel, think, and desire—in short, all that we are at a given moment.[4] In reality, it is this entire zone that constitutes our state. Moreover, we can say that states defined in this way are not distinct elements. They continue into each other in an endless flowing.[a]

And yet, since our attention has artificially distinguished and separated these states, it is obliged to then reunite them through some artificial connection. So, it imagines an amorphous, indifferent, and unchanging self [moi] upon which the psychological states that it itself had erected into independent entities would be threaded or strung together. Whereas in reality there is a flow of fleeting nuances that encroach upon each other, our attention sees rather separate and so-to-speak solid colors,

a   TN: *Ils se continuent les uns les autres en un écoulement sans fin.* I have translated Bergson's use of *se continuer* here as "to continue" so as to highlight the breaking down of borders between the states and to maintain his allusion to the "continuity" of change. It might also be rendered (a bit more freely) as "to merge" or "to fuse" together, since Bergson is perhaps also thinking of his distinction between juxtaposition (of external parts) and fusion (of interpenetrating moments). *Écoulement* is translated throughout as "flowing" or "flowing by."

juxtaposed like the various pearls of a necklace—it has no choice but to posit a thread, itself no less solid, able to hold the pearls together.[5] But if this colorless substrate is ceaselessly colored by what covers it over, then for us, in its indetermination, it is as if it did not exist at all. Now, we only ever perceive that which is colored, namely, psychological states. In fact, the "substrate" is not a reality. For our consciousness, it is a mere sign intended to serve as a constant reminder of the artificial character of the operation by which our attention juxtaposes one state after another, when there is rather a continuity that is unfolding.[6] If our existence was composed of distinct states that had to be synthesized together by an unchanging "self," there would be no durée for us.[7] A self that does not change does not endure [dure], and a psychological state that remains identical to itself so long as it is not replaced by the following state does not endure either.[a] This being the case, try as we might to string these states together one after another upon the "self" that supports them, these solids threaded together upon something solid will never equal the durée that flows. The truth is that we obtain merely an artificial imitation of inner life, a static equivalent that lends itself better to the requirements of logic and language precisely because real time has been eliminated from it. But when it comes to psychological life, such as it unfolds beneath the symbols that cover it over, we can easily see that time is its very fabric [étoffe].[b/8]

Besides, there is no fabric more resistant or more substantial.[9] For our durée is not one instant replacing another—if that were the case, only the present would ever exist, and there would be no continuation of the past into the actual, no evolution, and no concrete durée.[c/10] Durée is the continuous progression of the past, gnawing into the future and swelling up as it advances.[11] From the moment the past begins to grow continuously, so too does it begin to preserve itself indefinitely. Memory,

4

---

a TN: In order to maintain the connection with the noun durée, wherever Bergson uses the verb durer, I have opted for a form of the English verb "to endure."

b TN: ...le temps est l'étoffe même. Note that étoffe could also be translated as "stuff" or "material," but "fabric" resonates with Bergson's various metaphors relating to garments and tailoring, including the terms "ready-to-wear," "ready-made," "made-to-measure," etc.

c TN: ...actuel. Bergson usually uses the term actuel to mean "actual" (as opposed to "virtual"), but also employs it more colloquially to mean "current" or "present." Given the importance of the opposition between the actual and the virtual for understanding Bergson's metaphysics, I have tried to maintain this word wherever possible, sometimes writing "actual or present" where necessary.

as I have tried to establish elsewhere, is not a faculty for filing away recollections in a drawer or for recording them in a register.[a] There is no register and no drawer; properly speaking there is not even a faculty here since a faculty operates intermittently—when it wants to or when it can—whereas the piling up of the past upon itself continues without any breaks.[12] In reality, the past preserves itself, automatically.[b/13] It likely follows us around, at each instant, in its entirety. All that we have sensed, thought, and desired since our earliest childhood is there, leaning over the present that is about to join up with it, pressing against the door of consciousness that would prefer to leave it outside.[14] The cerebral mechanism is designed precisely so as to repress almost all of the past into the unconscious and to introduce into consciousness only that which is of a nature as to clarify the present situation, to aid the action that is being prepared, in short, to provide some *useful* work. At most, some non-essential memories manage to sneak their way through the half-open door, like contraband.[15] These messengers from the unconsciousness alert us to what we unknowingly drag along behind us. But, even when we have no distinct idea of it, we vaguely sense that our past remains present to us. After all, what are we, and what is our *character*, if not the condensation of the history that we have lived since our birth—and even before our birth since we carry prenatal dispositions with us? It may be true that we only think with a small part of our past, but it is with our entire past, including the original curvature [*courbure*] of our soul, that we desire, will, and act. Our past fully manifests itself to us through its thrust [*poussée*] and in the form of a tendency, although only a very small part of it becomes a representation.

This survival of the past makes it impossible for a consciousness to pass through the same state twice.[16] The circumstances might well be the same, but they no longer act on the same person since they take place at a new moment in the person's history. My personality, which is built up at each moment from accumulated experience, changes continuously.[17] By changing, my personality makes it impossible for a state—though it might well seem identical to itself on the surface—ever to repeat itself in depth. This is why our *durée* is irreversible. We can never relive a fragment of it because we would first have to erase the memory of all

---

a   Bergson's note: *Matière et mémoire*, Chapters II and III.
b   TN: *Le passé se conserve de lui-même, automatiquement.*

that followed. We could perhaps strike this memory from our intellect, but not from our will.

As such, our personality constantly sprouts, grows, and matures.[a] Each of its moments is something new added onto what came before. But let us go further: it is not just something new, it is something unforeseeable. My present or actual state is surely explained by what was there in me and by what was acting upon me a moment ago. In analyzing it, I would not find any other elements. But no intellect, not even a superhuman one, would have been able to foresee the simple and indivisible form that gives to these entirely abstract elements their concrete organization.[18] This is because foreseeing consists in projecting into the future what one has perceived in the past, or in imagining, as if in the future, a novel assemblage of already perceived but now differently arranged elements. But that which has never been perceived, and which is at the same time unique, is necessarily unforeseeable.[b] And this is precisely the case with each one of our states, taken as a moment in an unfolding history: each state is unique and could not have already been perceived, since it concentrates in its very indivisibility all that had previously been perceived along with what the present adds to this.[19] Each state is an original moment in a history that is itself no less original.

The completed portrait is explained by the physiognomy of the model, by the disposition of the artist, and by the colors laid out on the palette. And yet, even with the knowledge of what explains the portrait, no one, not even the artist, could have foreseen exactly what the portrait would end up being, since predicting this would be to produce it in advance of its being produced—an absurd and self-defeating hypothesis.[20] The same goes for the moments of our life, of which we are the artisans. 7 Each moment is a type of creation. And in the same way as the painter's talent is formed and deformed—or is, in any case, modified—under the influence of the works that he produces, so too is our personality modified by each one of our states as they emerge from us, they being the new form that we have just given to ourselves.[21] It is thus right to say that what we do depends on what we are; but it must be added that,

---

a TN: *Ainsi notre personnalité pousse, grandit, mûrit sans cesse.*
b TN: Here and in the following sentence, I have chosen "unique" as the translation for the French word *simple*, given that the context shows that Bergson intends *simple* as in "one of a kind," rather than as the opposite of "complex."

to a certain extent, we are what we do and we are continuously creating ourselves. This creation of the self by the self is all the more complete, for that matter, the more we reason about what we do.[22] For reason does not proceed here as it does in geometry, where the premises are impersonal and given once and for all, and where an impersonal conclusion is necessary. To the contrary, here the same reasons can command different persons—or even the same person at different moments—to carry out profoundly different though equally reasonable acts. True, these are not exactly the same reasons, since they are not reasons for the same person nor of the same moment.[23] This is why one cannot treat these reasons in *abstracto*, from the outside, as is done in geometry, and why one cannot resolve for another person [*autrui*] the problems they face in life. It falls to each person to resolve these problems from within and on their own.[24] But I need not go deeper into this here. We are simply looking for the precise sense that consciousness gives to the verb "to exist." We find that, for a conscious being, to exist is to change, to change is to mature, and to mature is to go on endlessly creating oneself.[a] Can we say the same of existence in general?[25]

[INORGANIC BODIES]

A material object chosen at random presents the opposite characteristics of the ones I have just enumerated. Whether it remains what it is or changes under the influence of an external force, we conceive of this change as a displacement of parts that themselves do not change. If these parts themselves suddenly dared to change, we would in turn divide them up into fragments. In this way, we will descend all the way down to the molecules that make up those fragments, to the atoms that constitute the molecules, to the generative corpuscles of the atoms, to the "imponderable" at the center of which the corpuscle was formed by some simple swirling vortex [*tourbillonnement*].[26] In short, we will push the division or the analysis as far as is necessary. And we will not stop until we have reached something immutable.

8

---

a TN: "To mature" is the translation for Bergson's use of the verb *se mûrir* ("to become ripe," "to mature," "to develop"), which continues the organic metaphors already at work in this section.

Now, we say that the composite object changes through the displacement of its parts. But when a part has left its position, nothing stops it from regaining that position. A group of elements that has passed through a particular state can thus always return to that state, if not on its own, then at least through the effect of an external cause that puts everything back into place. A state of a particular group can be repeated as often as one likes and, as a result, the group does not age. It has no history.

As such, here nothing is created; the form is no more created than is the matter. What the group will be is already present in what it is, so long as we understand "what it is" to include all of the points of the universe to which the group must be related.[27] A superhuman intellect could calculate, for any given moment of time, the position of any point of the system in space. And since the form of the whole is nothing more than the arrangement of its parts, future forms of the system are theoretically visible in its present configuration.

Our entire belief in objects, all the operations that we perform upon the systems that science isolates, indeed rest upon the idea that time does not bite into them. I have touched upon this question in a previous work.[28] I will come back to it again in the course of the present study.[29] For the moment, I will simply note that the abstract time t that science attributes to a material object or to an isolated system consists in nothing but a determinate number of *simultaneities* or, more generally, of *correspondences*, and that this number remains the same regardless of the nature of the intervals that separate the correspondences from each other. When speaking of brute matter, these intervals themselves are never at issue; or, if they are taken into consideration, this is only so as to find some new correspondences among them, between which, again, anything whatsoever might well take place. Common sense, which only deals with detached objects (just like science, which only considers isolated systems), places itself at the extremities of the intervals and not within the intervals themselves.[30] This is why we can imagine the flow of time becoming infinitely fast, such that the entire past, present, and future of material objects or of isolated systems would be spread out in space all at once—and yet there would be nothing to change in the scientific formulas nor even in the language of common sense. The number t would still signify the same thing. It would continue to count the same number of correspondences between the states of objects or of systems

9

and the points of the line, fully traced out, that would now be called "the course of time."

And yet, even in the material world, succession is an indisputable fact.[31] Although our reasoning with regard to isolated systems implies that the past, present, and future history of each system could be unfolded all at once and fully spread out like a fan, this history still develops gradually, as if it occupied a *durée* analogous to our own. If I want to make myself a glass of sweetened water, no matter what I do, I still must wait for the sugar to dissolve. This small fact is big with lessons.[a] For the time that I must spend waiting is no longer that mathematical time that could be applied just as well to the entire history of the material world, even if that history were spread out all at once in space. Rather, it coincides with my impatience, i.e., with a certain portion of my own *durée* that can be neither lengthened nor shortened at will. It is no longer something thought; it is something lived [*vécu*]. It is no longer a relation; it is something absolute.[32] What could this mean other than that the glass of water, the sugar, and the process of dissolving the sugar in the water are but abstractions, and that the Whole from which they have been cut out by my senses and by my understanding itself advances in the manner of a consciousness?[33]

Of course, the operation by which science isolates and closes off a system is not completely artificial. If it did not have an objective foundation, it would be impossible to explain why this operation is clearly indicated in some cases but impossible in others. We will see below that matter tends to constitute isolatable systems that can be treated geometrically.[34] I will even suggest that this tendency defines matter. But it is merely a tendency. Matter never makes it to the end of this movement, and the isolation is never complete. If science follows this movement to its end and fully isolates the systems in question, this is for the convenience of study. Science is aware that the supposedly isolated system remains subject to certain exterior influences. It simply leaves them to the side, either because it finds them weak enough to be ignored, or because it plans to take them into account later. It is no less true, however, that these influences are so many threads that connect this system

---

a   TN: *Ce petit fait est gros d'enseignements.* This could also be more freely translated as: "This small fact is highly instructive." For Bergson, this small fact promises to unlock an entire metaphysics of matter.

to a larger one, and this larger one again to a third that encompasses the first two, and so on, until we reach the most objectively isolated and the most independent system of all: the solar system as a whole. But even here the isolation is not absolute. Our sun radiates heat and light beyond the most distant planet. And, moreover, the sun itself moves, dragging with it the planets and their satellites in a determinate direction. The thread that attaches it to the rest of the universe is surely a tenuous one. And yet, the *durée* immanent to the whole of the universe is transmitted along this thread, all the way down to the tiniest particle of the world in which we live.

The universe endures. The deeper our study of the nature of time, the more we will come to understand that *durée* signifies invention, the creation of forms, and the continuous development of the absolutely new.[35] The systems that science delimits only endure because they are indissolubly tied to the rest of the universe. True, as I will show below, it is necessary to distinguish between two opposing movements within the universe itself: one that "descends," another that "ascends."[a/36] The first does nothing but unroll a ready-made roll.[b] This movement could in principle happen in an almost instantaneous way, such as when a spring is released. But the second, which corresponds to an inner work of maturation or of creation, essentially endures, and thereby imposes its rhythm on the first movement, which is inseparable from it.

Thus, nothing prevents us from attributing a *durée*—and thereby a form of existence analogous to our own—to the systems that science isolates, so long as those systems are reintegrated into the Whole. But those systems must indeed be reintegrated. And, *a fortiori*, the same would have to be said of the objects delimited by our perception. The distinct boundaries that we attribute to an object, and that give it its individuality, are nothing but the sketch of a certain kind of *influence* that we could exert at a certain point in space. When we perceive the surfaces and edges of things, this is actually the plan for our possible actions being reflected back to our eyes, as if by a mirror.[37] It you suppress this action,

11

---

a TN: *...l'un de "descente," l'autre de "montée"...* The distinction between a downward movement or "descent" and an upward movement or "ascent" [*montée*] is key to Bergson's emerging metaphysical argument: matter and intellect being associated with descent; spirit and intuition with ascent. Although Bergson sometimes varies his terms for this distinction, I have favored *descending* or *descent* versus *ascending* or *ascent* whenever possible.

b TN: *...dérouler un rouleau tout préparé.* Bergson will include *déroulement* ("unrolling," "unwinding," or "uncoiling") as one type of causality. Cf. below, 73–74.

the major pathways that perception traces out for it within the entanglement of the real and the individuality of the body are re-absorbed back
into the universal interaction that is surely reality itself.

[ORGANIC BODIES: AGING AND INDIVIDUALITY]

12   Now, we have thus far considered material objects chosen at random.
But are there not some privileged objects? I said that physical bodies are
cut out of the fabric of nature by a *perception* whose scissors, so to speak,
follow the dotted lines along which *action* might go.[a] But the body that
will accomplish this action—the body that, prior to accomplishing real
actions already projects upon matter the sketch of its virtual actions, the
one that need only turn its sensory organs toward the flow of the real in
order to cause it to crystallize into definite forms and thereby to create
all of the other bodies, in short, the *living body*—is this body a body like
the others?

Surely this body is also a portion of extended reality [*étendue*] tied to
the rest of extended reality, interdependent [*solidaire*] with the Whole,
and subject to the same physical and chemical laws that govern any
portion of matter whatsoever.[b] And yet, whereas the subdivision of
matter into isolated bodies is relative to our perception, and whereas
the constitution of closed systems of material points is relative to our
knowledge [*notre science*], the living body has been isolated and closed
off by nature itself.[c/38] The living body is composed of heterogeneous

a TN: *...les corps bruts...* Bergson usually reserves the adjective "brute" for matter as in raw
or brute matter. In this context, he is focusing on bodies as merely physical, the ones we
perceive and upon which we act.

b TN: Bergson uses *étendue* here and below in the technical sense in which it is used
in modern philosophy for referring to "extension" or "extended reality" (*res extensa* in
Descartes) versus the unextended or mental reality (*res cogitans* in Descartes). Bergson
also uses the term *solidaire* ("interdependent"), which is difficult to translate consistently
into English. Although the weaker meaning of "working together" is possible, Bergson
generally intends the stronger meaning of "interdependent."

c TN: Note that in French, especially in older or literary French, the word *science* can refer
not only to the set of knowledges of phenomena governed by laws and verified by experiment (captured well by the literal translation as "science" or the "positive sciences" in the
preceding paragraphs), but also more loosely to any "knowledge" insofar as it is somewhat precise or analytic. As such, when Bergson clearly means the kind of knowledge that
characterizes human intellect in general, rather than the specific scientific domains, I have
used "knowledge" while indicating the French when appropriate. Below, he will also use
"Science" in the sense of the intelligible structure of reality that exists in itself, according
to certain idealist traditions (cf. below, 320–22 and the Translator's Introduction).

parts that complement each other. It accomplishes a variety of functions that involve all of its parts. It is an individual—and this is something that cannot be said of any other object, not even of a crystal, since a crystal has neither heterogeneity of parts nor variety of functions.[39] Of course, it is not easy to determine, even in the organic world, what is an individual and what is not.[a] This is already a major difficulty in the animal kingdom; it becomes almost insurmountable when it comes to plants. The difficulty comes from profound causes, which I will explore more deeply in what follows.[40] We will see that individuality consists in an infinity of degrees and that nowhere, not even among humans, is it fully realized.[b] But this is no reason to refuse to see individuality as a property that is characteristic of life. Our inability to give a precise and general definition of individuality allows the biologist who attempts to proceed like a geometer to triumph far too easily here. A perfect definition only applies to a reality that is already made.[c] But vital properties are never fully realized; they are always on their way to being realized—they are less *states* than *tendencies*. And a tendency only brings about all that it aims at if it is not opposed by any other tendency—how could this ever occur in the domain of life, where, as I will show, there is always a reciprocal interaction of antagonistic tendencies?[41] In the case of individuality in particular, we can say that, while the tendency to individualize is present everywhere in the organic world, it is also opposed everywhere by the tendency to reproduce. For individuality to be perfect, it would be necessary that no detached part of an organism could live separately. But then reproduction would become impossible. After all, what is reproduction if not the reconstitution of a new organism from a fragment detached from the older one?[42] Individuality thus harbors its enemy at home. The very need it feels to perpetuate itself in time condemns it to never be complete in space. The biologist should always take these two

13

a TN: *...le monde organisé...* Bergson uses *organisé* ("organized") in the biological sense of vital or organic. Just below, the word *inorganisé* (literally, "unorganized") is used to mean "inorganic." Given the context and usage, I have opted for "organic" and "inorganic."

b TN: *...réalisé.* Note that for Bergson, generally speaking, the "possible" (or a plan) is "realized," whereas a "virtual" is "actualized." Cf. below, 159, 167, and editorial note 10 to Chapter I.

c TN: *...une réalité faite...* This is the first occurrences of the distinction that Bergson will insist upon between what is "made" or "ready-made" (*fait* or *tout fait*) and what is "being made" or "in-the-making" (*se faisant*). For more on this vocabulary, see the Translator's Introduction and editorial note 191 to Chapter III.

tendencies into account. As such, it is in vain that we ask them for a definition of individuality that could be formulated once and for all and applied automatically.

But too often we reason about the things of life in the way that we do about the modalities of brute matter. Nowhere is this confusion more apparent than in discussions about individuality. We are shown sections of a *Lumbriculus* [California Blackworm], each regenerating a head and from then on living as independent individuals, a Hydra whose parts become new Hydras, or the egg of a Sea Urchin whose fragments develop into complete embryos.[43] And then we are asked: so, where was the individuality of the egg, the Hydra, or the Worm?—But the fact that there are now several individuals does not mean that, just before, there was not really a single individual. I admit that having seen several drawers fall out of a piece of furniture, I no longer have the right to say that it was all one piece. But this is because the present state of this piece of furniture contains nothing more than what was there in its past, and so if it is now made up of several heterogeneous pieces, this must have been the case at the moment it was fabricated as well.[a] More generally, inorganic bodies—namely the ones we need in order to act and upon which we have modeled our way of thinking—are governed by this simple law: "the present contains nothing more than the past, and what is found in the effect was already there in its cause."[44] But if we assume that the distinctive trait of an organic body is to grow and to continuously modify itself, as even the most superficial observation will confirm, then there would be nothing surprising if such a body were at first *one* and then later *many*. This is precisely what happens in the reproduction of unicellular organisms: the living being divides into two halves, each one being a complete individual. True, in more complex animals, nature localizes the power to reproduce the whole in the so-called sexual cells, which are more or less independent.[45] But something of this power can remain diffused throughout the rest of the

14

---

a   TN: Bergson often refers to objects that have been made as *fabriqué* ("fabricated" or "manufactured"). Since the English term "manufactured" tends to imply making on a large scale and with industrial machinery, I have opted to use the more direct translation of "fabrication" or "fabricated" instead. Note that this is intended in the sense of "construct" or "manufacture" and not in the sense of "concoct" or "invent" with deceitful intent. This form of making is to be contrasted below with "organic development" [*organisation*] (cf. below, 93, 140–43).

organism, as the facts of regeneration prove. And some maintain that, in certain privileged cases, the faculty subsists in its entirety in a latent state and manifests itself at the first opportunity.[46] In fact, it is not necessary that an organism be unable to subdivide into viable fragments for me to have the right to speak of individuality. It is sufficient that this organism presented a certain systematization of its parts prior to the fragmentation and that this same systematization tends to be reproduced in the fragments once they are detached. Now, this is precisely what we observe in the organic world. Let us thus conclude that individuality is never perfect, that it is often difficult—and sometimes impossible—to say what is an individual and what is not, but that life no less manifests a search for individuality and that it tends to constitute naturally isolated and naturally closed systems.

A living being is distinguished in this way from everything that our perception or our knowledge [*notre science*] artificially isolates or closes off. As such, it would be wrong to compare a living being to an *object*. If we wanted to compare it to something in the inorganic world, the living organism would have to be likened not to a determinate material object but to the material universe as a whole. True, the comparison would no longer be very useful, since a living being is an observable being, whereas the whole of the universe is constructed or reconstructed by thought. But at least our attention would thereby be drawn to the essential characteristic of organic nature [*l'organisation*].[a] Like the universe as a whole, and like each conscious being taken separately, an organism that lives is a thing that endures. Its past, as a whole, is prolonged into its present, and remains present [*actuel*] and active there.[47] For how else could we understand that the organism passes through well-ordered phases and changes age—in short, that it has a history? If I consider my body in particular, I find that, like my consciousness, it matures little by little from childhood to old age; like me, my body ages. We might even say that maturity and old age are, properly speaking, nothing but attributes of my body; it is only through metaphor that I give the same name to the corresponding changes in my conscious self [*ma personne consciente*]. Now, if I transport myself from the highest to the lowest level on the

a TN: ...*le caractère essentiel de l'organisation.* Continuing the vocabulary of the organic as "organized" [*organisé*], Bergson sometimes uses the noun *organisation* to refer to the organic world or organic nature more generally. Thus, where he is intending the general process of *organisation*, I have used "organic development" or "organic nature."

scale of living beings, if I go from one of the most differentiated to one of the least differentiated living beings, from the multicellular organism that is man to the unicellular organism that is the Infusorian, I again find the same process of aging taking place even in this single cell.[a/48] The Infusorian is worn out after a certain number of divisions; and even

16   if by modifying its environment it is possible to delay the moment at which it must be rejuvenated through conjugation, that moment cannot be pushed back indefinitely.[b/49] True, between these two extreme cases in which the organism is completely individualized, a multitude of other cases could be found in which the individuality is much less marked and in which, although there is surely an aging taking place somewhere, we would be hard pressed to say what exactly is aging. To repeat, there are no universal biological laws that apply as such and automatically to every living being whatsoever.[50] There are only *directions* in which life launches out species in general. Each particular species, in the very act by which it is constituted, affirms its independence, pursues its various whims, deviates more or less from the line, and sometimes even goes back up the slope and seems to turn its back on the original direction. It would be easy to show that a tree does not age, since its outermost branches are always equally young and always equally capable of engendering new trees through cuttings. But in such an organism—which in fact is more a society than an individual—something ages, if only the leaves and the interior of the trunk.[51] And each cell, considered separately, evolves in a determinate way. *Wherever something is living, there is a register, open somewhere, in which time is being inscribed.*[c]

It will be said that this is but a metaphor—and in essence mechanism holds that every expression attributing to time an effective action and a reality of its own is merely metaphorical. It is useless for immediate observation to show us that the very foundation of our conscious existence is memory, that is, the prolongation of the past into the present, or again, an active and irreversible *durée*. And it is useless for reason to

17   prove that the more we leave aside objects and systems that have been

a TN: In English, infusoria is the term for the class of Protozoa of which an infusorian is a member.

b Bergson's note: Gary N. Calkins, "Studies on the Life-History of Protozoa," *Archiv für Entwickelungsmechanik der Organismen* 15, no. 1 (1902): 139–86.

c TN: The original of this striking conclusion reads: *Partout où quelque chose vit, il y a, ouvert quelque part, un registre où le temps s'inscrit.*

cut out or isolated by common sense or science, the more we are faced with a reality that changes as a whole in its inner dispositions, as if a memory that accumulates the past made it impossible for any backward movement. The mechanistic instinct of the mind is stronger than reason and stronger than direct observation. The metaphysician that we unconsciously carry within ourselves, and whose presence is explained (as we will see below) by the very place that man occupies in the ensemble of living beings, has his fixed requirements, his ready-made explanations, and his irreducible claims—all of which lead to the negation of concrete durée.[52] Change must be reducible to an arrangement or a rearrangement of parts, the irreversibility of time must be an appearance relative to our ignorance, and the impossibility of going backward nothing but man's lack of power to put things back in their places. That being the case, aging can no longer be anything other than the progressive acquisition or gradual loss of certain substances, or perhaps both at once. Time would have just as much reality for a living being as it does for an hourglass, where the upper chamber empties while the lower one is filled up, and where one can simply turn the apparatus again in order to put things back into their places.

True, there is little agreement on what is gained or what is lost between the day of birth and the moment of death. Some focus on the continual increase in the volume of protoplasm that occurs between the birth of the cell and its death.[a/53] A more plausible and more profound theory is the one that attributes the decrease to the nutritive substance enclosed in the "interior milieu"[54] where the organism regenerates, and the increase to the non-excreted residual substances that, by accumulating in the cell's body, end up causing it to "crust over."[b/55] But is it necessary to follow an eminent microbiologist in declaring that any explanation of aging that does not take phagocytosis into account is insufficient?[c/56]

18

a Bergson's note: Charles Sedgwick Minot, "On Certain Phenomena of Growing Old," in *Proceedings of the American Association for the Advancement of Science, 39th meeting, August 1890* (Salem: The Permanent Secretary, 1891): 271–89.

b Bergson's note: Félix Le Dantec, *L'individualité et l'erreur individualiste*, preface by Alfred Giard (Paris: Félix Alcan, 1898), 84ff.

c Bergson's note: Élie Metchnikov, "Revue de quelques travaux sur la dégénérescence sénile," *L'Année biologique: Comptes rendus annuels des travaux de biologie générale* 3 (1897): 249ff.; cf. from the same author, *Études sur la nature humaine: Essai de philosophie optimiste* (Paris: Masson, 1903), 312ff. [*The Nature of Man: Studies in Optimistic Philosophy*, trans. and ed. P. Chalmers Mitchell (New York and London: G. P. Putnam's Sons, 1903)].

am not qualified to respond to this question. But the fact that the two theories agree in affirming the constant accumulation or loss of a certain type of matter (whereas they have little in common when it comes to determining what is gained or lost) shows quite clearly that the framework for the explanation has been given *a priori*. This will become increasingly clear as we move forward in this study: when thinking about time, it is not easy to escape the image of the hourglass.

The cause of aging must be deeper. I reckon that there is an uninterrupted continuity between the evolution of the embryo and that of the complete organism. The thrust by which the living being grows, develops, and ages is the same one that caused it to pass through the phases of embryonic life. The embryo's development is a perpetual changing of form. Someone wanting to record all of the successive aspects would quickly be lost in an infinity of details, as happens whenever we are dealing with a continuity.[57] Life is the continuation of this prenatal evolution. The proof is that it is often impossible to say whether we are dealing with an organism that ages or with an embryo that continues to develop. This is the case, for example, with Insect or Crustacean larvae. Moreover, when it comes to organisms like our own, crises such as puberty or menopause, which bring about the complete transformation of the individual, are clearly comparable to the changes that take place during larval or embryonic life—and yet they are an integral part of our aging. Even if these crises happen at a specific age and within a relatively short period of time, no one will claim that they therefore arrive *ex abrupto*, from the outside, simply because one has reached a certain age (such as how the summons to obligatory military service arrives when one turns 20 years old). Of course, a change like puberty is being prepared at every moment since birth, and even since before birth, and the aging of the living being up until this crisis consists, at least in part, in this gradual preparation. In short, what is properly vital in aging is the imperceptible and infinitely divided continuation of the changing of form. Moreover, some phenomena of organic destruction unquestionably accompany this changing of form. These phenomena will be explained through a mechanistic account of aging. Such an account will record the facts of sclerosis, the gradual accumulation of residual substances, and the growing hypertrophy of the cell's protoplasm. But an inner cause is hidden beneath these visible effects. The development [*évolution*] of the living being, like that of the embryo, implies a continual registration of *durée*, a persistence of the past in the present, and consequently an appearance of at least an organic memory.[a]

The present state of a physical body [*un corps brut*] depends exclusively on what happened in the preceding instant. The position of the material points of a system defined and isolated by science is determined by the position of these same points at the moment that immediately preceded it. In other words, the laws that govern inorganic matter can be expressed, in principle, as differential equations in which time (in the sense in which the mathematician understands this word) plays the role of an independent variable. Is the same true for the laws of life? Does the state of a living body [*un corps vivant*] find its complete explanation in the state that immediately preceded it? Yes, if we agree *a priori* to assimilate the living body to the other bodies of nature and to identify it, for the sake of the analysis, with the artificial systems that the chemist, physicist, and astronomer analyze. But in astronomy, physics, and chemistry, the theory has a well-defined sense: it means that certain aspects of the present, the ones that are significant for science, can be calculated as a function of the immediate past. There is nothing like this in the domain of life. Here calculation can at best get a hold on certain phenomena of organic *destruction*. But when it comes to organic *creation*, that is, to the evolutionary phenomena that in fact constitute life, we cannot even imagine how they might ever be subject to a mathematical treatment. It will be said that this inability is merely the result of our ignorance. But our inability might also indicate that the present or actual moment of a living body does not find its *raison d'être* in the moment that immediately preceded it, and that instead we would have to add the entire past of the organism, its heredity, in short, the entirety of a very long history. In fact, this second hypothesis is the one that expresses the current state of the biological sciences, and indeed their direction. As for the idea that some superhuman calculator might be able to subject the living body to the same mathematical treatment it applies to our solar system, this idea gradually emerged from a certain metaphysics that has taken on a more definitive form since Galileo's discoveries in physics, but that—as I will show—has always been the natural metaphysics of the human

a TN: Note that Bergson here is using *évolution* in the sense of the "gradual development" of an individual living being. Where he is intending this meaning, rather than the evolution and diversification of species, I have occasionally opted for "development."

mind.[58] The apparent clarity of this metaphysics, our impatient desire to take it as true, the eagerness with which so many excellent minds accept it without any proof, in short, all of the seductions that it exercises on our thought should have made us wary of it. The attraction that we feel toward it proves well enough that it satisfies some innate inclination. But as I will show below, the intellectual tendencies that are innate today—tendencies that life must have created in the course of its evolution—are designed for something completely different than for providing us with an explanation of life.[59]

We run counter to this tendency the moment we want to distinguish between an artificial system and a natural system, between what is dead and what is living.[a/60] This tendency makes it such that we find it just as difficult to conceive of the organic as enduring as of the inorganic as not enduring. — But hold on! By saying that the state of an artificial system depends exclusively on its state in the previous moment, are you not introducing time, that is, are you not placing the system within durée? And moreover, this past that, according to you, is incorporated into the present moment of the living being, does organic memory not thereby contract it entirely into the immediately preceding moment, which, as a result, becomes the unique cause of the present state? — And yet, to speak in this way is to misunderstand the key difference that separates the concrete time along which a real system develops from the abstract time that enters into our speculations on artificial systems. What exactly do I mean when I say that the state of an artificial system depends on what it was in the immediately preceding moment? There is no instant, nor could there be one, that immediately precedes another instant, no more than there is a mathematical point that is contiguous with another mathematical point. The "immediately preceding" instant is, in reality, one that is linked to the present instant by the interval dt. Thus, all that I mean is that the present state of a system is defined by equations in which differential coefficients are introduced, such as $\dfrac{de}{dt}, \dfrac{dv}{dt}$, i.e., basically, *present* speeds and *present* accelerations. In

---

a TN: *...entre le mort et le vivant.* The masculine noun *le mort* usually refers to a particular dead person and *le vivant* is often translated as a particular "living being." Here, however, Bergson associates artificial or inorganic being with what is dead, versus natural or living systems taken generally. Cf. below, 166.

short, the present alone is in question here, though admittedly a present that includes its *tendency*. And, in fact, the systems that science works on are in an instantaneous present that is continuously renewed; they are never in real or concrete *durée* where the past is incorporated into the present.[61] When the mathematician calculates the future state of a system at the end of time t, nothing prevents him from imagining that, between now and then, the material universe vanishes completely only to suddenly reappear. The only moment that matters for him is the moment t that he has chosen—and this is something that will be a pure instant. That which flows by in the interval—i.e., real time—does not matter and cannot enter into the calculation. Even if the mathematician claims to place himself within this interval, it is always at a certain point or a certain moment, by which I mean that he transports himself to the extreme limit of a time t'. As such, it is no longer a question of the interval that leads up to T'. If he attempts to divide the interval into infinitely small parts by considering the differential dt, he thereby simply indicates that he will take into account accelerations and speeds, i.e., numbers that record tendencies and that allow for the calculation of the state of the system at a given moment. But this is still a question of a given moment, by which I mean one that is fixed, and not of flowing time. In short, *the world upon which the mathematician operates is a world that dies and that is reborn at each instant—that very world that Descartes had in mind when he spoke about continuous creation.*[62] And yet, if time is understood in this way, how could we conceive of an evolution, the very trait that is characteristic of life? Evolution itself implies a real continuation of the past in the present, a *durée* that is a link or a hyphen [*trait d'union*]. In other words, knowledge of a living being or of a *natural system* is one that bears upon the very interval of durée, whereas knowledge of a mathematical or *artificial system* only bears upon the extremity.

Continuity of change, conservation of the past in the present, true *durée*—the living being thus seems to indeed share these attributes with consciousness. Can we go even further and claim that, like conscious activity, life is invention and incessant creation?

[ON TRANSFORMISM AND THE WAYS OF INTERPRETING IT][a]

It is not my intention here to enumerate the various proofs in favor of transformism. I merely want to explain briefly why, in the present work, this theory will be accepted as a sufficiently rigorous and precise translation of known facts. The natural classification of organic beings already contains the seed of the idea of transformism. Indeed, the naturalist brings together organisms that resemble each other, then divides the group into subgroups within which the resemblance is even greater, and so on. Throughout the operation, the characteristics of the group appear as general themes upon which each of the subgroups can execute its particular variations.[63] Now, this is precisely the relation we find in the animal world and in the plant world between those that engender and those that are engendered: upon the tapestry that the ancestor transmits to its descendants, and that they all possess in common, each one places its own original embroidery. True, there are but slight differences between the descendant and its ancestor, and one might wonder whether a single living matter has enough plasticity to take on, in turn, forms as different as those of a Fish, a Reptile, and a Bird. But [scientific] observation provides a preemptive response to this question.[b] It shows that, up to a certain period in its development, the embryo of the Bird can hardly be distinguished from that of the Reptile; and it shows that the individual organism develops, throughout its embryonic life in general, by a series of transformations comparable to those by which, according to evolutionary theory, one species passes into another. A single cell, obtained through the union of a male cell and a female cell, accomplishes this work by dividing itself. Every day and before our very eyes the highest forms of life emerge from an extremely basic form. Experience thus establishes that, by way of evolution, the most complex was indeed able to emerge from the most basic. Now, did it actually emerge from it? Paleontology, despite the insufficiency of its

24

---

a TN: Bergson often uses the now less common term *transformisme* ("transformism") for the collection of theories of the evolution of species through some mechanism or process of transformation.

b TN: Bergson here simply writes *observation*, but the context indicates that he has in mind "scientific study," which is implied in the French term, more than simple or direct observation.

evidence, invites us to believe that this did occur, given that wherever it discovers the order of succession among species with some measure of precision, this order is precisely the one that observations drawn from embryogenesis and comparative anatomy would have led us to expect, and each new paleontological discovery brings with it a new confirmation of transformism.[64] In this way, the proof drawn from pure and simple scientific observation continues to strengthen the theory, while, on the other hand, experimentation is setting aside the objections one by one. For instance, this is how Hugo de Vries's somewhat curious experiments, which demonstrate that large variations can come about suddenly and be transmitted regularly, knock down some of the biggest difficulties facing the theory.[65] These experiments allow us to greatly shorten the time that biological evolution seemed to require. They also make us less demanding when it comes to paleontology. To sum up, the transformist hypothesis increasingly appears to be at the very least an approximate expression of the truth. The theory cannot be rigorously demonstrated, but just beneath the certitude offered by theoretical and experimental demonstration, there is that indefinitely increasing probability that stands in for evidence and that tends toward it as if toward its limit—this is the kind of probability that we find for transformism.[66]

Let us imagine, however, that transformism is wrong. Suppose that someone establishes, through reasoning or through experimentation, that species are in fact born through a discontinuous process that we currently know nothing about. Would the theory be undermined in terms of its most interesting and, for us, most important aspects? The classification would likely remain, at least in its broad outline. The current givens of embryology would also remain. And so too would the correspondence between comparative embryogenesis and comparative anatomy. As a result, biology would not only be able to, but even would have to, continue to establish the same relations and the same kinships between living forms that are posited by transformism today. True, this would be an ideal kinship and no longer a material lineage. Nevertheless, since the current givens of paleontology would also remain, it would still be necessary to acknowledge that the forms that admit of an ideal kinship in fact appeared successively and not simultaneously. Now, evolutionary theory—in terms of what is important to the philosopher—asks for nothing more. This theory consists above all in observing ideal kinship relations and claiming that, wherever we

25

find this so-to-speak *logical* relation of filiation between forms, there is also a relation of *chronological* succession among the species in which these forms materialize. This double thesis would remain in all cases. And consequently, it would remain necessary to assume an evolution somewhere: either in a creative Thought where the ideas of the various species would be engendered one after another, exactly as transformism claims that the species themselves are engendered on the Earth; or in a plan for the organic development [*organisation*] of life, immanent to nature, that would slowly unfold and according to which the logical and chronological relations of filiation among the pure forms would be precisely the ones that transformism presents to us as real relations of filiation among living individuals; or finally, in some unknown cause of life that would develop its effects *as if* the ones engendered the others. Thus, evolution would simply have been *transposed*, made to pass over from the visible into the invisible. Nearly everything that transformism tells us today would be preserved, although interpreted in a different manner. So, would it not be better to hold ourselves to the letter of transformism, such as it is put forward by the vast majority of scientists? If we hold aside the question as to what extent this evolutionary theory describes the facts and to what extent it symbolizes them, then there is nothing irreconcilable with the doctrines that it claimed to replace, even the ones that propose isolated creations, and with which it is generally held to be in opposition. This is why I reckon that the language of transformism now imposes itself upon every philosophy, in the same way that the dogmatic affirmation of transformism imposes itself upon science.

But as such, we must no longer speak of *life in general* as if this were a mere abstraction or a simple rubric under which all living beings are inscribed.[67] At a certain moment and at certain points in space, a clearly visible current was born: this current of life, passing through the bodies that it had organized [*organisé*] one after the other, passing from generation to generation, was divided among species and distributed to individuals without thereby losing anything of its force, and rather intensifying as it advanced.[a] We know that, according to [August] Weismann's theory of

26

---

a TN: Here again Bergson makes use of the connections between "organic" and "organized" bodies, such that *organisation* is the process in which life "organizes" bodies into organisms, into bodies with organs, into living bodies. Some of these connections are lost in English.

the "continuity of the germ-plasm [*plasma germinatif*]," the sexual elements of the generating organism would transmit their properties directly to the sexual elements of the engendered organism.[a/68] In this extreme form, the theory seemed questionable since only in exceptional cases are the sexual glands formed at the moment of the segmentation of the fertilized ovum. But, even if the cells that generate the sexual elements do not, in general, appear at the beginning of embryonic life, it is no less true that they are always formed out of the embryonic tissues that have not yet undergone any particular functional differentiation and whose cells are formed from unmodified protoplasm.[b/69] In other words, the genetic power of the fertilized ovum weakens to the extent that it is spread over the embryo's growing mass of tissue. But, while it is being diluted in this way, it concentrates anew something of itself at a certain special point, namely, in the cells from which the ova or the spermatozoa will be born. As such, we could say that even though the germ-plasm is not continuous, there is at least a continuity of genetic energy, this energy being spent only for a few moments (just long enough to give an impulsion to embryonic life) and then being recuperated as soon as possible into new sexual elements where, once again, it will bide its time.[70] Seen from this point of view, *life seems like a current that passes from germ to germ by way of the intermediary of a developed organism.*[71] Everything takes place as if the organism itself is nothing but an excrescence, a bud shot out by the previous germ as it works at continuing itself into a new germ. What is essential is the continuity of the progression that goes on indefinitely, an invisible progression upon which each visible organism rides during the short interval of time given to it to live.

27

[RADICAL MECHANISM: BIOLOGY AND PHYSICO-CHEMISTRY]

Now, the more we fix our attention upon this continuity of life, the more we will see organic evolution come closer to the evolution of a consciousness, where the past presses against the present and causes

---

a TN: Although Bergson sometimes varies the formulation he uses for this idea of the "germ-plasm," I generally follow the English translations of Weismann's terminology where possible.

b Bergson's note: Louis Roule, *L'embryologie générale* (Paris: C. Reinwald, 1893), 319.

a new form to spring forth from it, a form that is incommensurable with its antecedents. No one will deny that the appearance of a plant or an animal species is the result of some precise causes. But this must be taken to mean that if, after the fact, we knew the details of these causes, then we would be able to use them to explain the form that was produced; it was never a question of foreseeing the form in advance.[a/72]

Will the objection be made that we could have foreseen the form, if only we had known in all of their details the conditions in which the form was produced? But these conditions make up part of the body of the form and are even inextricably united with it, being as they are the characteristics of the moment in which life then finds itself in its history. How could we claim to know in advance a situation that is one of a kind, that has not yet happened, and that will never happen again? The future can only be foreseen when it comes to something that resembles the past or something that can be recomposed from similar elements to those found in the past. Such is the case for astronomical, physical, and chemical facts, i.e., for all those facts that are a part of a system where the supposedly unchanging elements are simply juxtaposed, where nothing happens other than changes of position, where there is no theoretical absurdity in imagining that things might be put back into their places, and where, consequently, the same overall phenomenon or at least the same elementary phenomena can be repeated.[73] But when it comes to an original situation—one that communicates something of its originality to its elements, that is, to the partial views that we take of it—how could we imagine this as given prior to its having taken place?[b/74] The most we can say is that this original situation is explained, after it has taken place, by the elements that the analysis discovers there. But what is true of the production of a new species is also true of the production of a new individual, and more generally of any moment whatsoever of any living form whatsoever. For, even if it is necessary that the variation attain a certain size and a certain generality so as to give birth to a new species, it is nonetheless continuously and imperceptibly taking place at

---

a   Bergson's note: The irreversibility of the series of living beings was clearly demonstrated by James Mark Baldwin, *Development and Evolution, including Psychophysical Evolution, Evolution by Orthoplasy, and the Theory of Genetic Modes* (New York and London: Macmillan, 1902), see in particular page 327.

b   Bergson's note: I emphasized this point in my book *Essai sur les données immédiates de la conscience*, 140–51. [TN: Bergson is likely in fact thinking of pages 137–49.]

each moment in every living being. And even sudden mutations (which are being discussed these days) are clearly only possible if a work of incubation, or better, of maturation, has taken place throughout a series of generations that themselves did not appear to be changing.[75] In this sense, it can be said that life, like consciousness, is creating something at every instant.[a/76]

    29

But our whole intellect rebels against this idea of the absolute originality and unforeseeability of forms. The essential function of our intellect, such as it has been shaped by the evolution of life, is to light the way for our behavior, to prepare for our action upon things, and to predict, for a given situation, the favorable or unfavorable events that might follow.[77] As such, the intellect instinctively picks out, within a given situation, that which resembles what is already known; it looks for what is alike, so as to be able to apply its principle that "like produces like."[78] This is precisely how common sense goes about foreseeing the future. Science then elevates this operation to its highest degree of rigor and precision, but it does not change its essential nature. Just like ordinary knowledge, science only retains the aspect of *repetition* that it finds in things. Even if the whole is original, science goes about analyzing it into elements or into aspects that are *approximately* the reproduction of the past. Science can only operate upon something that is assumed to repeat, on something that is, by definition, shielded from the action of *durée*. That which is irreducible and irreversible in the successive moments of a history escapes science. In order to conceive of this irreducibility and irreversibility, we must break with the habits of science, adapted as they are to the fundamental requirements of thought; we must do violence to the mind and

a Bergson's note: In his excellent book, *Essai sur le génie dans l'art* (Paris: Germer Baillière, 1883), Gabriel Séailles develops this double thesis that art prolongs nature, and that life is creation. We happily accept the second formulation, but is it necessary to understand creation (as the author does) as a *synthesis* of elements? Wherever the elements pre-exist, the synthesis that will be made of them is already given, virtually, since it is nothing other than one of their possible arrangements. A superhuman intellect would have been able to see this arrangement in advance, among all the other possible arrangements that frame it. I reckon, to the contrary, that in the domain of life, the elements have no real and separate existence. They are but multiple views the mind takes of an indivisible process. And this is why there is a radical contingency in that progression, an incommensurability between what precedes and what follows—in short, *durée*.

30   go counter to the natural inclination [*pente*] of the intellect.[a][79] But that is precisely philosophy's role.[80]

This is why, despite the fact that life evolves before our very eyes in a continuous creation of unforeseeable form, the idea still persists that form, unforeseeability, and continuity are nothing but pure appearances, reflecting so many gaps in our knowledge. We will be told that what appears to the senses as a continuous history could be decomposed into successive states. And that what gives us the impression of an original state is reduced, through analysis, down into elementary facts, where each one is the repetition of a fact that is already known. And finally, that what we call an unforeseeable form is nothing but a new arrangement of previous elements. The elementary causes that together have determined this arrangement are themselves but previous causes that repeat by adopting a new order. Knowledge of the elements and the elementary causes would have allowed us to sketch out, in advance, the living form, which is the sum and the result of them. After having reduced the biological aspect of phenomena down into physico-chemical factors, we will, if necessary, leap beyond physics and chemistry: we will go from masses to molecules, from molecules to atoms, from atoms to corpuscles. We must ultimately reach something that can be treated like a sort of solar system, something that can be treated astronomically. If you deny this, then you are questioning the very principle of scientific mechanism and declaring arbitrarily that living matter is not made up of the same elements as non-living matter. — To this I will respond that I do not question the fundamental identity of brute matter and organic matter.[81] The only question is whether the natural systems that we call living beings should be likened to the artificial systems that science cuts out of brute matter, or whether they should rather be compared to that natural system that is the universe as a whole. I happily admit that life is

31   some sort of mechanism. But is it the kind of mechanism whose parts can be artificially isolated within the whole of the universe, or is it the mechanism of the real as a whole? As I said above, the real as a whole

---

a   TN: ...*la pente naturelle de l'intelligence*... The word *pente* in this important and repeated formulation could be translated in several ways, including "incline," "inclination," "slope," or "bent." Since Bergson regularly uses *pente* in other images as well (cf., for example, below, 103, 209), I have opted for "inclination" whenever it has to do with the intellect and "incline" when it is a more physical or metaphysical image.

might well be an indivisible continuity; and if that is so, then the systems we cut out of this whole would not be its parts, properly speaking, but rather partial views taken of the whole.[82] And you will never achieve even the beginnings of a recomposition of the whole by laying these partial views out, end to end, no more than you will reproduce the materiality of an object by multiplying the photographs of it and by taking them from a thousand different angles.[83] The same goes for life and for the physico-chemical phenomena into which it is claimed life can be reduced. Analysis will surely discover an increasing number of physico-chemical phenomena in the processes of organic creation. And the chemists and the physicists will hold themselves to these alone. But it does not follow from this that chemistry and physics must give us the key to life.

A very small section of a curve is almost a straight line. And the smaller the section we consider, the more it will resemble a straight line. It will be said that at the limit one can choose to say either that it is a part of a straight line or that it is a part of a curve. At each of its points, in effect, the curve merges with its tangent. In this way, "vitality" is tangential at any one of its points to the physical and chemical forces; but these points themselves are, ultimately, nothing but the views of a mind that imagines some stopping points [arrêts] at various moments in the movement that generates the curve. In reality, life is no more made up of physico-chemical elements than is a curve composed of straight lines.[84]

In general, the most radical progress that a science is capable of accomplishing consists in inserting already acquired results into a new whole [ensemble], in relation to which they become instantaneous and immobile views or snapshots taken, every now and again, along the continuity of a movement.[85] This is, for example, the relationship between modern geometry and ancient geometry. Ancient geometry was purely static and operated upon shapes that were already described; modern geometry studies the variation of a function, i.e., the continuity of the movement that describes the shape. Of course, one way to have more precision is to eliminate all consideration of movement from our mathematical operations; but it is no less true that the introduction of movement into the genesis of shapes is the foundation of modern mathematics.[86] I reckon that were biology ever able to get as tight a grip on its object as the one mathematics has on its object, then biology would come to be to the physico-chemical composition of organic bodies what modern

32

mathematics has turned out to be to ancient geometry. The entirely superficial displacements of masses and molecules studied by physics and chemistry would become, when compared to this deeper, vital movement (a movement that is transformation and not merely translation), what the stopping point of a moving object is to its movement in space. And, as far as I can predict, the procedure by which we would go from the definition of a certain vital action to the system of physico-chemical facts that it implies would likely be analogous to the operation by which we go from the function to its derivative, from the equation of the curve (i.e., the law of the continuous movement by which the curve is generated) to the equation of the tangent that gives the direction of the curve at a particular instant. Such a science would be a *mechanics of transformation*, of which our *mechanics of translation* would be merely a particular case, a simplification, a projection on the plane of pure quantity. And just as there exists an infinity of functions having the same differential, these functions differing from each other by a constant, so too perhaps would the integration of the physico-chemical elements of a properly vital action only determine this action in part—a part would be left to indetermination. But we can hardly do more than dream of such an integration; that is, we do not claim that this dream can ever become a reality. By developing a certain comparison as far as possible, we have simply wanted to show how close our theory comes to pure mechanism and in what way it distinguishes itself from it.

We can, moreover, push this imitation of the living being by the inorganic quite far. Not only does chemistry carry out organic syntheses, but it is able to artificially reproduce the external appearance of certain facts of organic life [*organisation*], such as indirect cell division and the circulation of protoplasm. We know that the cell's protoplasm performs a variety of movements within its envelope. Then again, so-called indirect cell division takes place through extremely complex operations of which some have to do with the nucleus and others with the cytoplasm. These latter ones begin by the doubling of the centrosome, which is a small spherical body situated next to the nucleus. The two resulting centrosomes move away from each other, drawing toward themselves the cut and doubled sections of the filament of which the original nucleus was mainly composed. They thereby succeed in forming two new nuclei around which the two new cells are constituted, succeeding the original one. Now, it has been possible to imitate in broad outline and in external appearance at least some of these operations. If we pulverize some sugar or table salt, mix it with some very old oil, and then examine it through a microscope, we see a froth that has an alveolar structure that, according to certain theoreticians, resembles that of protoplasm, and in which movements take place that are remarkably similar to those of protoplasmic circulation.[a/87] If, in a froth of the same kind, the air is extracted from an alveolar, one can observe the formation of a cone of attraction analogous to those that form around the centrosomes so as to bring about the division of the nucleus.[b/88] It is even believed that the exterior movements of a unicellular organism can be explained mechanically, at least those of an Amoeba.[89] The displacements of an Amoeba in a drop of water would be comparable to the to-and-fro of a grain of dust in a room where the open doors and windows allow air currents to circulate. Its mass continuously absorbs certain dissolvable materials contained in the ambient water and sends back into the water certain other ones. These continuous exchanges, which are similar to those that take place between two containers separated by a porous partition, would create an everchanging vortex around the tiny organism. As for the temporary prolongations or pseudopods that the Amoeba seems to fashion for itself, they would be less sent out by the organism than drawn out of it through a sort of aspiration or suction by the ambient environment.[c/90] This mode of explanation is gradually extended to the more complex movements that the Infusorian itself executes with its vibratory cilia, which are, for that matter, probably little more than stabilized pseudopods.

Nevertheless, scientists are hardly in agreement regarding the value of explanations and of schemas of this kind. Some chemists have noted

34

a Bergson's note: Otto Bütschli, *Untersuchungen über mikroskopische Schäume und das Protoplasma: Versuche und Beobachtungen zur Lösung der Frage nach den physikalischen Bedingungen der Lebenserscheinungen* (Leipzig: Wilhelm Engelmann, 1892), First part [4–101; Otto Bütschli, *Investigations on Microscopic Foams and on Protoplasm: Experiments and Observations Directed Towards a Solution of the Question of the Physical Conditions of the Phenomena of Life*, trans. E. A. Minchin (London: Adam and Charles Black, 1894)].

b Bergson's note: Ludwig Rhumbler, "Versuch einer mechanischen Erklärung der indirekten Zell und Kerntheilung," *Archiv für Entwickelungsmechanik der Organismen (Roux's Archiv)* 3, no. 4 (1896): 527–623.

c Bergson's note: Gottfried Berthold, *Studien über Protoplasmamechanik* (Leipzig: Verlag von Arthur Felix, 1886), 102. Cf. the explanation proposed by Félix Le Dantec, *Théorie nouvelle de la vie* (Paris: Félix Alcan, 1896), 60.

that even by limiting ourselves to organic materials [l'organique], let alone going so far as the organism [l'organisé], science has so far reconstituted nothing but the waste products of vital activity; properly active or plastic substances continue to defy synthesis.[91] One of the most not-

35   able naturalists of our time has insisted upon the opposition between two orders of phenomena observable in living tissue: *anagenesis* on the one hand; *catagenesis* on the other. The role of anagenetic energies is to raise inferior energies up to their own level through the assimilation of inorganic substances. They construct the tissues. To the contrary, the very functioning of life (with the exception of assimilation, growth, and reproduction) is of the catagenetic order, the falling [*descente*] of energy, and no longer the raising [*montée*] of it. Physico-chemistry can only get ahold of these facts of the catagenetic order, which is to say that it deals with what is dead and not with what is living.[a/92] And surely the facts of the former kind seem to resist physico-chemical analysis, even if they are not, strictly speaking, anagenetic. As for the artificial imitation of the external appearance of protoplasm, must we accord this a real theoretical importance given that we have not yet settled the question of the physical configuration of this substance? It is still less a question of recomposing it chemically, at least for the moment. And finally, a physico-chemical explanation of the Amoeba's movements, and even more so for those of the Infusoria, seems impossible to those scientists who have closely observed these rudimentary organisms. They see the mark of an effective psychological activity even in these humblest manifestations of life.[b/93] But above all, what is most instructive is to see the extent to which the in-depth study of histological phenomena often discourages, rather than

36   strengthens, the tendency to explain everything through physics and

a Bergson's note: Edward Drinker Cope, *The Primary Factors of Organic Evolution* (Chicago: The Open Court Publishing Company, 1896), 475–84.

b Bergson's note: Émile Maupas, "Contribution à l'étude morphologique et anatomique des Infusoires ciliés," *Archives de zoologie expérimentale et générale*, 2nd series, 1 (1883): 471, 491, 518, 549 in particular; Paul Vignon, "Recherches de cytologie générale sur les épithéliums: L'appareil pariétal, protecteur ou moteur. Le rôle de la coordination biologique," *Archives de zoologie expérimentale et générale*, 3rd series, 9 (1901): 655. A profound study of the movements of the Infusoria and a penetrating critique of the idea of tropism has been recently offered by Herbert S. Jennings in his *Contributions to the Study of the Behavior of Lower Organisms* (Washington: The Carnegie Institution of Washington, 1904). The "type of behaviour" of these inferior organisms, such as Jennings defines it (pages 237–52), is incontestably of the psychological order.

chemistry. Such is the conclusion of the genuinely admirable book that histologist E. B. Wilson devoted to cell development: "[T]he study of the cell has on the whole seemed to widen rather than to narrow the enormous gap that separates even the lowest forms of life from the inorganic world."[a/94]

In sum, those who only pay attention to the functional activity of the living being are led to believe that physics and chemistry will provide the key to biological processes.[b/95] Indeed, they primarily deal with phenomena that continuously *repeat* in the living being, as if in a chemist's retort [*une cornue*].[c] This in part explains physiology's mechanistic tendencies. By contrast, those who concentrate their attention upon the minute structure of living tissues and on their genesis and evolution—histologists and embryologists on the one hand, naturalists on the other—are focused on the retort itself and no longer merely on its contents.[96] They find that this retort creates its own form through a *unique* series of acts that constitute a genuine history. And these scientists—histologists, embryologists, and naturalists—are much less likely than the physiologists to believe in the physico-chemical character of vital actions.

The fact is that neither of these two theories, neither the one that affirms nor the one that rejects the possibility of ever chemically producing an elementary organism, can invoke the authority of experimentation. They are both unverifiable: the first because science has still not made a single step toward the chemical synthesis of a living substance; the second because no conceivable means for proving the impossibility of a fact via experimentation exists.[97] But we have explained the theoretical reasons that prevent us from likening the living being, which is a system closed off by nature, to the systems isolated by science.[98] I admit that these reasons have less force when it comes to rudimentary organisms such as the Amoeba, which barely evolves. But they gain in force if we consider a more complex organism that accomplishes a

37

---

a   Bergson's note: Edmund Beecher Wilson, *The Cell in Development and Inheritance* [1896] (New York and London: Macmillan, 1897), 330. [TN: Bergson provided his own French translation of this passage, and included the original English citation in his footnote.]
b   Bergson's note: Albert Dastre, *La Vie et la Mort* (Paris: Ernest Flammarion, 1903), 43 [Albert Dastre, *Life and Death*, trans. W. J. Greenstreet (London: Walter Scott, 1911), 46].
c   TN: A "retort" is a glass container with a long neck used for distilling liquids in chemistry experiments.

regular cycle of transformations. The more that durée marks the living being with its imprint, the more the organism is distinguished from a pure and simple mechanism across which durée slides without penetrating. And the demonstration achieves its greatest force when we consider the evolution of life as a whole, from its humblest origins to its highest current forms, insofar as this evolution constitutes—through the unity and continuity of the living matter [matière animée] that supports it—a single indivisible history. Which is why I find it so difficult to understand why the evolutionary hypothesis is generally considered to be akin to the mechanistic conception of life. Of course, I do not claim to provide a mathematical and scientific refutation of this mechanistic conception. And yet, the refutation that I draw from considerations of durée (which is, in my opinion, the only refutation possible) acquires more rigor and becomes more convincing the more we place ourselves directly within the evolutionary hypothesis. I will have to develop this point below. But first, let me indicate more clearly the conception of life that I am moving toward.

As mentioned, mechanistic explanations are valid for the systems that our thought artificially detaches from the whole.[99] But when it comes to the whole itself and to the systems that, within this whole, are naturally constituted in its image, we cannot assume a priori that they will be explicable mechanistically, for in that case time would be useless and even unreal [irréel].[100] The essence of mechanistic explanations is, in effect, to consider the future and the past to be calculable as a function of the present, and to thereby claim that everything is given.[a/101] According to this hypothesis, past, present, and future would be simultaneously visible to a superhuman intellect capable of performing the calculation. As such, scientists who believed in the universality and in the perfect objectivity of mechanistic explanations adopted, consciously or unconsciously, a hypothesis of this kind. Pierre-Simon de Laplace already formulated this position with the greatest of precision:

> Given for one instant an intellect that could comprehend all the forces by which nature is animated and the respective situation of the beings

a TN: ...que tout est donné. This locution, which for Bergson captures the ontological assumptions of mechanism and the intellect more generally, resonates with his use of se donner in his critique of philosophical systems that "take for granted" or "give themselves everything in advance." Cf. "everything is given" in the Index.

who compose it—an intellect sufficiently vast to submit these data to analysis—it would embrace in the same formula the movements of the greatest bodies of the universe and those of the lightest atom; for such an intellect, nothing would be uncertain and the future, like the past, would be present to its eyes.[a/102]

And Emil Du Bois-Reymond writes:

Nay, we may conceive of a degree of natural science wherein the whole process of the universe might be represented by one mathematical formula, by one infinite system of simultaneous differential equations, which should give the location, the direction of movement, and the velocity of each atom in the universe at each instant.[b/103]

Thomas Henry Huxley, for his part, expressed the same idea in a more concrete form:

[The fundamental proposition of evolution is] that the whole world, living and not living, is the result of the mutual interaction, according to definite laws, of the forces possessed by the molecules of which the primitive nebulosity of the universe was composed. If this be true, it is no less certain that the existing world lay, potentially, in the cosmic vapour; and that a sufficient intelligence could, from a knowledge of the properties of the molecules of that vapour, have predicted, say the state of the Fauna of Britain in 1869, with as much certainty as one can say what will happen to the vapour of the breath in a cold winter's day.[c/104]

39

a Bergson's note: Pierre-Simon de Laplace, *Théorie analytique des probabilités* [1812], in *Œuvres complètes de Laplace*, vol. 7 (Paris: Gauthier-Villars, 1886), vi–vii. [TN: This slightly modified English translation is from Pierre-Simon de Laplace, *A Philosophical Essay on Probabilities*, trans. Frederick William Truscott and Frederick Lincoln Emory (New York: John Wiley & Sons, 1902), 4.]

b Bergson's note: Emil Du Bois-Reymond, *Über die Grenzen des Naturerkennens. Die sieben Welträthsel: Zwei Vorträge*, 5th ed. (Leipzig: Verlag von Veit, 1882). [TN: The English translation provided here is from: Emil Du Bois-Reymond, "The Limits of Our Knowledge of Nature," trans. J. Fitzgerald, *Popular Science Monthly* 5 (May 1874): 18.]

c [Bergson's missing note: Thomas Henry Huxley, "*The Natural History of Creation*–By Dr. Ernest Haeckel [Review of *Natürliche Schöpfungs-Geschichte*]," *The Academy* 1 (October 9, 1869): 13–14. Bergson provided his own French translation of this uncited passage.]

Such theories still speak of time in the sense that the word is uttered, but they hardly think of time itself. For time here is stripped of all efficacy, and, from the moment that it does nothing, it is nothing.[105] Radical mechanism implies a metaphysics in which the totality of the real is posited as a whole for all eternity and in which the apparent durée of things simply expresses the weakness of a mind that cannot know everything at once. And yet, for our consciousness (for that which is most unquestionable in our experience) durée is surely something else entirely.[a] We perceive durée as a current that we cannot follow back upstream. Durée is the ground of our being and, as we clearly sense, the very substance of the things that we are in communication with.[106] In vain do they hold up before our eyes the brilliance of the perspective of a universal mathematics: we cannot sacrifice experience to the requirements of a system.[107] That is why I reject radical mechanism.

[RADICAL FINALISM: BIOLOGY AND PHILOSOPHY]

But, and for the same reason, radical finalism appears to be just as unacceptable. The doctrine of finality, in its extreme form (such as we find it, for example, in Leibniz), implies that things and beings do nothing other than realize [réaliser] a previously drawn up program. But if there is nothing unpredictable, if there is no invention and no creation in the universe, then time once again becomes useless. As was the case for the mechanistic hypothesis, here again it is assumed that *everything is given*. Finalism thus understood is nothing more than a mechanism in reverse. It takes its inspiration from the same postulate, the only difference being that insofar as our finite intellects follow along the merely apparent succession of things, finalism places the light with which it claims to guide us out in front of us, whereas mechanism places it behind us. It substitutes the attraction of the future for the impulsion of the past. But the succession, along with the very movement by which the intellect follows it along, remains no less a pure appearance. According to Leibniz's doctrine, time is reduced to a confused perception, one that is relative to the human point of view and that would evaporate, like a fog that rises, for a mind placed at the center of things.[108]

a   TN: The formulations that Bergson uses here recall the very first lines of this chapter.

Yet unlike mechanism, finalism is not a doctrine with fixed lines. It contains as many inflections as one might like to give to it. When it comes to mechanistic philosophy, you either take it or leave it: it must be rejected the moment the smallest grain of dust, by deviating from the trajectory predicted by mechanics, manifests the slightest trace of spontaneity. To the contrary, the doctrine of final causes will never be definitively refuted. If one form of finalism is set aside, it will take on another form. Its principle, which is of a psychological nature, is very flexible.[109] It is so extensible, and therefore so broad, that the moment pure mechanism has been rejected, something of finalism has been accepted. The theory I will defend in this book thus necessarily participates, to a certain extent, in finalism. Hence the importance of clearly indicating what I will take from it and what I intend to leave behind.

Let me first say that the attempt to weaken Leibnizian finalism by breaking it up into an infinite number of pieces seems to me the wrong path to take. And yet, this is precisely the direction that the doctrine of finality took. They clearly sense that, even if the universe as a whole is the realization of a plan, it would be impossible to demonstrate this empirically. And they clearly sense that, even if we limit ourselves to the organic world, it is hardly any easier to prove that everything there is harmonious. When we examine the facts, they suggest quite the opposite. Nature sets living beings into conflict with each other.[110] Everywhere it shows us disorder alongside order, and regression alongside progression. But although the doctrine of finality cannot be affirmed of matter as a whole, nor of life as a whole, might it still not be true of each organism taken separately? Do we not observe in the organism an admirable division of labor, a marvelous interdependence between the parts, a perfect order in infinite complexity? In this sense, does not each living being realize a plan immanent to its substance? This theory ultimately consists in breaking the ancient conception of finality into pieces. It does not accept, and even happily makes a mockery of, the idea of an *external* finality by which living beings would be coordinated among themselves. It is absurd, they say, to suppose that the grass was made for the cow, and the lamb for the wolf. But they say there is an

41

internal finality: each being is made for itself, all of its parts come together for the greatest good of the whole and are intelligently organized in relation to that end. And for a long time, this was the classic conception of finality. Finalism shrunk down to the point of never encompassing more than one living being at a time. By making itself smaller, it probably thought that it was leaving less surface area open to attacks.

The truth is it opened itself up to even more attacks. As radical as this claim might appear to be, I suggest that finality is either external or it is nothing at all.

Indeed, imagine the most complex and most harmonious organism. Finalism will claim that all of its elements conspire for the greatest good of the whole. So be it. But we must not forget that in certain cases each element may itself be an organism, and by subordinating the existence of this smaller organism to the life of the larger one, we adopt a principle of external finality. As such, the idea of an always internal finality is self-defeating. An organism is composed of tissues, and each one lives for itself. The cells that make up the tissues also have a certain independence.[111] Strictly speaking, if there were a complete subordination of all of the individual's elements to the individual as a whole, then we could refuse to see them as organisms in themselves, reserving this name for the individual alone and thereby only speaking of internal finality. But everyone knows that these elements can possess a genuine autonomy. There is no need to even mention phagocytes, which take independence so far as to attack the organism that nourishes them, or germinal cells, which lead their own lives alongside somatic cells, for the facts of regeneration are enough to prove this autonomy.[112] In regeneration, an element or a group of elements suddenly demonstrates that, even though it is normally forced to occupy but a minor place and to accomplish merely a specific function, it was capable of doing much more, capable even, in certain cases, of taking itself to be the equivalent of the whole.[113]

This is the stumbling block of vitalist theories.[114] I will not criticize them (as is ordinarily done) for having responded to the question by merely restating the question. Perhaps the "vital principle" explains very little; at least it has the advantage of being a sort of sign of our ignorance that will remind us of it from time to time,[a][115] whereas mechanism invites us to forget our ignorance. But the truth is that the vitalist position is made very difficult by the fact that, in nature, there is no purely internal finality and no absolutely distinct individuality. The organic elements that enter into the composition of the individual themselves have a certain individuality and, if the individual is to have its own vital principle, then each of these elements will claim to have one as well. But, on the other hand, the individual itself is not so independent, not so isolated from everything else that we could accord it a "vital principle" of its own. An organism such as a higher Vertebrate is the most individuated of all the organisms. Nevertheless, if we recognize that this organism is nothing but the development of an ovum that was a part of its mother's body and a sperm that belonged to its father's body, and that the egg (i.e., the fertilized ovum) is a genuine hyphen [trait d'union] linking the two progenitors since it is common to their two substances, then we will see that every individual organism, even a human being, is but a simple bud that has grown upon the combined bodies of its two parents.[116] So where does the individual's vital principle begin, and where does it end? Step by step, we would have to go all

43

---

a   Bergson's note: In fact, there are two parts to contemporary neo-vitalism. On the one hand, there is the affirmation that pure mechanism is insufficient (a claim that carries a lot of weight when it comes from scholars such as Hans Driesch or Johannes von Reinke, for example). On the other hand, there are the hypotheses that this vitalism superposes upon mechanism (Driesch's "entelechies," or Reinke's "dominants," etc.). Of these two parts, the first is surely the more interesting one. See the excellent studies by Driesch (*Die Lokalisation morphogenetischer Vorgänge: Ein Beweis vitalistischen Geschehens* (Leipzig: Wilhelm Engelmann, 1899) [a shortened and revised version appears in: *The Science and Philosophy of the Organism*, vol. 1, *The Gifford Lectures Delivered Before the University of Aberdeen in the Year 1907* (London: Black, 1908)]; *Die organischen Regulationen: Vorbereitungen zu einer Theorie des Lebens* (Leipzig: Wilhelm Engelmann, 1901) [a shortened and revised version also appears in: *The Science and Philosophy of the Organism*, vol. 1, *The Gifford Lectures Delivered Before the University of Aberdeen in the Year 1907* (London: Black, 1908)]; *Naturbegriffe und Natururteile* (Leipzig: Wilhelm Engelmann, 1904) [an extensively revised version appears here: *The Science and Philosophy of the Organism*, vol. 2, *The Gifford Lectures Delivered Before the University of Aberdeen in the Year 1908* (London: Adam and Charles Black, 1908)]; *Der Vitalismus als Geschichte und als Lehre* (Leipzig: Johann Ambrosius Barth, 1905) [revised and rewritten for an English-speaking audience as: *The History and Theory of Vitalism*, trans. C. K. Ogden. London: Macmillan and Co., 1914]); and by Reinke (*Die Welt als That: Umrisse einer Weltansicht auf Naturwissenschaftlicher Grundlage* (Berlin: Gebrüder Paetel, 1899); *Einleitung in die theoretische Biologie* (Berlin: Gebrüder Paetel, 1901); *Philosophie der Botanik* (Leipzig: Johann Ambrosius Barth, 1905)).

the way back to its most ancient ancestors; as such, the individual's vital principle will turn out to be interconnected with them all, connected even with that of the tiny mass of protoplasmic jelly that is in all likelihood at the root of the genealogical tree of life. To a certain extent, by being joined to this most primitive ancestor, the individual would also be interconnected with every other individual that broke free along the diverging pathways of descent. In this sense, it could be said that the individual remains united with the totality of living beings through invisible links. In vain, then, do they claim to restrict the principle of finality to the individuality of the living being. If there is any finality in the living world, it encompasses the whole of life in a single indivisible embrace.[117] This life that is common to all living beings certainly presents many contradictions and many lacunae, and moreover, it is not so mathematically *one* that it is unable to allow each living being to individualize itself to a certain extent. It no less forms a single whole; and so, it is necessary to choose between a pure and simple rejection of finality and the hypothesis that coordinates not merely the parts of one organism within itself but also each living being with all of the others.

44

Pulverizing finality into a powder will not make it any easier to swallow. Either the hypothesis of a finality that is immanent to life must be rejected as a whole, or (as I believe) it must be modified in a completely different way.

The error of radical finalism, which was also the error of radical mechanism, is to have extended too far the application of certain concepts that are natural to our intellect. Originally, we only think in order to act. Our intellect was cast in the mold of action. Speculation is a luxury, whereas action is a necessity.[118] Now, in order to act, we begin by setting ourselves a goal; we make a plan and then turn to the details of the mechanism that will realize that plan. This final step is only possible if we know what we can count on. We must have extracted from nature the similarities that will allow us to anticipate the future. Thus, we must have applied, either consciously or unconsciously, the law of causality. Moreover, the more clearly the idea of efficient causality is formed in our minds, the more efficient causality takes the shape of a mechanical causality. And this latter relationship is, in turn, all the more mathematical insofar as it expresses a more rigorous necessity.[119] This is why we need to simply follow the inclination of our mind in order to become mathematicians.[120] But, on the other hand, this natural mathematics is nothing but the unconscious support for our conscious habit of linking the same causes to the same effects.[121] And the ordinary goal of this habit itself is to guide actions inspired by intentions or to direct combined movements so as to execute a model (which amounts to the same thing). We are born artisans just like we are born geometers, and in fact we are only geometers because we are artisans.[122] So, insofar as it was fashioned for the requirements of human action, human intellect is an intellect that proceeds simultaneously by intention and by calculation, by the coordination of means toward an end, and by the representation of mechanisms in increasingly geometric forms. Whether one imagines nature as an immense machine regulated by mathematical laws or whether one sees nature as the realization of a plan, in both cases one merely follows to the end two tendencies of the mind that are mutually complementary and that have their origin in the same vital necessities.

This is why radical finalism is so close to radical mechanism on a majority of points. Both doctrines are loath to recognize an unforeseeable creation of form taking place in the course of things, or even merely in the development of life.[123] Mechanism only considers reality in terms of similarities and repetition. As such, it is dominated by the law according to which in nature there is nothing but the same reproducing the same. The more the geometry that mechanism contains comes into view, the less it can admit that something is created, not even merely form. To the extent we are geometers, we reject the unforeseeable. Of course, we could embrace the unforeseeable to the extent that we are artists since art lives on creation and implies a latent belief in the spontaneity of nature.[124] But like pure speculation, disinterested art is a luxury. Long before being artists, we are artisans. And all fabrication, as rudimentary as it might be, lives on similarities and repetitions, like the natural geometry that serves as its footing. Fabrication works according to models that it intends to reproduce. And when it does invent, it proceeds (or at least imagines that it proceeds) through a new arrangement of already known elements. The principle of fabrication is that *the same is necessary for producing the same*.[125] In short, the rigorous application of the principle of finality, just like that of the principle of mechanical causality, leads to the conclusion that *everything is given*.[126] Because they are responding to the same need, the two principles make the same claim in their respective languages.

And that is why they agree again when it comes to completely doing away with time. Real durée bites into things and leaves the imprint of its teeth on them. If everything is in time, then everything changes inwardly, and so the same concrete reality never repeats. Thus, repetition is only possible in the abstract: what repeats is some aspect that our senses, and above all our intellect, have detached from reality, precisely because our action, toward which our intellect directs all of its effort, can only move among repetitions. As such, concentrating upon things that repeat, and exclusively preoccupied with soldering the same to the same, the intellect turns its back on the vision of time.[127] It loathes fluidity and solidifies everything that it touches. We do not think real time. But we live it because life overflows the intellect. The feeling that we have of our own evolution and of the evolution of all things in pure durée is there, forming around the intellectual representation (properly so-called) an indistinct fringe that fades off into the darkness.[a] Mechanism and finalism agree insofar as they only consider that luminous core that shines at the center. They forget that this core was formed out of the rest by way of condensation. And they forget that it would be necessary to use the whole—the fluid as much as, and perhaps even more than, what is condensed—in order to again catch hold of the inner movement of life.

In fact, if this fringe exists, as indistinct and vague as it might be, it should be even more important to the philosopher than the luminous core that it surrounds. For its presence allows us to affirm that the core is indeed a core, and that completely pure intellect is in fact a contraction, through condensation, of a more extensive power [puissance]. And, precisely because this vague intuition does not help us at all in guiding our action upon things, action being entirely localized at the surface of the real, we can assume that intuition no longer simply operates on the surface but rather in the depths.

The moment we break free of the frameworks in which radical mechanism and radical finalism lock up our thought, reality appears as an

***

a   TN: ...une frange indécise qui va se perdre dans la nuit. This striking image of the "fringe" or the atmosphere of intuition that surrounds the bright core or nucleus representing the intellect becomes an important one below. The image was implicitly introduced above, ix.

uninterrupted streaming forth of innovations where each one, having no more than sprung forth to make the present, has already receded into the past.[a] At that precise moment, reality falls under the gaze of the intellect, whose eyes are eternally looking backward.[128] This is already the case with our inner life. For each one of our actions, it is easy to find some antecedents of which this action will, in some sense, be merely the mechanical result. And it will also be said that each action is the accomplishment of an intention. In this sense, both mechanism and finality are everywhere in the evolution of our behavior. But if the action is one that concerns our person as a whole and is thereby genuinely ours, then it could not have been foreseen, even though it can be explained by its antecedents after it has been accomplished.[129] And even when realizing [réalisant] an intention, our action, which is a present and new reality, differs from the intention, which was only able to be a project for a repeating or a rearranging of the past. Thus, mechanism and finalism are here merely external views of our behavior. They extract from our behavior its intellectuality.[b] But our behavior itself slips between the two and extends much further. To repeat myself again, this is not to say that free action should be defined as capricious or irrational [déraisonnable].[130] To behave capriciously is to oscillate mechanically between two or several ready-made options and then, for all that, to settle upon one of them. This is not to have brought an inner situation to maturity, it is not to have evolved; rather, and as paradoxical as this claim might seem, it is to have forced the will to imitate the mechanism of the intellect. To the contrary, a behavior that is truly our own is one of a will that does not seek to mimic the intellect, but rather one that, by remaining itself (i.e., by evolving), and by way of gradual maturation, arrives at acts that the intellect will be able to indefinitely divide up into intelligible elements

48

a TN: ...jaillissement ininterrompu de nouveautés... By using the plural here of nouveautés, Bergson is thinking of the substantial new things that spring up at each moment, rather than simply "the new" in general. I have chosen "innovations" here to capture this, given possible misinterpretations with the English "novelties." The noun un jaillissement, translated here as "a streaming forth," comes from the verb jaillir, springing up or gushing forth, which also foreshadows the important image of a jet of water below (cf. below, 248–49).

b TN: Bergson's term intellectualité ("intellectuality") does not mean "intelligent" behavior per se, but rather any reality that has an intellectual character or fits the framework of an intellectual description, i.e., external, causal relations and systems shielded from real durée. This is the first occurrence of this term, which will become crucial in Chapter III.

without ever fully completing that division. The free act is incommensurable with the idea, and its "rationality" must be defined by this very incommensurability, which allows us to find in it as much intelligibility as we like.[131] Such is the character of our inner evolution. And it is probably the character of the evolution of life as well.

Our incorrigibly presumptuous reason takes itself to possess, by right of birth or of conquest, as innate or as acquired, all of the essential elements of the knowledge of truth. Even when it admits that it does not know the object that is presented to it, it believes that its ignorance only bears on the question of knowing which of its old categories is the one that works for the new object. Into which ready-to-open drawer will we put it? Which already tailored garment will we dress it in? Is it this, or that, or another thing? And the terms "this" and "that" and "another thing" are, for us, always things that are already conceived of and already known. We find it deeply repugnant to think that we might have to create, from scratch, a new concept or even a new method of thinking for a new object. And yet the history of philosophy is there; it shows the eternal conflict between systems, the impossibility of definitively forcing the real into these ready-to-wear garments (i.e., our ready-made concepts), and thus the necessity of making to measure.[132] Rather than going to this extreme, our reason prefers to announce, once and for all and with an arrogant modesty, that its knowledge is limited to what is relative and that the absolute is beyond its jurisdiction.[133] This preliminary declaration allows our reason simply to apply, without any scruples, its habitual method of thought and, on the pretense that it does not touch upon the absolute, to make an absolute decision with regard to all things. Plato was the first to set up the theory that to know the real consists in finding its Idea, that is, in forcing the real into a pre-existing framework that would already be available to us—as if we implicitly possessed universal science. But this belief is natural for the human intellect, obsessed as it is with always knowing into which old category it will catalog whatever new object comes along. It might be said that in a certain sense we are born Platonists, through and through.[a]

---

a   TN: ...*nous naissons tout platoniciens.* The sense of Bergson's claim here is that we are "naturally born" Platonists, or born as Platonists "through and through," in that Platonism has something to do with our natural way of thinking (i.e., the intellect), as the previous sentence shows. The adverb *tout* does not figure in the similar phrases just above, when he claims that we are born geometers and artisans (cf. above, 44–45).

Nowhere is the powerlessness of this method more clearly on display than in the theories of life. If Life—by evolving toward the Vertebrates in general, and toward man and the intellect in particular—was forced to abandon along the way many elements that were incompatible with this particular mode of organization, and, as I will show, to entrust them to other lines of development, then in order to grasp the true nature of vital activity, we will have to seek out and fuse together the totality of these elements with the intellect itself.[134] We will probably be aided in this task by the fringe of indistinct representation that surrounds our distinct representation, by which I mean our intellectual representation: after all, what could this useless fringe be if not that part of the evolving or source [principe] that has not been fully reduced down to the particular form of our organization and that has been smuggled along like contraband?[a] As such, we should look to this fringe for indications as to how we might dilate the intellectual form of our thought; it is there that we can find the élan necessary for raising ourselves up above ourselves.[b] Imagining the whole of life cannot consist in combining together the simple ideas deposited in us by life itself along the course of its evolution: for how could the part equal the whole, the contained equal the container, or a residue of the vital operation equal the operation itself?[135] And yet, this is precisely the illusion we create when we define the evolution of life as "a change from a homogeneous state to a heterogeneous state," or by any other concept that is obtained by combining together intellectual fragments.[c/136] We place ourselves at one of the endpoints of evolution, the principal one perhaps, but it is not the only one. Even

50

---

a   TN: This is the first time that Bergson uses the term *principe* ("principle") in this rather technical or philosophical sense, not to mean a fundamental proposition or a scientific theorem, but rather a fundamental "source" (and hence a metaphysical reality). To emphasize where he is aiming at this metaphysical sense, which is perhaps less apparent in the English word "principle," I have occasionally included "or source" in the text, as in the present case. Cf. below, 275, where Bergson even capitalizes the term to emphasize the metaphysical sense of the term in certain philosophical traditions.

b   TN: *...nous hausser au-dessus de nous-même.* As Arnaud François notes in the French edition, this somewhat paradoxical formulation resonates with Bergson's other claims regarding intuitive philosophy as a means for moving "beyond the human condition." Cf. the Index, "surpassing the human condition." The term *élan* here alludes both to the *élan vital* and to the idea of "energy" or "momentum" necessary for this self-transcendence. I have left it untranslated here to preserve this double allusion. Again, just below Bergson uses *élan* to refer to a certain "energy" (cf. below, 51).

c   [Bergson's missing note: Spencer, *First Principles*, 330.]

so, we do not consider everything that is contained there. We only retain of the intellect one or two of the concepts by which it expresses itself, and then we declare this part of a part to be representative of the whole, of something that even surpasses the consolidated whole, namely the evolutionary movement of which this "whole" is but the current phase! The truth is that it would not be too much to consider the whole of the intellect, and in fact this would not even be enough. It would still be necessary to bring the intellect together with what is found at each other terminal point of evolution. And it would be necessary to consider these diverse and diverging elements as so many extracts that are, or at least that were, in their most humble form, mutually complementary. This is the only way for us to get a sense for the real nature of the evolutionary movement—but again, this will only give us a sense of it since we would still be dealing with what has evolved, which is a result, and not with evolution itself, i.e., the action by which the result is obtained.

Such is the philosophy of life toward which I am moving. It claims to simultaneously go beyond both mechanism and finalism, but as I noted above, it is closer to the latter doctrine than to the former. It will be useful to emphasize this point, and to show in more precise terms where it resembles finalism and where it differs.

Like radical finalism, though perhaps in a form that is more vague, my philosophy of life will present the organic world as a harmonious whole.[137] But this harmony is far from being as perfect as they claimed. Life clearly allows for discordances because every species, and even every individual, retains only a certain *élan* from the overall impulsion of life, and tends to use this energy toward its own interests. Such is the nature of *adaptation*.[138] As such, the species and the individual think only of themselves, from which a potential conflict arises with the other forms of life. Harmony thus does not exist in fact but rather in principle—by this I mean that the original *élan* is a shared *élan*, and that the higher we climb back up, the more the diverse tendencies appear as complementary to each other.[139] Just like the wind that sweeps into a crossroads is divided up into diverging air currents, they are all still but one and the same gust of wind. Harmony, or better "complementarity," is only revealed in broad strokes, in tendencies rather than in states. Above all (and this is the point at which finalism is most gravely mistaken), harmony will be found behind rather than out in front.[140] It has to do with an identity of impulsion and not with a shared aspiration. In vain do they assign to life a goal, in the human sense of that word. To speak of a goal is to think of a pre-existing model that simply needs to be realized [*se réaliser*]. It is thus to assume that ultimately everything is given, and that the future could be read in the present.[141] It is to believe that life, in its movement and in its entirety, advances in the manner of our intellect, which itself is but an immobile and fragmentary view of life, a view that always naturally places itself outside of time. But life itself progresses and endures.[142] Of course, when glancing back at the pathway that has been traveled, we will always be able to trace the direction, to describe it in psychological terms, and to speak as if it unfolded as the pursuit of a goal. I myself will speak in this manner.[143] But when the pathway had yet to be traveled, the human mind had nothing to say, for the pathway was created through the act that traveled it, it being nothing other than the direction of that act itself. Thus, evolution must always admit of a psychological interpretation that is, from my point of view, the best explanation; but this explanation is only useful, indeed only meaningful, in a retroactive sense. The finalist interpretation, such as I will formulate it, must never be taken as an anticipation of the future. It is a certain vision of the past in light of the present. In short, the classical conception of finality posits at once too much and too little. It is both too broad and too narrow. By explaining life through the intellect, it excessively reduces the meaning of life: the intellect, at least such as we find it in ourselves, was fashioned by evolution along its pathway; it is cut out of a larger something; or rather, it is but the necessarily flat projection of a reality that has relief and depth. A true finalism would have to reconstitute this more comprehensive reality, or rather encompass it, if possible, in a single vision. But, on the other hand, precisely because this more comprehensive reality overflows the intellect (which is the faculty for linking the same to the same or for perceiving, and even producing, repetitions), this reality is surely creative, that is to say, productive of effects by which it dilates and goes beyond itself.[144] These effects were thus not given in it in advance and consequently it could not take them as ends, even though once they are produced they admit of a rational interpretation, like the interpretation of the fabricated object as the realization of a model. In short, the theory of final causes does not go far enough when it limits itself to placing the intellect into nature, and it goes too far when it assumes that the future pre-exists in the present in the form of an idea. Moreover, the second claim, which sins by excess, is the consequence of the first, which sins by not going far enough. The intellect, properly so-called, must be replaced by that more comprehensive reality of which the intellect is but a contraction. The future thereby appears as dilating the present. Thus, the future was not contained in the present in the form of a represented goal. And nevertheless, once it has been realized, the future will explain the present as much as be explained by the present, and perhaps even more so. That is, the future will have to be considered at least as much an end as a result, and perhaps even more so. Our intellect has the right to consider the future abstractly, from its habitual point of view, being itself an abstraction of the cause from which it emanates.[145]

True, the cause may thereby seem beyond our grasp. The finalist theory of life already evades any rigorous verification. What will happen, it will be asked, if we take finalism even further in one of its directions? And so here, after a necessary detour, we return to the question that I take to be essential: can facts be used to prove that mechanism is insufficient? I suggested that such a demonstration is only possible if we place ourselves directly within the evolutionary hypothesis.[146] The moment has come to establish that if mechanism is insufficient to account for evolution, then the way of proving this insufficiency is not by stopping at the classical conception of finality, and even less by restricting or attenuating it, but rather by going even further.[147]

*

[IN SEARCH OF A CRITERION]

Let me begin by indicating the principle of my demonstration. I said that from its origins, life is the continuation of a single and self-identical élan that is divided and shared among the diverging lines of evolution.[a/148] Something has grown or something has developed through a series of additions, all of which were creations. This development itself led to the dissociation of tendencies that themselves could not grow beyond a certain point without becoming incompatible with each other. In fact, there would be nothing to prevent us from imagining a single individual in which the entire evolution of life might be accomplished through

---

a TN: ...*s'est partagé*... This verb in French means both "to divide" and "to share," and Bergson's discussions of the *élan vital* rely upon both aspects. Where appropriate, I have included both words to preserve this double sense (cf. below, 88, 100).

a series of transformations spread across thousands of centuries.[149] Or again, in the absence of a single individual, we could imagine a plurality of successive individuals in a unilinear series. In both cases, evolution would have had but a single dimension, if we can express it this way. But in reality, evolution is accomplished through the intermediary of millions of individuals along diverging lines, where each line itself arrives at a crossroads from which new pathways radiate outward, and so on indefinitely. If my hypothesis is correct—that is, if the essential causes at work along these diverging pathways are psychological in nature—then they must preserve something in common despite the divergence of their effects, just like long-separated boyhood friends carry with them the same childhood memories.[150] Even though bifurcations took place, and even though lateral pathways opened up where the dissociated elements unfolded in an independent way, it is no less the case that the parts continue to move because of the original *élan* of the whole. Something of the whole must thus subsist in the parts. And this common element will make itself visible in a certain way, perhaps through the presence of identical organs in very different organisms. Imagine for a moment that mechanism were true: evolution would be accomplished by a series of accidents being added together, with each new accident, if it is advantageous, being preserved through selection and being added to the sum of the previously advantageous accidents represented by the actual form of the living being.[151] What chance will there be that, through two completely different series of accidents being added together, two completely different evolutions would arrive at similar results? The more two evolutionary lines diverge, the less probability that external accidental influences or internal accidental variations could have contributed to the construction of identical instruments [*appareils*].[a] This is especially the case if there had been no trace of these instruments at the moment the bifurcation took place. To the contrary, this similarity would be quite natural according to a hypothesis like the one we have put forward: we should rediscover, even in the most distant small streams, something

54

55

---

a TN: *...la construction d'appareils identiques.* The French word *appareil* could be translated in several ways, including "apparatus," "device," "machinery," "instrument," or even "system" (as in "the digestive system" for *l'appareil digestif*). I have chosen "instrument" here to help forefront how Bergson sees the organ as a relatively independent system or instrument serving a certain purpose. Indeed, Bergson himself makes this connection, writing: "*un instrument comme l'œil*" (below, 96; cf. 140–41).

of the impulsion received back at the source. If we could establish that life fabricates certain identical instruments through dissimilar means and along diverging lines of evolution, then pure mechanism would be refutable and finality, in the special sense in which I intend to use that term, would be demonstrable from a certain angle. Moreover, the force of the proof would be proportional to the degree of divergence between the lines of evolution selected, and to the degree of complexity of the similar structures that we find along them.[a]

Some will claim that the similarity of structure is due to the fact that life has evolved in identical general conditions. Despite the diversity of fleeting external influences and accidental internal variations, these durable external conditions would have imparted the same direction to the forces that constructed this or that instrument.[152] — Indeed, I am not unaware of the role that the concept of *adaptation* plays in contemporary science.[153] Of course, biologists do not all use this concept in the same way. For some, external conditions are capable of directly causing the variation of organisms in a definite direction, through physico-chemical modifications that they bring about in the living substance—such is Eimer's hypothesis, for example.[154] For others who are more loyal to the spirit of Darwinism, the influence of the conditions only operates indirectly by favoring, in the struggle for life [concurrence vitale], those members of a species that through the luck of birth were better adapted to the milieu. In other words, the former attribute a positive influence to external conditions; the latter attribute to them a negative action. According to the first hypothesis, this cause brings about the variations; according to the second, the cause only eliminates variations. But in both cases, the external cause is supposed to determine a precise adjustment of the organism to its conditions of existence. They will surely use this idea of shared adaptation in their attempt to explain, mechanistically, the very similarities of structure from which I believe the strongest arguments against mechanism can be drawn. This is why I must immediately indicate in general, and before going into the details, why the explanations that they would like to draw here from the concept of "adaptation" appear to be insufficient.

56

---

a   TN: ...*la vie fabrique certains appareils identiques*... It is worth noting again that Bergson uses "fabricate" in the sense of "manufacture" or "construct" rather than "concoct" or "invent (to deceive)."

Let me first note that of the two hypotheses just formulated, only the second one avoids equivocation.[155] Darwin's idea of an adaptation taking place through the automatic elimination of the poorly adapted is a simple and clear idea. But then again—and precisely because it attributes an entirely negative influence to the external cause guiding evolution—this approach already has significant difficulty accounting for the progressive and linear development of complex instruments [appareils] like the ones we will examine below.[156] What will become of this difficulty when the theory attempts to explain the identical structures of extraordinarily complex organs along diverging lines of evolution? A single accidental variation, as minimal as it might be, involves the action of a hoard of tiny physical and chemical causes.[157] The accumulation of accidental variations, such as is required to produce a complex structure, requires the convergence of, so to speak, an infinite number of infinitesimal causes. How could these entirely accidental causes reappear identically and in the same order at different points in space and time? No one will defend this claim, and even the Darwinian will say no more than that identical effects can come from different causes and more than one road leads to the same place. But we must not be fooled by a metaphor. The place where one arrives does not sketch out the form of the road that one has taken to get there, whereas an organic structure is the accumulation itself of the tiny differences that evolution had to go through to reach it. The struggle for life and natural selection cannot provide us any help in resolving this part of the problem, since we are not dealing here with what has disappeared; rather, we are simply looking at what has been conserved. Now, we see that identical structures are sketched out along independent lines of evolution through a gradual accumulation of effects that are added together. How could we suppose that accidental causes, coming about in an accidental order, could have repeatedly reached the same result given that the causes are infinite in number and the effect is infinitely complex?

The principle of mechanism is that *the same causes produce the same effects*.[158] True, this principle does not always entail that the same effects have the same causes. And yet, it does imply this consequence in the particular case in which the causes remain visible in the effect that they produce and where they are the constitutive elements of that effect. It is hardly out of the ordinary that two people out for a walk, leaving from different locations, and wandering through the countryside according to their individual whims, might eventually meet up. But it is entirely implausible that, by wandering about in this way, they might trace out identical and perfectly superposable lines. Moreover, this implausibility will be even greater if the pathways traveled from beginning to end involve more complex detours. And it will become an impossibility if the zigzags of the two walkers are of an infinite complexity. Now, how will this complexity of zigzags compare to the complexity of an organ where thousands of different cells are arranged in a certain order and where each cell is itself a kind of organism?[159]

So, let me turn to the second hypothesis to see how it might resolve the problem.[160] Here, adaptation will no longer consist simply in the elimination of the poorly adapted. Rather, it will be the result of the positive influence of external conditions that will have shaped the organism according to their own form. Clearly the similarity of the effects will here be explained through the similarity in the cause. It would seem that we are within a form of pure mechanism. But let us take a closer look. We will see that the explanation is merely verbal, that again we have been fooled by words, and that the artifice of this solution consists in taking the term "adaptation" to mean two entirely different things at the same time.

If I pour into the same glass, one after the other, first water and then wine, the two liquids will take the same form, and the similarity of the form comes from an identical adaptation of the contents to the container. Adaptation here clearly means mechanical insertion. This is because the form to which the matter adapts itself is already there, ready-made, and it imposed its own configuration upon the matter. But when we speak of the adaptation of an organism to the conditions in which it must live, where is the pre-existing form that awaits its matter? These conditions are not a mold into which life will insert itself and from which it will receive its form—when we reason in this way, we are fooled by a metaphor. There is no form as yet, and it will be up to life to create a form for itself that is appropriate to the conditions that are imposed upon it. Life will have to take advantage of these conditions; it will have to neutralize the disadvantages and make use of the advantages; in short, it will have to respond to external actions through the construction of a machine that does not resemble them at all. Here "to adapt" will no longer consist in *repeating*, but rather in *responding*, which is entirely different.[a] If there

a TN: *S'adapter ne consistera plus ici à répéter, mais à répliquer, ce qui est tout différent.*

is still adaptation here, it will be in the sense in which we might say that the solution to a geometry problem, for example, is adapted to the conditions set out in the problem.[161] I happily admit that adaptation, understood in this way, explains why different evolutionary processes bring about similar forms: the same problem surely calls for the same solution. But then it would be necessary to introduce, following the model of the solution of a geometry problem, an intelligent activity or at the very least a cause that behaves like an intelligent activity. But this will be to reintroduce finality, and indeed, this time it will be a finality that is much too loaded down with anthropomorphic elements. In short, if the adaptation that they are talking about is a passive and simple repetition in relief of what the conditions give in *intaglio*, then adaptation will never construct any of the things that they want it to construct. And if it is claimed that adaptation is active and capable of responding through a calculated solution to a problem that the conditions pose, then they go even farther than I do—and indeed too far, in my opinion—in the direction that I first indicated. But the truth is that they surreptitiously shift from one of these meanings to the other, and that they take refuge in the first meaning every time they are caught red-handed relying on finalism in their use of the second. Although it is really the second meaning that is most useful in current scientific practice, its philosophy is most often provided by the first. For each particular case, they speak as if the process of adaptation were an effort by the organism to construct a machine capable of making the most of the external conditions, and then they talk about adaptation in general as if it were the very imprint of the circumstances passively received by an indifferent matter.

[EXAMINATION OF THE VARIOUS TRANSFORMIST THEORIES THROUGH A PARTICULAR EXAMPLE]

But let us come to the examples. First, it would be interesting to establish here a general comparison between plants and animals. Indeed, how could we not be struck by the parallel developments that have been accomplished, on both sides, toward sexuality? Not only is fertilization itself identical in higher plants to what it is for animals, since in both cases it consists in the union of two partial nuclei that differ in their properties and in their structures prior to their union and that immediately thereafter become equivalent to each other.[162] But also, the preparation of the

59

sexual elements takes place under similar conditions on both sides: in essence, it consists in the reduction of the number of chromosomes and in the rejection of a certain quantity of chromosomal substance.[a/163] Nevertheless, plants and animals have evolved along independent lines, have been favored by dissimilar circumstances, and have faced different obstacles. So here we have two large series that have continued along while diverging. Along each series, thousands and thousands of causes came together to determine the morphological and functional evolution. And yet on both sides, these infinitely complex causes have culminated in the same effect. Moreover, beginning from this effect, one will hardly dare to say that this is a phenomenon of "adaptation": for how could we speak of adaptation, or appeal to the pressure of external circumstances, when the very utility of sexual generation is not apparent, when it could have been interpreted in an extremely diverse number of ways, and when some excellent minds see the sexuality of the plant, at least, as a luxury that nature could have simply gone without?[b/164] But I do not wish to weigh in on such controversial facts. The ambiguity in the term "adaptation," as well as the necessity of going beyond both the mechanistic perspective of causality and the anthropomorphic perspective of finality, will be clearer if we consider simpler examples. From the beginning, the doctrine of finality has taken advantage of the marvelous structure of the sense organs so as to liken the work of nature to that of an intelligent worker. Moreover, since these organs are found in a rudimentary state among lower animals, and since nature provides us with all of the intermediaries between the pigmented eyespot of the simplest organisms and the infinitely complex eye of the Vertebrates, one can very easily suggest that what determines the increasing perfection is the entirely mechanistic play of natural selection. In short, if there is a case in which it seems we are justified in invoking adaptation, this is the one. For, when it comes to the role and the significance of sexual generation, or to the relationship that ties it to the conditions in which

a Bergson's note: Paul Guérin, *Les connaissances actuelles sur la fécondation chez les Phanérogames* (Paris: A. Joanin, 1904), 144–48. Cf. Yves Delage, *L'hérédité et les grands problèmes de la biologie générale*, 2nd ed. (Paris: Schleicher Frères, 1903), 140–41.

b Bergson's note: Martin Möbius, *Beiträge zur Lehre von der Fortpflanzung der Gewächse* (Jena: Verlag von Gustav Fischer, 1897), 203–6. Cf. Marcus Hartog, "Sur les phénomènes de reproduction," *L'année biologique: Comptes rendus annuels des travaux de biologie générale* 1 (1895): 707–9.

it takes place, these remain debatable. But the relation between the eye and light is self-evident, and when we speak here of adaptation, we must know what we mean. So, if I can demonstrate the insufficiency of the principles invoked by either side of the debate in this privileged case, my demonstration will thereby immediately attain a fairly high degree of generality.

Consider the example that supporters of finality have always emphasized: the structure of an eye, such as the human eye. They have had no difficulty showing that in this extremely complex instrument, the elements are marvelously coordinated among themselves. In order for vision to take place, writes the author of a well-known book on *Final Causes*:

> It is necessary, first of all, that the solid membrane which constitutes the globe of the eye, and which is called the *sclerotic*, should become transparent in a point of its surface, to permit the luminous rays to traverse it; and this transparent part, which is called the *cornea*, must be found to correspond exactly with the opening of the orbit to the eye. [...] In the second place, there must be behind the transparent opening which permits the light to enter, convergent media. [...] And thirdly, there must be found at the extremity of this camera [obscura], and opposite the entrance, the *retina*.[a/165]

There must also be, "before the retina, and perpendicularly to it, an innumerable quantity of transparent cones, which allow to reach the nervous membrane only the light following the direction of their axis," etc., etc.[b]

In response to this, the supporters of final causes have been asked to place themselves within the evolutionist hypothesis. Indeed, everything appears marvelous if we consider an eye, such as our own, in which thousands of elements are coordinated in the unity of the function. But it would be necessary to consider the function at its origin, for the Infusorian, where the eye is reduced to the simple (almost purely chemical) light sensitivity of a pigmented spot. This function, which at the beginning was merely an accidental fact, was able—either directly,

62

---

a Bergson's note: Paul Janet, *Les causes finales* (Paris: Germer Baillière, 1876), 83 [*Final Causes*, trans. William Affleck, preface Robert Flint (Edinburgh: T & T Clark, 1878), 61–62].
b Bergson's note: *Ibid.*, 80 [59].

through an unknown mechanism, or indirectly, by the simple effect of the advantages that it procures for the living being and of the grip that natural selection was thereby offered—to bring about a slight complication of the organ, and this will have brought along with it a perfecting of the function.[166] In this way, they were able to explain the progressive formation of an eye as intricately designed as our own through an indefinite series of actions and reactions between the function and the organ, without having to introduce an extra-mechanistic cause.

The question is indeed difficult to resolve if it is first assumed as being between the function and the organ, as the doctrine of finality assumed and as mechanism does as well. This is because organ and function are two heterogeneous terms that nonetheless mutually condition each other to such an extent that it is impossible to say *a priori* whether— when it comes to identifying their relation—it would be better to begin with the former (as mechanism would have it) or with the latter (as the finalist theory would require).[167] But I believe that the discussion would take an altogether different turn if we first compared two terms of the same nature, an organ with an organ, and no longer an organ with its function. In this way, we might gradually move toward an increasingly plausible solution. And we would have all the more chance of reaching such a solution the more resolutely we place ourselves within the evolutionist hypothesis.

Consider side by side the eye of a Vertebrate and that of a Mollusk such as the Scallop.[a] In both, we find the same essential parts composed of analogous elements. The Scallop's eye includes a retina, a cornea, and a crystalline lens with a cellular structure like our own. We even find here that strange inversion of retinal elements, something that we do not in general find in the retinas of Invertebrates. Now, the origin of Mollusks may still be an open question, but, regardless of which opinion we adopt, everyone will agree that Mollusks and Vertebrates split from their shared branch well before the appearance of an eye as complex as the Scallop's. So where does their analogous structure come from?[168]

Let us put this question to the two opposing systems of evolutionary explanation, one after the other, beginning with the hypothesis of purely

a   TN: *...le Peigne...* Bergson uses a common name for a genus of large saltwater scallops (*Pecten*), which are bivalve mollusks of the *Pectinidae* family.

accidental variations, followed by that of a variation oriented in a definite direction under the influence of external conditions.[169]

In terms of the first hypothesis, we know it is presented these days in two quite different forms. Darwin spoke of very slight variations that would be added together through the effect of natural selection. He was not unaware of the evidence of sudden variations, but these "sports" (as he called them) would, according to him, only result in monstrosities incapable of perpetuating themselves.[170] Rather, he accounted for the genesis of species through an accumulation of imperceptible variations.[a/171] Many naturalists continue to hold this opinion. And yet, it seems to be losing ground to the opposite idea: that a new species is constituted quite suddenly, through the simultaneous appearance of several new characteristics that are quite different from the previous ones. This second hypothesis, which was already suggested by several authors (notably by Bateson, in a remarkable book[b/172]), has recently become very important and has acquired great force following Hugo de Vries's   64 striking experiments. This latter botanist, experimenting on the species *Oenothera lamarkiana* [Evening Primrose], obtained a certain number of new species after merely a few generations.[173] The theory he draws from his experiments is extremely interesting. Species, he suggests, would pass through alternating periods of stability and transformation. When the period of "mutability" occurred, they would produce unexpected forms.[c/174] I will not attempt to take sides between this theory

a Bergson's note: Charles Darwin, *Origine des espèces au moyen de la sélection naturelle, ou la lutte pour l'existence dans la nature*, trans. Edmond Barbier (Paris: C. Reinhold, 1887), 46 [*On the Origin of Species by Means of Natural Selection, Or the Preservation of Favoured Races in the Struggle for Life* [1859], 6th ed. (London: John Murray, 1872), 33].

b Bergson's note: William Bateson, *Materials for the Study of Variation: Treated with Especial Regard to Discontinuity in the* Origin of Species (London: Macmillan and Co., 1894), 567–75. Cf. William Berryman Scott, "On Variations and Mutations," *The American Journal of Science* 48, no. 287 (November 1894): 355–74.

c Bergson's note: Hugo de Vries, *Die Mutationstheorie: Versuche und Beobachtungen über die Entstehung der Arten im Pflanzenreich*, 2 vols. (Leipzig: Verlag von Veit, 1901–1903). Cf. Hugo de Vries, *Species and Varieties: Their Origin by Mutation*, ed. Daniel Trembly MacDougal (Chicago: The Open Court Publishing Company, 1905). Although the experimental basis for De Vries's theory was judged to be too narrow, the idea of mutation, or of sudden variation, still took on an important role in science. [TN: The English version of the two volumes of *Die Mutationstheorie* appeared as: *The Mutation Theory: Experiments and Observations on the Origin of Species in the Vegetable Kingdom*, trans. J. B. Farmer and A. D. Darbishire, vol. 1, *The Origin of Species by Mutation* (Chicago: The Open Court Publishing Company, 1909); vol. 1, *The Origin of Varieties by Mutation* (Chicago: The Open Court Publishing Company, 1910).]

and the theory of imperceptible variations.[175] I simply want to show that, whether small or large, the variations that are invoked here are incapable—if they are merely accidental—of accounting for a similarity of structure such as the one that we have indicated.[176]

[DARWIN AND IMPERCEPTIBLE VARIATION]

Let us first adopt the Darwinian theory of imperceptible variations by positing small, random differences that are continuously being added together. It must be remembered that all of the parts of an organism are necessarily coordinated among themselves. Regardless of whether the function is the effect or the cause of the organ, one thing is certain: the organ must be functioning for it to provide some benefit and for it to thereby give selection something to grasp onto.[177] That the delicate structure of the retina develops and becomes increasingly complex, that is, progresses, this fact, rather than favoring vision, would probably even disturb it, so long as the visual centers did not develop at the same time, along with various parts of the visual organ itself. If the variations are accidental, then we could hardly expect them to agree to take place in all the parts of the organ at the same time and in such a way that the organ continues to accomplish its function. Darwin understood this well, and it is one of the reasons that he assumed variations to be imperceptible.[a/178] Being very slight, a difference that accidentally appears at one point of the visual apparatus will not hinder the functioning of the organ. And that being the case, this first accidental variation can wait, so to speak, for complementary variations to come along to join with it and raise vision to a higher degree of perfection. Very well, but if the imperceptible variation does not hinder the functioning of the eye, it does not help it either, so long as the complementary variations have not yet occurred. In short, how could it be preserved by the effect of selection? Whether they like it or not, it will have to be argued that the slight variation was a sort of toothing-stone set down by the organism and reserved for a later construction.[b] This hypothesis, which is so little in keeping with

a   Bergson's note: Darwin, *L'origine des espèces*, 198. [TN: *On the Origin of Species*, 145. Note that Bergson regularly uses *variations insensible* ("imperceptible variations") for Darwin's discussions of "slight" and "gradual" variations.]

b   TN: A "toothing-stone" [*pierre d'attente*] is an architectural feature in which projecting stones or pieces of wood are left at the end or outer corner of a wall or on the side of a building so that another wall or building might eventually be built onto it.

Darwin's principles, already seems difficult to avoid when considering an organ that has developed along a single main line of evolution, such as the Vertebrate's eye. It becomes absolutely unavoidable when we note the similarity in the structure of the Vertebrate's eye and the Mollusk's eye. That is, how could we ever imagine that the same incalculable number of slight variations might take place in the same order along two independent lines of evolution, if indeed those variations were purely accidental? And how could they be preserved by natural selection and accumulated, along both lines—the very same variations in the very same order—when each variation, taken separately, had no utility?[179]

[DE VRIES AND SUDDEN VARIATION]

Let us turn to the theory of sudden variations and consider whether or not it solves the problem. It attenuates it, of course, on one point, but it aggravates it on another. I will have less difficulty understanding the similarity of the two organs if the actual form of the Mollusk's eye, like the Vertebrate's eye, had been set up through a relatively small number of sudden leaps, rather than if they were the composition of an incalculable number of infinitesimal similarities acquired successively. In both cases, chance is at work, but this new hypothesis does not ask chance to perform the miracle that it would have to accomplish on the former hypothesis. Not only is the number of similarities that I will have to add together more limited, but I also better understand why each of them would be preserved to be added to the later ones since this time the elementary variation is large enough to provide an advantage to the living being and thereby to open itself up to the play of natural selection. But here another problem arises, and indeed one that is no less daunting: how could all of the parts of the visual apparatus, being suddenly modified, remain so well coordinated among themselves that the eye continues to exercise its function? For from the very moment that this variation is no longer infinitesimal, the isolated variation of a single part of the eye would render vision impossible. Now it becomes necessary that all of the parts change simultaneously, and that each one consults with the others. I could admit that a mass of uncoordinated variations may have suddenly taken place among many less lucky individuals, that natural selection eliminated them, and that only the viable combination—i.e., the one capable of conserving and improving vision—survived.[180] But even then, this combination would have to have taken place. And, assuming that chance granted this favor once, how could it be claimed that it would repeat it throughout the history of a species in such a way as to bring about each time, and all of a sudden, new complications, wondrously ordered with each other and organized within the continuation of the previous complications? And above all, how could it be assumed that, through a series of simple "accidents," these sudden variations would take place identically along the pathway of two independent lines of evolution, in the same order, and requiring each time a perfect accord among elements that are more and more numerous and complex?

True, here they will invoke the law of correlation, to which Darwin himself appealed.[a][181] They will claim that a change is not localized to a single point in the organism, that it has necessary repercussions at other points. The examples used by Darwin have remained classic: white cats with blue eyes are generally deaf, hairless dogs have imperfect teeth, etc. Very well, but now is not the time to play around with the sense of the word "correlation." A collection of interdependent changes is one thing; a system of complementary changes is quite another, by which I mean changes that are coordinated among themselves so as to maintain and even to improve the functioning of an organ in more complex conditions.[182] The fact that an anomaly in the system of hair growth is accompanied by an anomaly in dentition does not call for a special principle of explanation: hair and teeth are similar formations,[b][183] and the same chemical alteration of the germ that blocks the formation of hairs will probably hinder that of the teeth as well.[184] There are probably similar kinds of causes that lead to the deafness of white cats with blue eyes. In these various examples, the "correlated" changes are merely interdependent (not to mention that in fact these are lesions, by which I mean reductions or suppressions of something, and not additions, which is quite different). But when they speak of "correlated" changes suddenly springing up in the various parts of the eye, the word is being used in an entirely new sense. Now it has to do with a collection of changes

---

a   Bergson's note: Darwin, L'origine des espèces, 11–12 [On the Origin of Species, 8–9].
b   Bergson's note: Regarding this homology between hair and teeth, see Alexander Brandt, "Ueber borstenartige Gebilde bei einem Hai und eine mutmaßliche Homologie der Haare und Zähne," Biologisches Centralblatt 18 (1898): 257–70.

that are not merely simultaneous and not merely linked together by   68
a shared origin, but rather coordinated together in such a way that
the organ continues to accomplish the same simple function and even
accomplishes that function better.[185] I admit that, strictly speaking, a
modification of the germ that influences the formation of the retina
would also act simultaneously upon the formation of the cornea, the
iris, the crystalline lens, the visual centers, etc., even though here we are
dealing with formations that are heterogeneous in a different way than
the hair and the teeth are.[186] But I cannot accept the part of the theory
of sudden variations that holds that all of these simultaneous variations
are accomplished in the direction of an improvement, or even merely a
maintaining, of vision—at least not so long as they refuse to posit a mysterious principle whose role would be to watch over the interests of the
function. But this would be to renounce the very idea of an "accidental"
variation. In reality, the two meanings of the word "correlation" often
interfere with each other in the biologist's mind, like the two meanings
of "adaptation."[187] And the merging of the two is more or less justified
in the field of botany, precisely where the theory of species formation
through sudden variation finds its most solid experimental foundation.
Indeed, for plants, the function is nowhere near as closely tied to the
form as it is for animals.[188] Profound morphological differences, such as
changes in the shape of the leaves, have no appreciable influence upon
the workings of their function and thereby do not create the need for
a system of complementary adjustments to allow the plant to remain
viable. The same is not true for animals, above all when we consider
an organ (such as the eye) that has both a highly complex structure
and a highly delicate function. It would be in vain here to identify the
variations that are simply interdependent with the variations that are, in
addition, complementary. The two meanings of the word "correlation"
must be carefully distinguished because we would commit a veritable   69
paralogism if we adopted one meaning in the premises of the argument
and the other in the conclusion. Nevertheless, this is precisely what is
done when the principle of correlation is invoked in the explanations
of the details to account for complementary variations, and when one
then speaks of correlation in general, as if it were merely any collection
of variations whatsoever provoked by any variation whatsoever in the
germ. They start by using the idea of correlation as found in current
science, much as a supporter of finality might. Then they tell themselves

that this is simply a convenient way of speaking, that they will correct it later, that they will come back to pure mechanism when they explain the nature of the principles and when they shift from science to philosophy.[189] Indeed, they do thereby come back to mechanism, but only on condition of giving the word "correlation" a new meaning, and one that is now improper for the details of the explanations they provided.[190]

[EIMER AND ORTHOGENESIS]

To sum up, if the accidental variations that determine evolution are imperceptible variations, then it would be necessary to appeal to a benevolent genius—the genius of the future species—to ensure that they are preserved and added together, since selection does not take care of this.[191] If, on the other hand, the accidental variations are sudden, then the previous function will only continue to work, or a new function will only replace the old one, if all of the changes that suddenly appear complement each other with regard to the accomplishment of a single act. Here again it will be necessary to appeal to a benevolent genius, this time in order to obtain the *convergence* of simultaneous changes, as before we had to do so to assure the *continuity in the direction* of successive variations. In neither case can we claim that the parallel development of complex identical structures along independent lines of evolution might be the result of a simple accumulation of accidental variations. I turn now to the second of the two major theories that I set out to examine. Let us assume that the variations are no longer the result of accidental and internal causes but rather of the direct influence of external conditions. And let us see how, on this hypothesis, one might account for the similarity of structure of the eye in two series that are independent from a phylogenetic point of view.

70

Even if Mollusks and Vertebrates have evolved separately, both have remained exposed to the influence of light. And light is a physical cause that brings about determinate effects. Acting in a continuous manner, light was able to produce a continuous variation in a consistent direction. Of course, it is unlikely that the eye of the Vertebrates and the eye of the Mollusks would have been constituted through a series of variations caused by mere chance. By supposing that light intervenes as an instrument of selection, so as to only allow useful variations to survive, there is no possibility that the play of chance—even if watched over in this way from the outside—would in both cases result in the same juxta-position of elements coordinated in precisely the same way.[192] But this would no longer be the case if we adopted the theory that light acts directly upon organic matter in order to modify its structure and to cause it to adapt, in some sense, to its own form. In this case, the similarity of the two effects would be explained simply by the identity of the cause. The increasingly complex eye would be something like the increasingly deep imprint of light upon a matter that, being organic, possesses a *sui generis* aptitude for receiving it.

But can an organic structure be compared to an imprint? We have already discussed the ambiguity of the term "adaptation."[193] The gradual complication of a form that better and better inserts itself into the mold of external conditions is one thing; the increasingly complex structure of an instrument that draws from these conditions an increasingly advantageous course of action is something completely different.[a] In the first case, matter is limited to receiving an imprint, while in the second it actively responds [*réagit*], that is, it solves a problem.[194] Of these two meanings of the word, it is clearly the second that is intended when it is said that the eye is better and better adapted to the influence of light. But they go more or less unconsciously from the second meaning to the first, and a purely mechanistic biology will attempt to bring into coincidence the passive adaptation of an inert matter, which undergoes the influence of the environment, and the active adaptation of an organism, which draws from this influence an appropriate course of action. Moreover, I admit that nature itself seems to invite our minds to merge these two genres of adaptation since nature usually begins with a passive adaptation where later it will have to construct a mechanism that will actively respond. For instance, in the case that I am focusing on here, it is incontestable that the first rudiments of the eye are found in the pigmented eyespot of lower organisms. This pigmented spot could very well have been physically produced by the action of the light itself, and we can observe an immense number of intermediaries between the pigmented eyespot and a complex eye, such as that of the

71

---

a TN: Throughout these pages, Bergson plays with the formulation *tirer parti de* ("to take advantage of" or "to make the most of"). Here he alludes again to the expression by using its component parts: *tirer* ("to draw") and *un parti* ("a course of action"). Cf. above, 58, 60, and particularly below, 72–73.

Vertebrates. — But, the fact that we go from one thing to another by degrees does not necessarily entail that the two things are of the same nature. From the fact that an orator first adopts the passions of his audience in order to then be able to take control of them, we cannot conclude that *to follow* is the same thing as *to lead*.[195] Now, it seems that the only way living matter can take advantage of the circumstances is by first passively adapting to them: wherever living matter must take hold of the direction of a movement, it begins by adopting that movement. Life proceeds by insinuation.[a][196] It was useless to show us all of the intermediaries between a pigmented spot and an eye: there will be no fewer intermediaries between these two than there are between a photograph and a device for taking photographs. Perhaps the photograph tends, little by little, in the direction of the photographic apparatus; but could light alone, a physical force, have been able to provoke this tendency and convert an impression that it has left behind into a machine capable of making use of that impression?

72

It will be objected that I am mistakenly introducing considerations of utility, and that the eye is not made for seeing; rather, it will be said, we see because we have eyes, the organ simply is what it is, and "utility" is a word by which we designate the functional effects of the structure. But when I say that the eye "takes advantage" of the light, I do not merely mean to say that the eye is capable of seeing; rather, I am alluding to very precise relations that exist between this organ and the apparatus for locomotion.[197] The retina of the Vertebrates extends into an optical nerve, which itself is continued through cerebral centers connected to motor mechanisms.[198] Our eye takes advantage of the light insofar as it allows us to use, through movements of reaction, the objects that we see as advantageous and to avoid the ones that we see as harmful. Now, it will hardly be difficult to convince me that if light indeed physically produced the pigment spot, then it could also physically determine the movements of certain organisms. For example, some ciliated Infusorians react to light. And yet, no one will claim that the influence of the light physically caused the formation of a nervous system, a muscular system,

---

a TN: *La vie procède par insinuation.* Bergson does not use "insinuation" in the negative or pejorative sense of "hinting at something nefarious," but rather in the older, neutral sense of "the insertion of something into the sinuosities or curves of something else." This characterization of how life inserts itself into matter reappears below, 99.

a skeletal system, all of which are connected to the visual apparatus in Vertebrates. In fact, to speak of the gradual formation of the eye, and even more so when the eye is linked to all that is inseparable from it, is already to introduce something completely different from the direct action of light. This is to implicitly attribute to organic matter a certain *sui generis* ability, namely the mysterious power to set up highly complex machines in order to take advantage of the simple stimulation whose influence it undergoes.

73

And yet, this is precisely what they claim they can do without. They hold that physics and chemistry give us the key to everything. Theodor Eimer's major work is instructive in this respect.[199] It is well known how much insightful effort this biologist has devoted to demonstrating that the transformation takes place through the continuous influence of the external upon the internal, in a well-defined direction, and not, as Darwin held, through accidental variations. Eimer's thesis is based upon some extremely interesting observations, beginning from the study of the progression of color variations of the skin of certain Lizards. Or again, the somewhat dated experiments performed by Georg Dorfmeister show that the same chrysalis, depending on whether it is submitted to cold or to heat, gave rise to quite different butterflies, so much so that they were long considered independent species (*Vanessa levana* and *Vanessa prorsa*); an intermediate temperature produces an intermediary form.[200] These facts can be placed alongside the significant transformations that are observed in a small Crustacean, the *Artemia salina*, depending on whether the salinity of the water in which it lives is increased or decreased.[a/201] In these various experiments, the external agent seems to clearly behave as a cause of transformation. But in what sense should we understand the word "cause" here? Without attempting an exhaustive analysis of the idea of causality, I will simply note that three completely different senses of this word are often confused.[202] A cause can act by *impulsion*, by *releasing* [*déclenchement*], or by *unrolling* [*déroulement*]. The billiard ball that is launched at another billiard ball determines its movement by *impulsion*. The spark

74

---

a Bergson's note: Recent observations seem to show that the transformation of the Artemia is a more complex phenomenon than was at first believed. For more on this, see Max Samter and Richard Heymons, "Die Variationen bei *Artemia salina* Leach. und ihre Abhängigkeit von äußeren Einflüssen," *Abhandlungen der königlich preussischen Akademie der Wissenschaften*, "Abhandlungen nicht zur Akademie gehöriger Gelehrter: Physikalische Abhandlungen," Anhang 2 (1902): 1–62.

that provokes the explosion of the powder acts as a *releasing*. The gradual relaxing of the spring that causes the phonograph to turn *unrolls* the melody that is inscribed upon the cylinder: if I take the melody that plays to be an effect, and the relaxing of the spring as the cause, then I will say that the cause proceeds here through an unrolling. These three cases are distinguished from each other by a larger or smaller degree of interdependence between the cause and the effect. In the first case, the quantity and the quality of the effect varies with the quantity and the quality of the cause. In the second, neither the quality nor the quantity of the effect varies with the quality and the quantity of the cause; the effect is invariable. Finally, in the third, the quantity of the effect depends upon the quantity of the cause, but the cause has no influence upon the quality of the effect; the longer that the cylinder turns under the influence of the spring, the longer will be the portion of the melody that I hear, but the nature of the melody or of the portion that I hear does not depend upon the action of the spring. In reality, then, the cause only *explains* the effect in the first case. For the other two, the effect is more or less given in advance and the antecedent invoked is, to different degrees, of course, the opportunity for it more than its cause. Now, do they take the word "cause" in the first sense when it is claimed that the salinity of the water is the cause of the Artemia's transformation, or that the degree of temperature determines the color and design on the wings of certain chrysalides when they become butterflies? Clearly not—causality here has an intermediary sense, somewhere between unrolling and releasing. And this is surely how Eimer understands it when he speaks of the "kaleidoscopic" character of variation.[a/203] Or again when he says that

75   the variation of organic matter takes place in a definite direction, just

---

a Bergson's note: Theodor Eimer, *Orthogenesis der Schmetterlinge: Ein Beweis bestimmt gerichteter Entwickelung und Ohnmacht der natürlichen Zuchtwahl bei der Artbildung, zugleich eine Erwiderung an August Weismann*, vol. 2 of *Die Entstehung der Arten auf Grund von Vererben erworbener Eigenschaften nach den Gesetzen organischen Wachsens* (Leipzig: Engelmann, 1897), 24; cf. Theodor Eimer, *Die Entstehung der Arten auf Grund von Vererben erworbener Eigenschaften nach den Gesetzen organischen Wachsens*, vol. 1, *Ein Beitrag zur einheitlichen Auffassung der Lebewelt* (Jena: Fischer, 1888), 53. [TN: For the former, a partial translation in English appears as: *On Orthogenesis and the Impotence of Natural Selection in Species-Formation*, trans. Thomas J. McCormack (Chicago: The Open Court, 1898), 33; the latter appears as: *Organic Evolution as the Result of the Inheritance of Acquired Characteristics According to the Laws of Growth*, trans. J. T. Cunningham (London: Macmillan and Co., 1890), 48.]

like inorganic matter crystallizes in definite directions.[a/204] And we can admit that, strictly speaking, this might be a purely physico-chemical process when it comes to changes in the color of the skin. But if we were to extend this mode of explanation to the case of the gradual formation of the eye in Vertebrates, for example, it would be necessary to posit that the physico-chemical nature of the organism is such that, in this case, the influence of light would cause it to construct a progressive series of visual instruments, each one extremely complex, and each one being capable not only of seeing but of seeing better and better.[b/205] What more indeed would the most ardent partisan of the doctrine of finality say to characterize this utterly peculiar physico-chemistry? And will the position of mechanistic philosophy not become even more untenable when it has been shown that the egg of a Mollusk cannot have the same chemical composition as that of a Vertebrate, that the organic substance that evolved toward the first of the two forms could not have been chemically identical to the one that took the other direction, and that nevertheless, under the influence of light, the same organ has been constructed in both cases?

The more we reflect upon this, the more we will see just how contradictory this production of the same effect by two different accumulations of an enormous number of small causes is to the principles invoked by mechanistic philosophy. We have focused our entire analysis thus far on an example taken from phylogenesis. But ontogenesis would have provided us with examples that are no less convincing. At each moment, and before our very eyes, nature arrives at identical results, often in neighboring species, through embryogenetic processes that are completely different. Observations of "heteroblasty" have multiplied in the past few years,[c/206] and it has been necessary to give up the almost classical theory of the specificity of embryonic layers.[207] To limit ourselves once again to our comparison between the eye of the Vertebrates and

76

---

a Bergson's note: Eimer, *Die Entstehung der Arten*, 25 [*Organic Evolution as the Result of the Inheritance of Acquired Characteristics*, 23].

b Bergson's note: Eimer, *ibid.*, 165ff. [TN: See Eimer, *ibid.*, 153ff.]

c Bergson's note: W. Salensky, "Heteroblastie," in *Proceedings of the Fourth International Congress of Zoology (Cambridge, 22–27 August 1898)*, ed. Adam Sedgwick (London: C. J. Clay and Sons, 1899), 111–18. Salensky coined the term "heteroblasty" to designate cases in which equivalent organs are formed, in the same locations, for related animals whose embryological origin is nevertheless different.

that of the Mollusks, we will observe that the retina of the Vertebrates is produced through an expansion emitted by the rudimentary brain of the young embryo. It is a veritable nervous center that was carried toward the periphery. To the contrary, the retina in Mollusks derives directly from the ectoderm, and not indirectly through the intermediary of the embryonic encephalon. Thus, the development of the same retina clearly takes place for humans and for the Scallop through two different evolutionary processes. But without having to go to the lengths of comparing two such distant organisms, we would reach an identical conclusion by studying certain curious facts of regeneration in a single organism. If a Newt's crystalline lens is removed, it is regenerated by the iris.[a/208] And yet, the original crystalline lens was constituted from the ectoderm, whereas the iris has its origin in the mesoderm. And even more so: if we remove the *Salamandra maculata*'s crystalline lens without touching its iris, then the regeneration of the lens is again accomplished beginning from the upper part of the iris; but, if the upper part of the iris itself is removed, then the regeneration begins in the interior or retinal layer of the remaining region.[b/209] As such, parts that are differently situated, differently constituted, and normally accomplishing different functions are capable of performing the same temporary replacements and of fabricating, when necessary, the same pieces of the machine. Here we clearly have one and the same effect obtained through various combinations of causes.

77

    Whether they like it or not, these theorists will have to appeal to an internal principle of direction in order to obtain this convergence of effects. Such a convergence appears impossible according to the Darwinian (and above all the neo-Darwinian) theory of imperceptible accidental variations, according to the hypothesis of sudden accidental variations, and even according to the theory that assigns definite directions to the evolution of various organs through some type of mechanical composition between external and internal forces. Thus, let us turn to the only current form of evolutionism that remains to be discussed: neo-Lamarckism.

a Bergson's note: Gustav Wolff, "Entwickelungsphysiologische Studien I: Die Regeneration der Urodelenlinse," *Archiv für Entwicklungsmechanik der Organismen* 1 (1895): 380–90.
b Bergson's note: Alfred Fischel, "Ueber die Regeneration der Linse," *Anatomischer Anzeiger: Centralblatt für die gesamte wissenschaftliche Anatomie. Amtliches Organ der anatomischen Gesellschaft* 14 (1898): 373–80.

[NEO-LAMARCKIANS AND THE INHERITANCE OF ACQUIRED CHARACTERISTICS]

It is well known that Lamarck attributed to the living being the faculty for variation according to the use or non-use of its organs, and also for transmitting the variation thus acquired to its descendants.[210] A certain number of biologists today are drawn to a doctrine of this kind.[211] The variation that is able to produce a new species would not be, according to them, an accidental variation inherent in the germ itself.[212] Nor would it be regulated by a *sui generis* determinism that would develop predetermined characteristics in a predetermined direction, independently of any concern for utility. The variation will be born of the very effort of the living being to adapt itself to the conditions in which it must live. Moreover, this effort could be nothing other than the mechanical exercise of certain organs, mechanically provoked by the pressures of external circumstances. But it could also involve consciousness and will, and it is in this sense that the American naturalist Cope, one of the most eminent representatives of this doctrine, seems to understand it.[a] 78 Thus, of all the current forms of evolutionary theory, neo-Lamarckism is the only one capable of allowing for an internal and psychological principle of development, though it does not necessarily appeal to this principle. And it is also the only evolutionary theory that seems to me to account for the formation of identical complex organs along independent lines of development. Indeed, they think that the same effort to take advantage of the same circumstances achieves the same result, and above all when the problem that is posed by the external circumstances is one of those problems that has but one solution.[213] It remains to be determined whether the term "effort" might not need to be taken in an even more profound and more psychological sense than any of the neo-Lamarckians suspect.

Indeed, a simple variation of size is one thing; a change in form is something altogether different. No one will object to the fact that an organ can be strengthened and grown through exercise. But this is a far cry from the progressive development of an eye such as the one found in Mollusks or Vertebrates. If this effect is attributed to the prolongation

a Bergson's note: Edward Drinker Cope, *The Origin of the Fittest: Essays on Evolution* (New York: D. Appleton and Company, 1887) and Cope, *The Primary Factors of Organic Evolution*.

of the passively received influence of light, then we fall back into the theory that we have just criticized. If, rather, an internal activity is indeed invoked, then this will be something completely different than what we ordinarily call an "effort," since effort has never been shown to produce the slightest complication of an organ, and yet it would have taken an enormous number of these complications, admirably coordinated among themselves, to go from the pigmented eyespot of the Infusorian to the eye of the Vertebrate.[214] Nevertheless, let us adopt this conception of the evolutionary process for animals: now, how could it be extended to the world of plants?[215] Here variations of form do not always seem to involve or to entail functional changes, and, even if the cause of the variation is of the psychological order, it is again difficult to call it an "effort" without greatly increasing the meaning of the word in a peculiar way.[216] The truth is that we must dig beneath the notion of effort itself and look for a deeper cause.

I believe that this is especially necessary if we wish to find a cause of regular hereditary variations. I do not intend to enter here into the details of the controversies surrounding the transmissibility of acquired characteristics; and even less would I presume to take too firm a stand on a question that is not within my area of competence. I cannot, however, completely avoid the question. For nowhere is it more obvious that philosophers today must not hold themselves to vague generalities, that they have an obligation to follow the scientists into the details of their experiments and to discuss with them the results.[217] If Spencer had begun by considering the question of the inheritance of acquired characteristics, his evolutionary theory would have likely taken an entirely different form.[218] If (as appears likely to me) a habit contracted by the individual is only transmitted to its descendants in extremely exceptional cases, then all of Spencer's psychology would have to be redone, and a good part of his philosophy would collapse. Let me clarify, then, how the problem appears to present itself, and in what direction it seems to me that a solution might be sought.

After having been adopted as a dogma, the transmissibility of acquired characteristics was rejected no less dogmatically, for reasons drawn a priori from the supposed nature of germ cells. It is well known how Weismann was led, through his hypothesis concerning the continuity of germ-plasm, to consider germ cells—ova and spermatozoa—as more or less independent of the somatic cells. Starting from this, it was claimed (and

many still claim) that hereditary transmission of an acquired characteristic would be something inconceivable. — But if by chance experiment
had shown that acquired characteristics are transmissible, this would   80
prove, by this very fact, that the germ-plasm is not as independent of
the somatic milieu as was claimed, and the transmissibility of acquired
characteristics would become *ipso facto* conceivable. This amounts to
saying that conceivability and inconceivability are useless in this kind of
situation, and that the question is simply a result of the experiment. But
here is precisely where the difficulty begins. The acquired characteristics
most often discussed are either habits or the effects of habit. And it is
rare that there is not some natural aptitude that serves as the foundation for the contracted habit. As such, we can always ask whether it is
indeed the habit acquired by the *soma* of the individual that is transmitted, or whether it is not rather the natural aptitude, which is anterior
to the contracted habit: this aptitude would have remained inherent
in the *germen* that the individual carried within itself since the aptitude
was already inherent in the individual and thereby in its germ. In this
way, nothing proves that the Mole became blind because it adopted the
habit of living underground; it may well have been because its eyes
were becoming atrophied that it was forced to condemn itself to a life
underground.[a][219] In this case, the tendency toward the loss of vision
would have been transmitted from germen to germen, without there
having been anything acquired or anything lost by the soma of the Mole
itself. The fact that the son of a fencing-master becomes an excellent
attacker much faster than his father does not necessarily entail that the
habit of the parent was transmitted to the child. After all, certain natural
dispositions that were increasing might have passed from the germen
that produced the father to the germen that produced the son, might
have grown along the way thanks to the influence of the primitive *élan*,
and might have thereby provided the son with a larger flexibility than
the father, without having to worry, so to speak, about what the father
actually did. The same holds for many of the examples cited from the   81
gradual domestication of animals. It is difficult to know whether it is the
contracted habit or rather a certain natural tendency that is transmitted,

a Bergson's note: Lucien Cuénot, "La nouvelle théorie transformiste: Jäger, Galton,
   Nussbaum et Weismann," *Revue générale des sciences pures et appliquées* 5, no. 3 (1894):
   74–79. Cf. Thomas Hunt Morgan, *Evolution and Adaptation* (New York and London:
   Macmillan, 1903), 356–57.

this tendency being the very reason that a particular species or certain members of a particular species were chosen for domestication. True, when we eliminate all of the questionable cases and all of the facts susceptible to multiple interpretations, hardly any absolutely incontestable examples of acquired and transmitted characteristics remain, other than Brown-Séquard's famous experiments, which have been repeated and confirmed by other physiologists.[a/220] By sectioning the spinal cord or the sciatic nerve in Guinea pigs, Brown-Séquard created an epileptic state that they transmitted to their descendants. Lesions to the same sciatic nerve, to the restiform body, etc., brought about a variety of disorders for Guinea pigs that their offspring could inherit, sometimes in a quite different form: exophthalmos, loss of toes, etc.[221] — But it has not been established that, in these various cases of hereditary transmission, there has been a genuine influence of the soma of the animal upon its germen. Weismann immediately objected that Brown-Séquard's procedure might have introduced certain specialized microbes into the Guinea pig's body, microbes that might have found an excellent nutritional source in the nervous tissues and that might have transmitted the disease by also penetrating the sexual elements.[b/222] Brown-Séquard himself had already rejected this objection.[c/223] But another, even more plausible one can be raised. Experiments conducted by Voisin and Peron have shown that epileptic fits are followed by the elimination of a toxic substance that, when injected into animals, is capable of producing convulsive episodes.[d/224] Perhaps the trophic disorders that followed the

82

---

a Bergson's note: Charles-Édouard Brown-Séquard, "Nouvelles recherches sur l'épilepsie due à certaines lésions de la moelle épinière et des nerfs rachidiens," *Archives de physiologie normale et pathologique* 2 (1869): 211, 422, and 497. [TN: Brown-Séquard's article is divided into three sections, which appear on pages 211–20, 422–38, 496–503.]

b Bergson's note: August Weismann, *Aufsätze über Vererbung und verwandte biologische Fragen* (Jena: Verlag Gustav Fischer, 1892), 376–78, and also August Weismann, *Vorträge über Deszendenztheorie*, vol. 2 (Jena: Verlag Gustav Fischer, 1902), 76. [TN: The English translations appear as: *Essays upon Heredity and Kindred Biological Problems*, ed. Edward B. Poulton, Selmar Schönland, and Arthur E. Shipley, 2nd ed. (Oxford: Clarendon Press, 1889), 320–23; and *The Evolution Theory*, trans. J. Arthur Thomson and Margaret R. Thomson (London: Edward Arnold, 1904), 67–68.]

c Bergson's note: Charles-Édouard Brown-Séquard, "Hérédité d'une affection due à une cause accidentelle: Faits et arguments contre les explications et les critiques de Weismann," *Archives de physiologie normale et pathologique*, 5th series, 4 (1892): 686–88.

d Bergson's note: Jules Voisin and Albert Peron, "Recherches sur la toxicité urinaire chez les épileptiques," *Archives de neurologie: Revue des maladies nerveuses et mentales* 24 (1892): 178–202; 25 (1893): 65–72. Cf. Jules Voisin, *L'épilepsie* (Paris: Félix Alcan, 1897), 125–33.

nerve lesions inflicted by Brown-Séquard were in fact brought about by the development of this convulsion-inducing poison. In this case, the toxin could have passed from the Guinea pig to its spermatozoon or to its ovum, and could have caused a general disorder in the development of the embryo that nevertheless might have only had visible effects at some particular point or other in the developed organism. Here things would take place much as they do in the experiments by Charrin, Delamare, and Moussu. In these experiments, they damaged the liver or kidney of Guinea pigs in gestation, and the lesion was transmitted to their offspring simply because the damage to the organ in the mother generated specific "cytotoxins" that also acted upon the corresponding organ of the fetus.[a][225] True, in these experiments it is the already formed fetus that was affected by the toxins, as was the case in a previous study by these same physiologists.[b][226] But other research by Charrin has shown that the same effect can be produced in the spermatozoa or the ova through an analogous mechanism.[c][227] In short, the inheritance of an acquired characteristic could be explained, in Brown-Séquard's experiments, through the effects of a toxin upon the germ. As localized as it might seem, the lesion could be transmitted by the same process as, for example, the defect of alcoholism.[228] But indeed, might it not be the same for all acquired characteristics that become hereditary? 83

In fact, there is one point upon which those who affirm and those who deny the transmissibility of acquired characteristics agree: namely, that certain influences, such as that of alcohol, can have a simultaneous effect upon the living being and upon the germ-plasm of which it is the keeper. In such a case, there is an inheritance of a defect, and everything happens *as if* the soma of the parent had acted upon its own germen, whereas in reality the germen and the soma have simply both suffered the action of the same cause. Supposing this to be the case, let

---

a Bergson's note: Albert Charrin, Gabriel Delamare, and Gustave Moussu, "Transmission expérimentale aux descendants des lésions développées chez les ascendants," *Comptes rendus hebdomadaires des séances de l'Académie des sciences* 135 (séance du 21 juillet 1902), 190–91; Morgan, *Evolution and Adaptation*, 257; Delage, *L'hérédité et les grands problèmes de la biologie générale*, 388.

b Bergson's note: Albert Charrin and Gabriel Delamare, "Hérédité cellulaire," *Comptes rendus hebdomadaires des séances de l'Académie des sciences* 133 (séance du 1er juillet 1901): 69–71).

c Bergson's note: Albert Charrin, "L'hérédité en pathologie," *Revue générale des sciences pures et appliquées* 7, no. 1 (1896): 1–7.

us admit that the soma can influence the germen, as is believed by those who hold that acquired characteristics are transmissible. Is not the most natural hypothesis to assume that things will happen in this second case as they did in the first, and that the direct effect of this influence of the soma will be a *general* alteration of the germ-plasm? If this were the case, then it would be an exception, and in some sense an accident, if the offspring's modification is the same as that of the parent. It would thus be like the inheritance of the defect of alcoholism: surely it passes from the father to his children, but it can take a different form for each child, and in none of them resemble what it was for the father. Let us call C the change that took place in the plasma, and C being for that matter either positive or negative, that is, representing either the gain or the loss of certain substances.[229] The effect will not reproduce its cause exactly; the modification of the germen provoked by a certain modification of a certain part of the soma will not produce the same modification of the same part of the new organism that is forming unless all of the other nascent parts of this new organism benefit from a sort of immunity to C.[230] In this case, the same part will thereby be modified in the new organism because the development of this part will alone turn out to be sensitive to the new influence. Nevertheless, it still might be modified in an entirely different way than was the corresponding part of the generating organism.

I would thus propose to introduce a distinction between the inheritance of the *divergence* [*écart*] and the inheritance of the *characteristic*. An individual who acquires a new characteristic thereby *diverges* from the form that it had and the form that would have been reproduced through the development of the germ cells (or more often, the half-germ cells) of which it is the keeper.[231] If this modification does not bring about the production of substances capable of modifying the germen, nor a general alteration of the nutrition able to deny the germen certain of its elements, then the modification will have no effect on the individual's descendants.[232] This is surely what most often takes place. But if, to the contrary, the modification has some effect, it is probably through the intermediary of a chemical change that it will have brought about in the germ-plasm. This chemical change might exceptionally bring about the original modification in the organism that the germ will produce, but it is just as likely—nay, even more likely—to do something else. In this latter case, the engendered organism might *diverge* from the normal type *as much* as the generating organism did, but it diverges from that form *differently*.[233] The engendered organism will have inherited the divergence but not the characteristic. In general, then, the habits contracted by an individual probably have no repercussions for its descendants. And when they do have an effect, the modification that takes place in the descendants might have no visible resemblance to the original modification. To me, at least, this is the most plausible hypothesis. In any case, until some counterproof is offered, and so long as the definitive experiments called for by an eminent biologist[a/234] have not been produced, we must hold ourselves to the actual results of observation. Now, even if we take the theory of the transmissibility of acquired characteristics in the best possible light, and even if we assume that the supposed acquired characteristic is not, in the majority of cases, simply the more or less late development of an innate characteristic, the facts show that hereditary transmission is the exception and not the rule. How, then, could we expect the development of an organ such as the eye from hereditary transmission? When we think of the enormous number of variations, all directed in the same direction, that must be assumed to accumulate, one after the other, so as to go from the pigmented spot of the Infusorian to the eye of the Mollusk and the Vertebrate, one will wonder how inheritance, such as we observe it, could have ever brought about this accumulation of differences, even assuming that individual efforts might have been able to produce each one of them in particular. In other words, neo-Lamarckism appears to us no more capable of resolving the problem than did the other forms of evolutionism.

85

[RESULTS OF THE ABOVE DISCUSSION][b]

By thus submitting the various current forms of evolutionary theory to a common test, and by showing that they all eventually run into the same insurmountable difficulty, it is not at all my intention to simply dismiss them all outright. Rather, each theory, being supported by a considerable number of facts, must be true in its own way. Each must correspond

---

a Bergson's note: Alfred Giard, *Controverses transformistes* (Paris: C. Naud, 1904), 147.
b TN: This section title is not included by Bergson in his Table of Contents, but it does appear as the running header starting near this point. The content of these paragraphs clearly indicates that he is here summarizing and concluding the four previous subsections.

to a certain point of view on the process of evolution. Moreover, perhaps a theory must hold exclusively to one particular point of view in order to remain scientific, i.e., in order to be able to give a precise direction to its research into the details. But each of these theories takes a partial view of a reality that must go beyond them all.[235] And that reality is the proper object of philosophy, which, since it does not aim at any practical application, is not at all constrained by the precision of science.[236]

86   So let me briefly indicate what each of the three major forms of evolutionary theory seem to contribute positively toward the solution of the problem, what each of them leaves to the side, and at what point it seems necessary to draw together this triple-effort so as to obtain a more comprehensive—although thereby more vague—idea of the evolutionary process.[237]

The neo-Darwinians are probably correct, I would say, when they teach that the essential causes of variation are differences inherent in the germ that the individual carries, and not in the activities this individual undertakes in the course of its career. It becomes difficult for me to follow these biologists, however, when they claim that the differences inherent in the germ are purely accidental and individual. I cannot help but believe that these differences are rather the development of an impulse that passes from germ to germ by way of individuals, that they are not therefore pure accidents, and that they can quite easily appear at the same time and in the same form in all of the representatives of a single species, or at least in a certain number of them. Moreover, the theory of *mutations* is already profoundly modifying Darwinism on this very point. It claims that, at a given moment, and after a long period of time has gone by, the entire species is seized by a tendency to change. As such, the *tendency to change* is not accidental. Of course, the change itself would be accidental if, as De Vries holds, the mutation operates in different directions for different members of the species.[238] But first it will be necessary to see if the theory can be confirmed in many other species of plants (De Vries only verified it for the species *Oenothera Lamarckiana*).[a/239] And even then, as I will explain in more detail below, it is not impossible

---

a   Bergson's note: Nevertheless, some analogous facts have been observed, always in the plant world. See Louis Blaringhem, "La notion d'espèce et la théorie de la mutation, d'après les travaux de Hugo de Vries," *L'année psychologique* 12 (1906): 95–112; Hugo de Vries, *Species and Varieties*, 655.

that the role played by chance in plant variation is larger than it is in animal variation, given that in the plant world function depends less strictly upon form.[240] Whatever the case may be, the neo-Darwinians are moving toward admitting that there are determinate periods of mutation. Thus, the direction of the mutation might be determinate as well, at least among animals, and to the extent I will indicate below.[241]

In this way, we arrive at something like Eimer's hypothesis, according to which the variations of different characteristics continue from generation to generation in definite directions. I find this hypothesis plausible, at least within the limits that Eimer himself imposes upon it. Of course, the evolution of the organic world must not be predetermined as a whole. I would suggest, to the contrary, that the spontaneity of life manifests itself in the organic world through a continual creation of forms succeeding other forms. But this indetermination cannot be complete; it must leave a certain role to determination.[242] An organ such as the eye, for example, would be constituted precisely through a continuous variation in a definite direction. Indeed, I cannot see how else one might explain the similarity of structure of the eye in species that have completely different histories. Where I must disagree with Eimer, however, is when he claims that the combination of physical and chemical causes is enough to ensure this result. To the contrary, I have tried to establish, by examining the specific example of the eye, that if there is indeed an "orthogenesis" at work here, then a psychological cause must intervene.[a]

Now, certain neo-Lamarckians do indeed appeal to a cause of a psychological nature. In my opinion, that is in fact one of the most solid points of neo-Lamarckism. But if this cause is nothing more than an individual's conscious effort, then it could only be operative in a relatively limited number of cases: at most, it would intervene in animals and not in the plant world.[243] Among animals themselves, this conscious effort would only act upon points that are directly or indirectly under the influence of the will. And even where it did act, I can hardly see how it might obtain a change as profound as an increase in complexity. At best, this might be conceivable if acquired characteristics were regularly transmitted and added together, but this kind of transmission seems to be the exception rather than the rule. A change that is hereditary and in

88

---

a TN: Recall that Eimer defines "orthogenesis" as "definitely directed evolution."

a definite direction, that will continue to accumulate and that will be coordinated with itself so as to construct an increasingly complicated machine, must surely be related to some sort of effort—but this will be an effort that is much deeper than individual effort and much more independent of the circumstances, an effort that is common to the majority of members of a single species, inherent to the germs that they carry rather than to their individual substance alone, and thereby assured of being transmitted to their descendants.

*

[THE *ÉLAN VITAL*]

This brings us back, after a long detour, to the idea from which I started, namely the idea of an *original élan* of life that passes from one generation of germs to the next through the intermediary of the mature organisms that form the link or hyphen between the germs.[a/244] This *élan*, preserving itself across the lines of evolution among which it is divided and shared, is the deep cause of the variations, or at least of those variations that are consistently transmitted, that accumulate, and that create new species. In general, once species have begun to diverge from a common source, they accentuate their divergence to the extent that they progress in their evolution. Nevertheless, if we accept the hypothesis of a common *élan*, then these species can and indeed must evolve identically at definite points. This is what we must now show more precisely in terms of the example that we have chosen: the formation of the eye in Mollusks and Vertebrates. This will, moreover, allow us to clarify the idea of an "original *élan*."

There are two equally striking aspects of an organ such as the eye: the complexity of structure and the simplicity of function.[245] The eye is composed of distinct parts, such as the sclera, the cornea, the retina, the crystalline lens, etc. The details of each one of these parts would go on to infinity. To speak only of the retina, we know it is composed of three superposed layers of nervous elements—multipolar cells, bipolar cells,

89

---

a   TN: Here Bergson slightly reformulates his earlier claim (cf. above, 27). There he described life as a *courant* ("current"), whereas now he suggests an *élan* ("thrust," "impulse," "force," or "momentum") of life.

and visual cells—each of which has its own individuality and surely itself constitutes an extremely complex organism. And this is but a simplified schema of the intricate structure of this membrane. The machine that is the eye is thus composed of an infinity of machines, each one being of an extreme complexity. And yet, vision is a simple fact. From the moment the eye opens, vision is at work. Precisely because the functioning is simple, the slightest distraction on the part of nature in the construction of the infinitely complex machine would have rendered vision impossible. This contrast between the complexity of the organ and the unity of the function is disconcerting for the mind.

A mechanistic theory will be any theory that presents to us the gradual construction of the machine through the influence of external circumstances, whether they intervene directly by acting upon the tissues, or indirectly by selecting the individuals that are best adapted.[246] And yet, no matter what form this theory takes, and even if we grant that it is somewhat valuable for understanding the details of the parts, it casts no light upon their correlation.

And so the doctrine of finality arises. This theory claims that the parts have been assembled according to a preconceived plan and in order to reach a goal. In this way, it likens nature's labor to that of the worker who also proceeds by assembling parts in order to realize [réaliser] an idea or to imitate a model. Mechanism will thus be correct to criticize finalism for its anthropomorphic character. But mechanism does not notice that it too proceeds according to the same method, simply truncating it. Sure, it has done away with the end that is pursued or the ideal model, but it too implies that nature works like the human worker, namely by assembling parts. And yet, a mere glance at the development of an embryo would have shown that life goes about its work in a completely different way. *Life does not proceed by association and the addition of elements but rather by dissociation and splitting* [dédoublement].[247]

Thus, we must move beyond both points of view, both mechanism and finalism, which are ultimately nothing more than points of view to which the human mind had been led by the spectacle of human work. But in what direction should we move beyond them? I said that even though the functioning of the whole is something simple, when the structure of an organ is analyzed, from one decomposition to the next, this process goes on to infinity. This contrast between the infinite

90

complication of the organ and the extreme simplicity of the function is precisely what should open our eyes to the right direction.

In general, when the same object appears from one side as simple and from the other as indefinitely composite, these two aspects are far from having the same importance, or better, the same degree of reality.[248] The simplicity belongs to the object itself, whereas the infinity of complexity belongs to the views that we take of the object when moving around it, to the juxtaposed symbols that our senses or our intellect use to conceive of the object, and, more generally, to elements of a *different order* with which we attempt to artificially imitate the object, but with which the object remains incommensurable, being of a different nature than these elements.[249] A talented artist has painted a figure on a canvas. We could imitate his painting with multicolored mosaic tiles. And the smaller, more numerous, and more varied in tone our tiles are, the better our reproduction of the curves and the nuances of the model will be. But it would take an infinity of infinitely small elements, presenting an infinity of nuances, to obtain the exact equivalent of this figure that the artist had conceived as a simple thing, that he wanted to transport to the canvas as a whole, and that is all the more complete to the extent that it appears as the projection of an indivisible intuition.[250] Now, imagine that our eyes were made in such a way that they could not help but see a mosaic effect in the painter's work. Or imagine that our intellect was designed such that the only way it could understand the appearance of the figure on the canvas was through a work of mosaic. Here we could simply speak of an assemblage of small tiles, and we would be within the mechanistic hypothesis. And we could add that, beyond the material used for the assemblage, there had to be a plan according to which the mosaicist worked—but this time we would be speaking like finalists. But in neither case would we reach the real process, since there were in fact no assembled tiles. It is the painting itself—by which I mean the simple act projected upon the canvas—that, by simply entering into our perception, is decomposed before our very eyes into thousands and thousands of tiny tiles that when recomposed will present an admirable arrangement. In the same way, the eye, with its marvelous complexity of structure, might be nothing but the simple act of vision insofar as it is divided up for us into a mosaic of cells, whose order seems marvelous to us once we have imagined the whole as an assemblage.

If I raise my hand from point A to point B, this movement simultaneously appears to me in two different ways. Sensed from within, it is a
simple and indivisible act. Viewed from the outside, it is the pathway
of a particular curve, AB. I will be able to identify in this line as many   92
positions as I like, and the line itself can be defined as a particular coordination of these positions among themselves. But the infinite number of
positions and the order that ties them together sprang forth automatically from the indivisible act by which my hand moved from A to B. Here,
mechanism would consist in seeing only the positions. Finalism would
only take their order into account. But both mechanism and finalism
would miss the movement, which is the reality itself.[251] In a certain
sense, the movement is *more* than the positions and *more* than their order
since it is enough that the movement be given in its indivisible simplicity for the infinity of successive positions and their order to be given
all at once, along with, in addition, something that is neither order nor
position, but rather what is essential: mobility.[252] But in another sense,
the movement is *less* than the series of positions and the order that ties
them together since, in order to be able to place the points in a certain
order, the order must first be conceived of and then realized using the
points. In other words, this requires a work of assemblage and it requires
intelligence, whereas the simple movement of the hand contains none
of that.[253] The movement is not intelligent, in the human sense of the
word, and it is not an assemblage, since it is not made up of elements.
The same is true of the relation of the eye to vision. In vision, there is
more than the eye's component cells and their reciprocal coordination;
in this sense, neither mechanism nor finalism go as far as is necessary.
But, in another sense, both mechanism and finalism go too far since
they attribute to nature the most formidable and Herculean works by
wanting to see it lift up an infinity of infinitely complex elements to
the simple act of vision, whereas nature had no more difficulty making
an eye than I had raising my hand. Nature's simple act is automatically
divided up into an infinity of elements that are coordinated according to
a single idea, just as the movement of my hand deposited beneath itself   93
an infinity of points that happen to satisfy a single equation.[254]

But this is precisely what is so difficult for us to understand because we
cannot stop ourselves from conceiving of organic development [*organisation*] as a kind of *fabrication*.[255] And yet to fabricate is one thing, to develop
organically [*organiser*] is something entirely different. The former is an

operation that is proper to man. It consists in assembling the parts of matter that have been cut out in such a way that we can fit them together and obtain a joint action from them. They are arranged, so to speak, around the action, which is already their ideal center. Fabrication thus goes from the periphery to the center, or, as the philosophers would say, from the many to the one.[256] By contrast, the work of organic development goes from the center to the periphery. It begins in a point that is nearly a mathematical point, and spreads out around this point through concentric waves that are continually growing.[257] The work of fabrication is all the more effective to the extent that it has at its disposal a larger quantity of matter. It proceeds by concentration and compression. By contrast, there is something explosive in the act of organic development. At the outset, it requires the smallest amount of place possible, a minimum of matter, as if the forces of organic development only reluctantly enter into space. The spermatozoon, which sets into motion the evolutionary process of embryonic life, is one of the organism's smallest cells—and moreover, it is but a tiny part of the spermatozoon that actually takes part in the operation.

And yet these are but superficial differences. By digging beneath them, I believe we will find a more profound difference.

The work that has been fabricated sketches out the form of the work of fabrication. By this I mean that the maker [le fabricant] finds in his product precisely what he had put there. If he wishes to build a machine, he will cut out the pieces one by one, then he will assemble them. Both the pieces and their assemblage will be visible in the completed machine. Here the whole of the result represents the whole of the work, and to each part of the work corresponds a part of the result.

Now, I admit that the positive sciences can and must proceed as if organic development were a work of this same kind. This is the only way for it to get a handle on organic bodies.[258] Indeed, their goal is not to reveal the essence of things but rather to provide us with the best means of acting upon them. Now, physics and chemistry are already advanced sciences, and living matter only lends itself to our action to the extent that we can deal with it through the procedures of our physics and of our chemistry. Organic development will thus only be able to be studied scientifically if the organic body has already first been assimilated to a machine.[259] The cells will be the parts of the machine; the organism will be their assemblage. And the elementary works, which have organized the parts, will be assumed to be the real elements of the work that had organized the whole. Such is the scientific point of view. The philosophical point of view is, in my opinion, entirely different.

For me, the whole of an organic machine still represents, strictly speaking, the whole of the organizing work (although this is only approximately true), but the parts of this machine no longer correspond to the parts of the work, for *the materiality of this machine no longer represents a collection of means that have been employed but rather a collection of obstacles that have been avoided*: the organic machine is a negation rather than a positive reality.[260] Thus, as I have shown in a previous study, vision is a power that could, *in principle*, reach an infinity of things that are nevertheless inaccessible to our gaze.[261] But such a vision could not be continued into action; it would be appropriate for a phantom, not for a living being.[262] A living being's vision is an effective vision that is limited to the objects upon which the being can act; it is a *canalized* vision, and the visual apparatus simply symbolizes the work of canalization.[263] As a result, the creation of the visual apparatus is no more explained by the assemblage of its anatomic elements than the digging of a canal could be explained by the contribution of the earth that made its banks. The mechanistic theory would consist in saying that the earth was brought in, one cartload at a time; finalism would add that the earth was not deposited randomly, that the carters followed a plan. But mechanism and finalism would both be wrong since the canal is built in an entirely different way.

More precisely, I compared the process by which nature constructed the eye to the simple act by which I raise my hand. But I assumed that the hand did not encounter any resistance.[264] Now imagine that, rather than moving through the air, my hand would have to move through iron filings that compress and resist to the extent that I advance. At a certain moment, my hand will have exhausted its effort and, at that precise moment, the iron filings will be juxtaposed and coordinated in a determinate form, namely that of the hand that has come to a halt along with a part of the arm. Now imagine that the hand and the arm remained invisible. The spectators will seek in the iron filings themselves, and in the forces internal to the mass, to find the reason for this arrangement.

95

Some of them will relate the position of each filing to the action that the neighboring filings exercise upon it: these will be the mechanists. Others will insist that an overall plan must have presided over the details of these elementary actions: they will be the finalists.[265] But in truth, there was simply an indivisible act, the action of the hand moving through the iron filings. The inexhaustible detail of the movement of the filings along with the order of their final arrangement express negatively, as it were, this undivided movement, this arrangement being the overall form of a resistance and not a synthesis of positive elementary actions. This is why, if we name the arrangement of the iron filings the "effect" and if we name the movement of the hand the "cause," then we can say that, strictly speaking, the whole of the effect is explained by the whole of the cause, but that the parts of the effect in no way correspond to the parts of the cause. In other words, neither mechanism nor finalism will work here, and so they will have to turn to a *sui generis* mode of explanation. Now, according to the hypothesis that I am proposing, the relation of vision to the visual apparatus would be more or less the same as the relation of the hand to the iron filings that sketch it out, that canalize it, and that limit its movement.[266]

The more considerable the effort of the hand, the farther it pushes into the interior of the iron filings. But at whatever point it stops, the filings immediately and automatically find their equilibrium and coordinate among themselves.[267] The same is true for vision and its organ. To the extent that the undivided act that constitutes vision advances more or less, the materiality of the organ is made up of a more or less considerable number of elements coordinated among themselves; and yet its order is necessarily complete and perfect.[268] Its order cannot be partial because, to repeat, the real process that gives birth to it does not have parts. This is something that neither mechanism nor finalism consider, and something that we also fail to watch out for when we stand in wonder before the marvelous structure of an instrument such as the eye. At the root of our wonder is still the idea that it *would have been possible* to realize [*réaliser*] *only part* of this order, and thus the complete realization of it is some sort of gift of grace. The finalists offer themselves this gift all at once, through the final cause; the mechanists claim to obtain it little by little through the effect of natural selection.[269] But both see something positive in this order and, consequently, in its cause, something that can be divided up, something that includes every possible degree of completion. In reality, the cause is more or less intense, but it can only produce its effect all at once and as fully completed. To the extent that it goes farther in the direction of vision, it will give the simple pigmented clusters of the lower organisms, or the rudimentary eye of the Serpula, or the already differentiated eye of the Alciopidae, or the marvelously perfected eye of a Bird—but all of these organs, whose complication is highly unequal, necessarily present an equal level of coordination.[270] Hence, no matter how distant from each other two species of animals might be, if the march toward vision has gone equally far in both cases, then both species will have the same visual organ because the form of the organ does nothing but express the extent to which the exercise of the function has been obtained.[271]

And yet, by speaking of a march toward vision, am I not simply going back to the ancient notion of finality? This would certainly be the case if this march required the conscious or unconscious representation of a goal to be reached. But the truth is that this march takes place in virtue of the original élan of life and is implied in this movement itself, which is precisely why we find it along independent lines of evolution. If someone now asked me why and how this march toward vision is implied in this movement, I would reply that life is, above all else, a tendency to act on brute matter. The direction of this action is surely not predetermined, and this is the reason for the unpredictable variety of forms that life sows along its pathway as it evolves. Rather, this action always presents, to a higher or lower degree, the characteristic of contingency; it implies at the very least some rudiment of choice. Now, a choice presupposes the anticipatory representation of several possible actions. Thus, the possibilities of action must be sketched out for the living being prior to the action itself. Visual perception is nothing other than this: the visible contours of bodies are the outline of our potential action upon them.[a] Vision will thus be found, to different degrees, among the most diverse animals, and it will manifest itself through the same complexity of structure everywhere where it has reached the same degree of intensity.

97

98

---

a Bergson's note: For more on this subject, see *Matière et mémoire*, Chapter I.

I have emphasized these similarities of structure in general, and the example of the eye in particular, because I had to define my own attitude by contrasting it with mechanism, on the one hand, and with finalism, on the other. I must now describe my position in itself and with more precision. I will do this by examining the diverging results of evolution, no longer in terms of the analogies they reveal but in terms of what they have that is mutually complementary.

II

THE DIVERGING DIRECTIONS OF LIFE'S EVOLUTION: TORPOR, INTELLECT, AND INSTINCT

[GENERAL IDEA OF THE EVOLUTIONARY PROCESS]

The movement of evolution would be a simple thing and we would have quickly determined its direction had life described a single trajectory, like the one followed by a solid cannonball shot from a cannon.[1] But here we are dealing with a shell that immediately exploded into fragments. And each of these fragments, being itself a kind of shell, in turn exploded into fragments destined to explode again, and so on for a very long time. We only perceive what is closest to us, namely the scattered movements of the pulverized explosions. We must begin from them to work back up, degree by degree, to the original movement.

When the shell explodes, its specific fragmentation is explained by both the explosive force of the powder that it contains and by the resistance that the metal opposes to that force. The same is true for the

99

DOI: 10.4324/9781315537818-3

fragmentation of life into individuals and into species. I believe that this fragmentation has to do with two series of causes: the resistance that life encounters from brute matter and the explosive force—resulting from an unstable equilibrium of tendencies—that life carries within itself.

[GROWTH]

The resistance of brute matter was the first obstacle that had to be avoided. Life seems to have succeeded by means of humility, by making itself very small and very insinuative, by tacking alongside the physical and chemical forces and even consenting to go along with them for part of the way, like the switch on the railroad track that momentarily adopts the direction of the rail from which it wants to detach.[a/2] When it comes to phenomena observed in the most elementary forms of life, we cannot say whether they are still physical and chemical, or whether they are already vital. In this way, life had to adopt the habits of brute matter so as to draw this mesmerized matter little by little over onto another track.[b] Thus, the very first living forms to appear were of an extreme simplicity. They were likely small masses of barely differentiated protoplasm, comparable from the outside to the Amoebas that are observed today, but with the addition of the incredible inner thrust [poussée intérieure] that was to raise them up to the higher forms of life.[3] It seems likely to me that, in virtue of this thrust, these first organisms would have tried to grow as much as possible; organic matter, however, has a limit of expansion that is quickly reached. Beyond a certain point, it splits apart rather than grows.[c] It surely must have taken centuries of effort

a TN: ...très petite et très insinuante... Bergson again uses a form of the word "insinuation" in the non-normative sense of an insertion in subtle and sinuous ways.

b TN: ...cette matière magnétisée... Here Bergson's plays between the literal image of matter being "magnetized," as in drawn over into life and "mesmerized" or "hypnotized." As Arnaud François notes here, Bergson uses these images of "fascinated" or "hypnotized" matter elsewhere (such as VOR, 275). I have chosen "mesmerized" here to maintain this important psychological aspect of Bergson's term. Cf. "hypnotism" in the Index.

c TN: ...elle se dédouble... Throughout this chapter, Bergson uses se dédoubler ("to divide" or "to split") and dédoublement ("dividing" or "splitting"). Where the context implies that Bergson has in mind less the phenomenon of division itself and more the result of the coming apart of the resulting parts or lines of development, I have opted for "splitting" or "to split apart." Otherwise, "division" has been used to maintain the resonance with cell division, etc. Cf. above, 90.

and prodigious subtleties for life to get around this new obstacle. Life managed to get an increasing number of elements—all ready to split apart—to somehow remain united. Through the division of labor, it tied these elements together with an unbreakable bond. The complex and quasi-discontinuous organism thereby functions just like a continuous living mass that had simply grown larger.

[DIVERGING AND COMPLEMENTARY TENDENCIES]

And yet, the true and deep causes of division were those that life carried within itself. For life is a tendency, and the essence of a tendency is to develop in the form of a burst, creating by the simple fact of its growth the diverging directions among which its *élan* will be divided and shared.[a] We observe this in ourselves in the evolution of that particular tendency that we call our character. Each of us, casting a retrospective glance back over our own history, will see that our childhood personality, though indivisible, collected together within itself a variety of persons who were able to remain merged together because they were in a nascent state—this indecision, so full of promises, is indeed one of the greatest charms of childhood.[4] But these interpenetrating personalities become incompatible as the child grows up, and since each one of us lives but a single life, we are forced to make a choice. In reality, we are continuously choosing, and we are also continuously abandoning so many things. The road we travel through time is littered with the debris of all that we began being, with all that we could have become. But nature, which an incalculable number of lives has at its disposal, is in no way forced to make such sacrifices. It preserves the diverse tendencies that bifurcated as it grew. And with them it creates the diverging series of species that will evolve separately.

Moreover, these series will perhaps be of unequal importance.[5] The author who starts a novel fills his hero with a multitude of things that he must abandon as he advances.[6] Perhaps he will take these things up

101

a   TN: *...en forme de gerbe...* The word *gerbe* is used for several images in French for the shape of something that begins together and opens out into a "burst," a "shower," a "spray," or a "bouquet." For instance, a "sheaf of wheat" or a "bouquet/spray of flowers." Given that it is also used for the "spray" of trajectories that might come from a cannon or the burst that results from a jet of water, I have opted for "burst" to maintain the resonances with other images used by Bergson below.

later, in other books, so as to compose with them new characters who will seem like extracts, or rather complements, of the first one. But these characters will almost always have something narrow about them when compared to the original character. The same is true for the evolution of life. The bifurcations along its path [trajet] have been numerous, but there are many dead ends alongside two or three main roads [grandes routes].[a] And among these main roads themselves, only one—the one that rises up through the Vertebrates to man—has been large enough to allow the great breath of life to pass through freely. This is the impression I have when comparing Bee and Ant societies, for example, to human societies. The former are admirably disciplined and united, but fixed; the latter are open to all kinds of advances, but are divided and in a constant battle with themselves.[7] The ideal would be a society always moving forward and always in harmony, but perhaps this ideal is not attainable; the two characteristics that should complement each other, that indeed were complementary in the embryonic state, become incompatible as they become more pronounced. If we can speak, more than just metaphorically, of an impulse [impulsion] toward social life, then it must be said that the bulk of the impulse has been carried along the line of evolution that ends in man, and that the rest has been collected together along the pathway that leads to the Hymenoptera.[b] In other words, Ant and Bee societies would thus present the complementary aspect to our own societies. But this would be nothing more than a manner of speaking. There was no specific impulse toward social life. There is simply the general movement of life, which creates ever-new forms along diverging lines. Should societies appear along two of these lines, they will have to manifest the divergence of their pathways at the same time as the community of the élan. In this way, they will develop two series of characteristics that we will find to be vaguely mutually complementary.[8]

[THE MEANING OF PROGRESS AND ADAPTATION]

The study of the evolutionary movement will thus consist in disentangling a certain number of diverging directions and evaluating the importance

---

a TN: For a discussion of the various translation decisions around Bergson's images of paths, roads, railroad tracks, and trajectories, see the Translator's Introduction.

b TN: Hymenoptera is the order of insects that includes bees, wasps, and ants.

of what has taken place along each of them—in short, in determining the nature of the dissociated tendencies and their various proportions. Then by combining these tendencies, we will obtain an approximation, or rather an imitation, of the indivisible motor principle from which their *élan* has emerged. That is, we will find in evolution something completely different from a series of adaptations to the circumstances (as mechanism claims) and something completely different from the realization of an overall plan (as the doctrine of finality would have it).

I do not question in any way the fact that adaptation to the milieu is the necessary condition of evolution. It is blatantly obvious that a species disappears when it does not bend to the conditions of existence that are imposed upon it. But it is one thing to recognize that external circumstances are forces that evolution must reckon with, and another thing altogether to maintain that they are the directing causes of evolution. This second theory is the one held by mechanism.[9] It absolutely rejects the hypothesis of an original *élan*, by which I mean that of an internal thrust [*poussée*] that could carry life to increasingly higher destinies by way of increasingly complex forms. And yet this *élan* is visible, and a simple glance cast over the various fossilized species shows that life could have done without evolving, or could have simply evolved within very strict limits, if it had resolved (and this would have been much more convenient for it) to become ankylosed in its primitive forms.[a] Certain *Foraminifera* have not varied since the Silurian Period.[10] Unchanging witnesses to the uncountable revolutions that have rocked our planet, the Lingula are today what they were during the earliest years of the Paleozoic Era.[11]

The truth is that adaptation explains the sinuosities of the evolutionary movement, but not the general directions of the movement, and even less the movement itself.[b/12] The road that leads to the village must surely climb the hills and descend the slopes; it *adapts itself* to the accidents of the landscape. But the accidents of the terrain are not the cause of the road and have not imposed its direction upon it. At each moment, they

a TN: *...s'ankyloser...* Bergson here uses a medical term that is more likely to be employed in a figurative sense in French than in English. Here it means "to become stagnated" as in "to become fixed or solidified" at a certain stage in the development of form.

b Bergson's note: This point of view regarding adaptation was noted by F. Marin in a noteworthy article on *On the Origin of Species* (F. Marin, "Sur l'origine des espèces," *Revue scientifique* (*Revue rose*), 4th series, 16, no. 19 (November 9, 1901): 580).

103

provide it with what is indispensable: the ground itself upon which the road sits. But if we consider the whole of the road and no longer each of its parts, then the accidents of the terrain appear as mere impediments or causes for delay since the road simply aimed at the village and would have preferred to be a straight line.[13] The same is true for the evolution of life and for the circumstances that it passes through, with the difference, however, that evolution does not sketch out a single route, that it commits itself to certain directions without thereby aiming at certain goals, and that, in the end, it remains inventive even in its adaptations.

And yet, if the evolution of life is something other than a series of adaptations to accidental circumstances, neither is it the realization of a plan. A plan is given in advance. It is represented, or at least is representable, prior to the details of its being realized. The complete execution of the plan can be pushed off into a distant future, even pushed back indefinitely, but the idea of it can no less be formulated, as of now, in terms that are actually given. To the contrary, if evolution is a continuously renewed creation, then as it goes along it creates not only the forms of life, but also the ideas that would allow an intellect to comprehend it and the terms that would serve to express it.[14] In other words, its future exceeds its present and could not be sketched out in the present as an idea.

Such is the first error of finalism; it leads to a second one that is even worse.

If life realizes a plan, then it should demonstrate an increasing harmony to the extent that it advances. For example, the house sketches out better and better the architect's idea insofar as the stone walls are built up. To the contrary, if the unity of life is entirely in the *élan* that pushes it along the road of time, then the harmony is not out in front, but rather behind. The unity comes from a *vis a tergo*: it is given at the beginning as an impulse, not placed at the end as an attraction.[a][15] Insofar as it communicates itself, the *élan* is more and more divided up. Life, to the extent that it progresses, spreads itself out into manifestations that surely owe to the community of their origin the fact that they are, in certain ways, mutually complementary, but that will be no less antagonistic and incompatible with each other. As such, the disharmony between

---

a   TN: A *vis a tergo* is literally a force acting from behind.

the species will become increasingly pronounced. Indeed, I have thus far merely indicated the essential cause of it. To simplify things, I have assumed that each species accepted the impulse received in order to transmit it to others, and that, in all the directions in which life evolves, propagation takes place in a straight line. In fact, there are species that come to a halt, while others move backward along the road they had traveled. Evolution is not simply a forward movement; in many cases we observe it coming to a sort of standstill and even more often we see it taking a detour or moving backward. This is in fact necessary, as I will show below, and the same causes that divide up the evolutionary movement make it such that, as it evolves, life too becomes distracted from itself, hypnotized by the form that it has just produced.[16] But this leads to an increasing disorder. Surely there is a progression [*progrès*], if by progression one means a continuous march in the general dir-ection determined by a first impulse.[17] But this progression is only accomplished along the two or three main lines of evolution, where increasingly complex and increasingly higher forms are sketched out. Between these main lines runs a multitude of secondary roads where the detours, the dead ends, and the backward movements are multi-plied. The philosopher who began by positing, in principle, that each detail is linked to an overall plan will, the day he decides to examine the facts, go from disappointment to disappointment. And since he had already put everything on the same level, not wanting to leave anything to accident, he will now come to believe that everything is accidental. We must begin, rather, by allowing accident its part, which is quite large. We must acknowledge that not everything in nature is coherent. We will thereby be led to determine the centers around which the inco-herence crystallizes. And this crystallization itself will clarify the rest: the main directions in which life moves by developing the original impulse will appear.[18] True, we will not witness the detailed accomplishment of a plan. Here we have something more, something better than a plan that is realized. A plan is an endpoint [*terme*] assigned to a work: it closes off the future whose form it sketches out. But out in front of the evolution of life, the doors of the future remain wide open. Life is a creation that, in virtue of an initial movement, continues without end. This movement makes the unity of the organic world—which is a fertile unity with an infinite richness—something greater than any intellect could dream up, for the intellect is but one of its aspects or one of its products.[19]

106

But it is easier to define the method than it is to put it into prac-tice. The complete interpretation of the evolutionary movement in the past, such as we conceive of it, would only be possible if the history of the organic world was fully written. We are far from having such results. The genealogies that are proposed for the various species are most often problematic. They vary according to the author in question or according to which theoretical perspective inspired them, and they give rise to debates that science in its current state cannot resolve. But a comparison of the various solutions among themselves will show that the controversy has more to do with the details than with the main lines. By following the main lines as closely as possible, we will be sure to not be led astray. Moreover, they alone are important here since I am not attempting (as a naturalist does) to discover the order of succession of the various species, but rather to define the principal directions of their evolution. And indeed, I am not equally interested in all of these directions—what interests me most particu-larly is the road that leads to man.[20] Thus, by following the main lines, one after another, we will not lose sight of the fact that it is a question of determining the relation of man to the whole of the animal kingdom, and the place of the animal kingdom itself in the whole of the organic world.

*

107   [THE RELATION OF THE ANIMAL TO THE PLANT]

To begin with the second point, let me start by saying that no specific characteristic distinguishes plants from animals. Attempts to rigorously define the two kingdoms have always failed. There is not one single prop-erty of plant life that is not found, to some degree, in certain animals; and not a single characteristic trait of the animal that cannot be observed in certain species, or at certain moments, in the plant world. So, it is under-standable that biologists enamored with rigor would have held the dis-tinction between the two kingdoms to be artificial. And they would have been correct had the definition to be made here indeed been like those in the mathematical and physical sciences, namely according to certain static attributes that the defined object possesses and that the others do not. But in my opinion, the kind of definition appropriate in the life sciences is quite different.[21] There is hardly any manifestation of life that does not contain—be it in a rudimentary, latent, or virtual state—the essential characteristics of the majority of the other manifestations. The difference is in the proportions. But this difference of proportion will suffice to define the group where it is encountered, so long as we can establish that it is not simply accidental and that the group, to the extent that it has evolved, has increasingly tended to *emphasize* these particular characteristics. In short, *a group will no longer be defined by the possession of certain characteristics, but rather by its tendency to accentuate them.* If we adopt this point of view, and focus less on states and more on tendencies, then we find that plants and animals can be defined and distinguished in a precise way, and that they indeed correspond to two diverging developments of life.[22]

This divergence first becomes pronounced in terms of the mode of nutrition. It is well known that the plant borrows the elements that are necessary to sustain its life (particularly carbon and nitrogen) directly from the air, the water, and the soil, and it takes them up in mineral form. To the contrary, the animal cannot acquire these same elem- 108 ents unless they have already been fixed for it in organic substances by plants or by animals that, directly or indirectly, owe them to the plants, such that ultimately it is the plant that nourishes the animal. Of course, this law is subject to many exceptions among plants. We do not hesitate to class among the plants the Drosera, the Dionaea, and the Pinguicula, which are all insectivore plants.[23] On the other hand, Fungi, which occupy such a large part of the plant world, feed themselves like animals; whether they be ferments, saprophytes, or parasites, they take their food from already formed organic substances.[24] So we cannot draw from this difference a static definition that would, in any case whatso- ever, automatically settle the question as to whether we are dealing with a plant or with an animal. But this difference can provide the beginnings of a dynamic definition of the two kingdoms, insofar as it marks the two diverging directions in which animals and plants have developed. It is a remarkable fact that Fungi, which are spread throughout nature and are so extraordinarily abundant, were unable to evolve further. Organically, they do not rise above the tissues that, in higher plants, are formed in the embryonic sac of the ovule and precede the germinative development of the new individual.[a][25] The Fungi are, so to speak, the runts of the plant world. Their various subspecies constitute so many dead ends, as if giving up the usual nutritional mode of plants brought them to a halt on the road of plant evolution.[26] As for the Drosera, the Dionaea, and the insectivore plants in general, they feed like other plants through their roots, and they too fix, in their green parts, the carbon in the carbonic acid contained in the atmosphere. The faculty for capturing, absorbing, and digesting insects is one that must have emerged late for them, and in entirely exceptional cases, when the soil was too poor to provide them with sufficient nutrition. In general, if we focus less on the presence of characteristics than on their tendency to be developed, and if we take as essential the tendency along which evolution was able to go on indefinitely, then it will be said that plants are distinguished from animals by their power to create organic material out of mineral elements that they draw directly from the atmosphere, from the soil, and from the water. But a second and already deeper difference is tied up with this first one.

The animal, not being able to directly fix the carbon and the nitrogen that are present all around, is obliged to seek out and feed upon the plants that have already fixed these elements or the animals who themselves have already taken them from the plant kingdom. As such, the animal is necessarily mobile. From the Amoeba who casts out its pseudopods at random so as to capture the scattered organic materials found in a drop of water, right up to the higher animals who possess sense organs for recognizing their prey, locomotive organs for going after it, and a nervous system for coordinating their movements with their sensations, animal life is characterized—in its general direction—by mobility in space.[27] In its most rudimentary form, the animal appears as a small mass of protoplasm enveloped, at most, by a thin albuminoid film that leaves it fully free to change its shape and to move. By contrast, the plant cell is surrounded by a cellulose membrane that condemns it to immobility. And so, from the lowest to the highest levels of the plant kingdom, we find these same increasingly sedentary habits—the plant has no need to move about, finding around itself, in the atmosphere, the water, and the soil in which it is placed, the mineral elements that it can directly appropriate. Of course, the phenomena of movement can

---

a Bergson's note: Gaston de Saporta and Antoine-Fortuné Marion, *L'évolution du règne végétal: Les cryptogames* (Paris: Germer Baillière, 1881), 37.

also be observed among plants. Darwin wrote a wonderful book on the movements of climbing plants. He studied the maneuvers that certain insectivore plants, such as the Drosera and the Dionaea, use to capture their prey. The movements of the leaves of the Acacia or of the Sensitive plant, etc., are well known.[28] Moreover, the to-and-fro movement of plant protoplasm within its envelope bears witness to its kinship with animal protoplasm. Inversely, we could mention phenomena of fixation analogous to those we find in plants among a large number of animal species (generally among parasites).[a/29] Here again we would be wrong to claim that fixity or mobility can be made into two characteristics allowing us to decide, through a simple inspection, whether we are in the presence of a plant or an animal. But in animals, fixity most often appears as a sort of torpor into which the species has fallen, as a refusal to evolve further in a certain direction; it is closely related to parasitism and comes along with characteristics reminiscent of plant life.[30] On the other hand, plant movements have neither the frequency nor the variety of animal movements. They ordinarily involve but one part of the organism, and almost never extend to the organism as a whole. In exceptional cases where a vague spontaneity is manifested in plant movement, it seems that we are witnessing an accidental awakening of an activity that is normally asleep. In short, even if mobility and fixity coexist in the plant world like they do in the animal world, the balance clearly swings toward fixity in the one case and toward mobility in the other. These two opposite tendencies are so clearly the guiding tendencies of the two evolutions that we might already use them to define the two kingdoms.[31] But fixity and mobility are, in turn, merely the superficial signs of tendencies that are even deeper.[32]

111

There is an obvious relationship between mobility and consciousness. Of course, consciousness in higher organisms appears to be interdependent with certain cerebral mechanisms [*dispositifs*]. The more the nervous system develops, the more the movements among which it can choose become numerous and precise, and the more lucid is the consciousness that accompanies them.[33] But the presence of a nervous system is not a necessary condition for this mobility, this choice,

---

a Bergson's note: For more on fixation and on parasitism in general, see Frédéric Houssay, *La forme et la vie: Essai de la méthode mécanique en zoologie* (Paris: Schleicher Frères, 1900), 721–807.

nor consequently this consciousness. The nervous system has simply canalized, in some determinate ways, and carried to a higher degree of intensity, a rudimentary and vague activity that was diffused throughout the mass of organic substance. The more we descend in the series of animals, the more the nervous centers become simplified and detached from each other; eventually the nervous elements disappear, submerged into the whole of a less differentiated organism.[34] But the same is true of all the other organs [appareils] and all the other anatomical elements; and it would be just as absurd to deny an animal consciousness because it has no brain as it would be to declare an animal incapable of feeding itself because it has no stomach.[35] The truth is that the nervous system is born, like all the other systems, from a division of labor. It does not create the function, it merely raises it to a higher degree of intensity and precision by giving it the double form of a reflex activity and a voluntary activity.[36] In order to accomplish a genuine reflex movement, a whole mechanism must be set up in the spinal cord or the medulla oblongata. In order to voluntarily choose between several determinate procedures, cerebral centers are necessary, that is, crossroads from which pathways emerge leading to motor mechanisms that have a variety of configurations and that are equally precise.[37] But, where a canalization into nervous elements has not yet taken place, let alone a concentration of nervous elements into a system, there is still something from which both reflex and voluntary action will emerge by way of a splitting [dédoublement], something that has neither the mechanical precision of the reflex nor the intelligent hesitations of the will, but that, participating in both to an infinitesimal degree, is simply an undecided reaction [réaction indécise] that is therefore already vaguely conscious.[38] In other words, the humblest organism is conscious to the extent that it moves itself freely. In relation to these movements, is consciousness here the effect or the cause? In one sense, it is the cause since its role is to guide the locomotion. But, in another, it is the effect since the motor activity maintains it, and consciousness atrophies, or better, goes to sleep, as soon as this activity disappears. Among Crustaceans such as the Rhizocephala, who must have previously presented a more differentiated structure, their fixity and their parasitism come along with the degeneration and the almost complete disappearance of the nervous system.[39] In such cases, given that the progression of organic development [organisation] had localized all of the conscious activity into nervous centers, we can conjecture that

112

consciousness is even weaker among animals of this sort than it is in much less differentiated organisms who have never had nervous centers, but who have remained mobile.

So how could the plant—fixed to the ground and finding its food right there—have ever been able to develop in the direction of conscious activity?[40] The cellulose membrane that surrounds the protoplasm in plants simultaneously immobilizes the simplest plant organism and separates it, for the most part, from those external stimuli that act upon the animal as irritants for its sensitivity and that prevent the animal from going to sleep.[a/41] Thus, the plant is generally unconscious. Here again it is necessary to resist radical distinctions. Unconsciousness and consciousness are not two labels that can be mechanically applied, the first to every plant cell, the other to all animals. If consciousness goes to sleep for the animal that has degenerated into an immobile parasite, inversely it in all likelihood awakens for the plant who has reconquered the freedom of its movements, and it awakens to the exact extent that the plant has reconquered that freedom. Nonetheless, consciousness and unconsciousness mark the directions in which the two kingdoms have developed, in the sense that to find the best specimens of consciousness among animals, it is necessary to *ascend* to the highest representatives of the series, whereas to discover likely cases of plant consciousness, one must *descend* as low as possible on the scale of plants, arriving for example at the zoospores of Algae, and more generally at the unicellular organisms that we could say seem to hesitate between the plant form and animality.[42] From this point of view, and to this extent, we can define the animal by sensibility and awakened consciousness, the plant by dormant consciousness and insensibility.[43]

To summarize, the plant directly fabricates organic substances out of mineral substances; this aptitude generally relieves it of the need to move and, thereby, of the need to sense. Animals, being obliged to seek out their food, have evolved in the direction of locomotive activity and consequently of a consciousness that is increasingly extensive and increasingly distinct.

Now, that the animal cell and the plant cell derive from a common source, and that the first living organisms would have oscillated between

a Bergson's note: Cope, *The Primary Factors of Organic Evolution*, 76.

113

114 the plant form and the animal form, participating in both at the same time, this hardly seems doubtful to me. Indeed, I have just shown that the characteristic tendencies of the evolution of the two kingdoms, although diverging, still coexist today, both in the plant and in the animal. The proportion alone differs. Ordinarily, one of the two tendencies conceals or stamps out the other, but in exceptional circumstances, the latter tendency breaks free and reconquers the place that it had lost. For the plant cell, mobility and consciousness are not so dormant that they cannot be reawakened when the circumstances allow it or require it. And, on the other hand, evolution in the animal kingdom has been continuously delayed, stopped, or even carried backward by the tendency toward plant life that it retained. As much as the activity of an animal species might appear to be full and overflowing, torpor and unconsciousness lie in wait for it. It can only maintain its role by exerting an effort, at the price of becoming fatigued.[44] Countless moments of weakness have taken place along the route upon which the animal has evolved, moments of degeneration that are usually linked to parasitic habits. These moments represent the various times animal life switched tracks toward plant life.[a] As such, everything leads us to the assumption that the plant and the animal descend from a common ancestor who combined the tendencies of both in a nascent state.

But the two tendencies that are reciprocally integrated in this rudimentary form became dissociated as they grew. This resulted in the plant world, characterized by fixity and insensibility, and the animals, characterized by their mobility and their consciousness. Moreover, there is no need to appeal to a mysterious force to explain this division. It is enough to point out that the living being naturally relies upon what is most convenient to it, and that plants and animals opted, respectively, for two different kinds of convenience in terms of how they procure the carbon and the nitrogen they need.[45] Plants continuously and mechanically draw these elements from an environment that constantly
115 provides them. Animals, through conscious and discontinuous action that is concentrated in certain instants, will seek out these bodies in the organisms that have already fixed them. Here we have two different ways of understanding work, or, if one prefers, laziness.[46] And from this

---

a TN: ...autant d'aiguillages sur la vie végétative. Here Bergson returns to the railroad metaphor to capture the sense of switching between tendencies or directions. Cf. below, 136.

it seems doubtful that we would ever discover nervous elements in the plant, no matter how rudimentary we might suppose them to be. In plants, what corresponds to the directing will of the animal is, I believe, the direction in which the plant inflects the energy from the sun's radiation when it makes use of it to break the bonds between carbon and oxygen in carbonic acid. And what corresponds to the sensibility of the animal is the highly specialized sensitivity [*impressionnabilité*] of the plant's chlorophyll to light. Now, a nervous system being above all else a mechanism that serves as an intermediary between sensation and volitions, the plant's true "nervous system" appears to me to be the mechanism, or rather the *sui generis* chemical activity, that serves as an intermediary between the sensitivity of its chlorophyll to light and the production of starch.[47] This amounts to saying that the plant must not have nervous elements, and that *the same élan that led the animal to develop nerves and nerve centers must have resulted, for the plant, in the chlorophyllian function.*[a/48]

This first glance at the organic world will allow us to determine, in more precise terms, what unites the two kingdoms as well as what separates them.

Suppose, as I suggested in the previous chapter, that at the foundation   116
of life there is an effort toward grafting the largest amount of indetermination possible onto the necessity of physical forces.[49] This effort cannot succeed in creating energy, or, if it does create some, then the quantity created remains below the level detectible by our senses and by our measuring instruments, that is, by our experience and our science.[50] Everything thus takes place as if the effort merely aimed at using, as best as possible, a pre-existing energy that it finds available to it. It has but one means of succeeding at this: it must obtain from matter a significant enough accumulation of potential energy that it will be able to obtain the work it needs to act by simply pulling a trigger at a given moment.[51]

---

a Bergson's note: Just as the plant, in certain cases, rediscovers the faculty for active movement that lies dormant within it, so too can the animal, in exceptional circumstances, return to the conditions of plant life and develop in itself an equivalent of the chlorophyllian function. The results of recent experiments by Maria von Linden seem to show that the chrysalis and the caterpillar of certain Lepidoptera, under the influence of light, fix the carbon of the carbonic acid in the atmosphere. Maria von Linden, "L'assimilation de l'acide carbonique par les chrysalides de Lépidoptères," *Comptes rendus hebdomadaires des séances et mémoires de la Société de biologie* 59, no. 2, séance du 23 décembre (1905): 692–94.

The effort itself possesses merely this power for releasing [déclencher]. And yet, although the work of releasing itself is always the same and always weaker than any given quantity whatsoever, it will be all the more effective to the extent that it causes a heavier weight to fall from a higher location, or, in other words, insofar as the sum of the accumulated and available potential energy is greater. Indeed, the Sun is the principal source of usable energy on the surface of our planet. The problem was thus as follows: make it such that, here and there on the surface of the Earth, the Sun's incessant outpouring of usable energy be partially and provisionally suspended, such that a certain quantity of that energy be stockpiled in the form of unused energy in some appropriate reservoirs from which it will later be able to flow forth at the moment, at the location, and in the direction desired.[52] The animal feeds on substances that are precisely reservoirs of this type.[53] Formed of very complex molecules that lock up a considerable sum of chemical energy in a state of potentiality, these substances are types of explosives, waiting for nothing other than a spark to set the stockpiled force free.[54] Now, it is probable that life at first tended to obtain, all at once, the fabrication of the explosive and the explosion that makes use of it. In this case, the same organism that had directly stockpiled the energy of solar radiation would have spent it on free movements in space.[a] And this is why we must assume that the first living beings attempted, on the one hand, to continuously accumulate energy borrowed from the Sun and, on the other hand, to spend that energy in a discontinuous and explosive manner through the movements of locomotion. Perhaps the Euglena, which are Infusorians that make use of chlorophyll, still symbolize today this primordial tendency of life, though in a restricted form that is unable to evolve.[55] Does the diverging development of the two kingdoms correspond to what we might metaphorically call the forgetting, by each kingdom, of one half of the program? Or, and this seems more likely, was the very nature of the matter that life encountered on our planet itself an obstacle to these two tendencies being able to evolve together in a single organism for very long? What is certain is that the plant relied above all on the first direction and the animal on the second. But if from the beginning the explosion was the goal of the fabrication of the explosive, then the

---

a TN: ...dépensée... Bergson here mixes the metaphors of explosives and of the financial activity of spending the resources that have been saved.

evolution of the animal, much more than that of the plant, indicates the fundamental direction of life in general.

The "harmony" between the two kingdoms and the complementary characteristics that they present thus ultimately came from the fact that they developed two tendencies that were at first melded together into a single one. The more the original and singular tendency grows, the more it finds it difficult to maintain in one and the same living being the two elements that complement each other in the rudimentary state.[a] This leads to a splitting [dédoublement]; it leads to two diverging evolutions. It also leads to two series of characteristics that are in conflict on certain points and are complementary on others, but which, whether they are complementary or opposed, always maintain a certain air of kinship between them. Whereas the animal evolved (though not without accidents along the way) toward an increasingly free and discontinuous spending of energy, the plant rather perfected its system of accumulating while remaining in place. I will not dwell on this second point. Let it suffice to say that the plant must have greatly benefited in turn from a new division, analogous to the one that took place between plants and animals. If the primitive plant cell had to fix, all by itself, both its carbon and its nitrogen, it was able to almost completely abandon this second function the day microscopic plants pushed exclusively in this direction—indeed, even specializing in diverse ways in this highly complicated work.[56] The microbes that fix the nitrogen of the atmosphere and those that, in turn, convert ammonic compounds into nitrogen compounds and these latter into nitrates, have performed for the whole of the plant world (through the same dissociation of a tendency that was originally singular) the same kind of service that the plants in general perform for the animals. If we created a special kingdom for these microscopic plants, then we might say that the microbes in the soil, the plants, and the animals together show the *analysis*—carried out by the matter that life had at its disposal on our planet—of all that life at first contained in the state of reciprocal implication.[b] Is this, properly

a   TN: ...s'impliquent les unes les autres...

b   TN: Note that Bergson is using *l'analyse* ("analysis") both here and in the following paragraph in the technical sense of "division into its essential or component parts." Of course, for Bergson the parts do not exist independently prior to the analysis, as he clarifies in the following paragraph. That is, the reciprocal implication in the original state is one where the subsequent independent tendencies did not yet, properly speaking, have an independent existence.

speaking, a "division of labor"? These words would not give a precise idea of evolution, such as I am imagining it. Where there is a division of labor, there is an *association* and there is also a *convergence* of effort.[57] To the contrary, the evolution that I am speaking of never takes place in the direction of an association, but rather in the direction of a *dissociation*—and never toward convergence, but rather toward the divergence of efforts. The harmony among terms that are mutually complementary on certain points is not, in my opinion, produced along the way through a reciprocal adaptation; to the contrary, it is only fully complete at the beginning. It derives from an original identity. It comes from the fact that, as it grows, the evolutionary process, which spreads outward in the form of a burst [*en forme de gerbe*], separates the terms from each other that at first were so complementary that they were fused together.[a]

Moreover, the elements that a tendency is divided up into do not necessarily all have the same importance, nor above all the same power to evolve. I have just identified three different kingdoms, if one can speak this way, in the organic world. Whereas the first only contains micro-organisms that have remained in the rudimentary state, animals and plants have taken flight for much higher destinies. Now, this is something that normally takes place when a tendency is analyzed. Among the diverging developments that are born from this division, some go on indefinitely, while others arrive at the end of their tether sooner or later. These latter developments do not come directly from the primitive tendency itself; rather, they come from one of the elements into which the primitive tendency had been divided. They are residual developments, actualized and deposited along the road by some truly fundamental tendency that itself continues to evolve. As for these truly fundamental tendencies, I believe that they bear a mark that allows them to be recognized.

This mark is like the trace that remains visible in each one of these tendencies of what was contained in the original tendency whose fundamental directions they represent. The elements of a tendency are not comparable to objects juxtaposed in space and separate from each other, but rather to psychological states, where each one, despite first being itself, nevertheless participates in others, and in this way contains, virtually, the entire personality to which it belongs.[58] There is no essential manifestation

119

---

a   TN: Cf. the Translator's note above, 100.

of life, I said, that does not present, in a rudimentary or virtual state, the characteristics of the other manifestations of life.[59] Reciprocally, when we encounter along one evolutionary line the memory, so to speak, of what develops along other lines, we must conclude that we are dealing with dissociated elements of a single, original tendency.[60] In this sense, plants and animals clearly represent the two major diverging developments of life. Insofar as the plant is distinguished from the animal by its fixity and by its insensibility, movement and consciousness remain dormant in it, like memories that could reawaken. Moreover, alongside these memories that are normally dormant, there are some that are awake and active. These are the memories whose activity does not hinder the development of the fundamental tendency itself. The following law can thus be formulated: *When a tendency is analyzed as it develops, each of the particular tendencies born in this way would like to preserve and develop everything in the primitive tendency that is not incompatible with the work in which it has specialized.*[61] This law would serve to explain precisely the fact that I dwelt upon at length in the previous chapter, namely the formation of identical complex mechanisms along independent lines of evolution. And this is probably the reason for certain deep analogies between plants and animals; sexual generation is perhaps merely a luxury for the plant, but it was a necessity for the animal, and the plant must have been carried toward this by the same *élan* that pushed the animal there, namely, a primitive or original *élan*, predating the division into the two kingdoms.[62] I will say the same about the tendency among plants toward an increasing complexity. This tendency is essential for the animal kingdom, which is motivated by the need for always more expansive and more effective action. But plants, condemned to insensibility and immobility, only present the same tendency because in the beginning they received the same impulse. Recent experiments show them to vary in random ways when the period of "mutation" arrives; whereas the animal must have evolved, I believe, in much more clearly defined directions.[63] But I will not dwell on this original division of life. I turn now to the evolution of animals, which I find especially interesting.

[THE SCHEMA OF ANIMAL LIFE]

I said above that what constitutes animality is the faculty for using a releasing mechanism to convert as large a sum as possible of accumulated potential energy into "explosive" actions.[64] In the beginning, the explosion takes place at random, unable to choose its direction; that is how the Amoeba launches its pseudopodic prolongations simultaneously in every which way.[65] But the higher we go in the animal series, the more we see that the very form of the body itself sketches out a certain number of well-determined directions along which the energy will travel. These directions are marked out by chains of nervous elements, placed end to end.[66] Now, the nervous element increasingly stands out from the barely differentiated mass of organic tissue. Thus, it can be assumed that the faculty for suddenly releasing accumulated energy is concentrated, from the beginning, in the nervous element and its appendages. True, every living cell continuously spends energy in order to maintain its equilibrium. The plant cell, in a drowsy state from the beginning, is entirely absorbed in this work of preservation, as if it took as an end what should have started out as merely a means. But for the animal everything converges toward action, i.e., toward the utilization of energy for movements of transportation.[a] Of course, each animal cell spends a good part of the energy at its disposal (and sometimes all of this energy) just to live; but the organism as a whole would like to direct as much of this energy as possible to the points at which the movements of locomotion are accomplished. This takes place such that wherever a nervous system exists, along with the sensory organs and the motor systems [appareils moteurs] that serve as its appendages, everything must take place as if the essential function of the rest of the body was to prepare for them—so as to transfer it to them at the desired moment—the force that they will free through a sort of explosion.[67]

The role of food for higher animals is in fact extremely complex. First, food serves to repair tissues. Next, it provides the animal with the warmth it needs to make itself as independent as possible from variations in the outside temperature. In this way, it preserves, maintains, and supports the organism in which the nervous system is inserted and on which the nervous elements must live.[68] But these nervous elements would have no raison d'être if this organism did not pass a certain amount of energy to be spent on to them—either directly to them or above all to the muscles

122

a TN: ...mouvements de translation. The French word translation generally only has the formal or technical meaning of its English equivalent, namely "the transportation of something" from one place to another. Bergson has this general sense in mind here, and not the process of traduction ["translation"] from one language to another.

that they operate. And one might even conjecture that in the end this is the essential and ultimate destination of the food. This does not mean that the largest part of the food is used in this work. A State might well have to make enormous expenditures in order to assure the collection of tax revenue; the sum that it will have available after the deduction of the costs of collecting it will perhaps be quite modest, but this sum is no less the *raison d'être* of the tax and for all that has been spent in order to obtain this revenue. The same goes for the energy that the animal demands from dietary substances.

Many facts seem to indicate that nervous and muscular elements occupy this place with regard to the rest of the organism. Let us take a quick look at the distribution of the dietary substances among the diverse elements of the living body. These substances divide into two categories: the first are quaternary or albuminoid; the others are ternary, including carbohydrates and fats. The albuminoids are specifically plastic, intended for rebuilding the tissues—although they can occasionally become energetic thanks to the carbon they contain. But the energetic function more particularly falls to the second category. These latter ones, deposited in the cell rather than incorporated into its substance, provide the cell with a potential energy (in the form of a chemical potential) that will be converted directly into movement or into heat.[69] In short, the main role for the former category is to repair or remake [*refaire*] the machine, while the latter provides it with energy.[70] It is natural that the albuminoids have no privileged location since all of the parts of the machine need maintenance. The same is not true of the energetic substances. Carbohydrates are distributed quite unequally, and this inequality of distribution appears to me to be highly instructive.

Carried along by the arterial blood in the form of glucose, these substances are then deposited in the form of glycogen in the various cells that form the tissues. We know that one of the principal functions of the liver is to maintain a constant glucose level in the blood, thanks to the reserves of glucose that the hepatic cells secrete.[71] Now, in this circulation of glucose and in this accumulation of glycogen, it is easy to see that everything takes place as if the entire effort of the organism was devoted to supplying the elements of muscle tissues and those of the nervous tissues with potential energy. The organism proceeds differently in the two cases, but the results are the same. In the case of the muscle tissues, it provides the cells with a considerable reserve, deposited in them in advance; indeed, the quantity of glycogen contained in the muscles is enormous by comparison to that which is found in the other tissues. To the contrary, in the nervous tissue, the reserve is much smaller (the nervous elements, whose role is simply to release the potential energy stockpiled in the muscle, never really need to provide much work at a given moment). But—and this is quite remarkable—this reserve is replenished by the blood the moment it is spent, such that the nerve is instantly recharged with potential energy. Muscle tissue and nervous tissue are thus clearly privileged: one insofar as it is supplied with a considerable reserve of energy; the other insofar as it is always provided with energy at the instant it needs it and to the exact measure it needs it.[72]

124

More specifically, the call for glycogen (i.e., for potential energy) comes from the sensorimotor system, as if the rest of the organism were there simply to transfer force to the nervous system and to the muscles that the nerves operate. Of course, when we think about the role that the nervous system (and even the sensorimotor system) plays as a regulator of organic life, we might wonder if, in this supposed exchange of goods between it and the rest of the body, it is actually more like a master that the body serves.[73] And indeed, considering the distribution of potential energy among the tissues, even in the static state so to speak, one will already be inclined toward this hypothesis. I think they will rally behind it once they reflect upon the conditions in which that energy is spent and restored. Indeed, imagine that the sensorimotor system was just a system like the others, that is, at the same rank as the others. Carried along by the whole of the organism, it will wait for an excess of chemical potentiality to be provided to it so as to accomplish its work. In other words, the production of glycogen itself would regulate the consumption of it by the nerves and the muscles. By contrast, imagine that the sensorimotor system is truly the master. The *durée* and the scope of its action will be independent, at least to a certain extent, of the glycogen reserve that it contains, and even of the reserve held by the whole of the organism. This system will provide the work, and the other tissues will have to organize themselves so as to bring potential energy to it.[a] And

---

a   TN: Here Bergson's use of *travail* is not in the technical sense of spending energy (which is how he uses it elsewhere), but rather in the sense of how a business provides or sets up work for its employees.

this is precisely how things take place, as demonstrated in particular by Morat and Dufourt's experiments.[a][74] If the glycogenic function of the liver depends on the action of the stimulating nerves that govern it, the action of these nerves is subordinated to that of the nerves that set the locomotive muscles into motion, in the sense that these muscles spend without calculation, consuming glycogen in this way, lessening the glucose in the blood, and ultimately forcing the liver, which would have had to transfer a part of its reserve into the depleted blood, to fabricate more. In short, everything thus begins with the sensorimotor system, and everything converges toward it, and one can say quite literally that the rest of the organism is at its service.[75]

Or again, consider what happens in a prolonged period of fasting. It is a remarkable fact that the brain in animals who have died of starvation is found to be pretty much intact, whereas the other organs have lost some part of their weight and their cells have suffered profound alterations.[b][76] It seems that the rest of the body sustained the nervous system right up to the point of death, treating itself as a mere means for which the nervous system would be the end.

To sum up, if we might abbreviate somewhat by including the cerebrospinal nervous system along with, in addition, the sensory apparatuses into which it extends and the locomotive muscles that it governs, under the name "sensorimotor system," then we can say that a higher organism is essentially a sensorimotor system installed upon systems for digestion, respiration, circulation, secretion, etc. The role of those systems is to repair it, to clean it, to protect it, to create for it a constant interior milieu, and, finally and above all, to pass to it the potential

125

126

a Bergson's note: *Archives de physiologie*, 1892. [Jean-Pierre Morat and E. Dufourt, "Consommation du sucre par les muscles: Origine probable du glycogène musculaire," *Archives de physiologie normale et pathologique*, 5th series, 4 (1892): 327–36; "Sur la consommation du glycogène des muscles pendant l'activité de ces organes," *Archives de physiologie normale et pathologique*, 5th series, 4 (1892): 457–64.]

b Bergson's note: Marie de Manacéïne, "Quelques observations expérimentales sur l'influence de l'insomnie absolue," *Archives italiennes de biologie* 21 (1894): 322–25. Analogous observations have been recently made of a man who died after 35 days of fasting. For more on this subject, see the summary in *L'Année biologique* from 1898, page 338, of a work in Russian by Tarakevich and Stchasny. [W. Podwyssozki, "Compte-rendu: *Les Modifications du système nerveux central et des organes internes dans un cas de mort d'un Homme par suite d'inanition pendant 35 jours* par L. Tarakevich et S. Stchasny," *L'Année biologique: Comptes rendus annuels des travaux de biologie générale* 4 (1898): 338.]

energy that will be converted into movements of locomotion.[a/77] True, the more the nervous function is perfected, the more the functions designed to support it have to develop and so the more demanding they become in themselves. To the extent that the nervous activity emerged from the protoplasmic mass in which it had been submerged, it must have called forth around itself activities of all types to support it: these activities could only be developed upon other activities, which imply again others, and so on indefinitely.[78] This is how the increasing complication of the functioning of higher organisms goes on to infinity. The examination of one of these organisms turns us around in circles, as if everything served as a means to everything else. Yet this circle surely has a center, namely the system of nervous elements stretched out between the sensory organs and the apparatus for locomotion.

I will not dwell here on a point that I have treated at length in a previous work.[79] Let me simply recall that the progression of the nervous system is accomplished simultaneously in the direction of a more precise adaptation of movements and in the direction of a larger latitude available to the living being for choosing among those movements.[80] These two tendencies can seem antagonistic, and indeed they are. Nevertheless, a nervous chain, even in its most rudimentary form, is able to reconcile them. On the one hand, it sketches out a well-defined line between two points on the periphery: one sensory and the other motor. It has thus canalized an activity that was at first diffused throughout the protoplasmic mass. But, on the other hand, it is composed of elements that are probably discontinuous. In any case, supposing that they anastomose with each other, they present a functional discontinuity since each one of them ends with a kind of crossroads where, most likely, the nerve impulse can choose its route.[81] From the humblest Monera to the most gifted

127

---

a Bergson's note: Cuvier had already said: "[T]he nervous system is, basically, the entire animal. The other systems are only there to serve it." (Georges Cuvier, "Sur un nouveau rapprochement à établir entre les classes qui composent le Règne animal," *Annales du Muséum d'histoire naturelle* 19 (1812): 73–84 [here, 76].) Of course, a large number of restrictions would have to be added to this formula, considering, for example, cases of degradation and of regression where the nervous system becomes secondary. And above all, it is necessary to join to the nervous system the sensory organs, on the one side, and the motor apparatuses on the other, between which it serves as an intermediary. Cf. Michael Foster, "Physiology: Part I.–General View," *Encyclopædia Britannica: A Dictionary of Arts, Sciences, and General Literature*, 9th ed. (Edinburgh: Adam and Charles Black, 1885), 19: 17.

Insects, to the most intelligent Vertebrates, the progress accomplished has above all been a progression of the nervous system with, at each stage, all of the creations and complications of the parts that this progression required.[82] As I have implied from the beginning of this book, the role of life is to insert indetermination into matter.[83] The forms that life will create along the course of its evolution are indeterminate, by which I mean unforeseeable. And these forms serve as the vehicle for an activity that is more and more indeterminate, by which I mean more and more free. A nervous system—with its neurons placed end to end in such a way that multiple pathways open up at the extremity of each one, and where just as many questions are asked—is a genuine *reservoir of indetermination*.[84] A simple glance at the whole of the organic world shows that the essential part of the vital thrust [*la poussée vitale*] has been devoted to the creation of systems of this kind. As for this thrust of life itself, a couple of clarifications are indispensable.

[THE DEVELOPMENT OF ANIMALITY]

It is important to keep in mind that the force that evolves through the organic world is a limited force that always attempts to go beyond itself and that always remains inadequate to the work that it strives to produce.[85] Failures to understand this point give birth to the errors and the puerilities of radical finalism. This doctrine imagines the whole of the living world to be a construction, and indeed a construction analogous to our own constructions. All of the parts would thus be arranged so as to ensure the best possible functioning of the machine. Each species would have its *raison d'être*, its function, and its purpose. Together the species would perform a great concert in which the apparent dissonances would serve merely as a means to bring out the fundamental harmony.[86] In short, everything in nature would take place as it does in the works of human genius, where although the result obtained might be modest, there would at least be a perfect match [*adéquation*] between the fabricated object and the work of fabrication.

128

Nothing of the sort takes place in the evolution of life. Here the disproportion between the work and the result is striking. From the lowest to the highest levels of the organic world, there is still but a single great effort; but most often this effort is cut short: sometimes paralyzed by the forces that oppose it; sometimes distracted from what it must do by what it is doing, absorbed by the form that it is in the process of becoming, and hypnotized by it as if gazing into a mirror.[87] Even in its most perfect works, where it seems to have triumphed over external resistances as well as over its own resistances, this effort remains subject to the materiality that it has had to take on. Each one of us can experience this in ourselves. Our freedom, in the very movements by which it affirms itself, creates the nascent habits that will suffocate it if it does not renew itself through a constant effort: automatism lies in wait. The liveliest thought is frozen in the phrase that expresses it. The word turns against the idea. The letter kills the spirit. And our most ardent enthusiasm, when it expresses itself in action, sometimes becomes so naturally fixed into the cold calculation of self-interest or vanity, the one so easily adopting the form of the other, that we might merge them together, doubt our sincerity, and negate kindness and love, if, that is, we did not know that the dead maintains the traits of the living at least for a while.

The deep cause of these dissonances lies in an irremediable difference of rhythm. Life in general is mobility itself, whereas the particular manifestations of life only receive this mobility reluctantly, and are always falling behind it. Life is always moving forward, whereas its manifestations would like to stay put.[88] Evolution in general would take place, as much as possible, along a straight line, whereas each particular evolution is a circular process. Like the eddies of dust picked up by the passing wind, living beings spin round and round, suspended upon the great gust of life. They are thus relatively stable and even imitate immobility so well that we treat them as things rather than as progressions, forgetting that the very permanence of their form is nothing but the sketch of a movement.[89] Nevertheless, sometimes the invisible gust of air that carries them along materializes, before our very eyes, in a fleeting apparition. We experience this sudden enlightenment when we encounter certain forms of maternal love, so striking and so touching in the majority of animals, and observable even in the care that the plant has for its seed. This love, in which some have seen the great mystery of life, will perhaps at least deliver to us life's secret. It shows each generation leaning over the one that will follow.[a] It allows us to catch sight of

129

---

a   TN: ...*chaque génération penchée sur celle qui la suivra.* Bergson's use of the adjective *penchée* ("leaning over") recalls how he speaks of the past leaning over the present in *Matière et mémoire*. Cf. *MM*, 167. For a similar usage, cf. also *MM*, 102.

how the living being is above all a place of passage, and that the essence of life is to be found in the movement that transmits it.

This contrast between "life in general" and the forms in which it manifests itself always presents the same character. We could say that whereas life tends to act as much as possible, each species prefers to exert the least amount of effort possible. Considered in terms of its very essence, namely as a transition from species to species, life is a continuously growing action. But each species that life passes through aims at nothing but its own convenience. It goes toward what requires the least amount of effort. Becoming engrossed in the form that it is about to adopt, it falls into a half-sleep in which it is unaware of almost all of the rest of life; it shapes itself with an eye toward the easiest possible exploitation of its immediate surroundings. In this way, the act through which life goes about the creation of a new form and the act by which this form takes shape are two different and often antagonistic movements. The first continues into the second, but it cannot be continued there without being distracted from its direction, like a jumper who, in order to clear an obstacle, is obliged to take their eyes off the obstacle and to focus on themselves instead.[90]

Living forms are, by definition, viable forms.[91] No matter how the adaptation of the organism to its conditions of existence is explained, this adaptation is necessarily sufficient from the moment that the species survives. In this sense, each one of the successive species described by paleontology and zoology was a *success* that life achieved. But things take on an entirely different appearance when we no longer compare each species to the conditions into which it is inserted, but rather to the movement that deposited it along its path. Often this movement veered off course, and often it even came to a full stop: what was meant to be merely a place of passage became rather an endpoint. From this new perspective, failure appears to be the rule, and success as both exceptional and always imperfect. As we will see, of the four main directions that animal life has embarked upon, two have led to dead ends and, along the two other lines, the effort has generally been disproportionate to the result.

We are missing the documents needed for reconstituting the details of this history. Nevertheless, we can still make out the main lines. I said that animals and plants must have split off from their common source relatively quickly—the plant falling dormant into immobility, and the

animal, by contrast, waking up more and more and marching toward
the conquest of a nervous system. It is likely that the effort of the animal
kingdom led to the creation of still very simple organisms, though
endowed with a certain mobility, and above all quite undefined in terms
of their form so as to lend themselves to all future determinations.[92]
These animals might well have resembled our Worms, the difference
being, however, that the Worms of today that we would compare them
to are the emptied and fixed copies of the infinitely plastic forms that
were brimming with an indefinite future and that were the common
source of the Echinoderms, Mollusks, Arthropods, and Vertebrates.

A danger was lying in wait for them, an obstacle that might well
have brought the rise of animal life to a halt. There is one particularity
that I cannot help but find striking when I glance back at the fauna of
primitive times, namely the imprisonment of the animal in a more or
less hard envelope that must have impeded and often even paralyzed its
movements. First, Mollusks had, more universally than today, a shell.[93]
Arthropods in general were furnished with a carapace—these were
the Crustaceans. The most ancient Fish possessed an extremely hard,
bony envelope.[a/94] The explanation of this general fact must be sought,
I believe, in a tendency of soft organisms to defend themselves from
one another as much as possible by making themselves impossible to
devour. Each species, in its act of self-constitution, goes toward whatever is most convenient for it.[95] Just as was true among the primitive
organisms, where certain ones became oriented toward animality by
renouncing the fabrication of the organic from the inorganic and by
rather borrowing ready-made organic substances from organisms that
were already directed toward plant life, the same is true among the
animal species themselves, where many animals manage to live off of
other animals.[96] An organism that is an animal, by which I mean mobile,
will indeed be able to benefit from its mobility to seek out defenseless
animals and to eat its fill, just as easily as it might do the same with
plants. In this way, the more the species made themselves mobile, the
more they became voracious and dangerous to each other. This should
have resulted in a sudden halt in the animal world as a whole in terms of

---

a Bergson's note: For more on these various points, see Albert Gaudry's work: *Essai de
paléontologie philosophique: Ouvrage faisant suite aux Enchaînements du monde animal dans
les temps géologiques* (Paris: Masson, 1896), 14–16 and 78–79.

the progress that had carried it to a more and more advanced mobility, for the Echinoderm's hard and calcareous skin, the Mollusk's shell, the Crustacean's carapace, and the ancient Fish's ganoid armor probably have, as their common origin, the effort of the animal species to protect themselves against enemy species.[97] But this armor, behind which the animal took shelter, impeded its movement and sometimes even left it immobile. If the plant renounced consciousness by enveloping itself in a cellulose membrane, the animal that encloses itself within a fortress or within a suit of armor also condemns itself to a half-sleep. The Echinoderms, and even the Mollusks, continue to live in this state of torpor even today.[98] Arthropods and Vertebrates were surely threatened by this as well. They managed to escape, and the current flourishing of the highest forms of life is a result of this happy occurrence.

In fact, along two pathways, we see life's thrust toward movement regain the upper hand. Fish exchanged their ganoid armor for scales. Long before that, the Insects appeared, also unencumbered of the armor that had protected their ancestors. In both cases, they compensated for the insufficiency of their protective envelope through an agility that allowed them to escape their enemies and also to take the offensive, that is, to choose the time and place of the encounter. We find a similar sort of progress that we see in the evolution of human weaponry. The first movement is to seek shelter; the second, which is best, is to make oneself as supple [souple] as possible for fleeing and above all for attacking— again, attacking being the most efficient means of self-defense.[a] In this way, the heavy hoplite was supplanted by the legionary; the armor-clad knight had to give way to the foot soldier free in his movements; and, in a general manner, in the evolution of life as a whole (as in that of human societies and individual destinies), the greatest successes are reserved for those who have accepted the greatest risks.[99]

Surely it was in the interests of the animal to render itself more mobile. As we said above with regard to adaptation in general, the transformation of species can always be explained by their particular interest.[100] In

133

---

a TN: Bergson uses here (and below) various forms of the word *souple* ("supple") and intends all of the rich senses of this word, from physically supple (as in not rigid or constrained, and hence "flexible" and "agile"), to supple in use (as in how the hand can be used in an indefinite number of ways), to supple in attitude or orientation (as in capable of adapting or quickly responding to a changing situation). The term appears on the first page of this book, above, v.

this way, we give the immediate cause of the variation. But one will often thereby only give the most superficial cause. The deep cause is the impulse that launched life into the world, caused life to split between plants and animals, oriented animality toward the suppleness of form, and at a certain moment, within an animal kingdom threatened by the prospect of drifting off to sleep, made it such that (at least at several points) they were reawakened and began to move forward again.

Along the two pathways where Vertebrates and Arthropods evolved independently, the development (if we disregard the backward movements resulting from parasitism or any other cause) above all consisted in a progression of the sensorimotor nervous system. They were looking for mobility, for suppleness, and (through a lot of trial and error, and not without first showing a tendency to exaggerate their mass and brute force) for a variety of movements. But this research is itself conducted in diverging directions. A simple glance at the nervous systems of Arthropods and Vertebrates alerts us to these differences. For the Arthropods, the body is formed of a more or less long series of juxtaposed rings; motor activity is then divided out among a variable and sometimes quite considerable number of appendages, each one having its specialization. For the Vertebrates, the activity is concentrated in only two pairs of limbs, and these organs accomplish functions that depend much less strictly on their form.[a/101] This independence becomes complete in the human, whose hand can execute any type of work whatsoever.[102]

At least, this is what we see. Now, behind what we see, there is what we can conjecture: two powers immanent to life, at first merged together, and that must have become dissociated as they grew.

To define these powers, we must examine the species that mark the high point[b] of the evolution of Arthropods and Vertebrates respectively. But how can that point be determined?[103] Here again, aiming for geometrical precision will lead us off course.[104] There is no unique and simple sign by which we can recognize one species as being more advanced than another along a single evolutionary line. There are multiple

a   Bergson's note: For more on this subject, see: Nathaniel Southgate Shaler, *The Individual: A Study of Life and Death* (New York: D. Appleton and Company, 1900), 118–25.
b   TN: ...*point culminant*... Although this term might be more directly translated as "end point," this would lose the sense in the French of reaching a "high point" or "peak," which resonates with Bergson's overall metaphysics of ascending/descending and of the movement of evolution as "rising up through" the various species.

characteristics that must be compared and weighed in each particular case so as to determine to what point they are either essential or accidental, and to what extent it is important to take them into account.

It can hardly be contested, for instance, that success is the most general criterion of superiority, the two terms being synonyms (at least up to a certain point). When it comes to living beings, we must understand success to mean an aptitude for developing within the most diverse environments and through the largest possible variety of obstacles, so as to cover the largest expanse of the earth as possible. A species that claims the entire earth as its domain is a truly dominant and, consequently, superior species.[a] Such is the case for humans, who will represent the high point of the evolution of Vertebrates.[105] But it is also the case for Insects, and in particular for certain Hymenoptera, in the series of Articulated organisms.[b] It has been said that Ants are the masters of the underground, like man is the master of the earth [sol].

On the other hand, a group of species that appears late might well be a group of degenerate ones, but for that to be the case, some special cause of regression must have intervened. In principle, this group would be superior to the group from which it derived since it would correspond to a more advanced stage of evolution. Now, man is probably the most recent comer among the Vertebrates.[c/106] And in the series of Insects, there is probably no more recent comer than the Hymenoptera, other than perhaps the Lepidoptera, a species that is likely a degeneration, being a veritable parasite for flowering plants.

In this way, and along different paths, we are led to the same conclusion.[107] The evolution of Arthropods attained its high point with the Insect, and in particular with the Hymenoptera, and the evolution of

a TN: Note that Bergson is here playing with the double meaning of *supérieure* for both higher in level and higher in value. I have generally translated this adjective as "higher" where it is primarily a question of level.

b TN: Bergson again here uses *les Articulés*, in the tradition of Cuvier for *Articulata* or "articulated organisms," and given the current context, more specifically for arthropods.

c Bergson's note: This point is contested by René Quinton, who considers carnivorous and ruminant Mammals, as well as certain Birds, as posterior to man (René Quinton, *L'eau de mer, milieu organique: Constance du milieu marin original, comme milieu vital des cellules, à travers la série animale* (Paris: Masson, 1904), 435). Let it be said, in passing, that our general conclusions, although very different from Quinton's, are not irreconcilable with them. For, if evolution has indeed been such as I am conceiving of it, then the Vertebrates must have tried to maintain themselves in the most favorable conditions for action, the very ones in which life at first placed itself.

the Vertebrates peaks in humans. Now, if it is observed that nowhere is instinct more developed than it is in the world of Insects, and that instinct is nowhere more marvelous than it is among Hymenoptera, then we will be able to say that the evolution of the animal kingdom as a whole, setting aside the steps back toward plant life, has been accomplished along two diverging pathways, one leading to instinct and the other to intelligence.[a][108]

136   Vegetative torpor, instinct, and intelligence: these are thus the elements that coincided in the vital impulse common to plants and animals, and that, along the course of their development through which they were manifested in the most unpredictable forms, dissociated from each other as a result of the simple fact of their growth. *The major error, the one that has been passed along since Aristotle and that has tainted the majority of philosophies of nature, is to see in plant life, in instinctual life, and in the life of reason three successive degrees of a single tendency that develops, whereas they are in fact three diverging directions of an activity that was divided up as it grew.*[109] The difference among them is not a difference of intensity, nor more generally of degree, but rather a difference of nature.

*

[THE MAIN DIRECTIONS IN THE EVOLUTION OF LIFE: TORPOR, INTELLECT, INSTINCT]

It is important to examine this point carefully. We have seen how plant life and animal life both complement each other and oppose each other. Now we must show that the intellect and instinct also oppose and yet complement each other. But let me first explain why it is tempting to see them as activities where the former would be superior to the latter, and would be superposed upon it, whereas in reality they are not things of the same order, nor things that might succeed one another, nor things that could be ranked.

This is because the intellect and instinct, having started out as interpenetrating, retain something of their common origin. Neither is ever

---

a   TN: It is worth recalling here that, in French, the word *intelligence* refers both to the quality of "intelligence" and to the faculty of knowledge, that is, "the intellect." For more on my usual translation of this term as "the intellect," see the Translator's introduction.

encountered in a pure state.[110] I said that the consciousness and mobility of the animal are dormant in plants, but could be awakened, and that the animal lives under the constant threat of switching tracks toward vegetative life. The two tendencies (that of the plant and that of the animal) were originally so thoroughly interpenetrating that there was never a complete break between them: the one continues to haunt the other; we find them everywhere mixed together; only the proportion differs. The same is true for the intellect and instinct. There is no intellect in which we do not discover traces of instinct, and above all no instinct that is not   137
surrounded by a fringe of intelligence.[111] This fringe of intelligence has been the cause of so many errors. From the fact that instinct is always more or less intelligent, it was concluded that the intellect and instinct are things of the same order, that between them there is but a difference of complexity or of perfection, and above all that one of them is expressible in terms of the other. In reality, they only accompany each other because they complement each other, and they only complement each other because they are different—what is instinctive in instinct being the opposite of what is intelligent in the intellect.

It will hardly be surprising if I insist upon this point. I consider it to be essential.

Let me first say that the distinctions that I am going to make here will be too cut and dried, precisely because I want to define what is instinctive in instinct and what is intelligent in the intellect, whereas every concrete instinct is mixed up with intellect, just as every real intellect is permeated with instinct. Moreover, neither intellect nor instinct lends itself to rigid definitions; they are tendencies, and not ready-made things [choses faites].[112] Finally, it must not be forgotten that, in the current chapter, we are considering the intellect and instinct as coming out of life, which deposits them along its way. Now, the life manifested by an organism is, in my opinion, a certain effort for obtaining certain things from brute matter.[113] Thus, it will hardly be surprising if what strikes us in instinct and in the intellect is the diversity of this effort, nor will it be surprising if we see in these two forms of mental activity [activité psychique] above all two different methods of acting upon inert matter.[114] This somewhat narrow way of picturing intellect and instinct will have the advantage of providing an objective means for distinguishing them. But then again, it will only give us the average position of intellect or of instinct in general, above and below which they both continuously   138

oscillate. This is why what follows should be taken merely as a schematic drawing in which the respective contours of intellect and instinct
will be more pronounced than they should be, a drawing in which we
have neglected the softening of those contours that results both from
the hesitancy [indécision] of each one of them and from their reciprocal
encroachment [empiétement] on each other. When it comes to a topic this
obscure, one must not spare any effort in moving toward the light. It
will always be easy to blur the forms later on, to correct what might have
been too geometrical in the drawing—in short, to substitute the suppleness of life for the stiffness of the schema.

To what date can we trace back the appearance of man on earth? To
the time when the first weapons and the first tools were fabricated. The
memorable debate that emerged over Boucher de Perthes' discovery in
the Moulin-Quignon quarry has not been forgotten.[115] The question
was whether we were dealing with genuine axes or rather fragments of
flint that had been broken accidentally. And indeed, no one doubted for
a moment that if they were indeed hatchets, then we were surely in the
presence of an intellect, and more particularly, of a human intellect. On
the other hand, imagine that we open a collection of anecdotes on the
intelligence of animals. We will find that alongside many acts that can be
explained through imitation, or through the mechanical association of
images, there are some that we do not hesitate to declare "intelligent"; at
the front of the line figure those that bear witness to a thinking of *fabrication*, either when the animal itself is able to fashion a brute instrument,
or when it uses an object fabricated by humans to its own advantage.
The animals that we classify as immediately below humans in terms of
intelligence—Monkeys and Elephants—are the ones that know how to
use an artificial instrument. Beneath them, but not far from them, we
place those that *recognize* a fabricated object: for example, the Fox who
knows perfectly well that a trap is a trap. In all likelihood, there is intelligence everywhere that there is inference. But inference, which consists
in a bending of past experience in the direction of present experience, is already the beginning of invention.[116] Invention becomes complete when it materializes in a fabricated instrument. The intelligence
of animals tends in this direction, as if toward an ideal. And even if it
ordinarily does not arrive at fashioning artificial objects and making
use of them, it prepares for this through the variations themselves that

139

it executes upon the instincts that are furnished by nature.[a] In terms of human intellect, it has not often enough been noted that mechanical invention was at first its essential activity, that still today our social life gravitates around the fabrication and utilization of artificial instruments, and that the inventions that mark the route of progress have also traced out its direction. It is difficult for us to see this in ourselves because the modifications of humanity ordinarily lag behind the transform-ations of its tools. Our individual and even our social habits survive long enough beyond the circumstances for which they were made that the deep effects of an invention only make themselves known when we have already lost sight of its novelty. A century has gone by since the inven-tion of the steam engine, and we are only now beginning to sense the depths of the shock it gave us. The revolution that it brought about in industry no less dramatically changed the relations among men. New ideas are emerging.[117] New feelings are in the process of being born.[118] Thousands of years from now, when the distance of the past has left nothing but the main outlines visible, our wars and our revolutions will count for very little, assuming that they are even remembered at all. But they will perhaps discuss the steam engine, along with all of the various inventions that come along with it, the way that we speak of bronze or of carved stone. That is, the steam engine will serve to define an age.[b/119] If we could rid ourselves of all arrogance, and if, in order to define our species, we held ourselves strictly to what history and prehistory present to us as the constant nature of man and of the intellect, then perhaps we would not say *Homo Sapiens*, but rather *Homo Faber*. In fact, *intellect, considered according to what appears to be its original procedure, is the faculty of fabricating artificial objects, in particular tools for making tools, and of indefinitely varying their fabrication.*

Now, does an unintelligent animal also possess tools or machines? Yes, of course, but here the instrument is a part of the body that uses it. And corresponding to this instrument is an *instinct* that knows how to use it. Perhaps it is necessary that all instincts consist in a natural faculty for using an innate mechanism. Such a definition would not apply to the instincts that Romanes called "secondary," and more than one "primary"

---

a   TN: ...*les variations mêmes qu'elle exécute sur les instincts*... Here Bergson begins to increas-ingly incorporate the image of "theme and variation." Cf. the Index.

b   Bergson's hote: Paul Lacombe has shown the essential influence that the great inventions have exercised upon the evolution of humanity. (Paul Lacombe, *De l'histoire considérée comme science* (Paris: Hachette, 1894), in particular pages 168–247).

140

instinct would escape it as well.[a][120] But this definition of instinct, like the provisional one that I gave of the intellect, at most identifies the ideal limit toward which the numerous forms of the defined object tend.[121] It has quite often been noted that the majority of instincts are the extension, or better the completion, of the work of organic development [*organisation*] itself. Where does the instinct's activity begin? Where does nature's end? It is impossible to say.[122] In the metamorphoses from the larva to the nymph to the mature insect, metamorphoses that indeed often require the larva to undertake some appropriate steps and to have a sort of initiative, there is no clear-cut line of demarcation between animal instinct and the organizing work of living matter. We could say equally well that instinct organizes the instruments that it will make use of or that organic development extends into the instinct that must use the organ. The most wonderous instincts in the Insect do little more than develop an organ's specific structure into movements, to such a point that, where social life divides the work between individuals and thereby imposes different instincts on them, we observe a corresponding difference of structure, such as the well-known polymorphism of Ants, Bees, Wasps, and certain Pseudoneuroptera.[123] As such, if we consider only the limit cases in which we find the complete victory of intellect or of instinct, we find an essential difference between the two: *perfected instinct is a faculty for using and even for constructing organic instruments; perfected intellect is the faculty for fabricating and for using inorganic instruments.*[b]

The advantages and disadvantages of these two modes of activity are clear to see.[124] Instinct finds the appropriate instrument within reach: this instrument—which fabricates and repairs itself, and which, like all works of nature, presents an infinite complexity of detail and a marvelous simplicity of function—immediately does what it is called upon to do at the desired moment, without any difficulty and with an often-admirable perfection. But then again, the instrument maintains a more or less invariable structure since it cannot be modified without a

---

a   [Bergson's missing note: George John Romanes, *Mental Evolution in Animals, with a Posthumous Essay on Instinct by Charles Darwin* (London: Kegan Paul, Trench, & Co., 1883).]

b   TN: ...*l'instinct achevé est une faculté d'utiliser et même de construire des instruments organisés; l'intelligence achevée est la faculté de fabriquer et d'employer des instruments inorganisés.* I have included the French original of this important conclusion because it provides a good example of Bergson's use of *organisé* in the context of his discussion of *organisation* ("organic development") as opposed to the intellect's work of "fabrication."

141

corresponding modification of the species. The instinct is thus necessarily specialized, being simply the use of a specific instrument for a specific purpose. To the contrary, the instrument that is intelligently fabricated is an imperfect instrument. It can only be obtained at the price of some effort. It is almost always quite tiresome to manipulate. But given that it is made of inorganic matter [*matière inorganisé*], it can take on any form whatsoever, serve any use whatsoever, free the living being of every new difficulty that arises, and confer upon the living being an unlimited number of powers.[a][125] Inferior to the natural instrument in terms of the satisfaction of immediate needs, it has all the greater advantage over the natural instrument when it comes to needs that are less pressing. Above all, the fabricated instrument has an effect upon the nature of the being who fabricated it since, by calling upon this being to exercise a new function, the fabricated instrument (being an artificial organ that extends the natural organism) confers upon the living being a richer organic development or organization [*organisation*], so to speak.[126] For every need that it satisfies, it creates a new need, and so rather than closing the circle of action in which the animal will mechanically move (the way instinct does), the fabricated instrument opens up an indefinite field to this activity, pushing it further and further and making it more and more free.[127] But this advantage of the intellect over instinct only appears much later, when, having elevated fabrication to its highest degree of power, the intellect already fabricates machines for fabrication. At the outset, the advantages and the inconveniences of the fabricated instrument and the natural instrument are so well balanced that it is difficult to say which of the two will assure the living being a larger authority over nature.

We could conjecture that the two began by being interpenetrating, that the originary psychical activity had something of them both at the same time, and that, if we could work back far enough into the past, we would find instincts closer to intellect than are the instincts of our Insects, and an intellect closer to instinct than is the intellect of our Vertebrates—elementary intellect and elementary instinct being, moreover, prisoners of a matter that they are not able to dominate. If the force

142

---

a   TN: Note that here and just below, Bergson continues to play with the multiple senses of *inorganisé* ("inorganic" and "unorganized") and *organisation* ("organization" and "organic development").

immanent in life were an unlimited force, it might well have developed both instinct and intellect indefinitely within the same organisms.[128] But everything seems to indicate that this force is finite, and that it exhausts itself quite quickly when it manifests itself.[129] This force finds it difficult to go very far in several directions at once. It has to choose. Now, it has a choice between two ways of acting upon brute matter. It can either pro-

143   vide this action *immediately* by creating for itself an *organic* instrument that it will work with, or it can contribute this action *mediately* by way of an organism that will itself fabricate it by making it out of inorganic nature, rather than naturally possessing the required instrument. From this we get intellect and instinct, which, although increasingly diverging as they develop, never fully separate from each other. Indeed, on the one hand, the most perfect instinct of the Insect is accompanied by some glimmers of intelligence, be they merely in the choice of place, moment, and building materials. When Bees exceptionally build their hive in the open air, they invent new and genuinely intelligent mechanisms in order to adapt to these new conditions.[a/130] But on the other hand, the intellect needs instinct more than instinct needs the intellect since fashioning brute matter already presupposes a high degree of organic develop-ment [*organisation*] in animals—the animal could not have reached these heights other than on the wings of instinct. And again, whereas nature clearly evolved toward instinct in the Arthropods, among almost all of the Vertebrates we witness the search for, rather than the mere blooming of, the intellect. Instinct still forms the substratum of their psychical activity, but the intellect is there too and aspires to replace instinct. The intellect does not manage to invent instruments here; but at least it attempts to do so by executing as many variations as possible upon instinct, which it would otherwise like to do without. The intellect only fully takes possession of itself in man, and this victory is confirmed by the very insufficiency of the natural means at man's disposal for defending himself against his enemies, or against cold and hunger.[131] This insufficiency, when we seek to decipher its meaning, takes on the value of a prehistoric document: it is the final notice of dismissal that

144   instinct receives from the intellect. But it is no less true that nature must

---

a   Bergson's note: Eugène-Louis Bouvier, "La nidification des Abeilles à l'air libre," *Comptes rendus hebdomadaires des séances de l'Académie des sciences* 142, séance du 7 mai (1906): 1015–20.

have hesitated between two modes of psychical activity: one being assured of immediate success, but limited in its effects; the other more uncertain, but whose conquests, should they manage to reach a certain independence, might be extended indefinitely.[132] Moreover, here again the greatest success was attained on the side where there was the greatest risk.[133] *Instinct and intellect thus represent two equally elegant but diverging solutions to one and the same problem.*[134]

True, this results in deep differences of internal structure between instinct and the intellect. I will only focus here on the ones that concern the present study. Thus, let me say that the intellect and instinct imply two radically different kinds of knowledge. But first, some clarifications are necessary regarding the subject of consciousness in general.

It has been asked to what extent instinct is conscious. I will respond that there is a multitude of differences and degrees, that instinct is more or less conscious in certain cases and unconscious in others. The plant, as we will see, has instincts; but it is doubtful that these instincts in plants are accompanied by feeling.[135] Even among animals, we rarely find complex instincts that are not unconscious in at least a part of their exercise. But here we must indicate a difference—one that is not noted often enough—between two kinds of unconsciousness: one that consists in a nil consciousness and one that is the result of a *nullified* consciousness.[a/136] Nil consciousness and nullified consciousness are both equal to zero; but the first zero expresses that there is nothing, whereas the second expresses that we are dealing with two equal and opposite quantities that compensate for each other and cancel each other out.[137] The unconsciousness of a falling stone is a nil consciousness: the stone has no feeling of its fall.[138] Is it the same for the unconsciousness of the instinct in extreme cases where the instinct is unconscious? When we mechanically accomplish an habitual action, or when the sleepwalker mechanically acts out [joue] his dream, unconsciousness might well be absolute; but this time it is the result of how the representation of the act is held in check by the execution of the act itself, this execution being so perfectly similar to the representation and inserting itself so precisely into it that no consciousness can overflow the action any longer.[139] *The representation*

145

---

a TN: *...celle qui consiste en une conscience* nulle *et celle qui provient d'une conscience* annulée.

is blocked by the action.[a] The proof is that, if the accomplishment of the act is stopped or impeded by an obstacle, consciousness can spring forth. Thus, consciousness was there, but it was neutralized by the action that filled up the representation. The obstacle did not create anything positive; it simply created a void, it enacted an unblocking or uncorking. This inadequacy between the action and the representation is precisely what we here call consciousness.[140]

By examining this point more in depth, we will find that consciousness is the light immanent to the zone of possible actions or virtual activity that surrounds the action actually being accomplished by the living being.[141] Consciousness means hesitation or choice. Where many equally possible actions are sketched out without any real action being accomplished (such as in an ongoing deliberation), consciousness is intense. Where the real action is the only action possible (such as in activities like sleepwalking, or, more generally, automatic ones), consciousness becomes nil. Representation and knowledge no less exist in this second case, so long as it turns out that we find there a collection of systematized movements in which the final one is already prefigured in the first one, and, moreover, if consciousness can spring forth from the shock of an obstacle.[142] From this point of view, *we could define the consciousness of the living being as an arithmetical difference between the virtual activity and the real activity. Consciousness measures the gap between representation and action.*

That being the case, we can assume that the intellect will be oriented more toward consciousness, and instinct more toward unconscious-
146   ness. For wherever the instrument to be used is organized by nature, the point of application provided by nature, and the result to obtain desired by nature, there but a tiny part is left to choice—as much as the consciousness inherent to representation attempts to free itself, it will thus be counter-balanced by the accomplishment of the act that serves as its counterweight, this being identical to the representation. Where consciousness does appear, it illuminates less the instinct itself than those frustrations to which the instinct is subjected. It is the *deficit* of the instinct, the distance from the act to the idea, that becomes consciousness. As a result, consciousness will be merely an accident. In essence,

---

a   TN: *...qu'aucune conscience ne peut plus déborder. La représentation est bouchée par l'action.* In these two sentences, Bergson is using the image of a bottle spilling out or overflowing, rather being stopped up (as in with a cork).

consciousness will only emphasize the instinct's initial step, the one that released the entire series of automatic movements. To the contrary, when it comes to the intellect, deficit is its normal state. Its very essence is to be subjected to frustrations. Given that its primitive function is to fabricate inorganic instruments, the intellect must choose in the face of a thousand difficulties the time and the place, the form and the matter for this work. And it will never be entirely satisfied because every new satisfaction creates new needs.[143] In short, although instinct and intellect both involve knowledge, in the case of instinct, knowledge is *enacted or played* and unconscious, in the case of the intellect, knowledge is *thought and conscious*.[a][144] But this is more a difference of degree than of nature. So long as we only focus on consciousness, we will fail to see what, from the psychological point of view, is the main difference between the intellect and instinct.

To reach the essential difference, we must—without getting distracted by the more or less brilliant light that illuminates these two forms of inner activity—turn our attention directly to the two profoundly distinct objects to which they are applied.

When the Botfly deposits its eggs on the legs or on the shoulders of the animal, it acts as if it knew that its larva must develop in the stomach of the horse, and that, by licking itself, the horse will transport the nascent larva into its digestive track.[145] When a Parasitoid Wasp strikes its victim at the exact points where the nervous centers are located, and in such a way as to immobilize without killing, it proceeds in the manner of a highly trained entomologist and a skilled surgeon combined.[146] And what must that little Beetle, the Sitaris, whose story has often been told, have known?[147] This Coleoptera deposits its eggs at the entrance of the underground tunnels that a particular species of Bee, the Anthophora, digs out.[148] The larva of the Sitaris, after a long time of lying in wait for the male Anthophora at the exit of the tunnel, clings to it, and remains attached to it until its "nuptial flight." It seizes this opportunity to go from the male to the female, and then it again waits patiently for her to lay her eggs. The larva then jumps onto the egg, which it will use as a support in the honey, as it devours the egg over a few days. Then,

147

---

a TN: ...*la connaissance est plutôt jouée et inconsciente dans le cas de l'instinct, plutôt pensée et consciente dans le cas de l'intelligence.* In *Matière et mémoire*, Bergson used this same distinction between *joué* ("acted out" or "enacted") versus *pensé* ("thought"), MM, 193.

setting itself up on the shell, it undergoes its first metamorphosis. Being now developed [*organisée*] in such a way as to float upon the honey, it consumes this store of food and becomes a nymph, then a mature adult insect. Everything takes place *as* if the Sitaris larva knows, from the moment it hatches: that the male Anthophora will come through the tunnel exit first; that the nuptial flight will provide it the opportunity of transporting itself to the female; that the female will carry it into a storehouse of honey capable of feeding it when it will have undergone a transformation; and that, leading up to this transformation, it will have devoured the Anthophora egg so as to feed itself, support itself on the surface of the honey, and also remove the rival that would have come out of the egg. And everything takes place *as* if the Sitaris itself knows that its larva will know all of these things. The knowledge, if this is indeed knowledge, is only implicit. It is exteriorized in precise steps rather than being interiorized in consciousness.[149] It is no less true that the Insect's behavior sketches out the representation of determinate things, existing or taking place at precise points in space and in time, that the Insect knows without having learned them.

148

Now, if we look at the intellect from the same point of view, we find that it too knows certain things without having learned them. But this knowledge is of a very different order. I certainly do not want to reanimate that old debate among philosophers on the subject of innate knowledge. Thus, I limit myself here to noting the point upon which everyone agrees, namely that the young child immediately understands things that the animal will never understand, and that in this sense the intellect, like instinct, is an inherited function and hence innate. But this innate intellect, although it is a faculty of knowing, knows no particular object. When the newborn looks for its mother's breast for the first time, thereby showing that he has a knowledge (unconscious, of course) of a thing that he has never seen, it will be said that this is instinctual and not intellectual knowledge precisely because the innate knowledge here is of a determinate object. As such, the intellect would not bring with it an innate knowledge of any object. And yet, if it knew nothing naturally, then it would not have anything innate at all. So, what might the intellect know, this faculty that is ignorant of all things? — Besides *things*, there are *relations*. The newborn child knows neither any determinate object nor any determinate property of an object; but the day that someone attributes a property to an object, or applies an attribute to a noun, in the

child's presence, the child will immediately understand what is meant. Thus, he naturally grasps the relation of the attribute to the subject. And we could say just as much of the general relation expressed by the verb, a relation so immediately conceived of by the mind that language can take it as understood, as happens in rudimentary languages that do not have verbs. The intellect thus naturally makes use of the relations of equivalent to equivalent, of contained to container, of cause to effect, etc.,   149
which are implied in every sentence that includes a subject, an attribute, and a verb (whether stated or implied). Can we say that the intellect has an *innate* knowledge of each one of these particular relations? It would be the logician's task to determine whether this many irreducible relations are to be found here, or rather if they might not be reduced to some even more general relations. But no matter how we analyze thought, we will always end up at one or at several general frameworks of which the mind possesses an innate knowledge since it makes natural use of them. Thus, let us say that *if we consider what instinct and intellect contain in terms of innate knowledge, we find that, in the first case this innate knowledge bears upon* things *and in the second upon* relations.

Philosophers distinguish between the matter of our knowledge and its form.[150] The matter is what is given by the faculties of perception, taken in its raw state. The form is the set of relations that are established between these materials so as to constitute a systematic knowledge. Can the form, without any matter, already be an object of knowledge? Yes, surely, on condition that this knowledge resembles less a possessed thing than a contracted habit, less a state than a direction. This knowledge will be, if you will, a certain natural fold of attention. The student who knows that someone is about to dictate a fraction draws a line before knowing what the numerator and the denominator will be; he thus has present in his mind the general relationship between the two terms, though he does not yet know either of them—he knows the form without the matter. The same is true for the frameworks, anterior to all experience, into which our experience will come to be inserted. Thus, let us adopt the terms here established by usage. I will give the distinction between the intellect and instinct the following, more precise formulation: *the intellect, in terms of what it has that is innate, is the knowledge of a* form, *whereas instinct involves the knowledge of a* matter.   150

From this second point of view, which is that of knowledge and no longer that of action, the force immanent to life in general again appears as a limited principle in which two different and even diverging ways of knowing at first coexist and interpenetrate.[151] The first immediately attains the determinate objects in their very materiality. It says: "here is what is."[a] The second attains no object in particular; it is but a natural power for relating one object to another, or one part to another, or one aspect to another—in short, it is a power for drawing conclusions when we possess the premises and going from what has been learned to what remains unknown. It no longer says: "this is."[b] It merely says: if the conditions are such and such, then such will be the conditioned. In short, the first kind of knowledge, instinctive in nature, might be formulated in what philosophers call *categorical* propositions, whereas the second kind of knowledge, intellectual in nature, is always expressed *hypothetically*.[152] Of these two faculties, the first initially seems to be somewhat preferable to the other. And if it were extended to an indefinite number of objects, then it would indeed be preferable.[153] But in fact, it only ever applies to one specific object, and even to a limited part of that object. At least it has an internal and full knowledge of that part of the object—not an explicit knowledge, but one implicit in the accomplished action.[154] The second, by contrast, naturally only possesses an external and empty knowledge; but by this very fact it has the advantage of providing a framework into which an infinity of objects will be able to find their place, one after the other. Everything happens as if the force that evolves through the living forms, being a limited force, had to make a choice between two kinds of limitation in the domain of natural or innate knowledge: one limitation having to do with the *extension* of knowledge, the other with its *comprehension*.[c] In the first case, know-

151   ledge will be able to be stuffed [*étoffée*] and full, but it will be restricted to a specific object; in the second, it no longer limits its object, but this is because, being merely a form without matter, it no longer contains anything. The two tendencies, at first implicated in each other, had to separate in order to grow. They each set out on their own to try their luck in the world. They have ended up in instinct and in intellect.

a   TN: ..."*voici ce qui est.*" Here Bergson uses a somewhat colloquial phrasing to illustrate the existential claim implicit in instinctual knowing. The verb "to be" should be understood in the strong sense of "here is what *exists*."

b   TN: ..."*ceci est.*" Again, this should be read in the strong sense of "this exists."

c   TN: ...*compréhension.* Bergson uses this term here in the logical sense of "inclusion" or "intension."

If we adopt the point of view of knowledge, and no longer that of action, then these are the two diverging modes of knowledge by which the intellect and instinct must be defined. But knowledge and action are here merely two aspects of one and the same faculty. Indeed, it is easy to see that the second definition is but a new form of the first.

If instinct is the faculty *par excellence* for using a natural organic instrument, then it must contain innate knowledge (though virtual or unconscious, it is true) both of this instrument and of the object to which it is applied. Instinct is thus the innate knowledge of a *thing*. But the intellect is the faculty for fabricating inorganic instruments, namely artificial ones. If (via the intellect) nature stops equipping the living being with the instrument that will be of use to it, this is so that the living being can, according to the circumstances, vary the instrument's fabrication. As such, the essential function of the intellect will be to sort out the means of dealing with any circumstance whatsoever. It will look for what might be most useful, that is, what will fit into the proposed framework. Essentially, intellect bears upon the relations between the given situation and the means of making use of it. What will be innate in intellect, then, is the tendency to establish relations, and this tendency involves a natural knowledge of certain very general relationships, a veritable fabric that will be tailored into more specific relationships by the activity of each intellect. Wherever this activity is oriented toward fabrication, knowledge necessarily bears upon relations. But the intellect's entirely *formal* knowledge has an incalculable advantage over instinct's *material* knowledge. Precisely because a form is empty, it can be filled, in turn and at will, by an indefinite number of things, even by things that are entirely useless.[a/155] Formal knowledge is thus not limited to what is practically useful, even though it is in view of practical utility that formal knowledge made its appearance in the world. An intelligent being carries in himself the means for going beyond himself.[b]

152

---

a TN: ...*qui ne servent à rien.* This could be translated as "to serve for nothing" or "to serve no purpose." In this paragraph, Bergson goes back and forth between forms of *utiliser* and *servir à* for characterizing relations of use and utility. I have privileged the translation by "use" to maintain this resonance.

b TN: ...*de quoi se dépasser lui-même.* The term *se dépasser* ("going beyond oneself") here is used in the sense of "transcending oneself." I have, as mentioned in the Translator's Introduction, only used the English word "transcend" where Bergson himself uses the verb *se transcender.* Cf. below, 184, 193.

And yet he goes beyond himself less than he would like, and also less than he imagines himself to do. The purely formal nature of the intellect deprives it of the ballast it would have needed to be applied to the objects that would be of the highest interest for speculation. Instinct, by contrast, would have the necessary materiality, but is incapable of going as far in terms of seeking out its object—instinct does not speculate. Here we touch upon the most important point for our current study. The difference between instinct and intellect that I will here indicate is the very one that my entire analysis thus far has tended to bring out. I would formulate it as follows: *There are things that the intellect alone is capable of looking for, but will never find by itself. Instinct alone could find these things, but it will never go looking for them.*[156]

*

[THE PRIMORDIAL FUNCTION OF THE INTELLECT]

It is necessary here to enter into some provisional details regarding the mechanism of the intellect. I said that the function of the intellect is to establish relations.[157] Let me define more precisely the relations that the intellect establishes. As long as we see the intellect as a faculty destined for pure speculation, our understanding of it will remain vague and arbitrary on this point. We are thereby reduced to taking the general categories of the understanding to be something absolute, irreducible, and inexplicable.[158] The understanding would have fallen from heaven with its form, just like each one of us is born with his face. Perhaps we can make this form more definite, but that is the most we can do, and there is no reason to ask why it is what it is rather than something else. As such, it will be explained that the intellect is essentially unification, that all of its operations have as their common goal to introduce a certain unity into the diversity of phenomena, etc. But to start with, the term "unification" is vague, less clear than "relationship" or even "thought," and it says no more than they do. In addition, one might wonder whether the function of the intellect was not rather to divide, much more than to unify.[159] And finally, if the intellect proceeds as it does because it wants to unify, and if it seeks unification simply because it has a need for this, then our knowledge becomes relative to certain mental requirements that might well have been entirely different than they are. After all, for a differently formed intellect, knowledge would have been different as well.[160] The intellect being no longer dependent on anything, everything thus becomes dependent upon it. And as such, because we placed the understanding too high, we end up placing the knowledge that it gives us too low. From the moment the intellect is a sort of absolute, its knowledge becomes relative. To the contrary, I take human intellect to be relative to the necessities of action. Posit the action, and the very form of the intellect can be deduced from it. Thus, this form is neither irreducible nor inexplicable. And, precisely because the form is not independent, we can no longer say that knowledge is dependent upon the form.[161] Knowledge ceases to be a product of the intellect and becomes, in a certain sense, an integral part of reality.[162]

Philosophers will respond that action takes place in an *ordered* world, that this order already has to do with thought, and that I am thereby begging the question when I explain the intellect through action, which in fact presupposes it. And they would be right, if the point of view I am adopting in the present chapter was to be my definitive point of view. In that case, I would be duped by an illusion like Spencer's. He believed that the intellect was sufficiently explained when it had been reduced down to the imprint left upon us by the general characteristics of matter—as if the order inherent in matter were not intelligence itself![163] To what extent, and through what method, philosophy might attempt a genuine genesis of the intellect at the same time as a genesis of matter is a question that I reserve for the next chapter.[164] For the moment, I am focused on a problem that is psychological in nature. I am asking: what portion of the material world is our intellect especially adapted to? Now, there is no need to choose a philosophical system in order to respond to this question. It is enough to adopt the point of view of common sense.[165]

154

Thus, let us begin from action and posit that in principle the intellect aims first and foremost at fabricating. Fabrication works exclusively upon brute matter, in the sense that even if it employs organic materials, it treats them like inert objects, not worrying about the life that has given them their form. Fabrication retains of brute matter hardly anything other than what is solid: the rest slips away by its very fluidity. Thus, if the intellect tends toward fabricating, then we can predict that whatever is fluid in the real will partially escape it, and that whatever is properly vital in the living being will entirely escape it. *Our intellect, such as it emerges from the hands of nature, takes the inorganic solid as its principal object.*

Were we to inspect the intellectual faculties, we would see that the intellect only feels at ease, only feels fully at home, when it is working on brute matter and in particular on solids.[166] And what is the most general property of brute matter? That it is extended, that it presents us with objects outside of other objects and, in these objects, parts outside of other parts.[167] Surely it is useful for us, in view of our later manipulations, to consider each object as divisible into arbitrarily cut out parts, each part itself being divisible again as we please, and so on to infinity. But above all else, in terms of our present manipulation, we must take the real object that we are dealing with, or the real elements we have divided it up into, as *provisionally fixed* and treat them as so many unities.[a/168] We allude to the possibility of breaking down matter as much as we like, and however we like, when we speak of the *continuity* of material extension. But this continuity obviously boils down to the faculty that matter leaves to us for choosing the mode of discontinuity that we will find in it. In general, it is always the mode of discontinuity, once chosen, that seems to us as effectively real and that fixes our attention because our present action is modeled upon it. As such, the discontinuity is thought for itself, it is thinkable in itself, and we conceive of it through a positive mental act, whereas the intellectual representation of continuity is rather a negative one, being at root nothing but the refusal of our mind (finding itself before any actually given system of decomposition whatsoever) of taking this one to be the only possible one. *The discontinuous is the only thing the intellect conceives of clearly.*

On the other hand, there is no doubt that the objects we act on are moving objects. But what matters to us is to know where the moving object is going, and where it is at a given moment along its path [*trajet*].[169] In other words, we focus above all on its actual or future positions, and not on the *progression* according to which it goes from one position to another—the progression that is the movement itself. In the actions that we accomplish and that are systematized movements, we fix our mind on the goal or on the signification of the movement, on its overall design—in short, on the immobile plan of its execution. What there is of actual moving reality in the action only interest us to the extent that

---

a TN: ...*provisoirement définitifs*... Bergson's formulation here could also be translated as "provisionally definitive" or "provisionally final." I have chosen "fixed" to try to capture the double sense of both "final" and "stable."

the whole movement might be advanced, slowed down, or blocked by some incident that takes place along the way. Our intellect turns away from mobility itself because it has no interest in paying attention to it. If our intellect had been designed for pure theory, then it would have installed itself within movement, for movement is surely reality itself, and immobility is never more than apparent or relative.[170] But the intellect is designed for something else entirely. Unless it does violence to itself, the intellect follows the opposite course: it always begins from immobility, as if that were the ultimate reality or the element; and when it wants to picture movement, it reconstructs it by juxtaposing immobile parts [immobilités].[171] This operation—which I will show to be illegitimate and dangerous on the speculative order (where it leads to dead ends and artificially creates unsolvable philosophical problems)— is easily justified when we refer back to what it is designed for.[172] The intellect, in its natural state, aims at a goal that is useful in terms of practice. When it substitutes juxtaposed immobile parts for the movement itself, it does not claim to reconstitute movement such as it really is—it simply replaces movement with a practical equivalent. Philosophers are the ones who make a mistake by transporting a method of thinking designed for action into the domain of speculation.[173] But I will come back to this point below.[174] Let it suffice to say for now that, in virtue of its natural disposition, our intellect focuses on what is stable and unchanging. *Immobility is the only thing our intellect conceives of clearly.*

Now, fabricating consists in cutting the form of an object out of a material. The most important thing of all is the form to be obtained. In terms of the matter, the most convenient one is chosen. But to choose it—to go looking for it among so many others—we must have tried, at least in our imagination, to give to every kind of matter the form of the conceived object. In other words, an intellect that aims at fabricating is an intellect that never stops at the current form of things, that does not consider that form to be definitive, and that, by contrast, holds every matter as able to be cut up at will. Plato compares the good dialectician to the skilled chef who cuts up the animal without breaking its bones by following the joints laid out by nature.[a/175] Indeed, an intellect that

157

---

a Bergson's note: Plato, *Phaedrus*, 265e [Plato, *Phaedrus*, trans. Alexander Nehamas and Paul Woodruff, in *Complete Works*, ed. John M. Cooper, with D. S. Hutchinson (Indianapolis: Hackett Publishing Company, 1997)].

always worked in this way would surely be an intellect turned toward speculation. But action and particularly fabrication require the opposite mental tendency. Action demands that we consider the current form of all things (even of natural ones) to be artificial and provisional, that our thought erase from the perceived object (even if it is organic and living) the lines that mark on the outside its inner structure—in short, that we consider its matter to be indifferent to its form. The whole of matter will thus necessarily appear to our thought as an immense fabric from which we can cut out what we like so that we can sew it together again as we please. It should be noted in passing that this is the power we affirm when we say that there is *space*, i.e., a homogeneous and empty milieu, infinite and infinitely divisible, lending itself indifferently to any mode of decomposition whatsoever.[176] A medium of this kind is never perceived; it is only conceived.[177] What is perceived is extension insofar as it is colored, resistant, and divided up according to the lines that sketch out the contours of real bodies or of their real elementary parts. But when we conceive of our power over this matter, that is, of our faculty for decomposing and recomposing it as we please, we project, behind real extension and *en bloc*, all of the possible decompositions and recompositions in the form of a homogeneous, empty, and indifferent space that subtends it.[178] This space is thus, above all else, the schema of our possible action upon things, even if things have a natural tendency (as I will explain below) to enter into a schema of this kind—it is a view adopted by the mind.[179] The animal probably has no idea of space, even when (like us) it perceives extended things.[180] Space is a representation that symbolizes the human intellect's tendency toward fabrication. But I will not linger over this point at the moment. Let it suffice to say that *the intellect is characterized by its indefinite power for decomposing according to any law whatsoever and of recomposing into any system whatsoever.*

158

I have enumerated some of the essential traits of human intellect.[181] But I have considered the individual in an isolated state, without taking social life into account. In reality, man is a being who lives in society. If it is true that human intellect aims at fabricating, we must add that it also joins together with other intellects, both for fabricating and for everything else.[182] Now, it is difficult to imagine a society in which the members did not use signs to communicate with each other.[183] Societies of Insects surely have a language, and this language must be adapted, like human language, to the necessities of communal living.[184] Language makes *collective action* possible. But the necessities of collective action for an ant colony are not at all the same as those for a human society. In Insect societies, there is generally some polymorphism, the division of labor is natural, and each individual is riveted to the function that it will perform by its physical structure.[185] In any case, these societies are based upon instinct, and consequently upon certain actions or fabrications that are more or less tied to the form of the organs. Thus, for example, if Ants have a language, then the signs that make up that language must be very limited in number, and, once the species has been constituted, each sign must remain invariably attached to a certain object or to a certain operation. The sign adheres to the thing that is signified. By contrast, in a human society, fabrication and action have a variable form, and, moreover, each individual must learn his role given that he is not predestined to that role by his structure.[186] There must be a language that allows us at any given moment to go from what we know to what we do not know. There must be a language whose signs—which cannot be infinite in number—can nevertheless be extended to an infinity of things. This tendency of the sign to be transported from one object to another is characteristic of human language. We see this in the small child, from the day he begins to speak. Immediately, and quite naturally, he extends the sense of the words he learns, making use of the most accidental comparisons or of the most distant analogies so as to detach and to transport elsewhere the sign that had been attached to an object in front of him. "Anything can designate anything else": such is the latent principle of childhood language. It was wrong to confuse this tendency with the faculty for generalizing. Animals themselves generalize, and indeed a sign, even if it is an instinctive one, always more or less represents a genus.[187] What characterizes the signs of human language is less their generality than their mobility. *The instinctive sign is adherent, the intelligent sign is mobile.*

Now, this mobility of words, designed to allow them to go from one thing to another thing, also permitted them to be extended from things to ideas. Of course, language could not have given the faculty of reflection to an intellect that was entirely externalized and incapable of folding back upon itself. An intellect that reflects is an intellect that had a surplus of force to spend beyond its practically useful efforts. This is

159

160

a consciousness that has already, virtually, won back control of itself. But it is still necessary for this virtuality to become actual.[a] Presumably, without language the intellect would have remained riveted to the material objects that it had some interest in considering. It would have lived in a state of somnambulism, outside of itself, and hypnotized by its work. Language greatly contributed to liberating it. The word, designed to go from one thing to another, is in effect essentially transferable and free. As such, it can be extended not only from one perceived thing to another perceived thing, but also from the perceived thing to the memory of this thing, from the precise memory to a more fleeting image, and from a fleeting but nevertheless still represented image, to the representation of the act by which one represents it, that is, to the idea.[188] In this way, the intellect, which had been looking to the outside, would suddenly find before its eyes a whole inner world, namely the spectacle of its own operations. Moreover, the intellect was merely waiting for this opportunity. It made use of the fact that the word itself is a thing in order to penetrate (being carried along by the word) into the interior of its own work. As much as its first vocation was to fabricate instruments, this fabrication is only possible through the use of certain means that are not precisely tailored to their object, that go beyond it, and that thereby allow the intellect to undertake a supplementary work, namely one that is disinterested.[189] From the day that the intellect, reflecting upon its own procedures, perceived itself as the creator of ideas, as the faculty of representation in general, there was no longer any object that it did want to have an idea of, even if that object was lacking any direct relation to practical action. This is why I said that there are things that intellect alone can look for.[190] That is, only intellect is concerned with theory. And its theory wants to embrace everything, not only brute matter (upon which it has a natural hold), but also life and thought.

We can guess by what means, by what instruments, and by what method the intellect will eventually approach these problems. Originally, the intellect was adapted to the form of brute matter. Even language, which allowed it to extend its field of operations, is made for designating

161

---

a TN: *Mais encore faut-il que la virtualité passe à l'acte.* Note here that the relationship is between the virtual and the actual, not between the possible and the real. Cf. editorial note 10 to Chapter I and the Translator's Introduction.

things and nothing but things. Because the word is mobile, because it makes its way from one thing to the next, the intellect at some point had to take hold of it *along the way*, even though it had not come to rest upon anything, in order to apply it to an object that is not a thing and that, having been hidden until then, was waiting for the safety of the word before coming out of the shadow and into the light.[a] But by covering this object, the word converts it again into a thing. In this way, even when the intellect no longer operates upon brute matter, it still follows the habits that it contracted in that operation: it applies the very same forms it used for inorganic matter. It is made for this type of work. This type of work is the only one that fully satisfies it. And this is what it expresses when it claims that this type of work is the only way to arrive at *distinctness* and at *clarity*.[191]

Thus, in order to think itself clearly and distinctly, the intellect will have to perceive itself in the form of discontinuity.[192] Concepts are indeed external to each other, in the manner of objects in space. And concepts have the same stability as objects, having been created according to that model. Taken together, concepts constitute an "intelligible world" that resembles the world of solids in its essential characteristics, but whose elements are lighter, more diaphanous, and easier for the intellect to manipulate than are the pure and simple images of concrete things. Indeed, concepts are no longer the perception of things, but rather the representation of the act by which the intellect fixes upon them. They are thus no longer images, but rather symbols. Our logic is the collection of rules that must be followed in the manipulation of symbols. Since these symbols are derived from a consideration of solids, and since the rules for the composition of these symbols among themselves hardly does anything other than translate the most general relations between solids, our logic is triumphant in the science that takes the solidity of bodies as its object of study, namely in geometry. Logic and geometry reciprocally engender each other, as we will see below.[193] Natural logic is the result of an extension of a certain natural geometry, one that is suggested by the general and immediately perceived properties of solids. In turn, scientific geometry, which indefinitely extends the knowledge of the external

162

---

a TN: Although Bergson here seems to be making a distinction between *objets* ("objects") and *choses* ("things"), he returns to using the terms more or less interchangeably in the following lines.

properties of solids, comes out of this natural logic.[a/194] Geometry and logic are rigorously applicable to matter. They feel at home there and can march along all by themselves.[195] But beyond this domain, pure reasoning needs to be watched over by good sense, which is something else entirely.[196]

In this way, all of the elementary forces of the intellect tend to transform matter into an instrument of action, that is, into an *organ*, in the etymological sense of the word.[b/197] Life, not content to simply produce organisms, wanted to give them inorganic matter itself as an appendage converted into an immense organ through the industriousness of the living being.[c] This is the first task that life assigns to the intellect. This is why the intellect invariably still acts as if it were fascinated by the contemplation of inert matter.[198] The intellect is life looking to the outside, going outside of itself, and adopting in principle the procedures of inorganic nature [*nature inorganisée*] so as to in fact direct them. Hence its astonishment when it turns toward the living being and finds itself confronted with organic nature [*organisation*]. No matter what else it does, it always reduces the organic to the inorganic because without reversing its natural direction and without twisting back upon itself, the intellect is unable to think true continuity, real mobility, reciprocal interpenetration—in sum, it is unable to think that creative evolution that is life itself.[199]

Is it a question of continuity? The aspect of life accessible to our intellect—and, for that matter, to our senses, which our intellect extends—is the one that offers a certain grip to our action. For us to be able to modify an object, we must perceive it as divisible and discontinuous. From the point of view of the positive sciences, an incomparable progress was achieved the day that organic tissues were reduced down into cells.[200] The study of the cell, in turn, revealed it to be an

163

---

a   Bergson's note: I will return to these points in the next chapter.
b   TN: Like the English word "organ," the etymology of the French word *organe* traces back to the Greek οργανον, meaning "instrument" or "device," and is also used for the "bodily organs."
c   TN: Here (and in the following pages) Bergson uses the term *notre industrie* (literally "our industry") in the double (and older) sense of having "skill," "cleverness," or "ingenuity," and of accomplishing manual work, and less the idea of large-scale mechanical production. I have chosen "industriousness" to capture both of these senses. The term also appears in the Introduction, see above, v, viii.

organism whose complexity seems to increase to the extent that it is examined more closely. The more science advances, the more it finds that the living being is made up of an increasing number of heterogeneous elements that are juxtaposed and external to each other. Does it thereby come closer and closer to life? Or does that which is properly vital in the living being not seem to retreat to the extent that we push more and more into the details of the juxtaposed parts? There is already a tendency emerging among scientists to consider the substance of the organism as continuous and the cell as an artificial entity.[a/201] And yet, assuming that this view ends up winning out, by deepening itself, it will merely lead to another mode of analyzing the living being, and consequently to a new discontinuity—even if this new discontinuity is perhaps less distant from the real continuity of life.[202] The truth is that this continuity cannot be thought by an intellect that gives itself over to its own natural movement. This continuity involves both the multiplicity of elements and the reciprocal penetration of all of them together, two properties that can hardly be reconciled in the field where our industriousness, and thereby our intellect, operates.

In the same way that we separate in space, we fix in time. The intellect is not at all made for thinking *evolution*, in the proper sense of the word, namely the continuity of a change that would be pure mobility.[203] I will not linger here on this point, which I intend to examine in a special chapter below.[204] Suffice it to say that the intellect conceives of becoming as a series of *states* where each state is homogeneous with itself and consequently does not change. And what if our attention is drawn to the internal change taking place in one of these states? We quickly decompose it into another series of states that will constitute, when taken together, its internal modification. These new states themselves will each be invariable; otherwise, if we notice their internal change, it will immediately resolve into a new series of invariable states and so on indefinitely. Here again, thinking consists in reconstituting, and we naturally reconstitute with elements that are given and hence with elements that are stable. So much so that, as much as we might try to imitate the mobility of becoming through the indefinite progression of our additions, becoming itself slips through our fingers just when we thought we had taken hold of it.[205]

164

---

a   Bergson's note: I will return to this point in Chapter III, 281. [TN: See rather, below, 259–61.]

Precisely because it always attempts to reconstitute, and to reconstitute using what is already given, the intellect allows what is new in each moment of a history to escape.[206] It does not admit the unforeseeable. It rejects all creation. Our intellect is only satisfied when specific antecedents lead to a specific consequence that can be calculated as a function of them. Or again, we understand when, in order to be attained, a specific end calls for specific means. In both cases, we are dealing with what is known being combined with what is known and, in general, with what is old being repeated. Our intellect is at ease in all of this. And whatever its object might be, it will abstract, separate, and eliminate so as to substitute for the object itself, if necessary, an approximate equivalent in which things will take place in this manner. But that each moment is a contribution, that something new continuously springs forth, that a form is born of which it will perhaps be said, once produced, that it is an effect determined by its causes, but of which it was impossible to take as foreseeable in terms of what it would be (given that here the causes, unique in their genre, are part of the effect, have come about at the same time as the effect, and are determined by it as much as they determine it)—all of this is something that we can feel within ourselves and divine through sympathy outside of ourselves, but that can neither be expressed in the language of the pure understanding nor be, strictly speaking, thought.[207] Of course, none of this will be surprising so long as we recall what our understanding was designed to do. The causality that it seeks and that it finds everywhere expresses the very mechanism of our industriousness, by which we indefinitely recompose the same whole out of the same elements, or by which we repeat the same movements in order to obtain the same results. For our understanding, the finality *par excellence* is the one we find in our industriousness, whereby we work according to a model that is given in advance, that is to say, one that is old or that is composed of known elements. As for invention, properly so-called, which is nonetheless the starting point for industriousness itself, our intellect is not able to grasp it in its *springing forth*, that is, insofar as it is indivisible, nor in its *genius* [*génialité*], that is, insofar as it is creative.[208] Explaining invention always consists in reducing what was unforeseeable and new into known or old elements arranged in a different order. The intellect no more admits complete novelty than it does radical becoming. In other words, here again the intellect allows an essential aspect of life to escape, as if it were not at all made for thinking such an object.

All of our analyses lead us back to this conclusion. But it was hardly necessary to go into so much detail in terms of the mechanism of intellectual work—it would have been enough to consider its results. We can see that the intellect, as skillful as it is in manipulating the inert, displays its clumsiness as soon as it touches on the living.[209] Whether it is a question of treating the life of the body or the life of the mind, the intellect proceeds with the strictness, the rigidity, and the brutality of an instrument that was not designed for such a use. The history of hygiene or pedagogy would tell us much about this.[210] When we think of the important, pressing, and constant interest that we have in conserving our bodies and elevating our souls, of the special capabilities that we all have for continuously experimenting on ourselves and on others, and of the palpable damage by which the imperfection of a medical or a pedagogical practice is manifested along with how much it costs us, we are amazed at the stupidity and above all the persistence of errors. We would easily discover their origin in our stubbornness in treating the living being like the inert and in thinking of all reality—as fluid as it might be—under the form of a definitively fixed solid. We are only at ease in the discontinuous, in the immobile, and among the dead.[a] *The intellect is characterized by a natural incomprehension of life.*

166

*

[THE NATURE OF INSTINCT]

By contrast, instinct is molded upon the form of life itself. Whereas the intellect treats all things mechanically, instinct proceeds organically, if one can speak this way. If the consciousness that sleeps within instinct reawakened, if it internalized itself in knowledge rather than externalized itself in action, if we knew how to interrogate it and if it were able to respond, then it would deliver over to us the most intimate secrets of life.

---

a TN: *...dans le discontinu, dans l'immobile, dans le mort.* As above, Bergson uses *le mort* ("dead beings" or "the dead") as a contrast to the general category of *le vivant* ("the living" or "living beings").

For instinct merely continues the work by which life organizes matter to such an extent that, as has often been shown, we cannot say where organic development [*organisation*] ends and where instinct begins.[211] When the chick breaks its shell with a strike of its beak, it acts by instinct; nevertheless, it is merely following out the movement that had carried it along though embryonic life. Inversely, in the course of embryonic life itself (and above all when the embryo freely lives in the form of a larva) many procedures are accomplished that must be related to instinct. The most essential of the primary instincts are thus really vital processes.[212] The virtual consciousness that accompanies these instincts usually only actualizes itself [*s'actualise*] in the initial phase of the act and leaves the rest of the process to be accomplished all by itself. This consciousness would merely have to open itself up more fully, and then deepen itself completely, in order to coincide with the generative force of life.

When we see thousands of cells in a living body working together toward a common goal, dividing up the task among themselves, each one living at once for itself and for the others, preserving itself, feeding itself, reproducing itself, and responding to threats of danger through appropriate defensive reactions, how could we not think of these as so many instincts? And yet, these are the cell's natural functions and the constitutive elements of its vitality.[213] Reciprocally, when we see that the Bees of a single hive form a system so tightly organized that none of the individuals can live separated from it for very long, even if they are given shelter and food, how could we not acknowledge that the hive is literally—and not metaphorically—a single organism and that each Bee is one cell united with the others through invisible connections? The instinct that animates the Bee is thus indistinguishable from the force that animates the cell, or perhaps instinct merely prolongs that force. In extreme cases such as this one, instinct coincides with the work of organic development [*organisation*].

Of course, there are clearly degrees of perfection in the same instinct. There is a large distance between the Bumblebee and the Honeybee, for example, and we would pass from one to the other by way of an immense number of intermediaries corresponding to just as many complications of the social life.[214] But the same diversity would be found in the functioning of histological elements belonging to different tissues, and more or less related to each other. In both cases, there are multiple variations executed upon the same theme.[215] Yet the constancy of the theme is no less apparent, and the variations do nothing more than adapt it to the diversity of the circumstances.

Now, in both cases, whether it is a question of an animal's instincts or of the vital properties of a cell, the same knowledge [*science*] and the same ignorance are apparent. Things take place as if the cell knew what was of interest to it in the other cells, and as if the animal knew what it would be able to use of the other animals, with everything else remaining in the shadows. It seems that life, from the moment it is contracted into a determinate species, loses contact with the rest of itself except on one or two points that interest the species that has just been born. Is it not obvious that life here proceeds like consciousness in general, that is, like memory?[216] We drag along behind ourselves, without noticing it, the totality of our past; but our memory only deposits into the present the two or three memories that will in some way complete our current situation.[217] The instinctive knowledge that a species possesses of another species on a certain specific point thus has its roots in the very unity of life, which is, to use an expression from an ancient philosopher, a whole that is sympathetic with itself.[a/218] It is impossible to examine certain specialized instincts of the animal and the plant—which are clearly born in extraordinary circumstances—without relating them to those seemingly forgotten memories that suddenly spring forth under the pressure of an urgent need.

Of course, an immense number of secondary instincts, along with many modalities of the primary instinct, can be scientifically explained.[219] Nevertheless, it is unlikely that science, with its current explanatory procedures, will ever arrive at a complete analysis of instinct.[220] This is because instinct and intellect are two diverging developments from a single principle that, in one case, remains within itself and, in the other, goes outside of itself and gets caught up in the use of brute matter. This continuing divergence bears witness to a radical incompatibility and to the fact that it is impossible for the intellect to reabsorb instinct. The

169

---

a [Bergson's missing citation: Plotinus discusses this idea of a whole that is "in a state of sympathy with itself" in *Ennead* IV. See Plotinus, *The Enneads*, ed. Lloyd P. Gerson, trans. George Boys-Stones, et al. (Cambridge: Cambridge University Press, 2018), 4.4.35 (page 456).]

essential nature of instinct cannot be expressed in the language of the intellect, nor, consequently, can it be analyzed.[a]

A person blind from birth, living among others who were also blind from birth, would not admit that it is possible to perceive a distant object without first having passed through the perception of all the intermediary objects.[221] Nevertheless, vision performs this miracle. True, we could say that the blind person is right insofar as vision—having its beginnings in the stimulation of the retina caused by the vibrations of the light—is in fact nothing other than a touching with the retina. And this is indeed the scientific explanation since the role of science is precisely to translate every perception into the language of touch.[222] But I have shown elsewhere that the philosophical explanation of perception must be of an entirely different nature, to assume that we can still speak here of "explanation."[b] But instinct too is a knowledge from a distance. Instinct is to the intellect what vision is to touch. Science can do nothing other than translate instinct into the language of the intellect; but it will thereby construct an imitation of instinct rather than penetrate into instinct itself.

Anyone who studies the ingenious theories of evolutionary biology will be convinced of this conclusion. The theories reduce down to two types, which moreover interfere with one another. Sometimes, according to neo-Darwinian principles, instinct is seen as a sum of accidental differences preserved through selection: some useful procedure, accomplished naturally by the individual in virtue of an accidental predisposition of the germ, would be transmitted from one germ to the next until such time as chance might come to add new improvements to it through the same procedure.[223] And sometimes instinct is turned into a degraded intellect: the action judged useful by the species or by some of its representatives would have engendered a habit, and the habit, having been transmitted through heredity, would have become an instinct. Of these two systems, the first has the advantage of being able to speak of hereditary transmission without giving rise to any major objections because the accidental modification that it places at the origin

170

a   TN: ...*ce qu'il y a d'essentiel dans l'instinct*. Bergson's phrase here could be more literally translated as "what is essential in instinct," but I have used "the essential nature of instinct" to emphasize the connection here with Bergson's section title.
b   Bergson's note: *Matière et mémoire*, Chapter I.

of the instinct would not be acquired by the individual but rather is inherent to the germ. On the other hand, this neo-Darwinian system is absolutely incapable of explaining instincts that are as sophisticated as those possessed by the majority of Insects.[a] Of course, these instincts must not have reached, all at once, the degree of complexity that they have today—they have surely evolved. But according to a hypothesis like the neo-Darwinian one, the evolution of the instinct could only take place through the progressive addition of new pieces that, in some way and by some fortuitous accidents, would have come to gear into the previous ones. Now, it is obvious that in the majority of cases the instinct was not able to perfect itself through a simple incremental growth. Each new piece in fact required, on pain of ruining everything, a complete reworking of the whole. How could we expect chance to undertake such a reworking?[224] I admit that there can be a hereditary transmission of the accidental modification of the germ, and that it can wait, so to speak, for other new accidental modifications to come along and make it more complicated.[225] I also admit that natural selection will eliminate all of those more complicated forms that are not viable. Still, in order for the life of the instinct to evolve, viable complications must take place. Now, these complications will only be produced if, in certain cases, the addition of a new element brings about the correlative change of all the old elements. No one will claim that chance can perform such a miracle. In one form or another, an appeal to intellect will be made. It will be assumed that through some more or less conscious effort the living being develops a higher instinct within itself. But then it will be necessary to accept that a contracted habit can become hereditary, and that it does so consistently enough to ensure an evolution. Such a thing is doubtful, to say the least. Even if we could relate animal instincts to a habit that is hereditarily transmitted and intelligently acquired, it still remains unclear how this mode of explanation could be extended to the plant world, where effort, even supposing that it is sometimes conscious, is never intelligent.[226] And yet, when we see the reliability and the precision with which climbing plants use their tendrils, or the marvelously

171

---

a TN: ...*des instincts aussi savants*... The adjective *savant* ("learned," "skillful," or "sophisticated") is also used as a noun in French for scientist or scholar, which resonates with how the instincts act *as if* they had the knowledge of a scientist. Cf. above 147; below, 174–75.

combined maneuvers executed by Orchids so as to get Insects[a] to pollinate them, how could we not think of these as so many instincts?[227]

This does not mean that we must completely abandon the neo-Darwinian theory, nor the neo-Lamarckian theory. The neo-Darwinians are likely correct when they claim that evolution takes place from germ to germ rather than from individual to individual, and the neo-Lamarckians are likely correct when they say that there is an effort at the origin of instinct (even if, as I believe, this is something quite different from an *intelligent* effort). But the neo-Darwinians are likely wrong when they turn the evolution of the instinct into an accidental evolution, and the neo-Lamarckians are likely wrong when they understand the effort from which instinct emerges to be an individual effort. The effort by which a species modifies its instincts and thereby modifies itself must be something much deeper, and must be something that solely depends on neither the circumstances nor the individuals. It does not depend merely upon individual initiative, even though the individuals surely collaborate with it; and it is not purely accidental, even though accident surely occupies a large place.[228]

172

Let us compare the various forms of the same instinct found in the various species of Hymenoptera. The impression we get is not always one of an increasing complexity being obtained through the successive additions of elements to each other, nor of a rising series of mechanisms arranged, so to speak, along the steps of a ladder. Rather, at least in a large number of cases, we will think of a planetary system: the diverse varieties of the instinct each taking off from a different point but all facing toward the same center, all making an effort in that same direction, and each one only approaching it to the extent of their means and to the extent to which the center point was illuminated for it.[b]

---

a Bergson's note: See the following two works by Darwin: *Les mouvements et les habitudes des plantes grimpantes*, trans. Richard Gordon, 2nd ed. (Paris: Reinwald, 1890) and *De la fécondation des Orchidées par les Insectes et des bons résultats du croisement*, trans. Louis Rérolle, 2nd ed. (Paris: C. Reinwald, [1891]). [The English originals are: *The Movements and Habits of Climbing Plants*, 2nd edition (London: John Murray, 1875); *The Various Contrivances by Which Orchids are Fertilised by Insects*, 2nd rev. ed. (London: John Murray, 1877).]

b TN: Bergson's term here is *circonférence*, which translates literally as "circumference" or simply "circle." The second part of the sentence, however, implies that Bergson is thinking of the *circonférence orbitale*, as in the movement of a planetary system such that the variations of species all circle metaphorically around the theme at various distances and speeds.

In other words, the instinct is everywhere complete, but it is more or less simplified and above all simplified in *diverse ways*. Moreover, where we see a linear gradation along which the instinct makes itself more complicated in one and the same direction, as if moving up the steps of a ladder, the species classed along such a linear series because of their instincts are far from always having kinship relations among themselves. In this way, the comparative study undertaken in recent years of the social instinct among the various Apidae establishes that the instinct of the Meliponines is intermediary in terms of complexity between the still rudimentary tendency of the Bombini and the consummate knowledge of our Bees. And yet, there can be no relation of kinship between the Bees and the Meliponines.[a/229] In all likelihood, the larger or smaller degree of complexity of these various societies does not come from a larger or smaller number of added elements. Rather, it is much more likely that we find ourselves before a certain *musical theme* that would first be itself   173
transposed, as a whole, into a certain number of keys, and then upon which, also as a whole, diverse variations would have been executed, some being very simple, others infinitely sophisticated.[230] As for the original theme, it is everywhere and nowhere. In vain does one hope to write it down in the language of representation—it was, in all likelihood, originally something *sensed* [*senti*] rather than something *thought*. We have the same impression when observing the paralyzing instinct of certain Wasps. It is well known that the various species of paralyzing Hymenoptera deposit their eggs in Spiders, Beetles, or Caterpillars who, having first been attacked by the Wasp in a sophisticated surgical operation, will continue to live immobilized for a certain number of days and will thereby provide the larvae with fresh food. In terms of the sting that these Wasps inflict upon the nervous centers of their victims so as to immobilize without killing, these various species of Hymenoptera adapt themselves to the various species of prey that each one respectively targets. The Scolia, which attacks a Rose Chafer Beetle larva, only stings it at one point, but the point in question is where a concentration of motor ganglia (and only motor ganglia) is found. The stinging of any other ganglia could lead to death and rotting, which must

---

a Bergson's note: Hugo von Buttel-Reepen, "Die phylogenetische Entstehung des Bienenstaates, sowie Mitteilungen zur Biologie der solitären und sozialen Apiden," *Biologisches Centralblatt* 23, no. 1 (1903): 108 in particular.

be avoided.[a][231] The Golden Digger Wasp [*Sphex à ailes jaunes*], which has chosen the Cricket as its victim, knows that a Cricket has three nervous centers that animate its three pairs of legs, or at least it acts as if it knows this.[232] It stings the insect first under its neck, then behind its prothorax, and finally where its abdomen begins.[b] The Ammophila Hirsuta inflicts nine successive stinger strikes to nine nervous centers of its Caterpillar, and finally grabs the head and bites it just enough so as to ensure paralysis without death.[c][233] The general theme is "the need to paralyze without killing"; the variations are subordinated to the structure of the subject being operated upon.[234] Of course, it is hardly the case that the operation is always executed to perfection. It has recently been shown that the Sphex Ammophila sometimes kills the Caterpillar rather than paralyzing it, and sometimes only partially paralyzes it.[d][235] But from the fact that instinct, like the intellect, is fallible, and that it too is susceptible to presenting individual divergences, it does not follow at all that the Sphex's instinct was acquired through some form of intellectual trial and error, as has been claimed. Even if we assume that over time the Sphex was able to recognize, one by one and through trial and error, the points on its victim that must be stung in order to immobilize it and the special treatment that needs to be inflicted upon the brain so that the paralysis takes hold without leading to death, how could we ever think that such specialized elements of such a precise form of knowledge might be reliably transmitted, one by one, through inheritance? If there were even a single, indisputable example of a transmission of this kind anywhere in our current experience, then the inheritance of acquired characteristics would not be questioned by anyone. But in fact,

174

---

a Bergson's note: Jean-Henri Fabre, *Souvenirs entomologiques: Études sur l'instinct et les mœurs des insectes*, troisième série [1886], 2nd ed. (Paris: Delagrave, 1890), 1–69.

b Bergson's note: Jean-Henri Fabre, *Souvenirs entomologiques: Études sur l'instinct et les mœurs des insectes*, première série [1879], 3rd ed. (Paris: Delagrave, 1894), 93–100. [TN: The chapter that Bergson cites here ("Les trois coups de poignard") immediately follows the one titled "Le Sphex à ailes jaunes," the species just mentioned (*ibid.*, 81–92).]

c Bergson's note: Jean-Henri Fabre, *Nouveaux souvenirs entomologiques: Études sur l'instinct et les mœurs des insectes* (Paris: Delagrave, 1882), 14–27. [TN: This is the second series of Fabre's *Souvenirs entomologiques*. The chapter in question is titled: "L'Ammophile hérissée."]

d Bergson's note: George W. Peckham and Elisabeth G. Peckham, *Wasps: Social and Solitary* (Westminster: Archibald Constable, 1905), 28–30.

the hereditary transmission of the contracted habit takes place in an imprecise and irregular fashion, assuming that it ever actually takes place at all.[236]

The entire difficulty comes from the fact that we want to translate the Hymenoptera's knowledge [science] into the language of the intellect. We are thereby forced to assimilate the Sphex to an entomologist who knows the Caterpillar in the same way that he knows all other things, that is, from the outside [du dehors], without having from that perspective any particular or vital interest in the thing in question.[237] As such, the Sphex would have to learn, one by one, like the entomologist, the positions of the Caterpillar's nervous centers—it would have to acquire at least a practical knowledge of these positions by experimenting with the effects of its stinging. But none of this would be required if we assumed rather a sympathy (in the etymological sense of the word) between the Sphex and its victim, a sympathy that would inform it from within [du dedans], so to speak, of the Caterpillar's vulnerability.[238] This feeling for its vulnerability might well owe nothing to external perception and might result from a simple present encounter between the Sphex and the Caterpillar, no longer considered as two organisms but rather as two activities. The feeling would express in a concrete form the relation of the one to the other. True, a scientific theory cannot appeal to these types of considerations. Such a theory must not place action prior to organic development [organisation], nor sympathy prior to perception and knowledge. And yet, to repeat, either philosophy has nothing to do here, or its role begins where the role of science ends.[239]

So, whether science turns instinct into a "compound reflex," or into an intelligently contracted habit that has become automated, or into a sum of small accidental advantages accumulated and fixed through selection, in all cases it claims to resolve instinct entirely either into intelligent processes or into mechanisms constructed one piece at a time, like the ones that our intellect combines.[240] Now, I happily admit that here science is playing its proper role. It will give us, in the absence of a real analysis of the object, a translation of this object into the language of the intellect. But how could we fail to notice that science itself invites philosophy to consider things from another angle? If our biology was still Aristotle's, if it took the series of living beings to be unilinear, if it showed life as a whole evolving toward intelligence and passing

175

through sensibility and instinct in order to do so, then we would have the right, as intelligent beings, to turn back toward anterior and, consequently, inferior manifestations of life and claim to grasp them within the frameworks of our intellect without deforming them. But one of the most obvious results of biology was to show that evolution has been accomplished along diverging lines. We find intellect and instinct in their more or less pure forms at the endpoints of two of these lines (the two principal ones). So why would instinct be resolvable into intelligent elements? And why, for that matter, into terms that are intelligible at all?[241] Is it not obvious here that thinking in terms of what is intelligent, or of what is absolutely intelligible, is to return to the Aristotelian theory of nature? Perhaps turning back to Aristotle was better than coming to a dead stop before instinct, as if before an unfathomable mystery. And yet, despite not belonging to the intellectual domain, instinct is not situated beyond the limits of the mind [esprit]. In phenomena of feelings, in unreflective sympathies and antipathies, we experience in ourselves, in a much vaguer form and one that is also much too penetrated by intellect, something like what must take place in the consciousness of an insect acting by instinct.[242] Evolution simply separated from each other, so as to develop them both entirely, elements that were interpenetrating at the origin. More precisely, the intellect is above all the faculty of relating one point of space to another, one material object to another. It applies to all things, but remains outside of them, and when it comes to understanding a deeper cause, it only ever perceives its distribution into juxtaposed effects. Whatever the force might be that is translated into the genesis of the Caterpillar's nervous system, we will never grasp it (using our eyes and our intellect) as anything other than a juxtaposition of nerves and nervous centers. True, we thereby reach the entire external effect of that force. The Sphex, for its part, surely only grasps very little of this external effect, namely the aspects that interest it; but at least it grasps it from within [du dedans]—in a manner completely different from a knowledge procedure—through an intuition (lived rather than represented) that likely resembles what we call, in our own case, a divining sympathy [sympathie divinatrice].

It is a remarkable fact the way that scientific theories of instinct go back and forth between seeing it as intelligent and seeing it as merely intelligible, that is, between the assimilation of the instinct to a "fallen" intellect and the reduction of the instinct to a pure mechanism.[a][243] Each of these two systems of explanation triumphs in its critique of the other: the first when it shows that instinct cannot be a pure reflex, the second when it says that instinct is something other than intellect, even one that has fallen into unconsciousness.[244] What could this mean other than that they are two symbolic systems equally acceptable from certain angles and, from others, equally inadequate to their object? The concrete explanation, no longer scientific but rather metaphysical, must be sought along an entirely different pathway, no longer in the direction of the intellect but rather in the direction of "sympathy."

Instinct is sympathy. If this sympathy could spread out its object and also reflect upon itself, it would give us the key to vital operations—just as the intellect, fully developed and properly disciplined, introduces us into matter. For—and I cannot repeat this enough—intellect and instinct are turned in two opposite directions: the former toward inert matter and the latter toward life. The intellect, through the intermediary of science (which is its own creation), delivers the secret of physical operations to us more and more completely; when it comes to life, it only provides us with a translation of life into the language of inertia (and indeed does not claim to provide us with anything more). It circles all around it and, adopting as many external views of it as possible, draws this object toward itself rather than entering into it.[245] But intuition—by which I mean instinct that has become disinterested, self-conscious, and capable of reflecting upon and indefinitely enlarging its object—takes us into the very interior of life.[246]   178

The fact that an aesthetic faculty exists in humans alongside normal perception already demonstrates that an effort of this type is not impossible.[247] Our eye perceives the traits of the living being, but as juxtaposed alongside each other and not as organized among themselves. The intention of life, that simple movement that runs through its lines, that ties them together, and that gives them meaning, escapes our perception. This is the intention that the artist aims to recapture by placing himself

---

a   Bergson's note: See, in particular, the following recent studies: Albrecht Bethe, *Dürfen wir den Ameisen und Bienen psychische Qualitäten zuschreiben?* Special reprint from the *Archiv für die gesammte Physiologie* 70 (Bonn: Verlag von Emil Strauss, 1898); and Auguste Forel, "Un aperçu de psychologie comparée," *L'année psychologique* 2 (1895): 18–44.

back into the interior of the object through a kind of sympathy and by breaking down, through an effort of intuition, the barrier that space interposes between himself and the model.[248] True, this aesthetic intuition (like external perception) only attains what is individual.[249] But we can imagine an inquiry oriented in the same direction as art that would take life in general as its object, the way that physics, by following to its end the direction marked out by external perception, extends individual facts into general laws. Of course, this philosophy will probably never obtain a knowledge of its object comparable to the one that science has of its object. The intellect remains the illuminated core around which instinct, even when enlarged and refined into intuition, forms but a vague nebulosity. But in the absence of knowledge properly so-called, which is reserved for pure intellect alone, intuition will be able to help us grasp what is insufficient here in the givens of the intellect and allow us to catch sight of the means of completing them. On the one hand, intuition will make use of the very mechanism of the intellect to show how the intellectual frameworks no longer find their precise application here, and, on the other hand, through its own work, intuition will suggest at the very least a vague feeling of what must replace these intellectual frameworks. In this way, intuition will be able to bring the intellect to recognize that life fully enters neither into the category of the many nor into that of the one, and that neither mechanical causality nor finality offer a sufficient translation of the vital process.[250] Then, through the sympathetic communication that it will establish between us and the rest of the living beings, through the dilation that it will obtain of our consciousness, intuition will introduce us into the domain of life itself, which is reciprocal interpenetration and indefinitely continuous creation. But if intuition goes beyond intellect in this way, the intellect will have been the source of the disturbance that made it rise up to the place it has reached. Without the intellect, intuition would have remained in the form of instinct, riveted to the specific object that interests it practically, and thereby externalized by that object into movements of locomotion.

I will attempt to show how the theory of knowledge must account for these two faculties—intellect and instinct—and also how, for having failed to establish a clear enough distinction between intuition and the intellect, it gets itself into inextricable difficulties, creating phantoms of ideas to which phantoms of problems cling.[251] We will see that the problem of knowledge, considered from this angle, is the same as the metaphysical problem, and that both thereby have to do with experience.[252] On the one hand, if the intellect is in accord with matter and intuition with life, then they must both be squeezed so as to extract from them the quintessence of their object; as such, metaphysics will be dependent on the theory of knowledge. But, on the other hand, if consciousness is divided in this way into intuition and intellect, this is because of the necessity of applying itself to matter and simultaneously following the current of life. In this way, the splitting of consciousness would be the result of the double form of the real, and so the theory   180 of knowledge would be dependent upon metaphysics. In truth, each of these two lines of inquiry leads to the other: they form a circle, and the only center this circle can have is the empirical study of evolution. It is only by observing consciousness coursing through matter, losing itself there and finding itself there again, dividing itself and reconstituting itself, that we will form an idea of the opposition between these two terms and, perhaps, of their common origin.[253] Moreover, by basing ourselves on this opposition between the two elements and on their common origin, we will surely bring the meaning of evolution itself more clearly into relief.

*

[LIFE AND CONSCIOUSNESS. THE APPARENT PLACE OF MAN IN NATURE]

That will be my goal in the next chapter.[254] But the facts that I have just examined would already suggest the idea of connecting life either to consciousness itself or to something that resembles consciousness.

I claimed above that, across the whole extension of the animal kingdom, consciousness appears to be proportional to the power of choice available to the living being.[255] Consciousness lights up the zone of virtualities that surrounds the action. It measures the gap between what is being done and what could be done. Examined from the outside, one might well take consciousness for a simple auxiliary of action, for a light that action kindles, a fleeting spark that would shoot out from the friction between the real action and possible actions. But it must be noted that things would take place precisely in the same way if consciousness were the cause rather than being the effect.[256] Or we could suppose that consciousness in principle covers an enormous field even when it comes to the most rudimentary animal, but that in fact it is compressed in a sort of vice. By giving the organism a choice among a greater number of actions, each development in the nervous centers would launch an appeal to the virtualities capable of surrounding the real, thereby loosening the vice and allowing consciousness to pass through more freely. According to this second hypothesis, as with the first one, consciousness would clearly be the instrument of action; but here it would be even more true to say that action is the instrument of consciousness, since the increasing complication of action with itself and the gearing together of the action with other actions would be the only possible means for an imprisoned consciousness to free itself. How can we choose between these two hypotheses? If the first one was true, then consciousness would sketch out precisely, and at each moment, the state of the brain: there would be a strict parallelism (to the extent that this is intelligible) between the psychological state and the brain state.[257] By contrast, according to the second hypothesis, there would surely be solidarity and interdependence between the brain and consciousness, but not a parallelism: the more complex the brain becomes, thus augmenting the number of possible actions among which the organism can choose, the more consciousness must overflow its physical concomitant. In this way, the memory of one and the same spectacle that both a dog and a man witnessed will, if the perception of it was the same, probably modify their respective brains in the same way; and yet, the memory must be something completely different in a human's consciousness than in a dog's consciousness. For the dog, the memory will remain imprisoned by the perception. It will only be reawakened when an analogous perception comes along to recall it by reproducing the same spectacle and it will thereby appear through a recognition—which is more *enacted* than *thought*—of the current perception, much more than through a genuine rebirth of the memory itself. To the contrary, man is capable of recalling the memory at will, at any moment whatsoever and independently of his current perception. He is not limited to merely enacting his past life; he imagines it and dreams it.[258] Given that the local modification of the brain to which the memory is attached is the same in both cases, the psychological difference between the two memories cannot be justified by some difference of detail between the two cerebral mechanisms, but rather by the difference between the two brains taken globally. By connecting a larger number of mechanisms with each other, the more complex of the two will have permitted consciousness to escape from the grips of them taken separately and to thereby gain its independence. In a previous work, I have attempted to prove (through the study of the facts that bring the relation between the conscious state and the brain state the most into relief, namely the facts of normal and pathological recognition, and in particular the forms of aphasia) that things indeed take place in this way and that the second hypothesis is the one that must be adopted.[a] But reasoning itself surely could have foreseen this conclusion. Indeed, I have shown how the hypothesis of an equivalence between the brain state and the psychological state rests upon a self-contradictory postulate and upon a confusion between two incompatible symbolisms.[b/259]

The evolution of life, examined from this angle, takes on a clearer meaning, even if it cannot be subsumed under a genuine idea. Everything takes place as if a vast current of consciousness had penetrated into matter, bearing with it (like every consciousness) an enormous multiplicity of interpenetrating virtualities. This current drew matter into organic development [*organisation*], but as a result the movement of this current was infinitely slowed down and infinitely divided up. Indeed, on the one hand, this consciousness had to fall off to sleep, like the chrysalis in the cocoon where it develops its wings, and on the other hand, the multiple tendencies it contained were shared out among the diverging series of organisms that, for that matter, externalized these tendencies into movements rather than internalized them into representations. During the course of this evolution, while some fell more and more deeply into sleep, others awakened more and more fully, and the torpor of the former organisms aided the activity of the latter ones. But this awakening was able to take place in two different ways. Life—that is, consciousness launched through matter—fixed its attention either on its own movement or on the matter it was passing through.[260] In this

183

---

a   Bergson's note: *Matière et mémoire*, Chapters II and III.

b   Bergson's note: "Le paralogisme psycho-physiologique," *Revue de métaphysique* 12, no. 6 (November 1904): 895–908. [TN: This article later appeared as "Le cerveau et la pensée: Une illusion philosophique," in *L'énergie spirituelle*, 191–210; and in English as: "Brain and Thought: A Philosophical Illusion," in *Mind-Energy: Lectures and Essays*, 231–55.]

way, it oriented itself either in the direction of intuition or in the direction of the intellect. At first glance, intuition seems much preferable to intellect since life and consciousness remain within each other. But the spectacle of the evolution of living beings shows that intuition could not go very far. On the side of intuition, consciousness found itself compressed to such a point by its envelope that it had to contract intuition down into instinct, i.e., to embrace only that tiny portion of life that interested it; and indeed, only embracing it in the dark, touching it without hardly even seeing it. From this side, the horizon was immediately closed off. To the contrary, consciousness establishing itself as intellect, that is, focusing first on matter, seemed in this way to externalize itself in relation to itself; but, precisely because it is adapted to external objects, it is able to circulate among them, to get around the barriers that it encounters, and to increase its domain indefinitely. Moreover, once liberated, it can fold back within itself and reawaken the virtualities of intuition that lie dormant within it.

From this point of view, not only does consciousness appear to be the motor principle of evolution, but also, among conscious living beings themselves, man comes to occupy a privileged place. Between animals and man, there is no longer a difference of degree, but rather a difference of nature. This is the conclusion that will emerge in the next chapter. For now, let me show how the preceding analyses already suggest that conclusion.[261]

184   It is worth noting the extraordinary disproportion between the consequences of an invention and the invention itself.[262] I said above that the intellect is modeled upon matter and that it initially aims at fabrication.[263] But does it fabricate simply for the sake of fabricating, or rather, does it pursue—involuntarily or even unconsciously—something entirely different? Fabricating consists in giving a form to matter, in making it supple and in folding it, in converting it into an instrument in order to make oneself its master.[a] Humanity benefits greatly from this mastery, much more than from the material results of the particular invention itself. Even if we draw an immediate advantage from the fabricated object, like an intelligent animal might do, and if even this advantage is the only thing that the inventor himself was seeking, this advantage is hardly anything when compared to the new ideas and the new feelings

a   TN: ...*informer la matière*... Bergson's formulation here means literally "to give form" or "to impart form" to matter.

that the invention can cause to spring forth on all sides, as if the essential effect of invention was to raise us up above ourselves and thereby to enlarge our horizon.[a] Here the disproportion between the effect and the cause is so large that it is difficult to hold that the cause *produces* its effect. True, the cause *releases* the effect, assigning it its direction.[264] But ultimately everything takes place as if the primary aim of the grip that the intellect has on matter was to *allow something to pass through* that matter otherwise stops.

The same impression emerges from a comparison between the human brain and the brains of animals. At first, the difference seems to be merely one of volume and complexity. But, judging by their functioning, surely there must be something else as well. For the animal, the motor mechanisms that its brain is able to set up, or, in other words, the habits that its will contracts, have no other objective and no other effect than to accomplish the movements that are sketched out in these habits or stored up in these mechanisms. But in humans, the motor habit can have a second result, one that is incommensurable with the first. One motor habit can hold other motor habits in check and, by thus taking control of the mechanical functioning [l'*automatisme*], free consciousness. We know that language occupies vast territories of the human brain. The cerebral mechanisms that correspond to words have the peculiarity that they can be geared into other mechanisms, those, for example, that correspond to the things themselves, or again be geared into each other. During this time, consciousness, which would have otherwise been carried off and submerged in the accomplishment of the act, takes hold of itself and frees itself.[b]

Thus, the difference must be more radical than a superficial examination might have led us to believe. It is the difference that we would find between a mechanism that absorbs all of our attention and a mechanism that we can use distractedly. The original steam engine, such as

185

a TN: ...*nous hausser au-dessus de nous-mêmes*... This formulation resonates with Bergson's various uses of *se dépasser* ("to go beyond or surpass oneself") and *se transcender* ("to transcend oneself"). Cf. "surpassing the human condition" in the Index.

b Bergson's note: Shaler, a geologist I have already cited [see above, Bergson's note to page 134], makes the following excellent claim: "When, however, we come to man, we appear to find the old bondage of the mind to the body swept away; and the intellectual parts develop with extraordinary rapidity, while the frame remains essentially unchanged." Nathaniel Southgate Shaler, *The Interpretation of Nature* [1893] (Boston and New York: Houghton, Mifflin and Company, 1899), 187.

Newcomen had designed it, required the presence of a person exclusively tasked with turning the valves, whether to introduce steam into the cylinder or to inject the cold spray designed to bring about condensation. The story is told that a child employed in this work, and quite bored with having to do it, had the idea of tying cords between the handles of the valves and the machine's balancing beam. From then on, the machine itself opened and closed the valves: it functioned all by itself.[265] Now, an observer who compared the structure of this second machine with that of the first (and not paying attention to the two children tasked with monitoring each one), would have found but a slight difference of complexity between the two machines. Indeed, this is all we can see when we only look at the machines themselves. But if we glance at the two children, we see that one is absorbed in the task of monitoring while the other is free to amuse himself as he pleases, and that, from this angle, the difference between the two machines is radical in that the first holds the attention captive while the second sets it free. This is the kind of difference that I believe would be found between the animal brain and the human brain.

186

To summarize, if we wanted to express this in the language of finality, we would have to say that consciousness, after having been obligated (so as to free itself) to split organic nature [organisation] into two complementary parts—plants and animals—looked for an escape route in the double direction of instinct and intellect. It did not find this escape route in instinct, and, on the side of intellect, only obtained it through a sudden leap from animal to man.[266] As a result, ultimately man would be the raison d'être of the entire organic development [organisation] of life on our planet.[267] And yet this would only be a manner of speaking. In reality, there is but a certain current of existence and the antagonistic current; the entire evolution of life comes from this. Now we must focus on the opposition between these two currents. Perhaps we will thereby discover their common source.[268] In this way, we will likely penetrate into the most obscure regions of metaphysics. But since the two directions that we must follow are clearly marked out—in the intellect, on the one hand, and in instinct and intuition, on the other—we need not worry about getting lost. The spectacle of the evolution of life suggests a certain conception of knowledge and a certain metaphysics that reciprocally imply each other. Once they have been made clear, this metaphysics and this critique will in turn cast some light on evolution as a whole.

III

ON THE MEANING OF LIFE, THE ORDER OF NATURE, AND THE FORM OF THE INTELLECT

[RELATION OF THE PROBLEM OF LIFE TO THE PROBLEM OF KNOWLEDGE. PHILOSOPHICAL METHOD]

In the first chapter, I drew a demarcation line between the inorganic and the organic. But I also indicated that the dividing up of matter into inorganic bodies is relative to our senses and our intellect, and that matter, when considered as an undivided whole, must be a flux rather than a thing.[1] I was thereby preparing the way for bringing together the inert and the living.

Moreover, I demonstrated in the second chapter that the same opposition is found between the intellect and instinct, with instinct being tuned to certain determinations of life and the intellect being modeled upon the configuration of brute matter.[2] But I also suggested that instinct and intellect stand out against a single background that we could call, for lack of a better term, Consciousness in general, and that must be coextensive with universal life.[3] As such, I hinted at the possibility

DOI: 10.4324/9781315537818-4

of engendering the intellect by starting from the consciousness that envelops it.[a]

The moment has now come to attempt a genesis of the intellect at the same time as a genesis of bodies—two obviously correlative undertakings, if it is true that the main outlines of our intellect sketch out the general form of our action upon matter and that the detail of matter is modeled upon the exigencies of our action. Intellectuality and materiality would have been constituted, in their details, through reciprocal adaptation.[4] Both would have been derived from a larger and higher form of existence. We will have to return them to that higher form of existence in order to witness them emerging from it.

Such an attempt will appear, at first glance, to be far more reckless than even the most audacious speculations of the metaphysicians.[5] This attempt would claim to go further than psychology, further than the cosmogonies, and further than traditional metaphysics because psychology, cosmology, and metaphysics begin by taking the intellect for granted in its essential nature, whereas here it is a question of engendering it in its form and in its matter.[6] In reality, this undertaking is much more modest, as I will show below. But let me first say how it differs from the others.

To begin with psychology, it must not be believed that it *engenders* the intellect when it follows its progressive development throughout the animal series. Comparative psychology teaches that the more intelligent an animal is, the more it tends to reflect upon the actions by which it uses things and, in this way, the closer it is to man. But, they continue, the animal's actions already adopted, on their own, the principal lines of human action; they already picked out the same general directions in the material world as we do; and they relied upon the same objects linked together by the same relations, such that animal intellect—even though it does not form concepts, properly speaking—already moves within a conceptual atmosphere. Absorbed at each moment by the acts and attitudes that emerge from it, drawn by them into the outside

a TN: ...*engendrer l'intelligence*... Bergson uses *engendrer* ("to engender" or "to generate") throughout this section to mean to generate something, but also to show or to demonstrate the origin of something. This should be understood as in direct opposition to philosophical systems (and particularly Kantian philosophy) that take for granted in advance [*se donner par avance*] the existence of the intellect rather than explaining its genesis. See "genesis" in the Index.

188

world and thereby externalizing itself in relation to itself, animal intellect surely enacts or plays [joue] its ideas rather than thinking them. But at the very least, this game already sketches out the broad schema of human intellect.[a] Thus, to explain human intellect through animal intellect consists simply in developing an embryo of humanity into the human. Comparative psychologists claim to show how a certain direction has been followed further and further by increasingly intelligent beings. And yet, from the moment that they posit the direction, they take the intellect for granted.

In a cosmogony such as Spencer's, the intellect is again taken for granted, and by the same stroke, matter is as well. We are shown matter obeying laws, as well as objects relating to other objects and facts relating to other facts according to constant relations; receiving the imprint of these relations and these laws, consciousness thereby adopts the general configuration of nature and establishes itself as intellect. But is it not obvious that the intellect itself is presupposed the moment objects and facts are posited?[7] A priori, and leaving aside all theories regarding the essence of matter, it is obvious that the materiality of a body does not stop at the point at which we touch it—the materiality of a body is present wherever its influence is felt.[8] Now, to take just one example, a body's attractive force influences the Sun, the planets, and perhaps the universe as a whole.[9] Moreover, the more physics advances, the more it erases the individuality of bodies and even of the particles that the scientific imagination first divided these bodies up into—bodies and corpuscles tend to dissolve into a universal interaction.[10] Our perceptions give us the sketch of our possible action upon things much more than the sketch of the things themselves. The contours that we find in objects simply mark out those aspects of them that we can reach and modify. The lines that we see traced out through matter are the very same ones we are called to move along.[11] Contours and routes increasingly stood out as the action of consciousness upon matter was being prepared—in short, as the intellect was being constituted. It is doubtful that animals that are constructed according to a different plan than us, such as a Mollusk or an Insect, cut up matter along the same joints.[12] It is not even necessary that they divide matter up into bodies. To follow the

189

190

---

a   Bergson's note: I have developed this point in *Matière et mémoire*, Chapters II and III. See, in particular, 78–80 and 169–86. [TN: See, rather, *MM*, 86–88, 170–85.]

indications of instinct, there is no need to perceive *objects*; it is enough to distinguish *properties*. To the contrary, even in its most humble forms, the intellect already aspires to make it such that matter acts upon matter. If matter, from some angle, allows itself to be divided up into agents and patients, or more simply into coexisting and distinct fragments, then the intellect will look at it from that angle. And the more that it busies itself with dividing, the more it will deploy in space (in the form of extension being juxtaposed with extension) a matter that surely tends toward spatiality, but whose parts are still nevertheless in a state of entanglement and of reciprocal interpenetration.[13] As such, the same movement that leads the mind to establish itself as intellect—that is, to structure itself into distinct concepts—also leads matter to be broken up into objects that are clearly external to each other. *The more consciousness is intellectualized, the more matter is spatialized.* In other words, when evolutionary philosophy imagines matter as cut up in space according to the very lines that our action will follow, it takes for granted, as ready-made [*toute-faite*], the very intellect that it claimed to engender.

Metaphysics engages in a similar kind of work (although it is subtler and more self-aware) when it deduces, *a priori*, the categories of thought. The intellect is compressed, reduced down to its quintessence, and forced to hold to a principle so simple that it might be believed to be empty—then they draw from this principle what had first been put into it as potentiality [*en puissance*].[14] In this way, they definitely show the coherence of the intellect with itself, define it, and give its formula, but they do not at all trace out its genesis. Even if a project like Fichte's is more philosophical than Spencer's (insofar as it shows more respect for the genuine order of things), it hardly takes us any further.[15] Fichte takes thought in its concentrated state and dilates it into reality. Spencer begins from external reality and condenses it back down into intellect. But in both cases, one must begin by taking the intellect for granted: either condensed or spread out, either grasped in itself through a direct vision or seen reflected in nature, as if in a mirror.

The agreement among the majority of philosophers on this point comes from the fact that they all agree in affirming the unity of nature and in conceiving of this unity under an abstract and geometrical form. They do not see, or they do not want to see, the rupture [*la coupure*] between the organic and the inorganic.[16] Some begin from the inorganic and, by making it increasingly interconnected with itself, claim to reconstitute the living being; others first posit life and then make their way toward brute matter through a skillfully managed *decrescendo*. But they all hold that in nature there is nothing but differences of degree—differences of complexity on the first hypothesis, differences of intensity on the second.[17] Once this principle is adopted, the intellect becomes as vast as the real since it is incontestable that whatever there is of the geometrical in things is entirely accessible to the human intellect. And if there is perfect continuity between geometry and the rest of the real, then everything else becomes equally intelligible and equally intelligent.[18] Such is the basic premise of the majority of [philosophical] systems. Anyone would be easily convinced of this through a simple comparison of doctrines that seem to have no point of contact or no common measure between them, such as those of Fichte and Spencer—two names that chance has just caused us to bring together.

There are two (correlative and complementary) convictions at the foundation of these speculations: that nature is one, and that the function of the intellect is to embrace nature as a whole. Given that the faculty of knowing is assumed to be coextensive with the totality of experience, it can no longer be a question of engendering that faculty. They take that faculty for granted and make use of it, like one makes use of vision so as to survey the horizon. True, there will be differences of opinion over the value of the result of this use: for some, the intellect encompasses reality itself; for others, it only grasps the phantom of reality. But whether phantom or reality, they all assume that the intellect grasps the totality of what is graspable.

This explains the exaggerated confidence that philosophy has in the powers of the individual mind. Whether it be dogmatic or critical, whether it accepts the relativity of our knowledge or claims to take up a position within the absolute, a philosophy is generally the work of one philosopher—a unique and global view of the whole. We can either take it or leave it.

The more modest philosophy that I am advocating is also the only one capable of being completed and of being improved. The human intellect, such as I am picturing it, is completely different from the one that Plato presents in the Allegory of the Cave. The intellect's function is no more to watch empty shadows pass by than it is to turn itself around to contemplate the brilliant sun. It has something else to do entirely. Yoked like an ox to a difficult task, we feel the play of our muscles and our joints, the weight of the cart, and the resistance of the soil—the function of the intellect is to act and to act knowingly, to enter into contact with reality and even to live that reality, but only to the extent that that reality concerns the work that is being accomplished and the furrow that is being plowed.[19] Nevertheless, we are bathed in a beneficent fluid from which we draw the very force we need to work and live.[20] From this ocean of life in which we are immersed, we ceaselessly aspire toward something, and we sense that our being, or at least the intellect that guides it, is formed there through a sort of local solidification. Philosophy can be nothing other than the effort to dissolve back into this whole.[21] The intellect, being reabsorbed back into its own principle or source [principe], will relive its own genesis in reverse. But such an undertaking can no longer be accomplished in a single stroke; it will necessarily be collective and gradual. It will consist in an exchange of impressions that, correcting each other and being layered upon each other, will end up dilating humanity in us and making it such that humanity itself transcends itself.[22]

[APPARENT VICIOUS CIRCLE IN THE PROPOSED METHOD]

But this method has against it the most deep-rooted habits of the mind. It immediately gives rise to the idea of a vicious circle. It will be said that my claim to go further than the intellect is in vain, "for how could you do this if not with the intellect itself? Everything that is illuminated within your consciousness is already intellect. You are within your thought, and you will never get outside of it. You can say, if you wish, that the intellect is capable of progress, and that it will see an increasing number of things more and more clearly. But do not speak of engendering the intellect, for again, it is with your intellect itself that you would have to accomplish this genesis."[a]

This objection comes to mind quite naturally.[23] But by following a similar reasoning, one could just as well prove the impossibility of acquiring any new habit whatsoever. The essence of reasoning is to enclose us within the circle of the givens. But action breaks the circle. If

---

a   TN: I have inserted these quotation marks to clarify that Bergson here adopts the voice of his interlocutor. Elsewhere he indicates this stylistic gesture with long dashes.

you had never seen a man swim, you would perhaps say that swimming is impossible, given that in order to learn to swim, he must begin by holding himself afloat in the water, and consequently he must already know how to swim.[24] Reasoning will always keep me rooted on solid ground, so to speak. But if I just fearlessly throw myself into the water, I will at first keep myself afloat on the surface of the water as best as I can by struggling against it, and little by little I will adapt myself to this new environment—I will learn how to swim. So there is indeed a kind of theoretical absurdity in wanting to know other than through the intellect. And yet, if we boldly accept the risk, then action will perhaps cut the knot that reasoning had tied and that reasoning will never be able to untie.[25]

194

Moreover, this risk will seem to diminish to the extent that we adopt the point of view that I am suggesting. I have shown that although the intellect emerged from a larger reality, there has never been a clean break between the two—an indistinct fringe subsists around conceptual thought that evokes its origin.[26] Even more, I compared the intellect to a solid core that was formed by way of condensation.[27] This core does not differ radically from the fluid that envelops it. It will only be reabsorbed into it because it is made of the same substance. The person who throws himself into the water, having only ever known the resistance of solid ground, would immediately drown if he did not struggle against the fluidity of the new environment. He must cling to whatever solidity the water still offers him, so to speak. On this condition alone can we eventually adapt ourselves to the inconsistent nature of the fluid. The same is true for our thought when it decides to make the leap.[28]

And yet, it must leap; that is, it must leave behind its usual environment. Reason, reasoning about its own powers, will never be able to extend them, even though such an extension, once accomplished, will not at all appear unreasonable.[29] Execute, if you will, thousands and thousands of variations on the theme of walking, you will never draw from this a rule for swimming. But get into the water, and, once you have learned to swim, you will understand that the mechanism for swimming is connected to the mechanism for walking. Swimming is an extension of walking, but walking would have never introduced you to swimming. In the same way, you can speculate as intelligently as you wish upon the mechanism of the intellect, but this method will never allow you to go beyond it. You will obtain something that is

195

increasingly complex, but not something superior, nor even something simply different. It is necessary to rush in headlong and, through an act of will, drive the intellect out of its home.[a]

The vicious circle is thus merely apparent. By contrast, I believe that it is real when it comes to every other way of philosophizing. This is what I would like to now show, briefly, and even if only to prove that philosophy cannot and indeed must not accept the relationship established by pure intellectualism between the theory of knowledge and the theory of the known, that is, between metaphysics and science.

[ACTUAL VICIOUS CIRCLE IN THE OPPOSITE METHOD]

At first glance, it might seem prudent to leave the study of facts to the positive sciences. Physics and chemistry will be responsible for brute matter; the biological and psychological sciences will study the manifestations of life. As such, the philosopher's task is clearly circumscribed. The facts and the laws are handed over to him by the scientist, and then either he seeks to go beyond them in order to reach the deeper causes, or he believes it impossible to go further and he proves this through the analysis of scientific knowledge itself. But in both cases, he has the same respect for the facts and relationships—such as science transmits them to him—that is owed to something that has already been proven. He will place atop this knowledge a critique of the faculty of knowing and, if need be, a metaphysics as well. As for the knowledge itself, in its materiality, he takes that to be the business of science and not of philosophy.[30]

But how could one not see that this supposed division of labor ends up muddling and confusing everything? The metaphysics or critique that the philosopher reserves as his own task are in fact received, readymade, from the positive sciences as already contained in the descriptions and analyses that the philosopher had left entirely for the scientist to take care of. Because he did not want to intervene from the beginning in questions of fact, when it comes to questions of principle, the philosopher finds himself reduced to formulating purely and simply, and in more precise terms, the unconscious and therefore inconsistent metaphysics and critique already sketched out in the very attitude that

196

a TN: ...hors de chez elle. Bergson here continues the metaphor of the intellect "feeling at ease" or "at home" among solids.

science has toward reality.[31] We must not be misled by an apparent analogy between natural things and human things. We are not here in the judicial domain, where the description of a fact and the judgment of that fact are two distinct things, for the simple reason that there is, above the fact and independent of it, a law decreed by a legislator. Here the laws are internal to the facts and relative to the lines that had been followed in cutting up the real into distinct facts. The appearance of the object cannot be described without prejudging its inner nature and its organization. Form can no longer be fully separated from matter, and so the philosopher—who began by reserving questions of principle for philosophy, and who wanted in this way to place philosophy above the sciences, like a final Court of Appeal above the crown courts and the appellate courts—will be led, degree by degree, to make of it merely a court of record, charged at most with drawing up in more precise terms the sentences that are handed down to it as already irrevocably rendered.[32]

The positive sciences are indeed the work of pure intellect.[33] Now, whether one accepts or rejects my conception of the intellect, there is one point that everyone will grant, namely that the intellect feels most at ease in the presence of inorganic matter. The intellect increasingly takes advantage of this matter by way of its mechanical inventions, and mechanical inventions become easier for it the more mechanically it thinks matter. The intellect bears within itself, in the form of a natural logic, a latent geometrical way of thinking [un géométrisme] that increasingly frees itself to the extent that it penetrates further into the depths of inert matter.[34] It is tuned to this matter, and this is why the physics and the metaphysics of brute matter are so close to each other. Now, when the intellect begins to study life, it necessarily treats the living like the inert by applying the same old forms to this new object and by transporting into this new domain the habits that had been so successful in the previous one. And it is right to do so, for on this condition alone will the living offer to our action the same hold that inert matter did. But whatever truth we arrive at in this way becomes entirely relative to our faculty of action. It is no longer anything more than a symbolic truth. It cannot have the same value as physical truth, being merely an extension of physics to an object that we agree a priori to conceive of only according to its external appearance. Philosophy's duty would thus be to intervene actively here, to examine the living without any ulterior motives for

197

practical use by freeing itself from the forms and habits that are properly intellectual. Philosophy's proper object is to speculate, that is, to see—so its attitude toward the living cannot be the same as the attitude that science adopts. Science's only aim is action, and, only being able to act through the intermediary of inert matter, it conceives of the rest of reality according to this single aspect.[35] What will happen if philosophy leaves biological facts and psychological facts to the positive sciences alone, as it had justifiably done with physical facts? It will accept *a priori* a mechanistic conception of the whole of nature, that is, an unreflective and even unconscious conception born of our material needs. It will accept *a priori* the doctrine of the simple unity of knowledge and the abstract unity of nature.[36]

198

From that point forward, philosophy's fate is sealed. The philosopher's only choice is between a metaphysical dogmatism and a metaphysical skepticism, which both ultimately rest upon the same postulate and neither of which add anything to the positive sciences.[37] The philosopher can hypostatize either the unity of nature or the unity of science (which amounts to the same thing) into a being that will be nothing since it will do nothing, into an ineffectual God who will simply summarize in himself all of the given, into an eternal Matter from whose center will pour out the properties of things and the laws of nature, or finally, into a pure Form that would seek to grasp an ungraspable multiplicity and that will be, as they wish, either the form of nature or the form of thought.[38] All of these philosophies will say (in their various languages) that science is right to treat the living like the inert, and that there is absolutely no difference of value and no distinction to be made between the results that the intellect obtains by applying its categories, regardless of whether it focuses on inert matter or whether it takes on life.[39]

And yet in many cases, we sense that this framework is cracking.[40] But since the philosophers did not begin by distinguishing between the inert and the living (the inert is adapted in advance to the framework into which it is inserted, the living is incapable of being held there except through some convention that eliminates its essential nature from it), they are forced to be equally suspicious of all that the framework contains. Metaphysical dogmatism, which built up the artificial unity of science into an absolute, is now succeeded by a skepticism or a relativism that will universalize and extend to all of the results of science the artificial character of certain ones among them.[41] In this way, philosophy will forever oscillate between the doctrine that holds absolute reality to be unknowable and the one that, in the conception of reality that it gives us, says nothing more than what science said.[42] For having wanted to prevent all conflict between science and philosophy, they will have sacrificed philosophy without science having gained much at all. And for having claimed to avoid the apparent vicious circle that would consist in using the intellect to go beyond the intellect, they will end up spinning around in a very real circle, the one that consists in laboriously rediscovering a unity in metaphysics that they had posited a priori at the outset, a unity that they had accepted blindly and unconsciously by the simple fact that they abandoned all experience to science and all of the real to the pure understanding.

199

Let us begin, rather, by drawing a demarcation line between the inert and the living. We will discover that the inert enters quite naturally into the frameworks of the intellect, that the living only lends itself to them artificially, and that, as such, we must adopt a special attitude toward the living and no longer examine it through the eyes of the positive sciences. Philosophy thereby invades the domain of experience. It gets mixed up in things that hitherto were none of its concern. Science, the theory of knowledge, and metaphysics will all find themselves directed toward the same terrain.[43] At first, this will result in a certain confusion among them. At first, all three will believe that they have lost something. But all three will end up profiting from the encounter.

Scientific knowledge could indeed pride itself on the fact that its claims were given a uniform value across the entire domain of experience. And yet, precisely because all of them were placed on the same level, all of them ended up being stained by the same relativity.[44] This will no longer be the case when we have begun by making the distinction that, in my opinion, is necessary here. The understanding is at home in the domain of inert matter.[45] Human action is essentially accomplished upon this matter, and action, as I said above, cannot move in the unreal.[46] In this way—as long as we only consider physics in its general form, and not in the details of its realization—we can say that physics touches the absolute.[47] To the contrary, it is merely by accident—by chance or by convention, if you will—that science gets a hold on the living analogous to the one it has on brute matter. Here the application of the understanding's frameworks is no longer natural. I do not wish to say that this application is no longer legitimate, in the scientific sense of the word. If science

200

is to extend our action on things, and if we can only act by using inert matter as an instrument, then science can and must continue to treat the living in the same way it treated the inert. But it will need to be understood that the more the intellect plunges into the depths of life, the more the knowledge it offers becomes symbolic, i.e., relative to the contingencies of action. Thus, philosophy will have to follow science onto this new terrain so as to place atop scientific truth a different kind of knowledge, one that will be called metaphysical.[48] From then on, all of our knowledge, whether scientific or metaphysical, is heightened.[49] For we exist, we move, and we live in the absolute.[50] Perhaps the knowledge that we have of the absolute is incomplete, but it is neither external nor relative.[51] We reach being itself, in its depths, through the combined and progressive development of science and philosophy.

By thus renouncing the artificial unity that the understanding imposes upon nature from the outside, we will perhaps rediscover nature's true, inner, and living unity. For our effort to go beyond pure understanding introduces us into something larger, something out of which our understanding has been cut and from which it must have detached itself. And given that matter is modeled upon [se règle sur] the intellect and that they are clearly in agreement with each other, the one cannot be engendered without also bringing about the genesis of other.[a] An identical process must have simultaneously cut matter and the intellect out of a fabric that contained them both. To the extent that we endeavor more and more to transcend pure intellect, we place ourselves more and more completely within that larger reality.[52]

201

*

[ON THE POSSIBILITY OF A SIMULTANEOUS GENESIS OF MATTER AND OF THE INTELLECT]

Let us concentrate, then, on what we have that is at once the most detached from the outside and the least penetrated by intellectuality. Let us seek, in the depths of ourselves, the point at which we feel most

a TN: ...on ne peut engendrer l'une sans faire la genèse de l'autre. Here we see Bergson again using the vocabulary of "engender" in the double sense of both "bringing about" and "demonstrating the genesis of." In addition, se régler sur has several senses and could also be translated here as "matter is adapted to the intellect" or "is in harmony" with it.

fully within our own life. We thereby plunge back into pure durée, a durée where the past that is always on the move continuously swells with the addition of an absolutely new present. But at the same time, we feel the spring of our will being tensed to its extreme limit.[53] Through a violent contraction of our personality upon itself, we must gather together our past (which is slipping away) so as to force it, compacted and undivided, into a present that it will itself create by being introduced into it. Rare are the moments in which we recover possession of ourselves to this extent; these moments are none other than our truly free actions.[54] And even then, we never entirely take hold of ourselves. Our feeling of durée, that is, the coinciding of our myself with itself, admits of degrees.[a] But, the deeper the feeling and the more complete the coinciding, the more the life it places us back into reabsorbs intellectuality (by going beyond it). Since the essential function of the intellect is to link together the same with the same, only facts that can be repeated are entirely adaptable to the framework of the intellect.[55] Now, the intellect probably gets, après coup, some hold on real moments of real durée by reconstituting the new state from a series of views or snapshots taken of it from the outside and that resemble as much as possible what is already known. In this sense, the state contains intellectuality as a "potentiality" [en puissance], so to speak. And yet, the state, being indivisible and new, overflows and remains incommensurable with that series of views.[56]

Now let us relax and interrupt the effort that pushes the largest possible part of the past into the present.[57] Were the relaxation complete, there would no longer be any memory or will; we never fall into this absolute passivity, no more than can we ever make ourselves absolutely free. But, at the limit, we catch sight of an existence made up of a present that endlessly begins anew—no longer any real durée, nothing but the instantaneous that continuously dies and is reborn.[58] Is this the existence of matter? Not exactly—since although analysis surely resolves matter into elementary vibrations whose shortest lengths are of an extremely brief duration (indeed nearly vanishing), they are not nothing [nulle]. Nevertheless, we can presume that physical existence tends in

202

---

a   TN: ...la coïncidence de notre moi avec lui-même. It is worth noting here that Bergson uses moi ("myself") rather than soi ("self"), which I have opted to preserve in the English despite the awkwardness it introduces. I have added the italics here for readability. It is his notion of the personal or deep self that Bergson has in mind here, as discussed in his other works. Cf. E, 93.

this second direction, just like psychical existence tends in the previous direction.

At the foundation of "spirituality" on the one hand, and "materiality" along with intellectuality on the other, there would thus be two processes that move in opposite directions, and we go from the first to the second by means of inversion, perhaps even by a simple interruption, if it is true that inversion and interruption must here be understood as synonymous terms (as I will show in detail below).[59] This presumption will be confirmed if things are considered from the point of view of extension and no longer simply from that of durée.

The more we become conscious of our progression in pure durée, the more we sense the various parts of our being interpenetrate and our entire personality becoming concentrated in a point—or better, in a sharp edge or tip that inserts itself into the future by continuously cutting into it. This is what free life and free action consist in. But instead, suppose we let ourselves go; rather than acting, suppose we dream. Our *myself* is suddenly broken up into pieces: our past, which up until then had gathered itself together in the indivisible impulse that it communicated to us, decomposes into thousands and thousands of memories that become external to each other.[60] They refuse to inter-
203   penetrate to the extent that they become increasingly fixed. In this way, our personality descends again in the direction of space. Moreover, it constantly mingles with space through sensation. But I will not insist here upon a point that I have developed elsewhere in depth.[61] Let me simply recall that extension admits of degrees, that every sensation is extensive to some degree, and that the idea of unextended sensations, artificially localized in space, is merely a view adopted by the mind that is suggested much more by an unconscious metaphysics than by psychological observation.

Of course, even when we let ourselves relax as much as we can, we merely take the first steps in the direction of the extended. But imagine for a moment that matter consists in this very same movement, just pushed even farther, and that physical reality is simply an inverted psychical reality.[a] We would then understand why the mind feels so at ease and moves about so naturally in space from the moment that matter suggests to the mind a more distinct representation of space.[62] The mind already

---

a TN: *...que le physique soit simplement du psychique inverti.* Note that Bergson's formulation is somewhat more elliptical than my translation, which I have lightly expanded for clarity.

had the implicit representation of this space in the very feeling that it gained of its own potential *relaxation* [*détente*], that is, of its possible *extension*. The mind rediscovers this space among the things, but it could have obtained it without them if its imagination had been powerful enough to push the inversion of its own natural movement to the end.[63] On the other hand, we would thus be able to understand why matter further accentuates its materiality under the gaze of the mind. Matter started by helping the mind re-descend its own incline; it provided the impulse. But once it has been launched, the mind continues along. The mind's representation of pure space is merely the *schema* of the point where this movement will come to an end. Once the mind possesses the form of space, it uses it like a net whose meshing can be made and unmade at will, a net that, when thrown across matter, divides it up according to the needs of our action.[64] As such, the space of our geometry and the spatiality of things are mutually engendered by the reciprocal action and reaction of two terms of the same essence that march in opposite directions. Space is not as foreign to our nature as we imagine, and matter is not as completely extended in space as our intellect and our senses represent it.

204

I have addressed the first point elsewhere.[65] In terms of the second, I will limit myself to noting that perfect spatiality would consist in a perfect exteriority of the parts in relation to each other, i.e., in a complete and reciprocal independence. But there is no material point that does not act upon every other material point.[66] If it is observed that a thing is genuinely wherever it is acting, then we will be led to say (as Faraday did) that all atoms interpenetrate and that each one of them fills the world.[a/67] According to such a theory, the atom (or more generally the material point) becomes a simple view adopted by the mind, namely the view we arrive at if we continue long enough the work by which we subdivide matter into bodies (which is a process entirely relative to our faculty of action). Nevertheless, it is undeniable that matter lends itself to this subdivision, and that by assuming matter to be divisible into parts that are external to each other, we construct a science that is sufficiently representative of the real.[68] It is also undeniable that, even if no system is completely isolated, science still finds the means of cutting up

a Bergson's note: Michael Faraday, "A Speculation touching Electric Conduction and the Nature of Matter, [Letter to the Editor]," *The London, Edinburgh, and Dublin Philosophical Magazine and Journal of Science*, series 3, 24, no. 157 (January–June 1844): 136–44.

the universe into systems that are relatively independent of each other, and that science does not thereby commit any appreciable error. What could this mean, if not that matter *extends itself* into space without being absolutely *extended* in space, and that by taking matter to be decomposable into isolated systems—by attributing to it clearly distinct elements that change in relation to each other without themselves changing (that are "displaced," let us say, without being altered), in short, by conferring upon matter the properties of pure space—we thereby transport ourselves to the end of the movement whose direction is merely sketched out by matter?

205

What Kant's "Transcendental Aesthetic" appears to have definitively established is that extension is not a material attribute comparable to the others.[69] Reason will not continue working indefinitely on the notions of heat, color, or weight: to know the modalities of weight or of heat, it will be necessary to regain contact with experience. This is not the case for the notion of space. Supposing that space is empirically given through vision and touch (and Kant never contested this fact), the concept of space is remarkable in that the mind, speculating on it with nothing but its own powers, *a priori* cuts out of it the figures whose properties the mind itself will *a priori* determine: experience, with which the mind had not maintained contact, nevertheless follows along behind it, throughout the infinite complications of its reasonings, and invariably shows it to be correct. These are the facts. Kant simply put them out there in broad daylight. But I believe that their explanation must be sought along an entirely different path than the one Kant followed.

The intellect, such as Kant presents it, is bathed in an atmosphere of spatiality to which it is just as inseparably united as the living body is to the air that it breathes.[70] Our perceptions only reach us after having passed through this atmosphere. There they are impregnated, in advance, by our geometry, such that our faculty of thinking merely rediscovers in matter the mathematical properties that our faculty of perception had first deposited in it. As such, we are guaranteed to see matter obediently submit to our reasoning; but this matter, insofar as it is intelligible, is our own creation [œuvre]. When it comes to reality "in itself," we know

206

nothing and will never know anything, since we only grasp it as refracted through the forms of our faculty of perception. The moment we claim to affirm one thing, the opposite affirmation immediately surges forth and is equally demonstrable and equally plausible: the ideality of space, proven directly by the analysis of knowledge, is indirectly proven by the antinomies to which the opposite thesis leads. Such is the driving principle of Kant's critique. It inspired Kant to offer a peremptory refutation of so-called "empiricist" theories of knowledge. His refutation is, in my opinion, definitive in terms of what it denies. But does it provide, in what it affirms, the solution to the problem?

Kant's approach takes "space" for granted as a ready-made form of our faculty of perception—a veritable *deus ex machina* of which we see neither how it arises nor why it is what it is rather than something else entirely. It takes the "things in themselves" for granted, and then claims that we can know nothing about them—so by what right then does it affirm their existence, even if taken as "problematic"?[71] If this unknowable reality projects into our faculty of perception a sensible manifold, capable of being inserted into it quite precisely, is not this reality, by that very fact, partially known?[72] And, by gaining a deeper understanding of this insertion, will we not be led, at least on one point, to assume a preestablished harmony between the things and our mind—a lazy hypothesis that Kant was right to want to do without?[73] In essence, because he did not distinguish between degrees of spatiality, Kant had to take ready-made space for granted—which is the source of his question as to how the "sensible manifold" is adapted to space.[74] The same reason led him to believe that matter is entirely developed into parts absolutely external to each other, and from there the antinomies arise in which one can easily see that the thesis and the antithesis presuppose the perfect coincidence between matter and geometrical space. But the antinomies disappear the moment we stop extending to matter what is true of pure space.[75] Finally, this is also what leads Kant to the conclusion that there are three (and only three) alternatives for theory of knowledge to choose between: the mind adapts itself to the things, the things adapt themselves to the mind, or it is necessary to assume a mysterious harmony between the things and the mind.[76]

207

But the truth is that there is a fourth option, one that Kant does not appear to have thought of: first, because he did not think that the mind extends beyond the intellect; second (and this is basically the same thing), because he did not attribute an absolute existence to *durée*, having *a priori* placed time on the same line as space.[77] This fourth solution would begin by considering the intellect as a special function of the mind, one that is essentially turned toward inert matter. Next, this solution would consist in recognizing that matter does not determine the form of the intellect, nor does the intellect impose a form upon matter, nor have matter and the intellect been put into agreement with each other through some sort of preestablished harmony. Rather, it will say that the intellect and matter have been progressively adapted to each other in order to finally arrive at a common form. *Moreover, this adaptation would be accomplished quite naturally because it is the same inversion of the same movement that creates both the intellectuality of the mind and the materiality of the things.*

From this point of view, the knowledge of matter that we are given through perception, on the one hand, and through science, on the other, might well appear to be approximate, but does not appear to be relative.[78] Our perception, whose role is to illuminate our actions, accomplishes a cutting up of matter that will always be too clean-cut, subordinated to practical demands, and consequently always needing revision.[79] Our science, which aspires to mathematical form, unnecessarily over-accentuates the spatiality of matter; thus, its schemas will be in general too precise, and indeed always needing to be redone. For a scientific theory to be final, the mind would have to be able to embrace the totality of things as a whole and locate them in precise relation to each other. But in reality, we are obliged to pose our problems one by one, in terms that are provisional for that very reason, in such a way that the solution to each problem will be indefinitely corrected by the solution that we will give to the problems that follow, and such that science as a whole is relative to the contingent order in which each of the subsequent problems were posed. In this sense, and to this extent, science must be considered to be conventional, but this is, so to speak, a conventionality in fact and not in principle.[80] In principle, the positive sciences bear upon reality itself, so long as they do not step outside of their proper domain, which is that of inert matter.[81]

Scientific knowledge, imagined in this way, ascends to a higher plane.[82] But as such, the theory of knowledge becomes an infinitely difficult undertaking, one that is beyond the power of pure intellect. Indeed, it is no longer sufficient to determine the categories of thought through a careful analysis; it is now a question of engendering them. In terms of space, it would be necessary to follow, through a *sui generis* effort of the mind, the progression, or rather the regression of the extra-spatial demoting itself down into spatiality. By first placing ourselves as high as possible within our own consciousness so as to then let ourselves gradually fall, we clearly have the feeling that our *myself* extends out into inert memories that are external to each other rather than being tensed into an indivisible and active will.[83] But this is merely a beginning. By sketching out this downward movement, our consciousness shows us the direction and allows us to catch sight of the possibility of continuing that movement right to the end—but our consciousness itself never goes that far. But then again, if we consider matter, which at first appears to coincide with space, we find that the more our attention focuses on it, the more the parts that we said were juxtaposed actually enter into each other, each one being subject to the action of the whole that itself is thereby somehow present.[84] In this way, even though it spreads out in the direction of space, matter never fully arrives there. From this we can conclude that matter continues much farther the very same movement that consciousness was able to sketch out in its nascent state within us. Thus, we hold the two ends of the chain, even though we are not able to grasp the other links.[85] Will they forever escape our grasp? It should be remembered that philosophy, as I am defining it, has not yet become fully self-conscious. Physics understands its role well when it pushes matter in the direction of spatiality. But has metaphysics understood its role when it purely and simply followed on the heels of physics with the chimerical hope of going farther in the same direction? To the contrary, would its proper task not be to go back up the incline that physics descends, to carry matter back to its origins, and to gradually constitute a cosmology that would be, if I can speak in this way, an inverted psychology?[86] From this new point of view, everything that appears *positive* to the physicist and geometer would become an interruption or an inversion of the true positivity, which would then have to be defined in psychological terms.

[THE GEOMETRY INHERENT IN MATTER]

Of course, if we consider the admirable order of mathematics, the perfect harmony among the objects it deals with, the logic immanent to numbers and shapes, and our certainty—despite the diversity and the complexity of our reasonings on the same subject—of always reaching the same conclusion, then we will hesitate to see a system of negations in properties that seem so positive, that is, the absence rather than the presence of a true reality.[87] But it must not be forgotten that our intellect,

210   which notices this order and admires it, is itself oriented in the very direction of the movement that ends at the materiality and spatiality of its
object. The more the intellect introduces complication by analyzing its
object, the more complicated is the order that it finds there. And because
they are moving in the same direction as the intellect, this order and this
complication necessarily give it the impression of a positive reality.

When a poet reads his verses to me, I can take enough of an interest
in him to enter into his thought, to insert myself into his feelings, and
to relive the simple state that he had broken up into phrases and words.[88]
As such, I sympathize with his inspiration; I follow it along through a
continuous movement that is, like the inspiration itself, an undivided
act. Now, it is enough that I relax my attention, that I release the tension
[détende] of what in me was tensed [tendu], for the sounds (which until
then were drowned out by the meaning) to appear to me distinctly, one
by one, in their very materiality.[89] For this to take place, I do not need to
add anything; it is enough for me to take something away. The more I let
myself go, the more the successive sounds will become individualized.
Just as the phrases were first decomposed into words, now the words
will become articulated into the separate syllables that I will perceive
one by one. Suppose we go even further in the direction of dreams: the
letters will separate from each other and I will see them marching by,
intertwined, upon some imaginary piece of paper. Here I will admire
the precision of their intertwinings, the marvelous order of their procession, and the precise insertion of the letters into syllables, the syllables
into words, and the words into sentences. The further I have gone in the
entirely negative direction of relaxation, the more I will have created
extension and complication; in turn, the more the complexity increases,
the more admirable the order that continues to reign undisturbed
among the elements will appear to be. Nevertheless, this complexity and
this extension do not represent anything positive: they express rather a
deficiency of will. And moreover, it is surely necessary that the order
increase along with the complication since it is nothing but an aspect
211   of it. The more we see, symbolically, parts within an indivisible whole,
the more the number of relations that the parts have among themselves
necessarily increases, since the same undividedness of the real whole
continues to hang over the growing multiplicity of symbolic elements
into which the scattering of the attention had decomposed it. A comparison of this kind will help us understand, to a certain extent, how

the same suppression of positive reality, the same inversion of an original movement can create all at once extension in space and the admirable order that our mathematics discovers there.[90] Of course, there is a difference between these two cases: the words and letters were invented by humanity through a positive effort, whereas space springs forth automatically, just as the remainder of a subtraction springs forth as soon as the first two terms are posited.[a/91] But in both cases, the infinite complication of the parts and their perfect coordination among themselves are created in the same stroke by an inversion that is, at root, an interruption—a diminution of positive reality.

[THE ESSENTIAL FUNCTIONS OF THE INTELLECT]

All of the operations of our intellect tend toward geometry, as if toward the endpoint where they will find their perfect completion. And yet, given that geometry is necessarily anterior to them (since these intellectual operations never fully reconstruct space and can do nothing other than take space for granted), it is obvious that a latent geometry, immanent in our idea of space, is the great mainspring of our intellect and sets it to work. We will be convinced of this by examining the two essential functions of the intellect: the faculty of deduction and the faculty of induction.

212

Let us begin with deduction. The same movement by which I trace a figure in space engenders its properties. These properties are visible and

---

a   Bergson's note: My comparison here simply develops the content of the term λογος ["word," "reason," "discourse"], such as Plotinus understands it. For on the one hand, Plotinus takes the λογος to be a generative and formative power, an aspect or a fragment of the ψυχή ["soul"]; on the other hand, he sometimes speaks of it as a *discourse*. More generally, the relationship that I am establishing in the present chapter between "extension" and "distension" [*distension*] is in some ways similar to the one posited by Plotinus (in passages that must have inspired Ravaisson) when he makes the extended, perhaps not an inversion of the original Being, but at least a weakening of its essence and one of the last stages of the procession. (See in particular, Plotinus, *The Enneads*, 4.3.9–11 and 3.6.17–18). And yet, ancient philosophy did not see what consequences followed from this in terms of mathematics, since Plotinus, like Plato before him, erected mathematical essences into absolute realities. Above all, ancient philosophy allowed itself to be tricked by the entirely external analogy between *durée* and extension. It treated *durée* like it treated extension, taking change to be a degradation of immutability, and the sensible to be a fall from the intelligible. As a result, as I will show in the next chapter, they developed a philosophy that misunderstood the real function and the real significance of the intellect.

tangible in this very movement, and it is in space that I sense and live the relation between the definition and its consequences, between the premises and the conclusion.[92] For all the other concepts whose ideas are suggested to me by experience, they can only be partially reconstituted *a priori*; the definition of those concepts will thereby be imperfect, and the deductions that include them—as rigorously as we link the conclusion to the premises—will participate in this imperfection. But when I trace out a rough sketch of a triangle in the sand, and when I begin to form the two angles at the base, I know with certainty and I understand absolutely that if these two angles are equal, then the sides will be as well, and that the figure will thus be able to be flipped over on itself without anything therein being changed. This is something I know long before having learned geometry. As such, prior to scientific geometry, there is a natural geometry whose clarity and whose self-evidence go beyond those of other deductions.[93] These other deductions have to do with qualities, and no longer with magnitudes. Thus, they are likely to have been formed on the model of the geometrical ones, and they must borrow their force from the fact that, from beneath the qualities, we see magnitude shining through in a confused way.[94] Note that problems of situation and of magnitude are the very first ones posed to our activity—they are the ones that the intellect (when externalized in action) resolves even before reflective intellect has appeared. The primitive person understands better than a civilized person how to evaluate distances, how to determine a direction, how to retrace in memory the often-complex pathway that he had followed, and to thereby follow a straight line back to his point of departure.[a/95] If the animal does not explicitly deduce anything, if it does not explicitly form concepts, then it does not have an idea of homogeneous space. You cannot take this space for granted without introducing, by the same stroke, a virtual geometry that will be reduced down, all by itself, into logic.[96] All of the repugnance that philosophers feel toward picturing things from this angle comes from the fact that the logical work of the intellect represents to them a positive effort of the mind or spirit [*esprit*]. But if spirituality is understood as a march forward toward ever new creations, toward

213

a   Bergson's note: H. Charlton Bastian, *Le cerveau: Organe de la pensée chez l'homme et chez les animaux*, vol. 1, *Les animaux* (Paris: Germer Baillière, 1882): 166–70. [The English original is: *The Brain as an Organ of Mind* (London: C. Kegan Paul & Co, 1880).]

conclusions that are incommensurate with the premises and that cannot be determined in relation to them, then it will have to be admitted that a thought that moves among relations of necessary determination, or through premises that contain their conclusion in advance, follows the opposite direction—the direction of materiality.[97] What appears to be an effort from the point of view of the intellect is actually in itself a giving up [un abandon]. From the point of view of the intellect, making geometry emerge automatically from space, or logic from geometry, is to beg the question; on the contrary, if space is the ultimate endpoint of the mind's movement of relaxation [détente], then we cannot take space for granted without thereby first positing logic and geometry, which are found along the pathway whose endpoint is pure spatial intuition.

It has not been noted often enough just how little deduction is able to accomplish in the psychological and moral sciences.[a/98] Here, beginning from a proposition that has been verified by the facts, one can only draw verifiable consequences to a certain extent or up to a certain point. An appeal to good sense, that is, to the continuous experience of the real, must soon be made in order to soften the consequences that have been deduced and to bend them back toward the sinuosities of life.[99] When it comes to moral or spiritual things [les choses morales], deduction only succeeds metaphorically, so to speak, and only to the exact measure in which the moral or spiritual is transposable into the physical, by which I mean translatable into spatial symbols. The metaphor never works for very long, no longer than the curve merges with its own tangent. How can we not be struck by what is strange, and even paradoxical, in this weakness in deduction? Here we have a pure operation of the mind [esprit] being accomplished through the force of the mind alone. It seems that if this operation should feel at home

214

a TN: In these lines, Bergson's use of the terms *sciences morales* and *le morale* are difficult to translate. Although *la morale* (the feminine noun) in French is relatively close to "ethics," "morality," or "moral code" in English, the masculine *le morale* refers more to a "state of mind," "morale," or "spirit" (as in, he is in "good spirits"). Moreover, during Bergson's era, *les sciences morales* loosely grouped together the human, psychological, ethical, and spiritual sciences, with this last term not necessarily being limited to religious concerns. Recall that *esprit* can be translated as either "mind" or "spirit." Nevertheless, translating *le morale* by "spiritual" would perhaps be misleading. I have thus maintained "moral" in the English, sometimes adding "and spiritual" to remind the reader to take "moral" in Bergson's broader sense of "human, spiritual, and psychological," and so including but not limited to ethical or moral questions.

and evolve with ease anywhere, it would be among mental or spiritual things, in the domain of the mind or spirit. But this is not at all the case, for that is precisely where deduction is immediately at the end of its powers.[a] On the contrary, in geometry, astronomy, and physics—where we are dealing with external things—deduction is all-powerful! In these physical sciences, observation and experience are probably necessary in order to arrive at the principle, i.e., to discover the aspect under which it was necessary to see things. But, strictly speaking, with some luck we might have immediately discovered that principle; and, as soon as we possess it, we deduce from it quite distant consequences that experience will always verify. What should we conclude from this, if not that deduction is an operation governed by the workings of matter, modeled upon the mobile articulations of matter, and, finally, implicitly given along with the space that underpins matter? Insofar as it rolls along in space or in spatialized time, deduction has nothing to do but to let itself go. It is durée that sticks a spoke in its wheels.

Deduction does not work, then, without some hidden idea of spatial intuition in the background. But the same could be said of induction. Of course, it is not necessary to think like a geometer, nor even to think at all, in order to expect a repetition of the same fact from the same conditions. Animal consciousness already does this work. Indeed, the living body itself, independently of all consciousness, is already constructed so as to extract the similarities that interest it from the successive situations it finds itself in and to thereby respond to the stimuli with appropriate reactions.[100] But mechanical anticipation and mechanical reaction, which take place in the body, are a far cry from induction properly so called, which is an intellectual operation. Induction is based on the belief that there are causes and effects, and that the same effects follow from the same causes.[101] Now, if we examine this double belief in more depth, here is what we find. First, it implies that reality can be decomposed into groups that can, for practical purposes, be taken as isolated and independent. If I bring water to boil in a saucepan placed on a burner, the operation and the objects that support it are, in reality, interdependent with an immense number of other objects and other

215

---

a   TN: ...au bout de son rouleau. This expression means to be "exhausted" or "out of resources." Since Bergson expresses the opposite situation in terms of "power" in the next line, I have chosen "at the end of its powers" to capture the idea.

operations—we would discover, step by step, that our solar system as a whole is implicated in what takes place at this point in space.[102] But to a certain extent, and for the particular goal I am pursuing, I can acknowledge that things take place as if the group "water-saucepan-hot burner" were an independent microcosm. That is the first claim I will make. Now, when I say that this microcosm will always behave in the same way and that the heat will necessarily cause the water to boil after a certain amount of time, I admit that here it is enough to take for granted a certain number of elements of the system for the system to be complete—the system automatically completes itself and I am not free to complete it in thought however I please.[a] If we posit the hot stove, the saucepan, and the water, along with a certain interval of *durée*, the boiling (which experience showed me yesterday to be what was missing from the system for it to be complete) will complete the system again tomorrow, or at any moment whatsoever, now and forever. What is the foundation of this belief? It should be noted that, depending on the situation, this belief admits of degrees of confidence, and when the microcosm under consideration only contains magnitudes, the belief takes on the character of an absolute certainty. Indeed, if I posit two numbers, I am no longer free to choose the difference between them. If I posit two sides of a triangle along with the angle between them, the third side springs forth all by itself and the triangle automatically completes itself. I can, anywhere and at any time, trace out the same two sides with the same angle, and the new triangles thus formed will obviously be able to be superposed upon the first—consequently, the same third side will have again come to complete the system. Now, if I have a "perfect" certainty when I am reasoning over pure spatial determinations, must I not assume that in other cases my certainty will be more or less perfect to the extent that they approach this limit case? Indeed, would it not be the limit case itself that shines through all the others[b/103] and that colors them, according to their degree of transparency, with a

216

---

a TN: *...pour que le système soit complet: il se complète automatiquement...* In this passage, Bergson uses the adjective *complet* (and the verb *se compléter*) in a way that resonates with his use of "closed" elsewhere. Just below, Bergson describes how, when two sides have already been posited, the triangle completes itself, as in "closes" its shape and becomes a complete (or closed) system.

b Bergson's note: I have developed this point in a previous work. See *Essai sur les données immédiates de la conscience*, 155–60 [TN: See, rather, *E*, 153–58].

more or less marked hue of geometric necessity? In fact, when I say that the water I place on the hot stove today will boil just like it did yesterday, and that this is an absolute necessity, I vaguely sense that my imagination transposes today's hot stove onto yesterday's, today's saucepan onto yesterday's, today's water onto yesterday's, the durée that flows by today onto the durée that flowed by yesterday, and that thereby the rest seems to necessarily coincide as well—all for the same reason that makes it such that the third sides of two superposed triangles will coincide if the first two already coincide with each other. But my imagination only proceeds in this way because it closes its eyes to two essential points. In order for today's system to be able to be superposed upon yesterday's system, it would be necessary that yesterday's system had waited for today's, that time came to a halt and that everything had become simultaneous with everything else. This is what takes place in geometry, but only in geometry. Induction thus implies, first, that in the physicist's world (like in the geometer's world), time does not count. But it also implies that qualities can be superposed on each other like magnitudes.[104] If I imaginatively transpose today's hot stove onto yesterday's, I certainly observe that its form remains the same; for this it is enough that the surfaces and the edges coincide. But what would the coincidence of two qualities be? And how could we superpose one quality upon the other to confirm that they are indeed identical? Nevertheless, I extend to this second order of reality everything that applies to the first. The physicist will later justify this operation by reducing, as much as possible, differences of quality to differences of magnitude. But, before any science at all, I am already inclined to assimilate qualities to quantities, as if I glimpsed a geometrical mechanism shining from behind the qualities.[a] The more complete this shining through, the more the repetition of the same fact under the same conditions will appear to me to be necessary. Our inductions are, for us, more or less certain to the precise extent to which we dissolve qualitative differences into the homogeneity of the space that subtends them, such that geometry is the ideal limit of our inductions as much as of our deductions. The movement that ends in spatiality deposits along its path the faculty of induction and the faculty of deduction, that is, intellectuality as a whole.

217

---

a   Bergson's note: *Essai sur les données immédiates de la conscience*, Chapters I and III, *passim*.

This movement creates these faculties in the mind. But it also creates the "order" that our induction (with the help of deduction) discovers in the things. This order, on which our action relies and where our intellect recognizes itself, seems marvelous to us. Not only do the same general causes always produce the same overall effects, but beneath the visible causes and effects, our science also discovers an infinity of infinitesimally small changes that are fit more and more precisely into each other the further the analysis is pushed—to the point that at the end of this analysis, matter would seem to be geometry itself.[105] True, here the intellect has every right to admire the growing order within the growing complexity; the order and complexity both have a positive reality for the intellect, since they are in the same direction as it. But things take on a different appearance when we consider the whole of reality as an undivided march toward successive creations. We would then discern that the complexity of material elements and the mathematical order that ties them together must automatically spring forth when a partial interruption or a partial inversion takes place within the whole.[106] Moreover, since the intellect is cut out of the mind or spirit [esprit] through a process of the same kind, it is tuned to this order and this complexity, and it admires them because it recognizes itself in them.[107] And yet, what is in itself admirable, and what really deserves to cause us wonder, is the continuously renewed creation accomplished by undivided reality as a whole as it advances. After all, no complication of the mathematical order with itself, as sophisticated as one imagines it to be, will introduce one atom of novelty into the world, whereas, once this power for creation is posited (and it does exist since we become aware of it in ourselves, at the very least, when we act freely), it need only become distracted from itself so as to relax [se détendre], to relax so as to extend out [s'étendre], and to extend out so that the mathematical order that presides over the arrangement of the elements thus distinguished, along with the rigid determinism that ties them together, will manifest the interruption of the creative act. Moreover, these elements and their order are nothing but this interruption itself.[108]

The particular laws of the physical world express this entirely negative tendency. Not one of them, taken individually, has any objective reality; each one is the work of a scientist who considered things from a certain angle, isolated certain variables, and applied certain conventional units of measurement.[109] And nevertheless, there is an approximately

218

219

mathematical order immanent to matter, an objective order to which our science gets closer and closer as it advances.[110] For if matter is a relaxation of the unextended into the extended, and thereby of freedom into necessity, then although matter might not fully coincide with pure homogeneous space, it is nonetheless constituted by the movement that leads there, and so is on the road to geometry. True, mathematically formulated laws will never completely correspond to matter. For that to be the case, matter would have to be pure space, and it would have to fully exit durée.

It is impossible to overemphasize what is artificial in the mathematical formulation of a physical law, and consequently what is artificial in our scientific knowledge of things.[a/111] Our units of measurement are conventional and strangers to the intentions of nature, if we can speak in this way. How could we ever assume that nature related all of the modalities of heat to the dilations of a single mass of mercury or to the changes of pressure of a single mass of air maintained at a constant volume? But even this is not saying enough. In general, *measuring* is an entirely human operation, which implies either a real or imaginative superposition of one object on another a certain number of times.[112] Nature did not dream up this superposition. Nature does not measure, nor does it count. And yet physics counts, measures, and relates "quantitative" variations together so as to obtain laws—and it is successful in doing so. Its success would be inexplicable if the constitutive movement of materiality was not the very same movement that—prolonged by us to its endpoint, i.e., to homogeneous space—ends up making us count, measure, and follow, in all of their respective variations, terms that are functions of one another. Moreover, to bring about this prolongation, our intellect need only prolong itself since it naturally goes toward space and toward mathematics, intellectuality and materiality being of the same nature and being produced in the same way.

Now, if the mathematical order were something positive, if there were laws comparable to our codes immanent to matter, then the success of our science would be something of a miracle. What chance would we have of actually discovering nature's standard and of isolating (so as to determine their reciprocal relations) precisely the variables that nature

220

---

a Bergson's note: Here I am alluding, above all, to the profound studies published by Édouard Le Roy in *Revue de métaphysique et de morale*.

would have chosen? But the success of a science in mathematical form would be no less incomprehensible if matter did not have all that was necessary for it to enter into our frameworks. Thus, only one hypothesis remains plausible: that mathematical order has nothing positive about it, that it is the form toward which a certain *interruption* tends of its own accord, and that materiality itself consists precisely in an interruption of this kind. It will thereby be understood that our science is contingent, relative to the variables that it itself has chosen, and relative to the order in which it has successively posed its questions, and that nevertheless it is successful.[113] It might have been completely different as a whole and yet still successful. This is precisely because there is no definite system of mathematical laws at the base of nature, and that mathematics in general represents simply the direction in which matter falls back [*retombe*]. You can place one of those little cork dolls whose feet are made of lead in any position whatsoever—lay it down, turn it on its head, toss it into the air—and it will always automatically return itself to its standing position. The same is true of matter: we can take it from any end and manipulate it in any way whatsoever, and it will always fall back into one of our mathematical frameworks because it is weighed down with geometry.[114]

221

*

[SKETCH OF A THEORY OF KNOWLEDGE BASED ON THE IDEA OF *DISORDER*]

But the philosopher will perhaps refuse to base a theory of knowledge on such considerations. He will find it repugnant because mathematical order, being a sort of order, will seem to him to contain something positive. We repeat in vain that this order is produced automatically by the interruption of the inverse order, that it is this very interruption itself. For nevertheless the idea still remains that *there might be no order at all*, and that the mathematical order of things, being a conquest over disorder, possesses a positive reality. By carefully examining this point, we will see the key role that the idea of *disorder* plays in the problems relative to the theory of knowledge.[115] The idea does not appear explicitly there, and that is why no one pays much attention to it. And yet, a theory of knowledge should begin from the critique of this idea, for if the main problem is to know why and how reality is submitted to an order, this is because the absence of every kind of order seems possible or conceivable. The realist and the idealist both believe themselves to be thinking about this absence of order: the realist when he speaks of the regulations that the "objective" laws actually impose upon a possible disorder in nature; the idealist when he presupposes a "sensible manifold" that would be coordinated—and so being at first without any order—under the organizing influence of our understanding.[116] The idea of disorder, understood in the sense of an *absence of order*, is thus the idea that must be analyzed first. Philosophy borrows it from everyday life. And it is undeniable that when in common parlance we speak of disorder, we are thinking about something. But what exactly are we thinking about?

222

We will see in the next chapter just how difficult it is to determine the content of a negative idea, as well as the illusions we open ourselves up to and the inextricable difficulties into which philosophy falls for not having undertaken this work.[117] Difficulties and illusions are normally the result of having accepted as final a manner of expressing oneself that is essentially provisional. They are the result of having transported a procedure designed for practical purposes into the speculative domain.[118] If I select at random a volume from my bookshelf, I might glance at it and return it to the shelf while saying: "These are not poetic verses."[a] But is that really what I saw while flipping through the book? Clearly not. I did not see, nor will I ever see, an absence of poetic verses. I saw prose. But since it is poetry that I want, I express what I find as a function of what I am seeking. Rather than saying "this is prose," I say "these are not poetic verses." Inversely, if I am suddenly inspired to read some prose but land upon a volume of poetry, I will exclaim: "This is not prose!" I thereby translate the perceptual givens (which show me poetic verses) into the language of my expectations and my attention (which are fixed upon the idea of prose and do not want to hear of anything else). Now, if Monsieur Jourdain were listening to me, he would surely infer from my two exclamations that prose and poetry are two forms of language reserved for books, and that these learned forms are superposed upon

---

a   TN: "*Ce ne sont pas des vers.*" Bergson here uses *vers* ("lines of poetry" or simply "verses") in opposition to *prose*. To help mark this distinction in English, I have added the adjective "poetic" where appropriate. Bergson's example is of course inspired by a famous scene from Molière's *Le bourgeois gentilhomme*, as Bergson makes clear below by naming the hapless character Monsieur Jourdain.

a brute language that is neither prose nor poetry. Speaking of this thing that is neither poetry nor prose, Monsieur Jourdain would also believe that he is thinking of that very thing. Nevertheless, this would be merely a pseudo-idea. But we can go even further: if Monsieur Jourdain were to ask his philosophy professor how the prose-form and the poetry-form are added onto something that possessed neither the one nor the other, or if he wanted to be given the theory, as it were, of the imposition of these two forms upon this simple matter, then this pseudo-idea   223 would create a pseudo-problem. His question would be absurd, and the absurdity would come from the fact that he will have hypostatized the simultaneous negation of poetry and prose into a common substrate of the two, forgetting that the negation of the one consists in the positing of the other.

Now, imagine that there are two species of order and that these two orders are opposites within one single genus. Imagine as well that the idea of disorder springs forth in our mind each time we encounter one species of order when we were seeking the other. The idea of disorder would thus have a clear meaning in the everyday practice of life. It would objectify, for the convenience of language, the feeling of disappointment that a mind has when it finds itself before an order different from the one it needs, an order for which it has no use and that, in this sense, does not exist for it. But this idea of disorder would not include any theoretical application. And now if we claim, despite all that, to introduce the idea of disorder into philosophy, we inevitably lose sight of its true meaning. The idea of disorder marked the absence of a certain order, but *to the benefit of another order* (which was of no interest to us). Only, since this idea is applied to each order in turn, and even though it continuously goes back and forth between the two, we take hold of it along its route, or rather, while it is up in the air, like the shuttlecock somewhere between the two rackets. We treat it as if it no longer indifferently represented either the absence of one order or the other, but rather the absence of both—something that is neither perceived nor conceived, something that is a mere verbal entity.[119] This was the birth of the problem of knowing how order imposes itself upon disorder, and of how form imposes itself upon matter.[120] By analyzing this subtilized idea of disorder, we would see that it does not represent anything at all; at the same time, the problems that have been raised around it would disappear.

True, we had to begin by distinguishing, even by presenting as opposites, two kinds of order that are normally mixed together. Given that this confusion created the main difficulties we find in the problem of knowledge, it will be quite useful to insist once again upon the traits by which the two kinds of order are distinguished from each other.

In general, reality is *ordered* [*ordonnée*] to the exact measure that it satisfies our thought. Order is thus a certain harmony between the subject and the object. Order is the mind or spirit [*esprit*] finding itself in things. But as I mentioned above, the mind can march in two opposite directions.[121] Sometimes it follows its natural direction: here it is a progression in the form of tension, continuous creation, and free action. Sometimes it inverts this direction, and this inversion, taken to its extreme, would lead to extension, the necessarily reciprocal determination of elements externalized in relation to each other, and ultimately geometrical mechanism. Now, whether experience appears to adopt the first direction, or whether it orients itself in the second, in both cases we say that there is order because the mind finds itself in both of these processes. The confusion between them is thus quite natural. In order to avoid this confusion, two different names would have had to have been given to the two kinds of order, and this is not easy given the variety and variability of forms that they adopt. The second kind of order can be defined by geometry, which is its extreme limit. But more generally, this is the kind of order in question every time we find a relation of necessary determination between causes and effects.[122] It evokes ideas of inertia, passivity, and automatism. As for the first kind of order, it surely oscillates around finality, but it would be impossible to define it by this because it is sometimes above and sometimes below finalism. In its highest forms, it is more than finality since we can say of a free action or of a work of art that they manifest a perfect order; nevertheless, free actions and works of art can only be expressed *après coup* and approximatively in the language of ideas.[123] Life as a whole, conceived of as a creative evolution, is something analogous: it transcends finality, if one understands finality to be the realization of an idea conceived, or at least conceivable, in advance.[124] The framework of finality is thus too narrow for life as a whole. But then again, this framework is often too broad for some particular expression of life taken separately.[125] In any case, we are still dealing with something *vital* here, and the entire present

study aims to establish that the vital is in the direction of the voluntary.[a]
One could thus say that this first kind of order is that of the *vital* or of
the *willed* [*voulu*], in opposition to the second, which is that of the *inert*
and of the *automatic*. Moreover, common sense instinctively distinguishes
between the two kinds of order, at least in extreme cases; it also instinctively brings them together. We might say that astronomical phenomena
manifest an admirable order, meaning thereby that they can be foreseen
mathematically. And, we will find an order that is no less admirable in
one of Beethoven's symphonies, which is the order of genius, of originality, and consequently of unforeseeability itself.[126]

And yet, the first kind of order only takes on such a distinct form as
an exception. In general, it presents itself through characteristics that
we have every interest in merging together with those of the opposite
order. It is quite certain, for example, that if we imagined the evolution
of life as a whole, we would be forced to pay attention to the spontaneity of its movement and the unforeseeable nature of its activities. But
what we encounter in our everyday experience is some particular determinate living being, some specific expressions of life, which *more or less*
repeat already known forms and facts. Indeed, the similarity of structure that we observe between that which engenders and that which is
engendered—a similarity that allows us to enclose an indefinite number
of living individuals within the same group—is for us the classical type
of the *generic*, with the inorganic genera seeming to take the living genera   226
as their model.[127] This is how the vital order, such as it is offered to us
in the experience that divides it up, presents the same character and
accomplishes the same function as the physical order: both cause our
experience to *repeat itself* and both allow our mind to *generalize*. In reality,
this character has completely different origins in these two cases, and
even opposite meanings. In the second case, geometrical necessity (in
virtue of which the same components give an identical outcome) is the
classical type, the ideal limit, and even the foundation of this character.
In the first case, this character involves rather the intervention of something that somehow manages to obtain the same effect, even though
the elementary (and infinitely complex) causes might be completely

---

a   TN: ...*dans la direction du volontaire*. Note that here *le volontaire* ("the voluntary") resonates
in French with both *volonté* ("the will") and *vouloir* ("to desire"). Arnaud François suggests
in his editorial notes that this alludes to a spiritual source that is both the source and principle of life and of materiality. Cf. below, 232–39.

different. I insisted on this second point above in Chapter I, where I showed how identical structures are encountered along independent lines of evolution.[128] But without having to go to such lengths, we can presume that the simple reproduction of the type of the ancestor by its descendants is already something completely other than the repetition of the same composition of forces that would yield an identical outcome.[129] When we think of the infinity of infinitesimal elements and infinitesimal causes that converge in the genesis of a living being, and when we realize that the absence or the deviation of a single one of them is enough to make it such that nothing works any longer, the mind's first move is to assume this army of tiny workers to be supervised by a skilled foreman—the "vital principle"—who could fix at each moment any errors committed, correct the effects of distraction, and put things back into their places.[130] We thereby attempt to translate the difference between the physical order and the vital order: the physical order makes it such that the same combination of causes gives the same overall effect; the vital order guarantees the stability of the effect even though there is some wavering in the causes. But this is merely a translation. When we reflect upon it, we find that there can be no foreman for the simple reason that there are no workers. The causes and the elements that physico-chemical analysis discovers are in all likelihood real causes and elements for the facts of organic destruction. As such, they are limited in number. But when we analyze vital phenomena, properly speaking, or the facts of organic creation, they open to us the prospect of an analysis that would go on to infinity. From this it can be inferred that the multiple causes and elements are here nothing but views taken by the mind attempting to create an imitation ever closer to the operation of nature, whereas the operation that is being imitated is itself an indivisible act.[131] The resemblance between individuals of a single species would thus have a completely different meaning and a completely different origin than the resemblance between the complex effects obtained by the same composition of the same causes. But in both cases, there is resemblance, and consequently a possible generalization. And since this possibility is all that interests us in practice, given that our daily life is necessarily an anticipation of the same things and the same situations, it was quite natural that this common character [the possibility for generalization], being essential from the point of view of our action, brought these two orders together despite an entirely internal diversity that itself was only

227

of interest to speculation. Hence the idea of a *general order of nature*, every-where the same, hovering over both life and matter.[132] And hence our habit of designating by the same word and of imagining in the same way the existence of *laws* in the domain of inert matter and the existence of *genera* in the domain of life.

[THE TWO OPPOSING FORMS OF ORDER: THE PROBLEM OF GENERA AND THE PROBLEM OF LAWS]

Moreover, it seems beyond any doubt that this confusion is at the origin of the majority of the difficulties raised by the problem of knowledge, as much for the ancients as for the moderns.[133] Indeed, with the gener-ality of laws and the generality of genera being designated by the same word and subsumed under the same idea, the geometrical order and the vital order were thereby merged together. Depending on the per-spective that was adopted, the generality of the laws was explained by that of the genera, or the generality of the genera by that of the laws. The two theories being defined in this way, the first is characteristic of ancient thought and the second belongs to modern philosophy.[134] And yet, in both ancient and modern philosophy, the idea of "generality" is an equivocal one that, in its extension and in its comprehension, brings together incompatible objects and elements. In both cases, two kinds of order—united merely by how they make our action on things easier—are grouped under the same concept. Two terms are brought together in virtue of an entirely external similarity, and although this perhaps justifies their being designated by the same word in terms of practice, it does not at all authorize us, in the domain of speculation, to merge them together under the same definition.

228

Indeed, the ancients did not wonder why nature submits itself to laws, but rather why it is ordered according to genera. The idea of a genus corresponds to an objective reality above all in the domain of life, where it translates an incontestable fact, namely heredity.[135] Moreover, there can only be genera where there are individual objects. Now, if organic being is cut out of the whole of matter by its very organic devel-opment [*organisation*], i.e., by nature, then it is our perception that breaks up inert matter into distinct bodies, guided by the interests of action and guided by the nascent reactions that our body sketches out, i.e., by the virtual genera that aspire to constitute themselves (as I have shown

229   elsewhere).[a] Genera and individuals are thus determined here through a semi-artificial operation that is entirely relative to our future action upon things.[136] Nevertheless, the ancients did not hesitate to place all genera on the same level, to attribute the same absolute existence to them all. Reality having thereby become a system of genera, the generality of laws had to be reduced to the generality of genera—in sum, to the expressive generality of the vital order. It would be interesting, in this respect, to compare Aristotle's theory of falling bodies to the explanation provided by Galileo.[137] Aristotle is solely preoccupied by the concepts "high" and "low," "proper place" and borrowed place, and "natural movement" and "forced movement"[b/138]: the physical law, in virtue of which the stone falls, expresses for him that the stone regains the "natural place" of all stones, namely the earth. For Aristotle, the stone is not fully a stone so long as it is not in its normal place. By falling back down to this place, it (like a living being that grows) aims at completing itself and thereby at fully achieving the essence of the genus "stone."[c] If this conception of the physical law were correct, the law would no longer be a simple relation established by the mind, and the subdivision of matter into bodies would no longer be relative to our faculty of perception: all bodies would have the same individuality as living bodies, and the universal laws of physics would express relations of real kinship between real genera. We know the kind of physics that emerged from this, and how—for having believed in the possibility of a unified and definitive science embracing the totality of the real and coinciding with the absolute—the ancients had to limit themselves to a more or less approximate translation of the physical into the vital.

230   But the same confusion is found among the moderns—the difference being that the relation between the two terms is inverted; that the laws are no longer related back to genera, but rather genera are related to laws; and that science, again being assumed here to be unified,

---

a Bergson's note: *Matière et mémoire*, Chapters III and IV.

b Bergson's note: See in particular: Aristotle, *Physics* IV, 215a2; V, 230b12; VIII, 255a2; and Aristotle, *On the Heavens*, IV, 1–5; II, 296b27; IV, 308a34 [Aristotle, *Physics*, trans. R. P. Hardie and R. K. Gaye, in *The Complete Works of Aristotle*, ed. Jonathan Barnes, the revised Oxford trans., vol. 1, Bolligen Series LXXI.2 (Princeton: Princeton University Press, 1995), 315–446; Aristotle, *On the Heavens*, trans. J. L. Stocks, in *ibid.*, 447–511].

c Bergson's note: Aristotle, *On the Heavens*, IV, 310a34: τὸ δ' εἰς τὸν αὐτοῦ τόπον φέρεσθαι ἕκαστον τὸ εἰς τὸ αὐτοῦ εἶδός ἐστι φέρεσθαι. ["...the movement of each body to its own place is motion toward its own form."]

becomes entirely relative, rather than fully coinciding with the absolute (as the ancients intended). The eclipsing of the problem of genera in modern philosophy is quite remarkable. Our theory of knowledge turns more or less exclusively on the question of laws: in one way or another, the genera will have to find some means of working with the laws. This is because our philosophy begins from the great astronomical and physical discoveries of the modern era. For philosophy, Kepler's and Galileo's laws have remained the ideal and unique type of all knowledge.[139] Now, a law is a relation between things or facts. More precisely, a mathematically formulated law expresses that a certain magnitude is a function of one or several other appropriately chosen variables. Indeed, there is already something contingent and conventional in the choice of variable magnitudes and in the dividing up of nature into objects and facts.[140] But even if we assume that the choice is fully indicated or even imposed by experience, still the law will no less remain a relationship—and a relationship essentially consists in a comparison. A law only has objective reality for an intellect who imagines several terms at the same time. This intellect might be neither mine nor yours, and so a science bearing upon laws might be an objective science, one that experience contained in advance and that we simply extract from experience.[141] But it remains no less true that comparison, even if it is not the work of any particular person, is at the very least carried out impersonally, and that an experience made up of laws, that is, of terms *related* to other terms, is an experience made up of comparisons, which, by the time we have gathered them together, must have already passed through an atmosphere of intellectuality. The idea of a science and of an experience as completely relative to human understanding is thus implicitly contained in the conception of a unified and integral science that would be composed of laws—Kant simply brought this out. But this conception is the result of an arbitrary confusion between the generality of laws and the generality of genera. Even though an intellect is necessary to determine these terms in relation to each other, it will be understood that the terms themselves might exist independently in certain cases.[142] And if, alongside the relationships between the terms, experience also showed us the independently existing terms (the living genera being something completely different from the systems of laws), then at least half of our knowledge would bear on the "thing in-itself," i.e., on reality itself.[143] This knowledge would be extremely difficult precisely because

231

it would no longer construct its object; rather, it would be obliged to submit itself to its object. And yet, no matter how little it gets its teeth into this object, this knowledge will have already bitten into the absolute itself. Let us go even further: the other half of knowledge would itself no longer be as radically or as definitively relative as certain philosophers claim if it could be established that it bears upon a reality of an inverse order, a reality that we always express in mathematical laws—in relations that imply comparisons—but that only lends itself to this work because it is weighed down by spatiality and consequently by geometry.[144] In any case, we find behind the relativism of the moderns the confusion between two kinds of order as much as it was there beneath the dogmatism of the ancients.

[DISORDER AND THE TWO ORDERS]

We have already said enough to indicate the origin of this confusion. It comes from the fact that the "vital" order, which is essentially creation, manifests itself less in its essence than in some of its accidents. These accidents imitate the physical and geometrical order. Just like the physical and geometrical order, they present repetitions that render generalization possible, and this is all that matters to us. There can be no doubt that life, as a whole, is an evolution, which is to say a continuous transformation. But life can only progress through the intermediary of the living beings who are its depositories. Thousands and thousands of nearly identical living beings must repeat each other in space and time so that the newness [la nouveauté] that they are working out might grow and ripen. Like a book that slowly moved toward a radically revised version by passing through thousands of printings, each comprised of thousands of copies. There is, however, a difference between the two cases in that the successive printings are identical, as are the copies printed during the same printing, whereas the representatives of the same species do not perfectly resemble each other, neither at the various points in space nor at the various moments in time. Heredity does not only transmit characteristics; it also transmits the élan in virtue of which the characteristics are modified—this élan is vitality itself.[145] This is why I claim that the repetition that serves as a basis for our generalizations is essential in the physical order, but only accidental in the vital order. The physical order is an "involuntary" or "automatic" order [ordre

"*automatique*"]; and although I will not say that the vital order is "voluntary," it is at least analogous to the "willed" order [*ordre "voulu"*].

Now as soon as we have a clear idea of the distinction between the
"willed" order and the "automatic" order, the ambiguity that the idea
of *disorder* lives on dissipates and, along with it, one of the principal difficulties of the problem of knowledge.

The main problem of the theory of knowledge is indeed to know
how science is possible, in short, why there is order and not disorder in
things.[146] Order exists, it's a fact. But then again, disorder, *which appears to
us to be less than order*, would also seem to exist in principle. The existence
of order would thus be a mystery to solve, or in any case a problem
to address.[147] To put it more simply, from the moment we attempt to
establish or justify order [*fonder l'ordre*], it is taken to be contingent, if not
in things then at least in the eyes of the mind. Something that was not
judged to be contingent would not require an explanation. If order did
not appear to us as a conquest over something else, or as an addition
to something (namely, to "the absence of order"), then ancient realism
would not have spoken of a "matter" to which the Idea would be added,
nor would modern idealism have posited a "sensible manifold" that
the understanding would then organize into nature.[148] And indeed, it is
incontestable that every order is contingent and conceived of as such.
But contingent in relation to what?

There is no doubt, in my opinion, as to the response to this question.
An order is contingent and appears to us as contingent in relation to the
inverse order, just as poetic verses are contingent in relation to prose
and prose in relation to poetic verses. But, just as all speaking that is
not prose is poetic verse and necessarily conceived of as poetic verse,
and just as all speaking that is not poetic verse is prose and necessarily
conceived of as prose, it is also true that every manner of being that is
not one of the two orders is the other and necessarily conceived of as
the other. But we are able not to pay attention to what we are conceiving
of, and only to catch sight of the idea that is really present to our mind
through a fog of affective states. It will be enough to consider how we
use the idea of disorder in everyday life to be convinced of this. When
I enter a room and judge it to be in "disorder," what do I mean? The
position of each object is explained by the involuntary or automatic
movements of the person who sleeps in the room, or by the efficient
causes, whatever they may be, that have caused each piece of furniture,

233

each piece of clothing, etc., to be in the place where they are: the room is in perfect order, in the second sense of the word.[a] But I am expecting the first kind of order: the order that a well-ordered person consciously puts into his life—in short, the willed order, not the automatic one. "Disorder" is the name I give the absence of this other order. In essence, everything that is real, perceived, and even conceived in the absence of one of the two orders is in fact the presence of the other order. But here the second order does not matter to me. *I am only interested in the first order,* and by saying that there is disorder, I express the presence of the second order as a function of the first instead of expressing it, so to speak, as a function of itself. Inversely, when we claim that we are imagining a chaos, i.e., a state of things in which the physical world no longer obeys laws, what are we thinking of? We are imagining facts that would *capriciously* appear and disappear. We begin by thinking of the physical universe such as we know it, with its causes and its effects perfectly proportional to each other. Then through a series of arbitrary decrees, we augment, diminish, and suppress so as to obtain what we call disorder. In reality, we have substituted something voluntary for the mechanism of nature. To the extent that we imagine the appearing or the disappearing of phenomena, we have replaced the "automatic order" with a multitude of elementary wills [*volontés élémentaires*].[149] Of course, in order for all of these tiny wills to constitute a "voluntary order," they must have accepted being directed by a higher will. And if we look closer, we will see that this is precisely what they do. Our will is there—it objectifies itself in each one of these capricious wills in turn, carefully avoids linking together the same with the same, carefully ensures that the effect not be proportional to the cause—in short, it is our will that causes a single intention to hover over the group of elementary volitions. So here too the absence of one of the two orders clearly consists in the presence of the other.

Now, by analyzing the idea of chance, which is closely related to the idea of disorder, we would find the same elements.[150] That the entirely mechanical play of causes that brings the roulette wheel to stop on a number causes me to win, and thereby operates like a benevolent genius

a TN: *...mouvements automatiques.* In these lines, Bergson's uses of the adjective *automatique* ("automatic," "mechanical," or "involuntary") continues the contrast with the opposite order, the "willed" order.

who cares about my interests, or that the entirely mechanical force of the wind breaks a shingle off from the roof and drops it on my head, and thereby acts in the manner of an evil genius conspiring against me personally, I nevertheless find in both cases a mechanism where I would have expected, or where it seems that I should have encountered, an intention.[151] That is what I express by speaking of *chance*. And when it comes to an anarchic world, where phenomena would follow each other at the whim of their caprice, I will again say that chance reigns, meaning that I encounter *wills*, or rather *decrees*, when I was expecting to find mechanism. This explains the rather singular flailing about [*ballottement*] of the mind when it attempts to define chance. Neither the efficient cause nor the final cause can provide it with the definition it seeks. Unable to steady itself, the mind oscillates between two ideas: the absence of a final cause and the absence of an efficient cause, with each definition sending the mind back to the other. Indeed, the problem remains irresolvable so long as one takes the idea of chance for a pure idea without any emotion being mixed in. And yet in reality, chance is simply the objectification of the state of mind of someone who, expecting one of the two kinds of order, encountered the other. Chance and disorder are thus necessarily conceived of as relative. As much as we might want to imagine them as absolute, we realize that we involuntarily go back and forth like a shuttle between the two kinds of order, passing into this one at the precise moment we were about to catch ourselves in the other. And so we realize that the supposed absence of all order whatsoever is in reality the presence of the two, along with the swinging back and forth of a mind that does not definitively come to rest on either. There can be no question of positing disorder as a substrate of order, neither in the things nor in our representation of the things, since disorder implies the two kinds of order and is made up of their combination.

236

But our intellect disregards all of that. With a simple *sic jubeo*, it posits a disorder that would be an "absence of order."[152] In doing so, it thinks about a word or a juxtaposition of words, but nothing more. Now, if the intellect seeks to place an idea beneath this word, it will find that disorder can surely be the negation of an order, but that this negation is in fact the implicit observation of the presence of the opposite order, an observation we close our eyes to because it does not interest us. Or we escape it by negating in turn the second order—that is, by essentially reestablishing the first order. So, how could we speak of an incoherent manifold that the understanding would organize? As much as one might claim that no one actually assumes this incoherence to be realized or even realizable, from the moment we speak of it, we believe we are thinking of it. Now, analyzing the idea that is actually present, we will only find there, once again, the disappointment of a mind standing before an order that does not interest it, or the oscillation of the mind between two kinds of order, or, finally, the pure and simple representation of an empty word that was created by attaching the negative prefix to a word that meant something. But we neglect to carry out this analysis. We omit it precisely because we do not think of distinguishing between two kinds of order that are irreducible to each other.

Indeed, I claimed above that every order necessarily appears as contingent.[153] If there are two kinds of order, then the contingent nature of order is explained: each form of order is contingent in relation to the other. Where I find the geometrical order, the vital one was possible; where the order that I find is vital, it could have been geometrical. But suppose for a moment that order is always of the same kind and simply contains degrees that go from the geometrical to the vital. If a given order continues to appear to me as contingent, and if it can no longer be contingent in relation to an order of another kind, then I will necessarily believe that it is contingent in relation to an *absence of itself*, that is, in relation to a state of things "where there would be no order at all." And I will believe that I am thinking of this state of things because it seems implicit in the contingent nature of order itself, which is an incontestable fact. In this case, I would place the vital order at the summit of the hierarchy, then the geometrical order as a diminution or as a lower level of complexity of the vital one, and finally, at the very bottom, the absence of order, incoherence itself, upon which order would be superposed.[154] This is why the word "incoherence" gives me the impression of a word that must actually signify something—if not something realized, then at least something thought. But if I notice that the state of things implied by the contingent nature of a given order is simply the presence of the opposite order, and if I thereby posit two opposite kinds of order, then I see for myself that no intermediary degrees between the two orders could be imagined, and that there is no descending from these two orders toward "the incoherent." Either "the incoherent" is simply a word lacking any meaning, or, if I give it a meaning, this is only by placing incoherence halfway between the two orders and not

237

underneath them both. What exists is not first the incoherent, then the geometrical, and then finally the vital. Rather, only the geometrical and the vital orders exist, then, by a wavering of the mind between the two, the idea of the incoherent emerges. To speak of an uncoordinated manifold on which order is imposed is thus to commit a veritable begging of the question, since by imagining something as uncoordinated, we in fact posit an order, or rather, we posit two of them.

*

[THE IDEAL GENESIS OF MATTER][a]

[Creation and Evolution]

This long analysis was necessary in order to show how the real can go from tension to extension and from freedom to mechanical necessity by way of inversion. It was not enough to establish that the relation between the two terms is suggested, simultaneously, by consciousness and by sensible experience.[155] It was necessary to prove that the geometrical order has no need of an explanation, being purely and simply the suppression of the inverse order. And for this, it was necessary to establish that this suppression is always a kind of substitution, and even that it is necessarily conceived of as such. But here the demands of practical life suggest a manner of speaking that misleads us, both in terms of what takes place in the things and in terms of what is present in our thought. I must now examine more closely the inversion whose consequences I have just described. So, what is the principle or source [principe] that simply needs to relax itself in order to extend itself out, the interruption of the cause here being equal to a reversal of the effect?[b]

238

I called this principle "consciousness," for lack of a better word.[156] But this is not the diminished consciousness at work in each one of

---

a   TN: Although not included in the Table of Contents, Bergson's running headers indicate quite clearly that the final part of this chapter is divided into two main sections: "The Ideal Genesis of Matter" (239–51) and "The Meaning of Evolution" (251–67). I have thus included these as section titles, taking the titles provided by Bergson in the Table of Contents to be subsections.

b   TN: Here Bergson undertakes a more explicit discussion of what he means by principe, not in the sense of a fundamental proposition or scientific theorem, but rather as a metaphysical "source." See above, 49, 187, and the Translator's Introduction.

us. Our personal consciousness is the consciousness of a certain living being who is placed at a certain point in space; and, even if it indeed goes in the same direction as its principle, it is continuously drawn in the inverse direction, obliged—though it marches forward—to look backward. This retrospective vision is, as I have shown, the natural function of the intellect and consequently of distinct consciousness.[157] For our consciousness to coincide with something of its principle, it would have to detach itself from what is *ready-made* and attach itself to what is *being made*.[a] Turning and twisting back upon itself, the faculty of *seeing* would have to become at one with the act of *willing* [*vouloir*].[158] This is a painful effort that we can accomplish suddenly by doing violence to nature, but that we cannot sustain for more than a few instants. In free action, when we contract all of our being in order to launch it forward, we have a more or less clear consciousness of the purpose and of the motives, and even, ultimately, of the becoming by which they are organized into an act. But pure will—the current that courses through this matter, imparting life to it—is something that we barely sense at all, something that, at most, we touch upon lightly in passing.[159] Let us try to situate ourselves within this pure will, if only for a moment: even then, we will only grasp an individual and fragmentary will.[160] To reach the principle or source of all life as well as of all materiality, it would be necessary to go even further.[161] Is this impossible? Certainly not—the history of philosophy bears witness to this possibility.[162] There is no enduring system that is not animated by intuition, at least in some of its parts.[163] The dialectic is necessary so as to put intuition to the test, and necessary so that intuition might be refracted into concepts and spread to other men. But the dialectic often merely develops the result of this intuition that goes beyond it. The fact is that these two procedures move in opposite directions: the same effort by which we link together ideas with other ideas causes to vanish the very intuition that the ideas had offered to store up. Once the philosopher has received the *élan* from the intuition, he is obliged to abandon the intuition itself and to rely upon himself alone to continue the movement, pushing one concept forward after another.[164] But very quickly he senses that he has lost his footing; a

239

---

a   TN: ...*il faudrait qu'elle se détachât du* tout fait *et s'attachât au* se faisant. Attempting to capture what is *durée* or becoming, Bergson's somewhat particular use of the term *se faisant* here could be translated as "what is being made" or "what is in-the-making."

new contact becomes necessary and the majority of what he had made will have to be unmade. In short, the dialectic is what ensures that our thought remains in harmony with itself; but through dialectic—which is nothing but a relaxation of intuition—many different harmonies are possible, and yet there is only one truth. Were intuition able to prolong itself beyond a few instants, it would guarantee not only the harmony between the philosopher and his own thought, but also a harmony of all the philosophers with each other. Such as it currently exists—fleeting and incomplete—intuition is that within each system that is more valuable than the system itself and that will survive the system. The aim of philosophy would be attained if this intuition could sustain itself, generalize itself, and above all reassure itself through some external points of reference so as to not go off course. For this, a continuous back-and-forth between nature and the mind is necessary.

240

When we place our being back into our will, and our will itself back into the impulse that it prolongs, we understand or we feel that reality is a perpetual growth, a creation that continues on endlessly.[165] Our will already performs this miracle.[166] Each human work that contains some invention, each voluntary act that contains some freedom, and each movement of an organism that manifests some spontaneity brings something new into the world. True, these are merely creations of form.[167] But how could they be anything else? After all, we are not the vital current itself; we are merely this current insofar as it is already weighed down by matter, that is, by congealed parts of its substance that it carries along the length of its course. Whether in the composition of a work of genius or in a simple free decision, as much as we might tighten the spring of our activity to its highest degree and thereby create what no pure and simple assemblage of matter could have given (after all, what juxtaposition of already known curves could ever equal the pencil stroke of a great artist?), it is no less the case that here too there are elements that preexist and survive their organization. But if a simple halt in the action that generates the form were able to constitute its matter (after all, are not the original lines sketched out by the artist themselves the fixing and, so to speak, the congealing of a movement?), then a creation of matter would be neither incomprehensible nor inadmissible.[168] For we grasp from within and live at each moment a creation of form; and it would be precisely there, in cases where the form is pure and where the creative current momentarily interrupts itself, that there is a creation of matter.[169] Consider all the letters of the alphabet that enter into the composition of everything that has ever been written. In order to write a new poem, we do not imagine that other letters spring forth and come to be added to the previous ones. But we understand perfectly well that the poet creates the poem and that human thought is thereby enriched. This creation is a simple act of the mind, and rather than continuing along, the action has only to come to a pause in a new creation in order for it to scatter itself out into words that themselves divide up into letters that in turn will be added on to the letters already in the world.[170] In the same way, the idea that the number of atoms making up the universe at a given moment might increase runs counter to our mental habits and contradicts our experience. But it is not inadmissible that a reality of an entirely different order—a reality that is to the atom what the poet's thought is to the letters of the alphabet—might grow through sudden additions. And the other side of every addition might well be a world that we then represent, symbolically of course, as a juxtaposition of atoms.[171]

[The Material World]

In fact, the mystery that spreads over the existence of the universe comes, in large part, from the fact that we want its genesis to have been accomplished in a single stroke, or rather that all matter be eternal.[172] Now, whether one speaks of creation or posits an uncreated matter, in both cases the totality of the universe is being considered. By closely examining this mental habit, we would find the presupposition that I will analyze in the next chapter: the idea (common to both materialists and their adversaries) that there is no truly effective or acting durée and that the absolute—whether it be matter or mind—can have no place within concrete time, that is, in the time that we feel to be the very fabric of our life.[173] It follows from this presupposition that everything is given, once and for all, and that it is necessary to posit for all eternity either the material multiplicity itself, or the act that creates this multiplicity given en bloc in the divine essence.[174] Once this presupposition is rooted out, the idea of creation becomes clearer since it merges with the idea of growth. But for this we must no longer speak of the universe in its totality.

And why would we speak of it that way? The universe is an assemblage of solar systems that we have every reason to believe are analogous to our own. Surely these systems are not absolutely independent of each other.[175] The warmth and light of our sun radiate beyond the most distant planet, and moreover, our solar system as a whole is moving in a definite direction, as if drawn that way.[176] So there is a link between worlds. But this link can be considered infinitely weaker by comparison to the interconnectedness that ties together the parts of a single world. As such, it is not merely artificial nor for simple convenience that we isolate our solar system; nature itself invites us to isolate it. Insofar as we are living beings, we depend on the planet we are on and the sun that nourishes it, but not on anything else. Insofar as we are thinking beings, we can apply the laws of our physics to our own world, and probably also extend them to every other world taken in isolation. But nothing indicates that our laws will still apply to the universe as whole, nor even that such an assertion has any meaning since the universe is not made, but rather is continuously being made.[a] The universe grows, perhaps indefinitely, through the addition of new worlds.

So, let us extend to the whole of our own solar system—though limiting ourselves to this relatively closed system, and to other relatively closed systems—the two most general laws of our science: the principle of the conservation of energy and that of degradation.[177] And let us see what the results of this will be. First, it must be noted that these two principles do not have the same metaphysical significance. The first principle is a quantitative law, and consequently is in part relative to the procedures we use for measuring. It says that in a system assumed to be closed, the total energy—i.e., the sum of kinetic and potential energy—remains constant. Now, if there were only kinetic energy in the world, or even if there were, beyond kinetic energy, one and only one kind of potential energy, then the artifice of measurement would not be sufficient to render this law artificial. The law of the conservation of energy would thus express that indeed a constant quantity of *something* is preserved.[178] But in fact, there are a variety of different types of energy,[b/179] and the measurement used for each one of them was clearly

243

a   TN: ...*se fait sans cesse*. Here Bergson's phrase could also be translated as "[the universe] makes itself continuously" or "is continuously in-the-making."

b   Bergson's note: For more on these qualitative differences, see Pierre Duhem's book: *L'évolution de la mécanique* (Paris: Librairie Scientifique A. Hermann, 1905), 197–208 [Pierre-Marie-Maurice Duhem, *The Evolution of Mechanics*, trans. Michael Cole (Alphen aan den Rijn: Sijthoff and Noordhoff, 1980), 105–15].

chosen in such a way as to justify the principle of the conservation of energy. Thus, the role that convention inherently plays in this principle is a significant one, although there is surely an interdependence among the variations of the different energies making up a single system, which is precisely what made possible the extension of the principle via conveniently chosen measurements. As such, if the philosopher applies this principle to the whole of the solar system, he must at the very least soften its contours. The law of the conservation of energy will no longer be able to express here the objective permanence of a certain quantity of a certain thing. Rather, it will express the necessity that each change that takes place be somewhere counterbalanced by a change in the opposite direction. In short, even if the law of the conservation of energy governs our entire solar system, it explains the relation of a fragment of this world to another fragment, and not the nature of the whole.[180]

Things are different when it comes to the second principle of thermodynamics.[181] In fact, the law of the degradation of energy does not essentially have to do with magnitudes. Of course, the idea was born of certain quantitative considerations in Carnot's work on the output of heat engines.[182] And of course, Clausius generalized the idea into mathematical terms and arrived at the conception of a calculable magnitude named "entropy."[183] These precisions are necessary for practical applications of the law. But the law itself would have remained vaguely formulable and might have ultimately been formulated in general terms, even if no one had ever dreamed of measuring the various energies of the physical world and even if no one had created the concept of energy. Indeed, the law essentially states that all physical changes tend to be degraded into heat and that heat itself tends toward a uniform distribution among bodies. In this less precise form, the law becomes independent of all conventions; it is the most metaphysical of the physical laws in that it shows us—without any interposed symbols and without any of the artifices of measurement—the direction in which the world marches.[184] It claims that visible and mutually heterogeneous changes become increasingly diluted into invisible and homogeneous changes, and that the instability to which we owe the richness and variety of the changes taking place in our solar system will little by little give way to the relative stability of elementary vibrations that will indefinitely repeat each other.[185] Much like a man who might preserve his strength but devote it less and less to actions, and who would ultimately end up

244

employing all of it in simply causing his lungs to breathe and his heart to beat.

Seen from this point of view, a world like our solar system appears to be using up, at each instant, something of the mutability that it contains. In the beginning, the possible utilization of energy was at its maximum; this mutability has gone on continuously diminishing. But where does it come from? It might initially be assumed that it comes from some other point in space, but the difficulty would merely be pushed back, and the same question could be asked of this external source of mutability. True, one could add that the number of worlds capable of passing mutability back and forth among themselves is unlimited, that the sum of mutability contained in the universe is infinite, and that, as such, there is no more a need to find the origin of it than there is a need to find its end. A hypothesis of this kind is as irrefutable as it is indemonstrable. But to speak of an infinite universe is to assume a perfect coincidence between matter and abstract space, and consequently an absolute exteriority of all of the parts of matter with regard to each other. Above we saw the judgment that must be made regarding such a theory and just how difficult it is to reconcile it with the idea of a reciprocal influence of all of the parts of matter upon each other, the very influence that is being appealed to here.[186] Finally, it might be assumed that the general instability comes out of a general state of stability, that the period in which we find ourselves, and during which the usable energy is diminishing, was preceded by a period in which the mutability was on the path of growth, and, moreover, that the alternatives of growth and diminution endlessly succeed each other. This hypothesis is theoretically conceivable, as has recently been rigorously demonstrated. But according to Boltzmann's calculations, its mathematical improbability goes beyond all imagination and hence practically amounts to an absolute impossibility.[a][187] In reality, the problem is unsolvable if we remain within the field of physics since the physicist is obliged to attach energy to extended particles and, even if he sees these particles as nothing but reservoirs of energy, thereby remains within space.[188] The physicist would betray his role were he to seek the origin of these

245

---

a Bergson's note: Ludwig Boltzmann, *Vorlesungen über Gastheorie*, vol. 2, *Theorie van der Waals'; Gase mit zusammengesetzten Molekülen; Gasdissociation; Schlussbemerkungen* (Leipzig: Verlag von Johann Ambrosius Barth, 1898), 253–60.

energies in an extra-spatial process. Nevertheless, the origin must, in my opinion, be sought precisely in some such extra-spatial process.

246   Are we considering, *in abstracto*, the extended in general? I said above that *extension* appears to be merely a *tension* that has been interrupted.[189] Are we thus attempting to grasp the concrete reality that fills this extension? The order that reigns there, and that is expressed by the laws of nature, is an order that must be born of itself when the inverse order is suppressed: this is precisely the suppression that would be produced by a relaxation of will.[190] In short, the direction in which this reality marches now suggests the idea of a thing *unmaking itself*; this is, beyond all doubt, one of the essential traits of materiality.[191] What else should we conclude from this, if not that the process by which this thing *makes itself* is directed in the inverse direction of physical processes, and thus is, by definition, immaterial? Our vision of the material world is of a weight that falls; no image drawn from matter, properly so called, will give us an idea of a weight that lifts itself up. But this conclusion will impose itself with even more force if we narrow our focus to concrete reality, if we no longer consider matter in general, but rather the living bodies within that matter.

Indeed, all of our analyses indicate that life is an effort to go back up the incline that matter descends. As such, our analyses allow us to catch sight of the possibility, even the necessity, of a process opposite that of materiality, a process that creates matter by simply being interrupted. Of course, the life that evolves on the surface of our planet is attached to matter. If that life were pure consciousness, or better, supra-consciousness, it would be a pure creative activity.[192] In fact, life is riveted to an organism that submits it to the general laws of inert matter. But everything takes place as if it does its best to break free of these laws. Life does not have the power to reverse the direction of physical changes as defined by Carnot's Principle. But at least life behaves exactly like a

247   force that, left to itself, would work in the inverse direction. Incapable of *halting* the march of material changes, life nevertheless manages to *slow it down*.[193] Indeed, the evolution of life continues an original impulse (as I have shown above).[194] This impulse—which determined the development of the chlorophyllian function in plants and the sensorimotor system in animals—carries life toward actions that are more and more effective through the fabrication and the use of explosives that are more and more powerful. Now, what do these explosives represent, if not a

stockpiling of solar energy whose degradation is thereby temporarily suspended at certain points into which it had flowed? The usable energy that the explosive contains will be spent, of course, at the moment of the explosion. But it would have been spent much earlier if an organism had not been there to halt its dissipation, to retain it, and to accumulate it. Life, such as we see it today, namely at the point to which it was led by a scission of mutually complementary tendencies that it contained within itself, is entirely dependent upon the chlorphyllian function in plants. When considered in its initial impulse and prior to any scission, life was a tendency to accumulate something in a reservoir (as above all the green parts of plants do), with an eye toward an instantaneous and effective expenditure (such as the ones animals accomplish), that would have simply dissipated without it.[195] Life is like an effort to lift up the weight that falls. True, it only manages to slow the fall. But at least it can give us an idea of what the lifting up of the weight would be.[a/196]

Imagine a container full of steam at a high pressure [tension], with some cracks here and there in the walls of the container through which steam is escaping in a jet. The steam shooting out into the air condenses almost entirely into droplets that fall, and this condensing and falling represent simply the loss of something, an interruption, a deficit. But a small part of the jet of steam subsists, uncondensed, for a couple of instants. This steam attempts to lift up the drops that fall; it can at most slow down their fall. In the same way, jets must ceaselessly shoot forth from an immense reservoir of life, where each one, falling back, is a world. The evolution of living species within this world represents what subsists of the initial direction of the original jet and of an impulse that continues along in the inverse direction of materiality. But we should

248

---

a   Bergson's note: In a book that is rich in facts and ideas, André Lalande shows us everything marching toward death, despite the momentary resistance that organisms appear to oppose to that movement (*La dissolution opposée à l'évolution dans les science physiques et morales* (Paris: Félix Alcan, 1899)). But, even on the side of inorganic matter, do we have the right to extend to the whole universe considerations drawn from the present state of our solar system? Alongside worlds that are dying, there are surely worlds that are being born. Moreover, within the organic world, the death of individuals does not at all appear to be a diminution of "life in general," nor as a necessity that life in general would have to suffer reluctantly. As I have mentioned more than once, life has never attempted to prolong the existence of the individual indefinitely, whereas, on many other points, it has made so many propitious efforts. Everything happens *as if* the death of the individual was willed or desired, or at the very least accepted, so as to allow for the greatest progress of life in general.

not rely too much on this comparison. It gives us but a weakened and even misleading image of reality since the crack, the jet of steam, and the lifting up of the droplets are necessarily determinate, whereas the creation of a world is a free act, and life within the material world, participates in this freedom. Rather, think of a gesture, such as an arm being raised. Then imagine that the arm, abandoned to itself, falls back down, and that nevertheless something of the will that animated it subsists in it, making every effort to raise it up again. This image of a *creative gesture unmaking itself* already gives us a more precise representation of matter.[a] In this way, we will see vital activity as what remains of the positive movement in the inversion of that movement, that is, *a reality that is making itself through the reality that is unmaking itself.*[b/197]

249   [On the Origin and the Destination of Life]

The idea of creation remains entirely obscure so long as we think (as we habitually do, and as the understanding cannot stop itself from doing) of a *thing* that creates and of *things* that would be created.[198] I will show the origin of this illusion in the next chapter.[199] This is a natural illusion for the intellect, which has an essentially practical function, having been made for showing us things and states rather than changes and actions.[200] But things and states are merely views or snapshots that our mind takes of becoming. There are no things; there are only actions.[201] More specifically, if I consider the world we live in, I find that the automatic and rigorously determinate evolution of this tightly woven whole is an action unmaking itself, and that the unforeseen forms that life cuts out within it—forms that are capable of prolonging themselves in unforeseen movements—represent an action that is being made. Now, I have every reason to believe that other worlds are analogous to our own and that things take place there in the same manner. And I know that these worlds are not all constituted at the same time since observation shows me, even today, nebulae in the process of being formed through concentration. If the same kind of action is being accomplished everywhere— an action that is sometimes unmaking itself and sometimes attempting

---

a TN: ...*un geste créateur qui se défait.*
b TN: ...*une réalité qui se fait à travers celle qui se défait.*

to remake itself—then I simply express this likely similarity between worlds when I speak of a center from which these worlds would shoot forth, like the sparks shooting out from an immense firework [*bouquet*], so long as I do not present this center as a thing, but rather as a continuity of shooting forth. God, thus defined, has nothing of the ready-made; he is unceasing life, action, and freedom.[202] Creation thus conceived is not a mystery; we experience it in ourselves the moment that we act freely. Of course, it is absurd to suggest that new things can be added onto the things that exist since the thing is the result of a solidification enacted by our understanding, and since there are never any things other than the ones that the understanding has constituted. To speak of things that create themselves would thus amount to saying that the understanding gives itself more than it gives itself—an affirmation that is self-contradictory, an empty and useless idea.[a] But that action grows as it advances, that it creates as it progresses, this is something we can all see when observing ourselves acting. Things are constituted when the understanding makes an instantaneous cut in a flux of this kind, at a particular moment; and the mysteries we face when we compare the cuts to each other become clear when we think back to the flux. In fact, the modalities of creative action, insofar as they continue along in the organic development [*organisation*] of living forms, are greatly simplified when we consider them from this angle. When confronted by the complexity of an organism and the quasi-infinite multitude of intertwining analyses and syntheses that it presupposes, our understanding recoils, disconcerted. We find it hard to believe that the pure and simple play of physical and chemical forces could create this marvel. And if a profound knowledge [*science*] is indeed at work, how are we to understand the influence exerted upon a matter without form by this form without matter?[203] But the difficulty is born from the fact that we have a static idea of ready-made material particles that are juxtaposed, and also a static idea of an external cause that would impose a sophisticated organization upon them. In reality, life is one movement, materiality is the inverse movement, and each of these two movements is simple—the matter that forms a world is an undivided flux, and the life that courses through that matter and that

250

a   TR: *...se donne plus qu'il ne se donne...* Here Bergson returns to the locution of *se donner* for "to give itself," "to presuppose," or "to take for granted."

cuts living beings out of it is also undivided.[204] Of these two currents, the second opposes the first, but the first still gains something from the second: between them results a modus vivendi, which is precisely organic nature [organisation].[205] To our senses and our intellect, this organic nature takes the form of parts entirely outside of parts in time and in space. Not only do we close our eyes to the unity of the élan that, passing through the generations, connects individuals to individuals, or species to species, and makes the series of living beings as a whole into a single immense wave coursing through matter, but each individual itself also appears to us as a mere aggregate: an aggregate of molecules and of facts. The reason for this would be found in the structure of our intellect, which is made for acting outwardly upon matter and only manages to do so by executing—within the flux of the real—some instantaneous cuts that become, in their fixity, indefinitely decomposable. Because the understanding only perceives the organism as a set of parts outside of parts, it must choose between two systems of explanation: either take the infinite complication (and, consequently, infinite sophistication) of organic nature to be an accidental assemblage, or relate organic nature back to the incomprehensible influence of an external force that would have grouped the elements together. But this complexity is the work of the understanding, and so too is this incomprehensibility. Let us no longer try to see through the eyes of the intellect alone, which only grasps what is ready-made and always observes from the outside, but rather see with the mind, by which I mean that faculty of seeing that is immanent to the faculty of acting and that somehow springs forth from the twisting back of the will upon itself.[206] Everything will be put back into movement, and everything will be resolved into movement. Where the understanding, operating on a supposedly fixed image of an action that is underway, showed us an infinite multiplicity of parts and an infinitely sophisticated order, we will catch a glimpse of a simple process, an action making itself through an action of the same kind that is unmaking itself, like the path cut by the last spark of a firework on its way through the falling debris of the sparks that are already extinguished.[207]

*

[THE MEANING OF EVOLUTION]

[The Essential and the Accidental in Vital Processes and in the Movement of Evolution]

From this point of view, the general considerations that I offered regarding the evolution of life will be clarified and completed.[208] I will be able to show more clearly what is accidental and what is essential in this evolution.[209]

The *élan of life* that I am discussing consists, in short, in a need for creation.[210] This *élan* is not capable of absolute creation since it encounters matter, i.e., the inverse movement of its own movement. But it takes hold of this matter, which is necessity itself, and attempts to introduce into it the largest possible amount of indetermination and freedom.[211] How does it go about doing this?

I said above that an animal that is higher in the series can be thought of, roughly, as a sensorimotor nervous system installed atop digestive, respiratory, and circulatory systems, among others.[212] The role of these systems is to clean, repair, and protect the sensorimotor system, and to make it as independent of the external circumstances as possible. But above all their role is to provide the sensorimotor system with the energy it will spend on movement. The increasing complexity of the organism is in theory (and notwithstanding the innumerable exceptions that result from the accidents of evolution) a result of the need to make the nervous system more complicated. Moreover, the complication of any one part of the organism brings with it many others, because this part itself must continue to live and any change at one point of a body has repercussions throughout. Thus, the complication might indeed go on to infinity, in all directions: but it is the complication of the nervous system that conditions the others, at least in principle, if not always in fact. Now, what exactly would the progression of the nervous system itself consist in? It would be a simultaneous development of automatic activity and voluntary activity, the former providing the latter with an appropriate instrument. So, in an organism like ourselves, a significant number of motor mechanisms are set up in the spinal cord and in the medulla, merely awaiting a signal to set free the corresponding action. In certain cases, the will itself works to set up the mechanism, in other cases it chooses which mechanisms to release, how to combine them, and when to release them.[213] The will of an animal is all the more

252

253

effective and all the more intense insofar as it has a choice among a larger number of these mechanisms, insofar as the crossroads where all of the motor pathways intersect is more complicated, or, in other words, insofar as its brain has reached a more significant stage of development. As such, the nervous system's development guarantees that the action is increasingly precise, increasingly varied, and increasingly efficient and independent. The organism behaves more and more like a machine for acting that rebuilds itself entirely for each new action, as if it were made of rubber and could, at any moment, change the form of all of its parts.[214] And yet, prior to the appearance of the nervous system, prior even to the formation of what might be properly called an organism, this essential property of animal life was already manifest in the undifferentiated mass of the Amoeba. The Amoeba changes shape in a variety of directions; its entire mass thus does what the differentiation of parts in the developed animal will localize into the sensorimotor system. Since it only accomplishes this activity in a rudimentary way, it has no need for the complication that we see in higher organisms. All that is needed here is for the auxiliary elements to transfer over to the motor elements the energy that is to be spent. The animal moves itself as a whole, and as a whole it procures energy by the intermediary of the organic substances that it assimilates. As such, whether we place ourselves at the bottom or at the top of the series of animals, we always find that animal life consists in: (1) procuring a certain stock of energy and (2) spending that energy in variable and unforeseeable directions by means of a matter that is as supple as possible.

Now, where does the energy come from? It comes from the food that is ingested, since food is a kind of explosive that simply waits for the spark to unleash the energy it is storing. Who fabricated this explosive? The food might be the flesh of an animal that will have fed on other animals, and so on; but, ultimately, we will arrive at plants. Only plants truly collect solar energy. Animals merely borrow that energy from them, either directly or by passing it back and forth among themselves. How did the plant store up this energy? Above all through the chlorophyllian function, that is, through a *sui generis* chemical reaction whose key we have not yet found, and that probably does not resemble the chemical reactions we see in our laboratories. The operation consists in making use of solar energy in order to fix the carbon of carbonic acid and, in this way, to store up this energy much like one would store up

254

the energy of a water carrier by employing him to fill up an elevated reservoir. Once the water has been elevated, it will be able to set into motion a mill or a turbine, however and whenever we wish.[215] Each fixed atom of carbon represents something like the elevation of a certain weight of water, or the tension of an elastic band uniting the carbon and the oxygen in the carbonic acid. The elastic will relax, the weight will fall back down, and the stored-up energy will finally reappear the day when, through a simple releasing [*déclenchement*], the carbon is allowed to rejoin its oxygen.[216]

As a result, life as a whole—animal life and plant life together—appears in essence to be an effort for accumulating energy and releasing that energy into flexible and deformable canals at the end of which it will accomplish an infinite variety of works. This is what the *élan vital*, passing through matter, wanted to obtain in a single stroke. And it might have succeeded had its power been unlimited, or if it could have received some help from the outside.[217] But the *élan* is finite and was given once and for all.[218] It cannot overcome all of the obstacles. The movement that it expresses is sometimes deflected, sometimes divided, but always contested, and the evolution of the organic world is simply the unfolding of this battle. The first great scission that must have taken place was between the two kingdoms, plants and animals, which thereby find themselves to be mutually complementary, without there necessarily having been an agreement established between them.[219] The plant does not accumulate energy for the animal, but rather for its own consumption. But as a result, its own expenditure is less discontinuous, less condensed, and less efficient than was required by the initial *élan* of life, essentially directed as it was toward free actions. A single organism could not simultaneously and with equal force sustain the two roles of gradual accumulation and sudden utilization.[220] This is why—all by themselves and without any external intervention, through the simple effect of the double tendency involved in the original *élan* and the resistance that matter opposes to this *élan*—some organisms lean in the first direction, others in the second.[221] Many other splittings follow from this first one.[222] Hence the diverging lines of evolution, at least in terms of their essential characteristics. But we must also consider regressions, halts, and accidents of all sorts. And above all it must be remembered that each species behaves as if the general movement of life came to a halt in it, rather than passing through it. The species thinks only of itself

255

and lives only for itself.[223] This is the source of the innumerable battles that take place in the theater of nature. And it is the source of a striking and shocking disharmony, though we must not hold the principle of life to be responsible for this disharmony.[224]

Contingency thus plays a large part in evolution. The majority of forms adopted, or rather invented, are contingent. So too is the dissociation of the primordial tendency into some complementary tendencies that create the diverging lines of evolution (that is, relative to the obstacles encountered in a given place and at a given moment). The halts and regressions are contingent; the adaptations, to a large extent, are as well. Only two things are necessary: (1) the gradual accumulation of energy; (2) an elastic canalization of this energy in variable and indeterminable directions, at the end of which are free actions.

This double result was obtained in a certain way on our planet. But it might have been obtained through entirely different means. It was not necessary that life set its focus primarily on the carbon molecule of carbonic acid.[225] Storing up solar energy was essential. But rather than asking the Sun to separate, for example, the oxygen atoms from the carbon atoms, life could have exposed to the Sun different chemical elements (at least in theory, and ignoring for the moment the perhaps insurmountable practical difficulties), elements that would have had to have been associated or dissociated through completely different physical means. And had the characteristic element of the organism's energetic substances been something other than carbon, the characteristic element of the plastic substances would have probably been something other than nitrogen.[226] The chemistry of living bodies would have been radically different from what it is. This would have resulted in living forms without any analogy to the ones that we know, and whose anatomy and physiology would have been entirely different. Only the sensorimotor function would have been preserved, if not in its specific mechanisms, at least in its effects. Life, then, probably develops on other planets, in other solar systems as well, in forms that are inconceivable to us and in physical conditions that would seem, from the point of view of our own physiology, absolutely abhorrent to life. If life essentially aims at capturing usable energy so as to spend it in explosive actions, it surely chooses, in each solar system and on each planet, like it does on Earth, the most appropriate means to obtain this result in light of the conditions that it is subjected to. At least this is what analogical reasoning

256

tells us, and to declare that life is impossible wherever it is subjected to different conditions than those found on the earth is an abuse of this type reasoning. The truth is that life is possible everywhere that energy descends the incline described by Carnot's Principle and where a cause, moving in the inverse direction, can slow down that descent. In other words, life is surely possible on all of the worlds that orbit the stars.[227] Let me go further: it is not even necessary that life be concentrated and take form in organisms properly so called, that is, in definite bodies that offer some ready-made (though still elastic) canals to the flowing away of energy. It is conceivable (though we are hardly able to imagine it) that energy might be held in reserve and then spent along variable lines coursing through a not-yet-solidified matter. All of the essential characteristics of life would be there since there would still be a slow accumulation of energy and a sudden relaxation. Between this vague and indistinct vitality and the clearly defined vitality that we know, there would hardly be any more difference than there is, in our psychological life, between the dream state and the waking state.[228] Such might well have been the condition of life in our own nebula before the condensation of matter had been accomplished, if it is true that life begins at the very moment when, through the effect of an inverse movement, the matter of the nebulous appears.

It is thus conceivable that life might have taken on an entirely different outward appearance and sketched out very different forms from the ones we know. With a different chemical substrate and under different physical conditions, the impulse would have remained the same, but it would have been divided up very differently along the way and, as a whole, another pathway would have been taken—perhaps a shorter road, perhaps a longer one. In any case, along the entire series of living beings, not one single endpoint would have been what it is. Now, was it necessary that there be a series and endpoints? Why was this singular *élan*   258 not imprinted upon a single body that could have continued evolving indefinitely?[229]

This question arises, of course, when we compare life to an *élan*. And life must be compared to an *élan* since there is no other image borrowed from the physical world that gives a better idea of it.[230] But this is only an image. Life is in fact of the psychological order, and the essence of the psychological is to envelop a confused plurality of interpenetrating terms.[231] In space, and only in space, a distinct multiplicity is undoubtedly possible: one point is absolutely exterior to another point. But pure and empty unity or oneness [unité] is itself only encountered in space: this is the unity of a mathematical point.[a] Abstract unity and abstract multiplicity are either determinations of space or categories of the understanding, whichever you prefer, given that spatiality and intellectuality are adapted to each other.[232] But something that has a psychological nature does not quite fit into space, nor can it fully enter into the frameworks of the understanding. Is my person, at a given moment, one or multiple? If I declare that it is one, inner voices rise up and protest, namely the voices of sensations, feelings, and representations among which my individuality is divided. But if I make it distinctly multiple, my consciousness rebels just as strongly; it claims that my sensations, feelings, and thoughts are but abstractions that I execute upon myself, and that each one of my states involves all of the others.[b/233] Thus, I am—and here I must adopt the language of the understanding since the understanding alone has a language—a multiple-unity [unité multiple] and a singular-multiplicity [multiplicité une].[c/234] But unity and multiplicity are merely views or snapshots taken of my personality by an understanding that imposes its categories on me. I fit into neither the one nor the other, nor into both at once, even though the reuniting of the two could give an approximate imitation of this reciprocal interpenetration and continuity that I find in the depths of myself. Such is my inner life, and such is also life in general. If life is comparable to an impulse or an élan when it is in contact with matter, then when conceived of in itself, life is an immensity of virtuality, a mutual encroachment of thousands and thousands of tendencies that will, nevertheless, only be "thousands and

259

a TN: Note that Bergson uses *unité* ("unity," "oneness," or "singularity") in the mathematical sense of a singular entity equivalent to the number one or of a self-contained individual thing without parts. *Unité* should not here be read as the unification of parts, but rather as the uniqueness of identity.

b TN: ...*distinctement multiple*. Note the resonance with the above term *multiplicité distincte* ("distinct multiplicity"), in which we have a whole with well defined and self-contained parts, in contrast to "confused" or "indistinct" multiplicities, where the parts are interdependent and interpenetrating.

c Bergson's note: I have developed this point in an article titled "Introduction à la métaphysique," *Revue de métaphysique et de morale* 11, no. 1 (January 1903): 1–25. [TN: See IM. Regarding Bergson's terms *unité multiple* and *multiplicité une*, these intentionally paradoxical formulations (hence his interjection about language) are difficult to translate. The phrase could also be translated in a circuitous way as "*a unity that is a multiplicity and a multiplicity that is one (or singular)*."

thousands" once they have been externalized in relation to each other, that is, spatialized. Life's contact with matter decides this dissociation. Matter actually divides up what was only virtually multiple, and, in this sense, individuation is in part the work of matter and in part the effect of what life bears within itself.[235] As such, of a poetic feeling—working itself out into distinct stanzas, distinct verses, and distinct words—we can say it contained this multiplicity of individuated elements and that nevertheless the materiality of language creates this multiplicity.[236]

But the simple inspiration, which is the whole of the poem, courses through the words, the verses, and the stanzas. In the same way, life still circulates among the dissociated individuals. The tendency to individuate is everywhere resisted and, at the same time, completed by an antagonistic and complementary tendency to associate, as if the multiple-unity of life, being drawn toward multiplicity, made all the more effort to pull back into itself.[237] No sooner is a part detached than it tends to reunite itself, if not with all the rest, then at least with what is closest to it. This leads to a certain equilibrium between individuation and association across the entire domain of life. Individuals are juxtaposed in a society; but the society, having hardly even been formed, would like to meld the juxtaposed individuals into a new organism so as to become itself an individual that can, in turn, become an integral part   260 of a new association.[238] At the lowest degree of the scale of organisms, we already find genuine associations, namely microbial colonies. And, if we are to believe a recent study, within these associations there is a tendency to individuate through the constitution of a core or nucleus.[a/239] The same tendency is found one step higher on the scale, among the Protophytes that, having just left the parent cell by means of division, remain united with each other through the gelatinous substance that covers their surface, much like the Protozoa that begin by intertwining their pseudopods and ultimately become fused together.[240] The so-called "colonial" theory of the genesis of higher organisms is well known. The Protozoa, each constituted as a single cell, would have formed aggregates by juxtaposing themselves; these aggregates, by coming together in turn, would have created aggregates of aggregates. In this

---

a   Bergson's note: W. Podwyssozki, "Compte-rendu: *Sur la structure des colonies bactériennes par S. Serkovski," L'année biologique: Comptes rendus annuels des travaux de biologie générale* 4 (1898): 317.

way, increasingly complex and increasingly differentiated organisms would have been born through the association of elementary and barely differentiated organisms.[a/241] In this extreme form, the theory invited some serious objections; and the idea that polyzoism is an exceptional and abnormal occurrence appears to be increasingly confirmed.[b/242] But it is no less true that things happen *as if* every higher organism were born from an association of cells that divided up the work among themselves. In all likelihood, it is not the cells that made the individual by way of association; rather, it is the individual who made the cells by way of dissociation.[c/243] But even this reveals that the social form haunts the genesis of the individual, as if the individual could only develop on condition of dividing up its substance into elements themselves having an appearance of individuality and united among themselves by an appearance of sociability. There are numerous cases where nature seems to hesitate between the two forms, and seems to ask itself whether it should constitute a society or an individual. In such cases, the slightest impulse is enough to tip the balance to one side or the other.[244] If we take a relatively large Infusorian, such as the Stentor, and if we cut it into two halves with each containing a part of the nucleus, each of the two halves regenerates an independent Stentor.[245] But if the division is not completed, leaving behind some protoplasmic connection between the two halves, then they perform, each on its own, movements in perfect synergy, such that here it is enough that a thread be either maintained or cut for life to adopt the social form or the individual form. As such, in rudimentary single cell organisms, we already see that the apparent individuality of the whole is the composition of an undefined number of virtual individualities that are associated virtually. But the same law is manifest from the bottom to the top of the series of living beings. And this is what we express when we say that unity and multiplicity are categories of inert matter, that the *élan vital* is neither pure unity nor pure

261

---

a   Bergson's note: Edmond Perrier, *Les colonies animales et la formation des organismes*, 2nd ed. (Paris: Masson, 1898).

b   Bergson's note: Yves Delage, *L'hérédité et les grands problèmes de la biologie générale*, 97. Cf., from the same author, "La conception polyzoïque des Êtres," *Revue scientifique (Revue rose)*, 4th series, 5, no. 21 (May 23, 1896): 641–53.

c   Bergson's note: This is the theory defended by Kunstler, Delage, Sedgwick, Labbé, etc. One can find an account of it, along with bibliographic references, in the work by Paul Busquet, *Les êtres vivants: Organisation–évolution* (Paris: Georges Carré et C. Naud, 1899).

multiplicity, and that if the matter into which it passes requires that it opt for one of the two, then its choice will never be definitive—it will forever leap from one to the other.[246] There is nothing accidental, then, in the evolution of life in the double direction of individuality and asso-ciation. This evolution comes from the very essence of life.

The march toward thought or reflection [réflexion] is also essential.[247] If my analyses are correct, then consciousness, or better supra-consciousness, is at the origin of life.[248] Consciousness or supra-consciousness is the firework whose extinguished debris falls into matter; and consciousness is also what subsists of the firework itself, coursing through the debris and lighting it up as organisms.[249] But this consciousness, which is a *need for creation*, only shows itself to itself where creation is possible.[250] Consciousness goes to sleep when life is condemned to automatism; it awakens the moment that the possibility of a choice is reborn. This is why, in organisms lacking a nervous system, consciousness varies in pro-portion to the power the organism has for locomotion and deformation. And, for animals with a nervous system, consciousness is proportional to the complication of the crossroads where the so-called sensory pathways and motor pathways intersect, i.e., the brain. But how should we under-stand this interdependence between the organism and consciousness?

262

[Humanity]

I will not insist here on a point I have developed at length in a pre-vious work.[251] I will merely recall that the theory according to which consciousness would be attached to certain neurons, for example, and would free itself from their workings like a sort of phosphorescence, can surely be accepted by the scientist with regard to the detail of the analysis; it is a convenient way of expressing oneself. But that is all it is. In reality, a living being is a center of action.[252] It represents a certain sum of contingency being introduced into the world, a certain quan-tity of possible actions—a quantity that varies among individuals and, above all, among species.[253] An animal's nervous system sketches out the flexible lines along which its action will run (even though the potential energy to be released is accumulated in the muscles rather than in the nervous system itself). Its nervous centers indicate, through their devel-opment and configuration, the more or less extensive set of choices that the organism will have among actions that are themselves more or less numerous and complicated.[254] Now, given that the awakening of consciousness for a living being is more complete insofar as a larger latitude of choice is left to it, and insofar as a more considerable sum of action is bestowed upon it, it is clear that the development of consciousness will appear to be modeled upon the development of the nervous centers. Moreover, every state of consciousness being (from a certain angle) a question posed to motor activity and even the beginnings of a response, there is no psychological fact that does not also involve the coming into play of the cortical mechanisms. Everything will thus appear to happen as if consciousness sprang forth from the brain, and as if the detail of conscious activity were modeled upon that of cerebral activity.[255] In reality, consciousness does not spring from the brain; rather, the brain and consciousness correspond because they both equally measure—the one through the complexity of its structure, the other through the intensity of its wakefulness—the quantity of choice that the living being has at its disposal.

The psychological state says more than the cerebral state does precisely because a cerebral state merely expresses what there is of nascent action in the corresponding psychological state.[256] As I have attempted to show elsewhere, a living being's consciousness is interdependent with its brain, in the sense in which a sharp knife is interdependent with its edge: the brain is the sharp tip by which consciousness penetrates into the compact fabric of events, but it is no more coextensive with consciousness than the sharp edge is coextensive with the knife.[257] So we cannot conclude from the fact that two brains are very similar, such as the brain of an ape and that of a human, that the corresponding consciousnesses are comparable or even commensurate with each other.[258]

And indeed, perhaps the two brains are less similar than is often assumed. How could we not be struck by the fact that the human is capable of learning any activity whatsoever, fabricating any object whatsoever, in short, acquiring any motor habit whatsoever, whereas the faculty of combining new movements is strictly limited for the most gifted animals, even for apes.[259] Here is where we find the cerebral characteristic of humans. Like all brains, the human brain is designed for setting up motor mechanisms and for allowing us to choose, at a given moment, the one that we will set into motion through a trigger mechanism. But it differs from other brains insofar as the number of mechanisms that it can set up, and consequently the number of triggers that it can choose among, is indefinite. Now, from the limited to the unlimited, there as much distance as there is from the closed to the open. This is not a difference of degree, but rather a difference of nature.

Radical too is the difference, therefore, between animal consciousness (even the most intelligent animals) and human consciousness. For consciousness corresponds exactly to the power of choice that the living being has at its disposal; it is coextensive with the fringe of possible action that surrounds real action.[260] Consciousness is synonymous with invention and freedom. Now, for animals, invention is never more than a certain variation upon the theme of its routine. Imprisoned within the habits of the species, the animal perhaps manages to enlarge those habits through its individual initiative. But it only escapes automatism for an instant, just long enough to create a new automatism. The doors of its prison close again as soon as they were opened. By pulling on its chain, the animal succeeds merely in lengthening it a bit. With man, consciousness breaks the chain. In man, and in man alone, consciousness liberates itself. The entire history of life up until then had been the history of an effort by consciousness to lift up matter, and of consciousness being more or less completely crushed by the matter that fell back on it. The project was paradoxical—if, that is, one can speak here (other than metaphorically) of a project and of an effort. It was an attempt to create out of matter, which is necessity itself, an instrument of freedom, to fabricate a mechanism that could triumph over mechanism, and to make use of nature's determinism in order to slip through the gaps in the net that it had set.[261] But consciousness, everywhere other than in man, gets caught by the net whose gaps it had wanted to slip through. Consciousness remained captive of the very mechanisms that it itself had set up. The automatism, which consciousness claimed to draw in the direction of freedom, coils back around it and drags it down. Consciousness does not have the force to free itself, because the energy it had stored up for actions is almost entirely employed in maintaining the infinitely subtle and essentially unstable equilibrium to which it had brought matter. But man does not merely maintain his machine: he is able to make use of it as he pleases. He owes this ability, of course, to the superiority of his brain, which allows him to construct an unlimited number of motor mechanisms, to continuously set up new habits atop the old ones, and, by dividing automatism against itself, to conquer

265

it.[262] He owes it to his language, which provides consciousness with an immaterial body into which it can be incarnated and that thereby frees it from having to impose itself exclusively upon material bodies that would first be carried off by the flux, and that would soon after be swallowed up. He owes it to social life, which stores and preserves the efforts that have been made (in the way that language stores thought), establishes in this way an average level to which individuals must first rise up, and, through this initial incitement, prevents the mediocre ones from falling off to sleep and pushes the best to rise up even higher.[263] But our brain, society, and language are merely the diverse and external signs of a single and self-same inner superiority. They each announce, in their own way, the unique and exceptional success that life achieved at a given moment of its evolution.[264] They translate the difference of nature, and not merely of degree, that separates man from the rest of animality. They allow us to conjecture that if, taking off from the vast springboard upon which life launched its élan, all of the other animals fell back down, finding the rope too high, then man alone cleared the obstacle.

In this very specific sense, man is the "endpoint" and the "goal" of evolution. Life, I said, transcends finality and all other categories.[265] It is essentially a current launched through matter, drawing from it whatever it can. Thus, properly speaking, there was no project and no plan. Moreover, it is obvious that the rest of nature is not simply for the sake of man: we struggle like the other species; we struggled against the other species.[266] In short, if the evolution of life had run into different accidents along its route, and if, thereby, the current of life had been otherwise divided up, we would have been both physically and morally or spiritually quite different from what we are.[a/267] For these various reasons, it would be wrong to consider humanity, such as we have it before our eyes, as pre-formed within the movement of evolution. It cannot even be said that humanity is the result of evolution as a whole, since evolution is accomplished along several diverging lines, and, even if the human species is at the extreme endpoint of one of those lines, other lines have been followed with other species at their endpoints. I consider humanity to be the *raison d'être* of evolution in a very different sense.

266

---

a TN: Note here and just below that Bergson uses the adverb *moralement* ("morally" or "psychologically") not in the sense of "ethically," but rather in the sense in which he understands the term *sciences morales*, i.e., relating to the spiritual, the psychological, and the intellectual (as opposed to the material). Cf. the translator's note to page 213.

From my point of view, life in general appears to be an immense wave that spreads forth from a center and that, along nearly the totality of its circumference, comes to a halt and is converted into an oscillation in place. But at one single point, the obstacle has been overcome and the impulse passes through freely. This is the freedom registered in the human form. Everywhere else, consciousness finds itself running into a dead end; in man alone was consciousness able to continue along its pathway. Man thus continues the vital movement indefinitely, although he does not himself drag along all that life carried in itself. Other tendencies included in life made their way along other lines of evolution, and since everything is interpenetrating, man probably preserved something of these, though only very little of them. *Everything takes place as if an indecisive and vague being, a being we could call, if you will, man or super-man, had attempted to realize himself and yet was only able to do so by abandoning a part of himself along the way.*[268] These waste products are represented by the rest of animality, and even by the plant world, at least in what the plant world possesses that is positive and superior to the accidents of evolution.

267

From this point of view, the disharmonies offered up by the spectacle of nature are attenuated in a very particular way. The whole of the organic world becomes something like the humus in which man, or some other being resembling man morally or spiritually, was to grow. Animals, as distant from us as they may be, and as much as they may be enemies of our species, have been no less useful traveling companions to whom consciousness was able to pass off the heavy weight that it was carrying, and who made it possible for consciousness to rise up, in man, to the heights from which it sees an unlimited horizon open up before itself.

True, consciousness did not only abandon *en route* a cumbersome piece of luggage. It was also forced to give up some valuable goods. In man, consciousness is primarily intellect. It could have, or (so it seems) it should have also been intuition. Intuition and intellect represent two opposite directions of conscious work: intuition marches in the same direction as life; intellect goes in the inverse direction and thereby finds itself quite naturally attuned to the movement of matter.[269] A complete and perfect humanity would be one in which these two forms of conscious activity reach their full development.[270] Now, we can conceive of many possible intermediaries between that perfect humanity and our own, corresponding to all the imaginable degrees of intellect and intuition. And here we find the role of contingency in the mental structure of our species. A different evolution might have led to an even more intelligent humanity or to a more intuitive one. In fact, in the humanity to which we belong, intuition is almost completely sacrificed to the intellect. It seems that in order to conquer matter and in order to reconquer itself, consciousness had to use up the better part of its force.[271] This conquest, within the specific conditions in which it was accomplished, required consciousness to adapt itself to the habits of matter and to concentrate all of its attention upon them—in short, to determine itself more specifically as intellect.[272] Nevertheless, intuition is there too, though it is vague and above all discontinuous. Intuition is a nearly extinguished lamp that is only rekindled here and there and for hardly more than a few instants. But in the end, it is rekindled wherever a vital interest is at stake. Intuition casts a flickering and weak light upon our personality, our freedom, the place that we occupy in the whole of nature, our origin, and perhaps even our destiny, but this light no less pierces the obscurity of the night in which the intellect abandons us.[273]

Philosophy must seize upon these fleeting intuitions, which only cast their light upon their object from farther and farther away, first to support them and then to expand them and link them together. The more it advances in this work, the more it realizes that intuition is mind itself and, in a certain sense, life itself. The intellect is cut out of this by a process that mimics the one that engendered matter.[274] This is how the unity of mental life appears. We only recognize this by placing ourselves within intuition in order to go from intuition to the intellect, since we will never be able to reach intuition by beginning from the intellect.[275]

[The Life of the Body and the Life of the Mind]

Thus, philosophy introduces us into spiritual life [la vie spirituelle]. And it shows us, at the same time, the relationship of the life of the mind [la vie de l'esprit] to the life of the body.[276] The major error in spiritualist doctrines was to have believed that by isolating spiritual life from all the rest, by suspending spiritual life as high up in space as possible, high above the earth, they were sheltering it from all attacks—as if these doctrines were not thereby simply exposing spiritual life to the risk of being mistaken for nothing more than a mirage![277] Of course, the spiritualists are right to listen to consciousness when it proclaims human freedom—but the intellect is there too, and it claims that the cause determines its effect, that the same brings about the same, that everything is repeated and everything is given.[278] They are right to believe in the absolute reality of the person and in his independence vis-à-vis matter—but science is there too, and it demonstrates the interdependence of conscious life and cerebral activity.[279] They are right to attribute to man a privileged place in nature, and to hold that the distance between the animal and man is infinite—but the history of life is there too, and it shows us the genesis of species by way of gradual transformation and thereby seems to reintegrate man into animality. When some powerful instinct proclaims the likely survival of the person, they are right not to shut their ears to its voice—but if "souls" exist in this way, capable of an independent life, then where do they come from? And when, how, and why do they enter into this body that we see, with our own eyes, emerging quite naturally from a mixed cell borrowed from the bodies of its two parents?[280] All of these questions will remain unanswered, and a philosophy of intuition will be merely the negation of science, sooner or later to be swept away by science, so long as it does not resolve itself to see the life of the body where it truly is, namely on the pathway that leads to the life of the mind.[281] But as such, this philosophy will no longer be concerned with some determinate living beings. Life as a whole, beginning from the initial impulse that launched it into the world, will appear to the philosophy of intuition as an ascending wave that is opposed by the descending movement of matter. Across the majority of its surface, and at various heights, the current is converted by matter into a vortex that turns in place. But at one single point, the current passes freely, drawing along with it the obstacle that will weigh upon its progress but that will not bring it to a halt. Humanity is found at that point; that is our privileged situation. Moreover, this rising wave is consciousness, and, like all consciousness, it envelops innumerable interpenetrating virtualities for which the categories of unity and multiplicity, having been made for inert matter, are hardly appropriate.[282] Only the matter that this wave sweeps along with it, and into whose interstices it inserts itself, can divide the wave up into distinct individualities. The current thus continues along, passing through human generations and subdividing itself into individuals. This subdivision was vaguely suggested in it, but without matter, it would have never become pronounced.[283] This

270

is how souls are continually being created that, nevertheless, in a certain sense, preexisted.[284] They are nothing other than the small streams into which the great river of life divides itself up, flowing through the body of humanity. The movement of a current is distinct from that through which it flows, even though it must necessarily adopt its various sinuosities.[285] Consciousness is distinct from the organism that it animates, even though it must suffer certain of its vicissitudes. Possible actions, whose rough outlines are contained in a state of consciousness, receive, at each moment, the beginnings of a performance in the nervous centers; in the same way, the brain accentuates, at each moment, the motor organization of the state of consciousness. But that is the extent of the interdependence between consciousness and the brain; the fate of consciousness is not tied to the fate of the cerebral matter.[286] In short, consciousness is essentially free, it is freedom itself; but it cannot move through matter without coming to rest upon it and without adapting itself to it. I call this adaptation "intellectuality," and the intellect, turning itself back around toward acting consciousness, that is, free consciousness, naturally forces it into the frameworks into which it customarily sees matter being inserted. The intellect will thus always see freedom in the form of necessity; it will always neglect the aspect of novelty or of creation inherent to the free act; it will always substitute, for the action itself, an artificial and approximate imitation obtained by combining the old with the old and the same with the same.[287] In this way, from the perspective of a philosophy that attempts rather to reabsorb the intellect into intuition, many difficulties vanish or subside.[288] But such a doctrine does not just facilitate speculation. It also gives us more force for acting and for living.[289] Because through this philosophy we no longer feel isolated in humanity, and humanity no longer seems isolated in the nature that it dominates.[290] Just as the smallest grain of dust is interdependent with our solar system as a whole, swept along with it in this undivided movement of descent that is materiality itself, so all organic beings—from the most humble to the most elevated, from the very first origins of life right up to the time in which we find ourselves, and in all places and all times—do nothing but render visible an impulse that is unique, the inverse of the movement of matter, and in itself indivisible.

271

All living beings stand together, and all yield to the same incredible thrust. The animal leans on the plant, man rides atop animality, and humanity as a whole, in space and in time, is an enormous army that gallops alongside each of us, out in front of us and behind us, in a rousing forward charge capable of overpowering all resistances and of clearing so many obstacles, perhaps even death.[291]

IV

THE CINEMATOGRAPHIC MECHANISM OF THOUGHT AND THE MECHANISTIC ILLUSION.[a/1] A GLANCE AT THE HISTORY OF SYSTEMS. REAL BECOMING AND FALSE EVOLUTIONISM

a Bergson's note: The part of this chapter that discusses the history of systems, and Greek philosophy in particular, is but a very succinct summary of the positions that I developed in detail in my lectures at the Collège de France between 1900 and 1904, and notably during a course titled *Histoire de l'idée de temps* (1902–1903). In those lectures, I compared the mechanism of conceptual thought to that of the *cinematographic projector* [*cinématographe*]. I believe it possible to take up this comparison again here.

DOI: 10.4324/9781315537818-5

[SKETCH OF A CRITIQUE OF SYSTEMS FOUNDED ON THE ANALYSIS OF THE IDEAS OF NOTHINGNESS AND IMMUTABILITY. EXISTENCE AND NOTHINGNESS]

Two theoretical illusions that we have continually encountered along the   272
way remain to be examined in themselves. Thus far, I have considered
only their consequences and not their principle or source.[a]/2 The examination of these two illusions will be the focus of the present chapter.
This will provide the opportunity to set aside certain objections, to dissipate certain misunderstandings, and above all to more clearly define,
by contrasting it with other philosophies, a philosophy that sees durée as
the very fabric of reality.

Whether matter or mind, reality has shown itself to be a perpetual
becoming. Reality is either making itself or unmaking itself, but it is
never something made.[b]/3 This is the intuition we have of the mind
when we pull aside the veil that interposes itself between us and our
own consciousness.[4] This is also what the intellect and the senses themselves would show of matter, were they to obtain an immediate and   273
disinterested representation of it. But, preoccupied above all else by
the necessities of action, the intellect, like the senses, limits itself to
taking a certain number of instantaneous and thereby immobile views
or snapshots of the becoming of matter. Consciousness, modeling itself
in turn upon the intellect, looks at what is already made in inner life
and only vaguely senses inner life being made. In this way, the moments
that interest us and that we have gathered up along the way are detached
from durée. We retain only these detached moments. And we are right to
do so, so long as action is our only concern. But when we are speculating
about the nature of the real, if we continue to look at it the way our practical interests require, we become incapable of seeing true evolution or
radical becoming.[5] Of becoming, we only see states; of durée, we only see
instants; and even when we are speaking about durée and becoming, we
are thinking about something else. This is the more striking of the two

---

a TN: Here Bergson again uses *principe* ("principle") in all of its senses, as in "fundamental
truth of X," as in "law" or "theoretical statement," and (primarily) as in metaphysical
"source" or "origin." Cf. below, 275 and the Translator's Introduction.
b TN: *Elle se fait ou elle se défait, mais elle n'est jamais quelque chose de fait.* For more on
the translation of Bergson's uses of *se faire* ("being made," "making itself," or "in the
making"), see the Translator's Introduction and editorial note 191 to Chapter III.

illusions that I want to examine in this chapter. It consists in believing that we will be able to think the unstable through the intermediary of the stable, or the moving through the immobile.

The other illusion is closely related to the first. It has the same origin. It too comes from the fact that we transport a procedure designed for practice into the realm of speculation. All action aims at obtaining an object we feel deprived of or creating something that does not yet exist. In this very specific sense, action fills a void and goes from the empty to the full, from an absence to a presence, from the unreal to the real. Moreover, the unreality in question here is purely relative to the direction in which our attention is engaged, for we are immersed in realities and cannot escape. It is simply that when the present reality is not the one we were looking for, we speak of the *absence* of this second reality where we observe the *presence* of the first. In this way, we express what we have as a function of what we would like to obtain. Nothing could be more legitimate in the domain of action. And yet, for better or worse, we preserve this manner of speaking and thinking when we speculate about the nature of things independently of the interest they might have for us. This is how the second of the two illusions that I mentioned is born, and I will examine it first. This second illusion, like the first, comes from the static habits our intellect contracts when it is preparing our action on things. Just as we pass through the immobile in order to arrive at the moving, we also make use of the empty in order to think the full.

We already encountered this illusion along the way when I broached the fundamental problem of knowledge.[6] As I said, the question is to know why there is order and not disorder in things. But this question only makes sense if one assumes that disorder, understood as an absence of order, is possible, imaginable, or conceivable. Now, there is no reality [*réel*] other than order; but, since order can take two forms, and since the presence of one consists, if you will, in the absence of the other, we speak of disorder each time that we are confronted with the one that we were not looking for. The idea of disorder is thus an entirely practical idea. It refers to a certain disappointment of a certain expectation, and it does not designate the absence of all order, but merely the presence of an order that is of no current interest. If one were to try to absolutely and completely negate order, they would find themselves jumping indefinitely from one kind of order to the other, and would realize that the supposed suppression of both of them actually implies the presence of the two. Finally, if they simply move on, if having made a decision they close their eyes to this movement of the mind and to all that it presupposes, then they are no longer dealing with an idea, and nothing remains of disorder but the word itself. As such, the problem of knowledge is complicated, and perhaps even rendered insoluble, by the idea that order fills a void and that its actual presence is superposed upon its virtual absence. We go from absence to presence, from empty to full, by virtue of the fundamental illusion of our understanding. I have already indicated one consequence of this error in the previous chapter.[7] As I already began to suggest, we will never definitively overcome this error until we confront it head on. We must look it directly in the face, in itself, and in the radically false conception that it involves of negation, void, and nothingness.[a/8]

["THE IDEA OF NOTHINGNESS"]

Philosophers have hardly paid any attention to the idea of nothingness. And yet, this idea is often the hidden mechanism or the invisible motor of philosophical thinking. From the moment reflection first awakens, it is the idea of nothingness that pushes to the forefront—and places directly under the gaze of consciousness—the problems that cause anxiety and the questions that we cannot focus on without being overcome by vertigo.[9] Having barely begun to philosophize, I already wonder why I exist. And when I become aware of the interdependence that connects me to the rest of the universe, the difficulty is merely pushed back since now I want to know why the universe exists. And if I then relate the universe back to some immanent or transcendent Principle that supports it or that creates it, my thought only rests easy in this principle for a few moments. The same question is again asked, this time in all of its fullness and generality: how does it come about and how can we understand that something exists? Even here, in this current book, when I defined matter as a kind of descent, this descent as the interruption of an ascent, and this ascent itself as a growth, or when I ultimately placed a Principle of

a Bergson's note: The analysis that I provide here of the idea of nothingness was previously published in *Revue philosophique* (November 1906). ["L'idée de néant," *Revue philosophique de la France et de l'étranger* 62, no. 11 (1906): 449–66. TN: I have added a paragraph break and the title of this article as a section title here to mark off this previously published material, which continues until page 298 below.]

creation at the foundation of things, the same question looms up: how and why does this principle exist, rather than nothing?[10]

276   Now, if I push these questions aside in order to go directly to what is hiding behind them, here is what I find. Existence appears to be a conquest over nothingness. I say to myself that there might not have been (and even should not have been) anything there at all, and so I am amazed that something is indeed there. Or perhaps I imagine all of reality as if extended across nothingness, like a carpet on the floor: nothingness was there first, then being came along as an addition. Or again, if something had always existed, then nothingness must have always served as its substrate or as its receptacle, and so nothingness would be eternally anterior to it.[11] Even if a glass has always been full, the liquid that fills it no less fills a void. In the same way, being might well have always been there: the nothingness that is filled up and, as it were, stuffed by being no less preexists being—if not in fact, then at least in principle. In short, I cannot rid myself of the idea that the full is embroidered upon the canvas of the void, that being is superposed upon nothingness, and that there is less in the idea of "nothing" than there is in the idea of "something." This is the source of the entire mystery.

This mystery must be solved. And if durée and free choice are to be placed at the foundation of things, then solving the mystery becomes all the more urgent. After all, the contempt that metaphysics holds for every reality that endures comes precisely from the fact that metaphysics does not arrive at being without first passing through "nothingness," and from the fact that it does not think that an existence that endures is strong enough to vanquish inexistence and to establish itself. This is the primary reason metaphysics tends to attribute a *logical* existence to true being, and not a psychological or physical existence.[12] For such is the nature of a purely logical existence, namely, it seems to be self-sufficient and to establish itself simply as a result of the force that is immanent to the truth. If I ask why bodies or minds exist rather than nothing, I find no response. But it seems quite natural to me that a logical principle (such as $A = A$) has the power of creating itself and of triumphing

277   over nothingness for all eternity.[13] The appearance of a circle traced in chalk on a blackboard is something that needs to be explained: by itself, this entirely physical existence does not have what it takes to vanquish inexistence. But the "logical essence" of the circle, that is, the possibility of tracing it according to a certain law—in short, its definition—is

something that seems eternal to me. This logical essence has no place and no date since nowhere and at no moment did the drawing of a circle begin to be possible.[14] Let us assume, then, that the principle upon which all things depend and by which all things manifest an existence is of the same nature as the definition of the circle, or as the axiom A = A. In this case, the mystery of existence evaporates since the being that is at the foundation of all things thus establishes itself in eternity, in the same way as logic itself establishes itself. Of course, this will cost quite a large sacrifice, for if the principle of all things exists in the manner of a logical axiom or a mathematical definition, then the things themselves must emerge from this principle like the applications of an axiom or the consequences drawn from a definition, and so there will no longer be any place, neither among the things nor at the level of their principle, for any efficient causality understood as free choice. Such are precisely the conclusions of Spinoza's doctrine, or even of Leibniz's, for example—and indeed, such was the genesis of their doctrines.[15]

If we could establish that the idea of nothingness (in the sense in which we understand it when we oppose it to the idea of existence) is a pseudo-idea, then the problems that it raises in its vicinity would become pseudo-problems.[16] The hypothesis of an absolute that could act freely and that could endure eminently would no longer be shocking.[17] The way would be opened up for a philosophy closer to intuition—a philosophy that would no longer require common sense to make the same sacrifices.[18]

Let us see, then, what we are thinking about when we speak of nothingness. Thinking about nothingness consists in either imagining it or conceiving of it. Let us examine what this image or this idea could be.   278
I'll begin with the image.

I will close my eyes, stop my ears, and extinguish one by one the sensations that come to me from the external world. There, it is done—all of my perceptions vanish, and the material universe disappears into silence and darkness.[19] I nevertheless subsist, and I cannot stop myself from subsisting. I am still there, along with the organic sensations that come from the periphery or from the interior of my body, with the memories my past perceptions have left behind, and even with the entirely positive and full impression of the void that I have just created around myself. How could I suppress all of that? How could I eliminate myself? I might well be able to set aside my memories and forget even my most immediate past. But I will still have, at the very least, the consciousness of my present, even if reduced to its most extreme poverty, that is, to the actual state of my body.[20] Nevertheless, let me attempt to do away with this consciousness itself. I will progressively weaken the sensations that my body sends to me. There, now they are on the verge of being extinguished, and now they are extinguished—they disappear into the darkness into which all things were already lost. But no! At the very moment my consciousness is extinguished, another consciousness lights up. Or rather, it was already lit up, since it had sprung forth a moment earlier so as to witness the disappearance of the first one. Indeed, the first consciousness could only disappear for another one, and in relation to another one. I can only witness myself annihilated if I have already resuscitated myself through a positive act, even if that act is involuntary and unconscious. So despite my best efforts, I always perceive something, either from the outside or from the inside. When I no longer know anything of the external objects, this is because I take refuge in the consciousness that I have of myself; if I abolish this internal object, then this abolishing [*abolition*] itself becomes an object for an imaginary self who, this time, perceives the self that disappears as an external object. External or internal, there is always an object that my imagination is imagining. True, it can go from one to the other, and imagine in turn a nothingness of external perception and a nothingness of internal perception—but it cannot imagine the two at the same time, given that the absence of the one ultimately consists in the exclusive presence of the other.[21] And yet, from the fact that two relative nothingnesses are imaginable, one after the other, we erroneously conclude that they are imaginable together. The absurdity of this conclusion should be obvious since we cannot imagine a nothingness without being aware, at least vaguely, that we are imagining it, that we are acting, that we are thinking, and that something, therefore, still subsists.

Thus, thought never actually forms the image (properly speaking) of a suppression of everything. The effort by which we attempt to create this image ends up simply causing us to oscillate between the vision of an external reality and the vision of an internal reality. In this back-and-forth of our mind, oscillating between the outside and the inside, there is a point situated at an equal distance between the two where it seems to us that we no longer see the one and that we do not yet see the other. That is where the image of nothingness is formed. In reality, having arrived at the point where the two terms are adjacent, we actually see them both. Indeed, the image of nothingness thus defined is an image that is full of things. It includes at once the image of the subject and that of the object, with, in addition, a perpetual leaping from one to the other and the refusal of ever definitively coming to rest upon one of the two. Clearly this is not the nothingness that we could place opposite being, and place before being or underneath it, since this nothingness already includes existence in general.

But some will object that if the visible or latent idea of nothingness intervenes in the philosophers' reasonings, it does so not in the form of an image, but in the form of an idea. Admitting that we do not imagine an abolishing of everything, they will claim rather that we can conceive of such an abolishing. As Descartes said, we understand a polygon with a thousand sides, even though we cannot see it in our imagination: it is enough that we can clearly imagine the possibility of constructing it.[22] The same would be true for the idea of abolishing of all things. They will say that the procedure for constructing this idea could not be simpler. Indeed, every single object in our experience can, in turn, be assumed to be abolished. Let us extend this abolishing of a first object to a second, then to a third, and so on for as long as you like: nothingness is simply the limit toward which this operation tends. And nothingness, defined in this way, is clearly the abolishing of everything. —That is the theory. Examining it in this form is enough to show the absurdity that it harbors.

An idea that the mind builds up from separate pieces is in fact only an idea, properly speaking, if the pieces are capable of coexisting together. It would be reduced to a mere word if the elements that we bring together to compose it drove each other away to the extent that they are assembled. When I have defined the circle, I can easily imagine a black circle or a white circle, a circle in cardboard, in iron, or in brass, a transparent circle or an opaque one—but I cannot imagine a square circle because the law for generating the circle excludes the possibility of delimiting this figure with straight lines. In this way, my mind can surely imagine abolishing any existing thing whatsoever—but if the abolishing of anything whatsoever by the mind were an operation whose mechanism involved working on a part of the Whole, and not on the Whole itself, then the extension of such an operation to the totality

280

of things might well become something absurd, or self-contradictory. And perhaps the idea of abolishing everything would present the same characteristics as the idea of a square circle—it would no longer be an idea, but rather merely a word. Let us examine, then, the mechanism of this operation more closely.

281   In fact, the object we suppress is either external or internal; it is either a thing or a state of consciousness. Consider the first case. I abolish, in thought, an external object. "There is no longer anything" at the place it used to be. — No longer anything of this object, that is certain, but another object took its place, since there is no absolute void in nature. Nevertheless, let us suppose that an absolute void is possible. This is not the void I am thinking of when I say that the abolished object leaves its place unoccupied, since this claim has to do with a *place*, i.e., with a void delimited by precise borders, which is a kind of *thing*. The void I am speaking of is thus, at root, merely the absence of the determinate object that was here and that now finds itself elsewhere, an object that, insofar as it is no longer in its previous place, leaves behind, so to speak, the void of itself. A being endowed with neither memory nor foresight would never utter the words "void" or "nothingness" here: he would simply express what is and what is perceived. Now, what is and what is perceived is always the *presence* of one thing or another, never the *absence* of anything whatsoever. Absence only exists for a being capable of remembering and anticipating. He remembered an object and expected to perhaps encounter it again; he finds another object there instead, and expresses the disappointment of his expectation—an expectation itself born from memory—by saying that he no longer finds anything, and that he has come up against nothingness. Even if he did not actually expect to encounter the object, then it is a possible expectation of finding this object (and again the eventual disappointment of this expectation) that he translates by saying that the object is no longer where it was. What he is actually perceiving, or what he succeeds in actually thinking, is the presence of the previous object in a new place or of a new object in the previous place. The rest—everything expressed negatively by words such as "nothingness" or "void"—is less thought than emotion, or, to be more precise, the emotional coloring of thought.

282   The idea of abolishing or of partial nothingness thus takes shape here, in the course of the substituting of one thing for another, the moment that this substitution is thought by a mind that would have preferred to

maintain the previous thing in the place of the new thing, or that, at the very least, conceives of this preference as a possibility. The idea implies a preference on the subjective side, a substitution on the objective side, and is nothing but a combination, or rather an interference, between this feeling of preference and this idea of substitution.

Such is the mechanism of the operation through which our mind abolishes an object and is able to imagine a partial nothingness in the external world. Now let us see how it imagines this within itself. The phenomena that we observe in ourselves are still phenomena that take place, and clearly not ones that do not take place. I experience a sensation or an emotion, I conceive of an idea, I make a resolution. My consciousness perceives these facts, which are so many *presences*, and at every moment, some facts of this kind are present to me. I can surely think of an interruption in the course of my inner life; I can imagine that I am sleeping without dreaming or even that I have ceased to exist. But at the very instant that I make this assumption, I conceive of myself and imagine myself watching over my slumber or surviving my own annihilation. I only give up perceiving myself from within in order to take refuge in the external perception of myself. In other words, here again the full is always succeeded by the full, and an intellect that was nothing but intellect—one that had neither regret nor desire, one that modeled its movement upon the very movement of its object—would not even be able to conceive of an absence or a void.[23] The conception of a void is born here when consciousness, lagging behind itself, remains attached to a memory of a previous state when another state is already present.[24] This idea is but a comparison between what is and what could or should be, between something full and something full. In short, whether it has to do with a void of matter or a void of consciousness, *the idea of a void is always a full idea, one that, under analysis, resolves into two positive elements: the distinct or confused idea of a substitution and the experienced or imagined feeling of a desire or a regret.*

It follows from this double analysis that the idea of absolute nothingness, understood in the sense of an abolishing of everything, is a self-destructive idea, a pseudo-idea, a mere word. If the suppression of one thing consists in replacing it by another, if thinking the absence of one thing is only possible by the more or less explicit representation of the presence of something else, and finally, if above all else abolishing means substituting, then the idea of "abolishing everything" is as absurd as the idea of a square circle. This absurdity is not immediately obvious

283

because there is no single object that we cannot imagine abolished. So from the fact that there is nothing to stop the suppression in thought of each thing, one by one, we conclude that it is possible to imagine them all suppressed as a whole. We fail to see that the suppression of each thing, one by one, consists precisely in gradually replacing each one by another, and that as a result the absolute suppression of everything involves a genuine contradiction in terms since this operation would consist in the destruction of the very condition that makes it possible.

But the illusion is tenacious. Though the suppression of one thing in fact consists in substituting another thing for it, some will not conclude, or are unwilling to conclude, that abolishing something in thought implies that thought substitutes a new thing for the previous thing. They will admit that a thing is always replaced by another thing, and even that our mind cannot think the disappearance of an external or internal object without imagining—true, in an indeterminate and confused form—that another object is substituted for it. But they will add that the representation of a disappearance is the representation of a phenomenon that takes place in space or at the very least in time, that it therefore also involves the evoking of an image, and that it was a question, precisely here, of breaking free of the imagination in order to appeal to pure understanding.[25] Let us no longer speak of disappearance or abolishing, they will say, for those are physical operations. We should no longer imagine that object A is abolished or absent—we should simply say that we think of it as "nonexistent." To abolish the object is to act upon it in time and perhaps also in space. As a result, this would be to accept the conditions of spatial and temporal existence, and to accept the interdependence that links one object to all the others and that blocks it from disappearing without immediately being replaced.[26] But, they will claim, we can free ourselves from these conditions: through an effort of abstraction. We simply need to evoke the idea of object A, all by itself, agree to first consider it as existing, and then, by the stroke of some intellectual stylus, cross out that clause. The object will then be, by our own decree, nonexistent.

So be it. Let us purely and simply cross out that clause. It must not be believed that our stroke of the stylus is self-sufficient and that it itself is separable from the rest of the things. It surely brings along with it, for better or worse, all that we claimed to cut off from ourselves. Indeed, let us compare the two ideas of object A, first assumed to be real and then this same object assumed to be "nonexistent."

The idea of object A assumed to be existent is nothing other than the pure and simple representation of object A, since we cannot imagine an object without thereby attributing to it a certain reality. There is absolutely no difference between thinking an object and thinking that object as existent—Kant fully clarified this point in his critique of the ontological argument.[27] That being the case, what does it mean to think object A as nonexistent? To imagine object A as nonexistent cannot consist in subtracting the attribute "existence" from the idea of the object, since, again, the representation of the existence of the object is inseparable from the representation of the object and is identical to it. Thus, to imagine object A as nonexistent can only consist in *adding* something to the idea of this object—we add the idea of an *exclusion* of this particular object from the current or actual state of reality in general. To think object A as nonexistent is first to think of the object (and therefore to think of the object as existing), then to think that another reality (one that is incompatible with the object) supplants it. Only it is useless to explicitly imagine this other reality—we have no need to worry about what that reality is. It is enough for us to know that this other reality chases away object A, the only object that interests us. That is why we think of the expulsion rather than of the cause that expels. But this cause is no less present to the mind. It is there in an implicit state, given that what expels is inseparable from the expulsion, just as the hand that moves the stylus is inseparable from the stroke of the stylus that crosses something out. The act by which one declares an object to be unreal thus posits the existence of the real in general. In other words, imagining an object as unreal cannot consist in denying it every kind of existence, since the representation of an object is necessarily the representation of the object as existing. Such an act consists simply in declaring that the existence attributed to the object by our mind, and inseparable from its representation, is a wholly ideal existence, the existence of a merely possible.[28] But the ideality or simple possibility of an object only makes sense in relation to a reality that chases this object that it is incompatible with into the region of the ideal or the merely possible. Assuming the stronger and more substantial existence is abolished, it is the weaker and

285

attenuated existence, that of the merely possible, that becomes reality
itself. As a result, you will no longer be imagining the object as nonexistent. In other words, and as strange as the claim may sound, *there is*
more, and not less, in the idea of an object conceived of as "not existing" than in the idea
of this same object conceived of as "existing," since the idea of the object as "not existing"
is necessarily the idea of the object as "existing" with, in addition, the representation of an
exclusion of this object by the current or actual state of reality taken as a whole.

But it will be claimed that our representation of the nonexistent is not
yet sufficiently freed of all imaginative elements—that it is not yet sufficiently negative. They will say: "It doesn't matter that the non-reality of
one thing consists in its expulsion by other things. We do not want to
know anything about that at all. Are we not free to direct our attention
wherever we want and however we want? So, after having evoked the
representation of an object and having thereby assumed, if you insist, that
the object exists, we merely place a "not" next to our affirmation and
this will be enough for us to think of the thing as nonexistent. This is an
entirely intellectual operation, independent of what happens outside the
mind. Let us think of anything whatsoever or of everything, then let us
place in the margins of our thought the "not" that prescribes the rejection
of what the thought contains: we will have ideally abolished all things
by the simple fact of decreeing that they be abolished." — At root, all of
the difficulties and errors here come from this power that is supposedly
inherent to negation. Negation and affirmation are imagined to be exactly
symmetrical. And negation is imagined to be self-sufficient, just like
affirmation. Were that the case, negation would have, like affirmation, the
power to create ideas, the only difference being that these would be negative. By affirming one thing, then another thing, and so on indefinitely,
I form the idea "Everything"; in the same way, by negating one thing, then
the others things, and finally by negating Everything, we would arrive at
the idea of Nothing. But it is precisely this assimilation that seems arbitrary. We fail to notice that although affirmation is a complete act of the
mind that can constitute an idea, negation is never more than half of an
intellectual act whose other half is implicit or, rather, postponed into an
indeterminate future. And we also fail to see that although affirmation is
an act of pure intellect, an extra-intellectual element enters into negation,
and that negation owes its specific character precisely to the intrusion of
this foreign element.

To begin with the second point, let us note that negating always consists in setting aside a possible affirmation.[a][29] Negation is merely an attitude taken by the mind vis-à-vis a possible affirmation. When I say "this table is black," I am clearly speaking about this table—I saw it as black, and my judgment translates what I saw. But if I say "this table is not white," I surely do not express something that I perceived since I saw black and not the absence of white. Thus, at root, my judgment does not apply to the table itself, but rather to the judgment that would declare the table to be white. I judge a judgment and not the table. The proposition "this table is not white" implies that you might believe it to be white, that you did believe it to be white, or that I was about to believe it to be white—I am warning you or alerting myself that this judgment is to be replaced by another (which, it is true, I leave indeterminate). In this way, whereas the affirmation applies directly to the thing, the negation only targets the thing indirectly by way of an interposed affirmation. An affirmative proposition translates a judgment applied to an object; a negative proposition translates a judgment applied to a judgment. Thus,   288 *negation differs from affirmation, properly so called, in that negation is an affirmation to the second degree: it affirms something with regard to an affirmation that itself affirms something about an object.*

But the first consequence of this is that negation is not the act of a pure mind, that is, of a mind detached from all motives, placed before the objects, and not concerned with anything else.[30] From the moment we negate, we are giving a lesson to others or to ourselves. We are taking a real or possible interlocutor to task for making a mistake; we are putting them on notice. He affirmed something; we are warning him that he will have to affirm something else (without specifying, however, the affirmation that must be substituted for the first one). As such, there is no longer simply one person in the presence of one object; there is, opposite the object, one person speaking to another person, debating

---

a   Bergson's note: Kant writes: "[I]n regard to the content of our cognition in general [...] negative judgments have the special job solely of preventing error." (Kant, *Critique of Pure Reason*, A709/B737). Cf. Christoph Sigwart, *Logik*, vol. 1, *Die Lehre vom Urtheil, vom Begriff und vom Schluss*, 2nd ed. rev. ed. (Fribourg-en-Brisgau: Mohr, 1889), 150–204. [TN: For an English translation of Sigwart, see: Christoph Sigwart, *Logic*, vol. 1, *The Judgment, Concept, and Inference*, trans. Helen Dendy, 2nd ed. (London: Swan Sonnenshein/ New York: Macmillan, 1895).]

with them and helping them at the same time; the beginnings of society. Negation targets someone and not merely something, as was the case for the purely intellectual operation. Negation is essentially pedagogical and social. It corrects, or rather it warns, and the person warned or corrected might for that matter be, through a sort of splitting, the very same person who is speaking.

Let this suffice for the second point. We now arrive at the first. I said that negation is never more than half of an intellectual act whose other half is left indeterminate. If I utter the negative proposition "this table is not white," I mean by this that you should substitute another judgment for your judgment that "the table is white." I am warning you, and the warning has to do with the need to make a substitution. True, I say nothing about what you should substitute for your affirmation. This might be because I do not know the actual color of the table; but it might also be, and perhaps this is the more important factor, because the color white is the only color that we are interested in at the moment. So I merely have to tell you that another color should be substituted for white without having to tell you which one. Thus, a negative judgment is clearly a judgment indicating that there is a need to substitute one affirmative judgment for another affirmative judgment, without thereby specifying the nature of this second judgment, sometimes because one does not know it, and more often because it is not of any current interest insofar as our attention is focused solely upon the material of the former judgment.

As such, every time I place a "not" next to an affirmation, every time that I negate, I accomplish two well-defined acts: (1) I take an interest in what another person affirms, or in what he was about to say, or in what another myself might have said and who needs to be warned; (2) I announce that a second affirmation, whose content I do not specify, will have to be substituted for the one that I find before me. But in neither of these two acts will we find anything other than affirmation. The *sui generis* character of negation comes from the superposition of the first upon the second. It is thus in vain that we attribute the *sui generis* power of creating ideas to negation, ideas that would be symmetrical to those created by affirmation and merely oriented in the opposite direction. No idea comes out of negation, since it has no other content than that of an affirmative judgment that it judges.

To be more precise, consider an existential judgment (and no longer an attributive judgment). If I say "object A does not exist," the first thing

I mean is that someone might believe object A exists. After all, how could someone think of object A without thinking of it as existing, and to repeat, what difference could there be between the idea of object A existing and the pure and simple idea of object A?[31] Thus, by simply saying "object A," I attribute to it a kind of existence, even if only the existence of a merely possible, of a pure idea.[32] And as a result, in the judgment that "object A does not exist," there is first an affirmation to the effect that "object A did exist," or that "object A will exist," or more generally that "object A exists at least as a merely possible." Now, when I add the words "does not exist," what could I mean if not that if we take it further, if we erect this possible object into a real object, then we will be mistaken, and that the possible I am speaking about is excluded from current or actual reality as being incompatible with it? Judgments that posit the nonexistence of a thing are thus judgments that formulate a contrast between the possible and the actual (that is, between two kinds of *existence*: one in thought and the other observed) in cases in which one person, real or imaginary, mistakenly believed that a certain possible had been realized. In place of this possible, there is a real that differs from it and that drives it away. The negative judgment expresses this opposition, but it expresses it in an intentionally incomplete form because it is addressed to a person who is assumed to be exclusively interested in the indicated possible and who will not be concerned with knowing what type of reality has replaced that possible. The expression of the substitution is thus obliged to be truncated. Rather than affirming that a second term has been substituted for the first one, the attention, which was initially directed toward the first term, remains fixed on it and on it alone. And, without straying from the first one, we implicitly affirm that a second term replaces it by saying that the first "does not exist." As such, we will be judging a judgment rather than judging a thing. We will be warning others or warning ourselves of a possible error rather than bringing forth a positive piece of information. If you suppress every intention of this type, give back to knowledge its exclusively scientific and philosophical character, and suppose, in other words, that reality comes to inscribe itself upon a mind that only pays attention to things and that takes no interest in people, then you will only ever affirm that such or such a thing is and never that a thing is not.[33]

So, what leads us to stubbornly persist in placing affirmation and negation along the same line and in endowing them with equal objectivity?

290

291

Why is it so hard for us to recognize the part of negation that is subjective, artificially truncated, relative to the human mind, and above all relative to social life? The reason for this is likely that negation and affirmation are both expressed through propositions, and that every proposition, being formed by *words* that symbolize *concepts*, is relative to social life and human intellect.[34] Whether I say "the ground is wet" or "the ground is not wet," in both cases the terms "ground" and "wet" are concepts more or less artificially created by the human mind, by which I mean concepts extracted from the continuity of experience by the mind's free initiative. In both cases, these concepts are represented by the same conventional words. In both cases, we can even say that the proposition ultimately aims at a social and pedagogical goal, since the first propagates a truth and the second warns against an error. If we look at things from this point of view, which is the perspective of formal logic, then affirmation and negation are in fact two symmetrical acts, where the first establishes a harmonious relationship between a subject and an attribute, and the second establishes a disharmonious one. — But is it not obvious that the symmetry here is entirely external and the resemblance merely superficial? Imagine language abolished, society dissolved, and among humans all intellectual initiative and every faculty for splitting oneself or for judging oneself atrophied. The wetness of the ground would no less remain, capable of being inscribed automatically in sensation and of sending a vague representation to this dazed intellect. The intellect would thus still make an affirmation, though in implicit terms. And, as a result, the essence of affirmation is not to be found in distinct concepts, nor in words, nor in the desire to spread the truth around oneself, nor in the desire to improve oneself. But this passive intellect—who mechanically follows along a step behind the course of experience, neither getting out in front of the course of the real, nor lagging behind it—would not have even the slightest inkling of the act of negating.[35] This intellect could not even receive the imprint of negation, since, again, although whatever exists can come to be recorded, the nonexistence of the nonexistent is not recorded. For such an intellect to be able to negate, it must first wake itself up from its torpor, formulate the disappointment of a real or possible anticipation, correct an actual or possible error, and finally decide to teach others or itself.

This may be more difficult to see in the example that I have chosen, but the example will therefore be more instructive and the argument all

the more convincing. It will be said that if wetness can be recorded automatically, then the same must be true of non-wetness since the dry (just
as well as the wet) can provide impressions to our sensibility that in turn
transmits them to the intellect as more or less distinct representations.
In this sense, the negation of wetness will be just as objective, just as
purely intellectual, and just as detached from all pedagogical intention
as affirmation. — But if they would just take a closer look, they would
see that the content of the negative proposition "the ground is not wet"
and the content of the affirmative proposition "the ground is dry" are
completely different. The second proposition implies that one knows the
dry, that one has experienced the specific sensations (for example, the
tactile or visual sensations) that are at the foundation of this representation. The first proposition requires nothing of the sort: it could just as
easily be formulated by an intelligent fish who had only ever perceived
wetness. Of course, this fish would have raised itself up to the distinction
between the real and the possible, and would have to be concerned with
getting out in front of the errors of his fellow fish, who surely consider
the wet conditions in which they actually live to be the only possible
ones. If you hold strictly to the terms of the proposition "the ground   293
is not wet," you will find that it signifies two things: (1) that someone
might believe that the ground is wet; and (2) that the wetness is in fact
replaced by a certain quality, x. This quality is left in a state of indetermination because either we do not have positive knowledge of it, or it
has no current interest for the person to whom the negation is being
addressed. Thus, negating always consists in presenting a system of two
affirmations in a truncated form: one affirmation being determinate
and bearing upon a certain possible, the other being indeterminate and
referring to an unknown or indifferent reality that supplants this possibility. The second affirmation is contained virtually in the judgment
that we apply to the first affirmation, this judgment being the negation
itself. And the subjective character of negation comes precisely from the
fact that when we observe a replacing, our observation only considers
what is being replaced and is not concerned with what is doing the
replacing. What has been replaced now only exists as a conception of
the mind. In order to continue to see what has been replaced and, consequently, to continue to speak of it, we must turn our back on reality,
which flows from the past to the present, from back to front.[36] This is
what we do when we negate. We observe the change, or more generally,

the substitution, much like a passenger would see the path of the car if he were to look backward and only wished to know at each moment the point where he had ceased to be. This passenger would never determine his current position except in relation to the one that he had just left behind, rather than expressing it in and for itself.

To summarize, for a mind that would purely and simply follow the thread of experience, there could be no void, no nothingness, not even a relative or partial nothingness, and no possible negation.[37] Such a mind would see facts following other facts, states following other states, and things following other things. At each moment, it would simply note the things that exist, the states that appear, and the events that take place. It would live in the actual [l'actuel] and, if it were capable of judging, it would never affirm anything other than the existence of the present [le présent].

Let us equip this mind with memory, and above all with the desire to dwell on the past.[38] Let us give it the faculty for dissociating and for distinguishing. Such a mind will no longer merely note the actual state of the reality that is passing by. It will conceive of this passage as a change, and thereby as an opposition between what had been and what is. And since there is no essential difference between a past that one remembers and a past that one imagines, this mind will quickly work its way up to the representation of the possible in general.[39]

As such, this mind will switch itself over to the track heading toward negation.[a] And above all, it will be on the point of imagining a disappearance. Nevertheless, it will not quite reach that point. To represent something as having disappeared, it is not enough to perceive an opposition between the past and the present; one must also turn one's back on the present, dwell upon the past, and think the opposition of the past with the present solely in terms of the past, without making the present appear there.[40]

The idea of abolition is thus not a pure idea: it implies that we regret the past or conceive of it as regrettable, and that we have some reason to dwell upon it. The idea is born when the phenomenon of substitution is cut in two by a mind that only considers the first half of it, because it is only interested in that half. If you suppress all interest and all emotion,

---

a TN: *Il s'aiguillera ainsi sur la voie de la négation*. Here Bergson recalls the railroad metaphors developed above. See, for instance, above, 100.

all that remains is the reality that passes by, along with the indefinitely renewed knowledge of its present state that it imprints in us.

Now there is but a single step from abolition to negation, which is a more general operation. Rather than imagining the opposition between what is and what was, now we just have to also imagine the opposition between what is and everything that could have been. And we must express this opposition in terms of what could have been and not in terms of what is—we must affirm the existence of the actual while only looking at the possible.[41] The formula that we obtain in this way no longer expresses merely the individual's disappointment. It is designed to correct an error or to warn against an error, which we take rather to be another person's error. In this sense, negation has a pedagogical and social character.

Now, once negation has been formulated, it very much appears to be symmetrical with affirmation. So, it seems that if affirmation affirmed an objective reality, then negation must affirm an equally objective and, so to speak, equally real non-reality. This assumption is both wrong and right. It is wrong because the negation could never make itself objective in what it contains of the negative. It is right, however, in that the negation of a thing implies the implicit affirmation of its replacement by another thing, which is systematically left to the side. But the negative form of negation benefits from the affirmation that is at its core: overlapping with the body of the positive reality to which it is attached, this phantom makes itself objective.[42] In this way, the idea of a void or a partial nothingness is formed, a thing no longer replaced by another thing, but rather replaced by a void that it itself leaves behind, i.e., by its own negation. Moreover, since this operation can be carried out on any thing whatsoever, we imagine it to be performed on each thing, one after the other, and finally on all things *en bloc*. We thereby obtain the idea of "absolute nothingness" [*néant absolu*]. If we now analyze this idea of Nothing [*Rien*], we find that it is at root the idea of Everything with the addition of a movement of the mind that continuously leaps from one thing to another, refuses to remain in place, and concentrates all of its attention on this refusal by never determining its actual position other than in relation to the one that it has just left behind. Thus, the idea of Nothing is an eminently comprehensive and full idea, as full and comprehensive as the idea of Everything, an idea to which it is very closely related.

295

296   So how could the idea of Nothing be opposed to that of Everything? Is it not obvious that this is to oppose the full to the full, and that the question as to "why something exists" is thereby a meaningless question, a pseudo-problem raised with regard to a pseudo-idea? And yet, we must once again explain why this phantom of a problem so stubbornly haunts the mind. In vain we show that in the idea of "abolishing the real" there is nothing more than the image of all realities driving each other away, indefinitely and in a circle. In vain we add that the idea of nonexistence is nothing other than that of the expulsion of an imponderable existence, or of a "merely possible" existence, by a more substantial existence, which would be the true reality.[43] In vain we find something extra-intellectual in the *sui generis* form of negation, the negation being the judgment of a judgment, a warning issued to another person or to oneself, such that it would be absurd to attribute to it the power of creating representations of a new kind, namely ideas without any content.[44] For despite all of this, the conviction still persists that there is nothingness prior to the things, or at least somehow beneath them. If we looked for the reason for this persistence, it would be found precisely in the emotional, social, and, most generally, practical element that gives negation its specific form. I said above that the most significant philosophical problems are born when the forms of human action venture beyond their proper domain.[45] We are made for action as much and even more than we are for thinking—or better, when we follow the movement of our nature, we think in order to act.[46] Thus, we should not be surprised that the habits of action influence or color our habits of representation, and that our mind always perceives things in the very order in which we have become accustomed to picturing them when

297   we set about acting upon them. Now, as I noted above, every human action unquestionably has its starting point in some dissatisfaction and, by this very fact, in a feeling of absence.[47] We would not act if we did not set ourselves a goal, and we only go looking for something because we sense that we are deprived of it. As such, our action proceeds from "nothing" to "something," and its very essence is to embroider "something" upon the canvas of the "nothing." The truth is that the nothing in question here is less the absence of a thing than it is the absence of a usefulness. If I lead a visitor to a room that I have not yet furnished, I warn him that "there is nothing in it." I know, of course, that the room is full of air; but since we do not sit on air, the room truly contains

nothing that currently counts as something for the visitor and myself. In general, human work consists in creating usefulness; and, so long as the work is not yet done, there is "nothing"—that is, nothing of that which we wanted to obtain. In this way, our life is spent filling voids that our intellect conceives of under the extra-intellectual influence of desire and regret, and under the pressure of vital necessities. And, if we understand a void as an absence of usefulness and not as an absence of things, then we can say, in this entirely relative sense, that we constantly move from the empty to the full. Our action marches in precisely this direction. And our speculation cannot stop itself from doing the same. It quite naturally moves from relative meaning to absolute meaning since it operates upon the things themselves and not upon the usefulness that they have for us. This is how the idea that reality fills a void is implanted in us, along with the idea that nothingness, understood as an absence of everything, preexists all things, at least in principle if not in fact. This is the illusion that I attempted to dissipate by showing that the idea of Nothing, if we claim to see here the idea of an abolition of all things, is an idea that destroys itself and reduces down to being a mere word—and that if, by contrast, it is genuinely an idea, then we find in it just as much material as we find in the idea of Everything.[48]

298

*

[BECOMING AND FORM]

This long analysis was necessary in order to show that *a self-sufficient reality is not necessarily a reality foreign to* durée.[49] If we (consciously or unconsciously) pass through the idea of nothingness in order to arrive at the idea of Being, then we arrive at a Being that is a logical or mathematical essence, and therefore non-temporal. And that being the case, a static conception of the real becomes inevitable: everything appears to be given in a single stroke and for all eternity.[50] But we must become accustomed to thinking Being directly, without making a detour and without first directing ourselves to the phantom of nothingness that interposes itself between us and Being. We must endeavor here to see for the sake of seeing, and no longer for the sake of acting. The Absolute is then revealed to be very close to us and, to a certain extent, within us. The Absolute is psychological in essence, and not mathematical or logical.[51] It lives with us. And like us—though in certain ways infinitely more concentrated and infinitely more condensed into itself—the Absolute endures.[52]

But do we ever think real durée? Here again it will be necessary to take possession of it directly.[53] We will never meet up with durée by taking a detour: it is necessary to place ourselves directly within it. This is what the intellect most often refuses to do, habituated as it is to thinking moving reality [penser le mouvant] through the intermediary of the immobile.[54]

The role of the intellect is indeed to preside over actions.[55] Now, in action we are above all concerned with the result—so long as the goal is achieved, the means are of little importance to us. As a result, we are entirely focused on the end to be realized, most often trusting ourselves to it so that, beginning as an idea, the end becomes an action.[a][56] This is also why the endpoint where our activity will come to rest is the only one our mind explicitly pictures. The constitutive movements of the action itself either escape our consciousness or merely come to consciousness in a confused way. Consider a very simple act, such as raising one's arm. Where would we be if we had to imagine in advance all of the elementary contractions and tensions that this movement involves, or had to perceive them, one by one, while each is taking place?[57] The mind immediately transports itself to the goal, i.e., to the schematic and simplified vision of the act taken as already accomplished [supposé accompli]. Then, if no antagonistic representation neutralizes the effect of the first one, the appropriate movements come along all by themselves to fill in the schema, as if somehow drawn in by the void formed by its interstices. The intellect thus only thinks of the goals to be achieved by the activity, that is, the various rest points. Our activity is transported by a series of leaps, from one accomplished goal to another, or from one rest point to another. During this time, our consciousness turns away as much as possible from the movement being accomplished in order to focus solely on the anticipated image of the accomplished movement.

Now in order for the intellect to imagine the result of the act that is taking place as immobile, it must also perceive the milieu in which this

---

a TN: The translation I have chosen here emphasizes the "becoming," already mentioned by Bergson in the section title. The phrase devienne acte might also be translated as "is actualized."

result is contained as immobile. Our activity is inserted into the material world. If the material world appeared to us as a perpetual flowing, then we could never assign a fixed endpoint to any of our actions.[58] We would sense each one of them being dissolved the entire time they were being accomplished, and we could never anticipate a future that is always fleeting. In order for our activity to leap from one *act* to another *act*, matter must also pass from one *state* to another *state*, since action must insert its results into a state of the material world in order to be accomplished. But does matter really appear in this way?

We can assume *a priori* that our perception is organized so as to grasp matter from this angle.[59] Sensory organs and motor organs are in fact coordinated with each other. Now, the former symbolize our faculty of perceiving, while the latter symbolize our faculty of acting.[60] As such, the organism reveals, in a visible and tangible form, the perfect accord between perception and action. Thus, if our activity always aims at a *result* into which it is momentarily fitted, then our perception must not retain of the material world, at any given moment, more than a *state* into which it provisionally places itself. Such is the hypothesis that comes to mind. It is easy to see that experience confirms it.

From the very first glance directed toward the world, and even before we pick out any *bodies*, we already distinguish *qualities*. One color follows another color, one sound follows another sound, one resistance follows another resistance, and so on. Each of these qualities, taken separately, is a state that seems to persist as such, immobile, while waiting to be replaced by another. And yet, analysis shows that each of these qualities resolves into an enormous number of elementary movements.[61] Whether we see these elementary movements as vibrations or imagine them in any other way, one thing is certain: every quality is in fact changing.[62] Moreover, it is in vain that we search beneath this change for the thing that changes. It is only provisionally that we attach movement to a moving thing, and so as to satisfy our imagination.[63] The moving thing forever slips away from the scientific gaze; science never deals with anything other than mobility. In the smallest perceptible fraction of a second, in the quasi-instantaneous perception of a sensible quality, there might be trillions of oscillations that repeat. The permanence of a sensible quality consists in this repetition of movements, just as the persistence of life is made up of successive palpitations. The primary function of perception is precisely to grasp, through a work of condensation, a

300

series of elementary changes in the form of a quality or a simple state. The greater the force for acting allotted to an animal species, the greater (in all likelihood) the elementary changes that its faculty of perceiving concentrates into one of its instants.[64] And in nature, there must be a continuous progression from the beings that vibrate almost in unison with the ethereal oscillations, right up to those who immobilize trillions of these oscillations in even the shortest of their simple perceptions. The former sense hardly anything other than movements; the latter perceive qualities. The former let themselves be more or less taken in by the gearing together of things; the latter react, and the tension of their faculty of acting is surely proportional to the concentration of their faculty of perceiving.[65] This progression continues right through humanity itself. One is more a "man of action" to the extent that he is able to embrace a larger number of events in a single glance. This is also why we either perceive successive events, one by one, and let ourselves be carried along by them, or rather grasp them *en bloc* and dominate them.[66] In short, the qualities of matter are but so many stable views or snapshots that we take of its instability.

Now, we delimit bodies within the continuity of sensible qualities. Each one of these bodies in fact changes at each moment. First, each body resolves into a group of qualities, and each quality, I said, consists of a succession of elementary movements. And yet, even if one envisions the quality as a stable state, the body is still unstable in that it incessantly changes qualities. The living body is the body *par excellence*, the one that we are the most justified in isolating within the continuity of matter because it constitutes a relatively closed system. Moreover, it is for this body that we cut all of the other ones out of the whole.[67] And yet life is an evolution. We concentrate a period of this evolution into a stable view that we call a form and, when change has become considerable enough to vanquish the convenient inertia of our perception, we say that the body has changed form. But in reality, the body is always changing form. Or rather there is no form, since form is essentially immobile and reality is movement. What is real is the continuous change of form; *form is but a snapshot taken of a transition*. Thus, here again, our perception manages to solidify the flowing continuity of the real into discontinuous images. When the successive images do not differ too much from each other, we consider them all to be the growth or the diminution of a single *average* image, or as the deformation of this image in different directions.[68] And we are thinking of this average when we speak of the *essence* of a thing, or of the thing itself.

Finally, constituted things show on their surface, through their changes of situation, the deep modifications that are taking place within the Whole.[69] That is when we say that they *act* upon each other. Of course, this action appears to us in the form of a movement. And yet, we turn our attention away from the mobility of the movement as much as possible. As I said above, what interests us is the immobile sketch of the movement and not the movement itself.[70] If it is a question of a simple movement, then we wonder *where* it is going. We imagine it at each moment according to its direction, that is, according to its provisional goal. If it is a question of a complex movement, then we want to know, above all else, *what* takes place, *what* the movement does—i.e., the result that is obtained or the intention that presides over it. Examine carefully what you have in mind when you speak of an action on the way to being accomplished. I am happy to admit that the idea of change is there, but it is hidden in the shadows. What appears in full light is the immobile sketch of the action taken as already accomplished. This, and this alone, is how the complex act is distinguished and defined. It would be very difficult for us to imagine the movements inherent to the actions of eating, drinking, fighting, etc. It is enough for us to know, in a general and indeterminate way, that all of these acts are movements. Once this has been established, the only thing we attempt to picture to ourselves is the *overall plan* of each one of these complex movements, that is, their underlying *immobile sketch*. Here again knowledge is directed toward a state rather than a change. Thus, this third case is the same as the previous two. Whether it is a question of a qualitative movement, an evolutionary movement, or a movement in extension, the mind is organized so as to take stable views or snapshots of instability.[71] As I have just shown, the mind thereby arrives at three kinds of representations: (1) qualities, (2) forms or essences, and (3) actions.

303

Three categories of words correspond to these three ways of seeing: *adjectives*, *nouns*, and *verbs*. These are the primordial elements of language. Adjectives and nouns thus surely symbolize *states*. But the verb too, if we remain focused on the well-illuminated part of the representation that it evokes, hardly expresses anything else.

If we now wanted to give a more precise characterization of our natural attitude toward becoming, here is what we would find. Becoming is infinitely varied. The becoming that goes from yellow to green does not resemble the one that goes from green to blue—these are different qualitative movements. The becoming that goes from the flower to the fruit does not resemble the one that goes from the larva to the nymph and from the nymph to the mature insect—these are different evolutionary movements. The action of eating or drinking does not resemble the action of fighting—these are different movements in extension. And these three kinds of movement themselves (qualitative, evolutionary, and in extension) are profoundly different. The artifice of our perception—

304   like that of our intellect and our language—consists in extracting from these highly varied becomings a single representation of becoming in general, an indeterminate becoming, a mere abstraction that by itself says nothing and that we rarely think of.[72] To this idea (which is always the same and, for that matter, always obscure and unconscious), we then adjoin, in each particular case, one or several clear images that represent *states* and that serve to distinguish all of the becomings from each other. We substitute for the specificity of the change itself this composition of a specific and determinate state with general and indeterminate change. An indefinite multiplicity of diversely colored becomings passes before our eyes, so to speak. We manage to see simple differences of color, i.e., different states, underneath which would flow by, in obscurity, a becoming that is always and everywhere the same and invariably colorless.

Imagine that we want to reproduce on a screen a lively scene, such as a military parade. A first way of going about it would be to cut out shapes with articulating joints to represent the soldiers, to cause each one to move in the manner of a march (a movement that is variable from one individual to the next, even though it is common to the human species), and to project the whole thing onto a screen. This little game would cost an enormous amount of work and, moreover, we would obtain but a fairly mediocre result. After all, how could such a game reproduce the suppleness and the variety of life? Now, there is a second way of proceeding that is much easier and at the same time much more effective. This would be to take a series of snapshots of the regiment as it passes by and to project these snapshots onto the screen such that they replace each other in quick succession. This is what the cinematograph does.[73] With a series of photographs that each represent the regiment in an immobile attitude, the projector reconstitutes the mobility of the regiment that passes by. True, if we were dealing with the photographs by themselves, then as much as we might look at them, we would never see them come to life—we will never make movement out of immobile parts [immobilités], not even out of immobile parts indefinitely juxtaposed with each other.[74] For the images to come to life, there must be movement somewhere. And indeed, movement surely exists here: it is in the projector. Because the filmstrip unwinds, causing the various photographs of the scene to follow one after the other, in turn, each actor in this scene regains his mobility: he strings together all of his successive attitudes on the invisible movement of the cinematographic filmstrip. To summarize, the procedure consisted in first extracting an impersonal, abstract, and simple movement, a *movement in general*, so to speak, from all of the movements belonging to all of the figures; then, in placing this movement into the projector; finally, in reconstituting the individuality of each particular movement through the composition of this anonymous movement with the individual attitudes [of the actors]. Such is the artifice of the cinematograph. And such is also the artifice of our knowledge. Rather than attaching ourselves to the inner becoming of things, we place ourselves on the outside of them in order to artificially recompose their becoming.[75] We take quasi-instantaneous views or snapshots of the reality that passes by, and, given that they are characteristic of this reality, we can simply string them together—along a becoming that is abstract, uniform, invisible, and situated at the foundation of the apparatus of knowledge—in order to imitate what is characteristic in this becoming itself.[76] In general, this is how perception, intellection, and language proceed. Whether it is a question of thinking, expressing, or even simply perceiving becoming, we hardly do anything other than set into motion a sort of inner cinematograph. We might thus summarize all of this by saying that the *mechanism of our ordinary knowledge is cinematographic in nature*.

It is impossible to doubt the entirely practical character of this operation. Each one of our actions aims to insert our will into reality in some way.[77] Between my body and the other bodies, there is an arrangement comparable to the one found among the pieces of glass that compose a kaleidoscopic image.[78] Our activity goes from an arrangement to a rearrangement; each action surely marks a new turn of the kaleidoscope, but we pay no attention to the turning and focus entirely on the new image. The knowledge of the operation of nature that our activity takes for granted must therefore be exactly symmetrical to the interest that it takes in its own operation. In this sense, we could say (if saying so were not to misuse a certain kind of comparison) that the *cinematographic character of our knowledge of things comes from the kaleidoscopic character of our adaptation to them.*

The cinematographic method is thus the only genuinely practical one since it consists in modeling the general style of knowledge on the general style of action, while assuming that the detail of each act will be modeled in turn on the detail of knowledge.[79] For action to be always illuminated, the intellect would have to always be present there; but for the intellect to accompany the movement of the action in this way and to maintain its direction, it must begin by adopting the rhythm of action.[80] Since action, like every pulsation of life, is discontinuous, knowledge will also be discontinuous.[81] The mechanism of the faculty of knowing was built according to this plan. But can the intellect, given that it is essentially practical, be used, just as it is, for speculation? Let us see what happens when we attempt to use the intellect to follow reality in its various detours.

I have taken a series of views or snapshots of the continuity of a particular becoming and I have linked them together through a notion of "becoming" in general. Of course, I cannot stop there. It is impossible to imagine something that cannot be made determinate, so I only have a verbal knowledge of "becoming in general." Just as the letter x designates a certain unknown thing, whatever it might be, my concept of "becoming in general" (which is always the same) here symbolizes a certain transition of which I have taken some snapshots. This concept tells me nothing about the transition itself. I will thus try to focus entirely on the transition itself to find out what takes place between two snapshots. But since I apply the same method, I arrive at the same result; a third view will simply be inserted between the two others. I will forever start over and forever merely juxtapose views to other views, never obtaining anything else. The application of the cinematographic method will thus end up in a perpetual starting over, one in which the mind—never finding satisfaction and never seeing anywhere to come to rest—probably persuades itself that its own instability imitates the very movement of the real. But even if the mind, working itself up to the point of dizziness, ends up creating for itself an illusion of mobility, its operation has not actually moved it forward one single step since this method always leaves the mind just as far from its goal. In order to move forward with moving reality, we would have to place ourselves back within that reality. Place yourself within that which is changing and you will immediately grasp both change itself and the successive states in which it *could*, at any moment, be immobilized.[82] But you will never reconstitute movement from these successive states, seen from the outside as real immobile parts (and no longer as virtual immobile parts).[83] Depending on the situation, you might call them "qualities," "forms," "positions," or "intentions." You can multiply them as much as you like, and thereby continuously bring two consecutive states closer together.[84] But when it comes to the intermediary movement between the two states, you will always experience the same disappointment as the child who tries to crush some smoke by pressing together his two open hands.[85] The movement will slip away into the interval because every attempt to reconstitute change from states involves the absurd proposition that movement is made up of immobile parts.

Philosophy noticed this absurdity the very moment it opened its eyes.[86] Zeno of Elea's arguments, though formulated with a very different intention, in fact say nothing else.[87]   308

Shall we consider the flying arrow? According to Zeno, the arrow is immobile at each instant, since it would only have the time to move— that is, to occupy at least two successive positions—if it is granted at least two instants. At any given moment, the arrow is thus at rest at a given point. Immobile at each point along its path, the arrow is thus immobile during the entire time that it moves.

Yes, provided we assume the arrow can ever *be* at one point of its path.[88] Yes, provided the arrow, which belongs to moving reality [*le mouvant*], ever coincided with a position, which belongs to immobile reality [*l'immobilité*]. But the arrow never *is* at any point of its path. The most we can say is that the arrow could be at a certain point, in the sense that it passes through that point and would be free to stop there. True, if it had stopped there, it would have remained there, and at that point we would no longer be dealing with a movement. The truth is that if the arrow leaves point A in order to land at point B, then the movement AB is, as a movement, just as simple and just as indecomposable as is the tension in the bow that launches it.[89] Just as the shrapnel, exploding prior to reaching the ground, extends an indivisible danger across the zone of the explosion, so too the arrow that goes from A to B deploys, in a single stroke, its indivisible mobility, although this time across a certain extension of *durée*. Imagine an elastic band that you stretch from A to B; can you divide up its extension?[90] The flight of the arrow is this extension itself: it is just as simple and, like it, indivisible. It is a single and unique leap. You focus on a point C in the interval that is traversed, and you say that at a certain moment the arrow was at C. But if it had been there, this means it would have stopped there, and now you would no longer have the flight from A to B, but rather two flights, one from A to C, the other from C to B, with an interval of rest in between. By definition, a single movement is an entire movement between two stopping points: if there are intermediary stopping points, then it is no longer a single movement. In essence, the illusion comes from the fact that the movement, *once completed*, has deposited along its path an immobile trajectory along which we can count as many immobile parts as we like.[91] From this we conclude that at each instant the movement, *while being accomplished*, deposited beneath itself a position with which it coincided. But we thereby fail to see that the trajectory is created in a single stroke, even though it takes a certain amount of time, and that even though we can divide at will the trajectory once it has been created, we could not divide its creation, which is not a thing but an act in progress.[92] To assume that the moving object is at a given point of its path amounts to making a cut with scissors at precisely that point, thereby cutting the path in two and substituting two trajectories for the single trajectory that we were first considering. It is to distinguish two successive acts where, by definition, there is only one. In short, it is to transport into the very flight of the arrow everything that might be said about the interval that the arrow flies through; it is to accept, *a priori*, the absurdity that movement coincides with the immobile.

I will not dwell on Zeno's three other arguments here. I have examined them elsewhere.[93] I will simply note that they too involve applying the movement to the length of the line traveled and then assuming that what is true of the line is also true of the movement. For example, the line can be divided into as many parts as we like, of any size that we like, and yet it is still the same line. From this it will be concluded that we have the right to imagine the movement divided up however we like and that it remains the same movement. In this

way, we obtain a series of absurdities that all express the same fundamental absurdity. And yet, the possibility of applying the movement to
the line traveled only exists for an observer who—remaining outside   310
of the movement and picturing at each moment the possibility of a
stopping point—claims to recompose the real movement out of these
possible immobile parts. This possibility disappears the moment we
adopt, in thought, the continuity of real movement, the continuity we
are all conscious of when we raise our arm or take a step forward. Here
we clearly sense that the line traveled between two stopping points
describes a single and indivisible stroke, and that it would be in vain
to attempt to make, within the movement that traces this line, some
divisions corresponding one by one to divisions arbitrarily chosen in
the traced line. The line traveled by the moving object lends itself to
any mode of decomposition whatsoever because it has no inner organization. But every movement is inwardly articulated.[94] A movement
is either a single indivisible leap (which could, of course, occupy a
very long durée) or a series of indivisible leaps. You must either take
into account the articulations of this movement or give up speculating
about its nature.

When Achilles chases the tortoise, each one of his steps (as well as
those of the tortoise) must be treated as indivisible.[95] After a certain
number of steps, Achilles will have overtaken the tortoise. Nothing
could be simpler. If you want to further divide up the two movements,
go ahead and distinguish on both sides, in both Achilles's path and
the tortoise's, some sub-multiples of steps taken by each—but be sure to
respect the natural articulations of the two paths. As long as you respect
those natural articulations, no difficulty will arise because you will be
following the indications of experience. But Zeno's trick consists in
recomposing Achilles's movement according to an arbitrarily chosen law.
In a first leap, Achilles would arrive at the point where the tortoise was;
in a second leap, he reaches the point to which the tortoise had moved
while Achilles was completing the first leap, and so on. This being the
case, Achilles would indeed always have a new leap to make. But it goes
without saying that Achilles manages to catch the tortoise in a com-   311
pletely different way. The movement that Zeno considers would only
be the equivalent of Achilles's movement if we could treat movement in
the same way that we treat the interval traversed, i.e., as decomposable

and recomposable at will. And the moment we subscribe to this first absurdity, all of the other ones come along with it.[a][96]

Moreover, nothing would be easier than to extend Zeno's argument to qualitative becoming and to evolutionary becoming. And again, we would find the same contradictions. We understand that the child becomes an adolescent, then a mature adult, and finally an elderly man when we realize that here vital evolution is the reality itself. By contrast, childhood, adolescence, maturity, and old age are in fact merely views taken by the mind—they are the *possible stopping points* imagined by us, from the outside, along the continuity of a progression. But suppose we take childhood, adolescence, maturity, and old age to be integral parts of the evolution itself—they become *real stopping points*, and since juxtaposed moments of rest will never equal a movement, we will no longer understand how evolution is possible. How could something that is being made be reconstituted from something that is already made? For example, once childhood has been posited as a *thing*, how could we go from childhood to adolescence given that (by definition) childhood alone has been given? If we take a closer look, we see that our habitual way of speaking, modeled as it is upon our habitual way of thinking, leads us into genuine logical impasses. We grapple with them without much worry because we vaguely sense that we would always be able to free ourselves from them. To do so, it would be enough to simply renounce our intellect's cinematographic habits. When we say that "the child becomes a man," we must avoid overanalyzing the literal sense of

312

a Bergson's note: In other words, I do not consider Zeno's sophistry to be refuted by the fact that the geometrical progression $a \left(1 + \frac{1}{n} + \frac{1}{n2} + \frac{1}{n3} + \ldots \text{ etc.}\right)$, where a designates the initial gap between Achilles and the tortoise and n designates the relation between their respective velocities, has a finite sum if n is greater than 1. On this point, I refer the reader to Évellin's argument, which I take to be decisive (see François Évellin, *Infini et quantité: Étude sur le concept de l'infini en philosophie et dans les sciences* (Paris: Baillière, 1880), 63–97. Cf. François Évellin, "Letter to Ribot, April 22, 1881," *Revue philosophique de la France et de l'étranger* 11, no. 5 (May 1881): 564–68). The truth is that mathematics—as I have tried to prove in a previous work—only operates upon lengths and nothing else. It was thus forced to find some tricks so as to introduce the divisibility of the line that is traveled into the movement itself (which is not a length), and then to re-establish the harmony between experience and the idea of a "movement-length" (a concept that goes against experience and that is full of absurdities), that is, of a movement *applied to* its own trajectory and considered to be, like it, arbitrarily decomposable.

the expression. For we would discover that when we posit the subject "child," the attribute "man" is not yet appropriate for him, and that, when we announce the attribute "man," it already no longer applies to the subject "child." The reality, which is the *transition* from childhood to maturity, has slipped through our fingers.[97] We have nothing but the imaginary stopping points "child" and "man," and we are very close to saying that one of these stopping points is the other, in the same way that Zeno's arrow is (according to Zeno) at each of the points along its path. The truth is that, were language here modeled upon the real, we would not say that "the child becomes a man," but rather that "there is becoming from the child to the man." In the first proposition, "becomes" is a verb that has an indeterminate sense, designed to mask the absurdity that we fall into by attributing the state "man" to the subject "child." Here the verb behaves more or less like the always identical movement of the cinematographic filmstrip, a movement hidden within the apparatus whose role is to superpose the successive images upon each other so as to imitate the movement of the real object. In the second proposition, "becoming" is a subject. It comes into the foreground. Becoming is the reality itself that is in question: childhood and maturity are no longer anything but virtual stopping points or mere views taken by the mind. Here we are dealing with objective movement itself, and no longer with a cinematographic imitation of it. But only the first way of expressing ourselves conforms to our linguistic habits. To adopt the second, we would have to free ourselves from the cinematographic mechanism of thought.

313

It would be necessary to completely set aside this cinematographic mechanism of thought in order to dissipate, all at once, the theoretical absurdities that emerge from the question of movement. Everything is obscure and contradictory when it is claimed that a transition can be fabricated out of states. The obscurity dissipates and the contraction subsides the moment we begin by placing ourselves alongside the transition in order to distinguish in it a certain number of states by mentally making a series of transversal cuts. This is because there is *more* in the transition itself than in the series of states (i.e., of possible cuts), and because there is *more* in the movement than in the series of positions (i.e., of possible stopping points). It is just that the first way of looking at it conforms to the procedures of the human mind, whereas the second requires rather that we work back up the incline of our intellectual habits.[98] Should we be surprised, then, if philosophy at first retreated when confronted with such a task? The Greeks had confidence in nature, confidence in the mind left to its natural inclination, and above all confidence in language, insofar as it naturally externalizes thought. Rather than blaming the attitude that thought and language adopt toward the course of things, the Greeks preferred to blame the course of things itself.[99]

*

[THE PHILOSOPHY OF FORMS AND ITS UNDERSTANDING OF BECOMING. PLATO AND ARISTOTLE. THE NATURAL INCLINATION OF THE INTELLECT]

This is what the philosophers of the Eleatic School did, and rather unceremoniously. Since becoming shocks the habits of thought and fits rather poorly into the frameworks of language, they declare it to be unreal. They saw nothing but pure illusion in spatial movement and change in general. This conclusion could have been attenuated without changing the premises by saying that although reality changes, it should not change. Experience puts us into the presence of becoming—that is sensible reality. But intelligible reality, the reality that should exist, is even more real, and it will be said that this reality does not change. Beneath qualitative becoming, evolutionary becoming, and becoming in extension, the mind must seek out what is resistant to change: the definable quality, the form or essence, and the end.[100] This was the fundamental principle of the philosophy that developed throughout classical antiquity—the philosophy of Forms, or, to employ a term that remains closer to the Greek, the philosophy of Ideas.

Indeed, the word εἶδος, which I translate here as "Idea," has this triple meaning. It designates: (1) the quality, (2) the form or the essence, and (3) the goal or the intention [dessein] of the act being accomplished, which is at root the sketch [dessin] of the act taken to be already accomplished. *These three points of view are those of the adjective, the noun, and the verb, and they correspond to the three essential categories of language.*[101] After the explanations that I gave above, I might (and perhaps I should) translate εἶδος by "view" or rather by "moment."[102] For the εἶδος is the stable view or snapshot taken of the instability of things: it is the quality that is a moment of becoming, the form that is a moment of evolution, the *essence* that is the average form above and below which the other forms line up as alterations of it, and finally, the *intention* that inspires the act being accomplished that is nothing other than, as I said above, the anticipated *sketch* [*dessin*] of the accomplished act.[103] Bringing things back to the Ideas thus consists in resolving becoming into its principal moments, each one being, moreover, by definition shielded from the laws of time and somehow gathered up into eternity. In other words, we arrive at the philosophy of Ideas when we apply the cinematographic mechanism of the intellect to the analysis of the real.

But the moment we place unchanging Ideas at the foundation of moving reality, an entire physics, an entire cosmology, and an entire theology necessarily follow. Let us pause over this point. I do not intend to summarize here, in a mere few pages, a philosophy as complex and as comprehensive as that of the Greeks. But, since I have just described the cinematographic mechanism of the intellect, it is important to show the particular representation of the real that results from the play of this mechanism. I believe that this representation is precisely the one found in ancient philosophy. The main lines of the doctrine that developed from Plato to Plotinus, passing through Aristotle (and even, to a certain extent, through the Stoics), has nothing accidental or contingent about it, that is, nothing that could be chalked up to philosophical fancy. Rather, they sketch out the vision of universal becoming that a systematic intellect will obtain when it only looks at becoming through the views or snapshots taken, every now and then, of its flowing by. Such that even today we continue to philosophize in the manner of the Greeks, and continue to reach many of their general conclusions (without even knowing them in advance), to the exact extent that we remain loyal to the cinematographic instinct of our thought.

I said above that there is *more* in a movement than there is in the successive positions attributed to the moving object, *more* in a becoming than in the forms that are successively passed through, and *more* in the evolution of the form than in the forms realized one after the other.[104] As such, philosophy will be able to deduce the terms of the second type from those of the first, but not the terms of the first type from those of the second. Speculation thus should begin with the first type. But the intellect reverses the order of the two terms and, on this point,

315

ancient philosophy proceeds in the same way as the intellect. It thus sets itself up in the immutable, and only considers Ideas. And yet there is becoming—it's a fact.[105] How, having posited immutability alone, could they draw change out of it? This could not be done through the addition of something else, since, by definition, nothing positive exists outside of Ideas. It would thus have to be through a diminution. The following postulate necessarily lies at the foundation of ancient philosophy: there is more in the immobile than in moving reality [le mouvant], and one goes from immutability to becoming by way of diminution or attenuation.

316

Thus, something negative, or at least something equal to zero, will have to be added to the Ideas in order to obtain change. Plato's "non-being" and Aristotle's "matter" consist in precisely this—a metaphysical zero that, placed alongside the Idea (the way the arithmetic zero is placed alongside the one) multiplies it in space and in time.[106] Through this zero, the immobile and simple Idea is refracted into an indefinitely propagated movement. In principle, there should only be immutable Ideas immutably fitting together. In fact, matter comes along to add its void to the Ideas and, by the same stroke, starts up universal becoming. Matter is that elusive nothing that, by slipping in between the Ideas, creates endless agitation and eternal restlessness, like a suspicion insinuates itself between two hearts in love. Demote the immutable ideas, and you obtain, by the same stroke, the perpetual flux of things. The Ideas or Forms are surely the whole of intelligible reality, that is, of the truth, insofar as they collectively represent the theoretical equilibrium of Being.[107] As for sensible reality, it is an indefinite oscillation from one side to the other of this equilibrium point.

This leads to a particular conception of durée, as well as to a particular conception of the relation of time to eternity, that are found throughout the philosophy of Ideas. To anyone who places themselves within becoming, durée appears as the very life of things, as the fundamental reality. In that case, the Forms, which the mind isolates and stores up in concepts, are merely views or snapshots taken of changing reality.[108] They are moments collected along the course of durée and, precisely because the thread that tied them to time has been cut, they no longer endure. They tend to merge with their own definition, i.e., with the artificial reconstruction and the symbolic expression that is their

317   intellectual equivalent. They enter into eternity, if you like; but what they have of the eternal is no longer anything other than what they have

of the unreal. — To the contrary, if becoming is approached through the cinematographic method, then the Forms are no longer views taken of changing reality; they become the constitutive elements of it and represent all that is positive in becoming. Eternity no longer hovers over time, like an abstraction; eternity becomes the foundation of time, like a reality. And this is precisely the attitude adopted by the philosophy of Forms or Ideas on this question. It establishes between eternity and time the same relation as the one between the piece of gold and some spare change—change so small that the payment continues indefinitely without the debt ever being fully paid. Yet one could free oneself from the debt, in a single stroke, with the gold piece itself.[109] This is what Plato expresses in his magnificent rhetoric when he says that God, not being able to make the world eternal, gave it Time, the "moving image of eternity."[a/110]

This also leads to a particular conception of extension, which is at the foundation of the philosophy of Ideas, although it was not set out quite as explicitly. Again, let us first imagine a mind that places itself back along the course of becoming and adopts its movement.[111] Each successive state, each quality, and each Form will ultimately appear to this mind as a simple cut that thought makes within universal becoming. Such a mind will discover that the form is essentially extended since it is inseparable from the becoming in extension that has materialized it along the course of its flowing by. Each form thereby occupies space in the same way that it occupies time. But the philosophy of Ideas marches in the opposite direction. It begins from the Form, seeing it as the very essence of reality. It does not obtain the form by taking a view or snapshot of becoming; rather, it takes the forms as given for all eternity.[b] And from the perspective of this immobile eternity, durée and becoming could be nothing but a degradation. Form posited in this way, as inde-   318 pendent of time, is no longer the form found in perception; rather, it is a *concept*. And, since a reality of the conceptual order no more occupies extension than it does durée, it must be that the Forms reside outside of space in the same way that they hover above time. Thus, according to

a Bergson's note: Plato, *Timaeus*, 37d [Plato, *Timaeus*, trans. Donald J. Zeyl, in *Complete Works*, ed. John. M. Cooper, with D. S. Hutchinson (Indianapolis: Hackett Publishing Company, 1997), 37d].

b TN: ...*elle se donne des formes dans l'éternel.*

ancient philosophy, space and time necessarily have the same origin and the same value. It is the same diminution of being that is expressed by a distension in time and by an extension in space.[112]

Extension and distension therefore simply express the gap between what is and what should be. From the perspective adopted by ancient philosophy, space and time are nothing other than the field posited by an incomplete reality—or rather, a reality that has gotten lost outside of itself—so that it can run about looking for itself. But it will be necessary to admit here that the field is created to the extent that the running about advances, and that the running about in some sense deposits the field beneath itself. Imagine that you move a pendulum away from its equilibrium point, which is a simple mathematical point: an unending oscillation takes place along which points are juxtaposed to other points and instants follow other instants.[113] The space and time that are born in this way have no more "positivity" than the movement itself. They represent the gap between the pendulum's artificially given position and its normal position, i.e., *what it is missing* in order to regain its natural stability. Now return the pendulum to its normal position: space, time, and movement all retract back down into a mathematical point. In the same way, human reasonings continue along in an endless chain, but they would be completely and instantly reabsorbed into the truth grasped by intuition since their extension and their distension are merely a gap, so to speak, between our thought and the truth.[a] The same holds for extension and for *durée* in relation to the pure Forms or Ideas. The sensible forms are out in front of us, always ready to regain their ideality, yet always blocked by the matter that they carry within themselves, that is, by their inner void, by the interval they leave between what they are and what they should be. They are continuously on the verge of taking hold of themselves and continuously in the process of losing themselves. They are condemned by an unavoidable law to fall back down, like Sisyphus's boulder, just when they were about to reach the summit, and this law, which has launched them into space and time, is nothing other than the very constancy of their original insufficiency. The alternation between generation and decline, the evolutions that

---

a   Bergson's note: I tried to disentangle what is true and what is false in this idea in terms of *spatiality* (see Chapter III [particularly 201–9]). This idea seems to me to be radically false when it comes to *durée*.

are continuously being reborn, the circular movement of the celestial spheres that is indefinitely repeated: all of this simply represents a certain fundamental deficit that is materiality itself.[114] Fill in this deficit—in a single stroke you suppress space and time, the indefinitely renewed oscillations around a stable equilibrium that is forever sought but never attained. Things enter back into each other. What was distended in space is contracted again into pure form. And the past, present, and future retract back into a single moment, which is eternity.

This amounts to saying that physical reality is simply logical reality corrupted.[a] This proposition sums up all of the philosophy of Ideas. And it is also the hidden principle of the philosophy that is innate to our own understanding. If immutability is more than becoming, then the form is more than change, and it is through a veritable fall that the logical system of Ideas, rationally subordinated and coordinated among themselves, is scattered out into a series of physical objects and events that are accidentally placed one after the other. The generative idea of a poem is developed into thousands of imaginings, which are them-   320 selves materialized into phrases and deployed in words. And the more we descend from the immobile idea, all rolled up into itself, to the words that unroll it, the more place is left for contingence and choice. After all, other metaphors expressed by other words might well have sprung up; but one image was called forth by another image, one word by another word. All of these words now run along, one after the other, seeking in vain to render, all by themselves, the simplicity of the generative idea. Our ear only hears the words; it thus only perceives accidents. But our mind, through successive leaps, jumps from the words to the images and from the images to the original idea, and in this way goes back up from the perception of words—mere accidents provoked by other accidents—to the conception of the Idea that establishes itself.[115] This is how the philosopher proceeds when considering the universe. Experience places before the philosopher's eyes phenomena that also pass by, one after another, in an accidental order determined by the circumstances of time and place. This physical order, which is a veritable

---

a TN: *...le physique est du logique gâté.* Here Bergson uses the masculine forms of the words *physique* and *logique*, indicating that he does not mean physics or logic, but rather physical and logical nature or reality. This could also be translated as "the physical is merely the logical corrupted."

degeneration of the logical order, is nothing other than the fall of logical reality into space and time. But the philosopher, moving back up from the percept to the concept, sees all the positive reality that the physical contained being condensed into the logical. The philosopher's intellect, disregarding the materiality that distends being, grasps being in itself in the immutable system of Ideas. This is how Science is obtained; and Science will appear complete and ready-made the moment we place our intellect back in its proper place, thereby correcting the gap that separated it from the intelligible. According to this perspective, science is not a human construct. It is anterior to our intellect, independent of it, and is genuinely generative of the things.[116]

And indeed, had we held the Forms to be mere views or snapshots taken by the mind of the continuity of becoming, then they would have been relative to the mind that conceives of them and would not have an independent existence. On this approach, the most that could be said is that each of these Ideas is an ideal. But we have adopted the opposite perspective here, and so the Ideas must exist by themselves. Ancient philosophy could not escape this conclusion. Plato formulated it and it was in vain that Aristotle tried to free himself from it. Since movement is born of the degradation of the immutable, there could be no movement—and consequently no sensible world—if immutability were not realized somewhere. Moreover, having begun by refusing the Ideas an independent existence and yet finding himself unable to take it away from them, Aristotle pressed them together, gathered them up into a ball, and placed above the physical world a single form that became the Form of Forms, the Idea of Ideas, or finally, to use his expression, the Thinking of Thinking.[117] This is Aristotle's God—necessarily immutable and outside of what takes place in the world, since he is merely the synthesis of all the concepts into a single concept. True, none of these many concepts could exist independently, as such, within the divine unity—it is in vain that we look for Plato's Ideas within Aristotle's God. But it is enough to imagine Aristotle's God himself being refracted, or simply inclining toward the world, for Platonic Ideas to immediately pour forth out of him, these Ideas being implicit in the unity of his essence—like the rays of sunlight shining forth from the sun, though the sun did not contain them in advance. It is surely this *possibility of a pouring forth of* Platonic Ideas from the Aristotelian God that is represented in Aristotle's philosophy by the active intellect, the νοῦς that is called ποιητικός—or, in other words, by what is essential, though unconscious, in human intellect. The νοῦς ποιητικός is integral Science, posited all at once, and is what the intellect, being conscious and discursive, is condemned to reconstruct with difficulty, piece by piece.[118] As such, there is in us, or rather behind us, a possible vision of God, as the Alexandrians would say, a vision that is always virtual and never actually realized by conscious intellect.[119] In this intuition, we would see God spread out into Ideas. This intuition "makes all things,"[a] playing the same role in relation to the discursive intellect (which moves in time) that the unmoved Mover himself plays in relation to the movements of the sky and to the course of things.

We would thus discover, immanent within the philosophy of Ideas, a *sui generis* conception of causality that must be brought out into the light because it is the same one that each of us will arrive at when, in order to work back up to the origin of things, we follow the natural movement of the intellect to its conclusion. True, ancient philosophers never explicitly formulated this conception. They limited themselves to drawing out its consequences and, in general, they indicated points of view upon it rather than presenting it for itself. Sometimes, for instance, they speak of an *attraction*, and sometimes of an *impulsion* exerted by the prime mover on the whole world.[120] The two views are found in Aristotle: he shows the movement of the universe to be an aspiration of things toward divine perfection and consequently an ascension toward God, whereas elsewhere he describes this movement as the effect of God's contact with the first sphere and thus as descending from God toward things.[121] Moreover, I believe the Alexandrians were simply following this double indication when they spoke of procession and conversion: everything derives from the first principle or source and everything also aspires to return to it.[122] But these two conceptions of divine causality can only be identified with each other if they are both referred back to a third, a

322

323

---

a Bergson's note: Aristotle, *De Anima*, 430a14: καὶ ἔστιν ὁ μὲν τοιοῦτος νοῦς τῷ πάντα γίνεσθαι, ὁ δὲ τῷ πάντα ποιεῖν, ὡς ἕξις τις, οἷον τὸ φῶς· Τρόπον γάρ τινα καὶ τὸ φῶς ποιεῖ τὰ δυνάμει ὄντα χρώματα ἐνεργείᾳ χρώματα. ["And in fact thought, as we have described it, is what it is by virtue of becoming all things, while there is another which is what it is by virtue of making all things: this is a sort of positive state like light; for in a sense light makes potential colours into actual colours." Aristotle, *On the Soul*, trans. J. A. Smith, in *The Complete Works of Aristotle*, ed. Jonathan Barnes, the revised Oxford translation, vol. 1, Bolligen Series LXXI.2 (Princeton: Princeton University Press, 1995), 684.]

conception that we take to be fundamental and that alone makes intelligible not only why and in what sense things move in space and time, but also why there is space and time, why there is movement, and why there are things.

I would formulate this conception—which increasingly shines through from beneath the reasonings of the Greek philosophers to the extent that we go from Plato to Plotinus—as follows: *The positing of a reality implies the simultaneous positing of all of the intermediary degrees of reality between it and pure nothingness.*[123] This principle is self-evident if we consider numbers. We cannot posit the number 10 without simultaneously positing the existence of the numbers 9, 8, 7..., etc., in short, the existence of every interval between 10 and zero.[124] But here our mind passes naturally from the sphere of quantity to that of quality. If a certain perfection is given, it seems that the whole continuity of degradations is also given—between this perfection, on the one hand, and the nothingness that we imagine ourselves capable of conceiving, on the other hand. Let us posit, then, Aristotle's God—the thinking of thinking, i.e., thought turning in a circle [*faisant cercle*], transforming itself from subject into object and from object into subject through an instantaneous, or better, eternal circular process.[125] Again, since nothingness seems to establish itself, and since, these two extremities being given, the interval between them is also given, it follows that all the degrees descending from being, i.e., from the divine perfection, all the way down to "absolute nothing," will be automatically realized, so to speak, the moment that God has been posited.

So let us travel along this interval from high to low. First, the slightest diminution of the first principle is enough for being to be precipitated into space and time, but the *durée* and the extension represented by this first diminution will be as close as possible to the unextended and eternal nature of the divine. As such, we must imagine this first degradation of the divine principle as a sphere turning in place, imitating through the perpetuity of its circular movement the eternity of the *circulus* that is divine thought. It thereby creates its own place and, as such, place in general[a] since nothing contains it and since it does not change place. It

324

---

a Bergson's note: Aristotle, *On the Heavens*, II, 287a12: τῆς ἐσχάτης περφορᾶς οὔτε κενόν ἐστιν ἔξωθεν οὔτε τόπος ["outside the farthest circumference there is neither void nor place"]. Aristotle, *Physics*, IV, 212a34: τὸ δὲ πᾶν ἔστι μὲν ὡς κινήσεται ἔστι δ' ὡς οὔ. Ὡς μὲν γὰρ ὅλον, ἅμα τὸν τόπον οὐ μεταβάλλει, κύκλῳ δὲ κινήσεται· τῶν μορίων γὰρ οὗτος ὁ τόπος ["the whole will be moved in one sense, but not in another. For as a whole it does not simultaneously change its place, though it will be moved in a circle; for this place is the place of its parts"].

also creates its own *durée* and, thereby, *durée* in general since its movement is the measure of all other movements.[a] Next we will see the perfection decrease, degree by degree, all the way down to our sublunary world where the cycle of generation, growth, and death imitates the original *circulus* one last time by corrupting it.[126] Understood in this way, the causal relationship between God and the world appears as an attraction when looked at from below and as an impulsion or as an action through contact when looked at from above, since the outer sphere is an imitation of God and since the imitation is the reception of a form.[127] Thus, depending on whether we look in one direction or in the other, we perceive God either as the efficient cause or the final cause. And yet, neither of these two relationships is the definitive causal relationship. The true relationship is the one found between two members of an equation in which the first member is a unique term and the second is a summation of an indefinite number of terms. It is, if you will, the relation of the piece of gold to its spare change, provided that the change is automatically revealed the moment that the piece of gold is presented. Only in this way will we understand that Aristotle demonstrated the necessity of an unmoved prime mover not by relying on the idea that the movement of things must have had a beginning, but rather by positing that this movement could not have a beginning and must never come to an end. If movement exists, or, in other words, if the spare change can be counted up, this is because the piece of gold exists somewhere. And if the summation continues forever, never having begun, this is because the single term that is its eminent equivalent is eternal. A perpetuity of mobility is only possible if it is propped up by an eternity of immutability that it unrolls in a chain that is without beginning and without end.

325

Such is the final word of Greek philosophy. I had no intention of reconstructing it *a priori*. It has multiple origins. It is attached by invisible threads to all of the fibers of the ancient soul. The attempt to deduce it from a single principle would be in vain.[b] But, if we eliminate everything in Greek philosophy that comes from poetry, religion, and social

---

a Bergson's note: Aristotle, *On the Heavens*, I, 279a12: οὐδὲ χρόνος ἐστὶν ἔξω τοῦ οὐρανοῦ ["there is also no place or void or time outside the heaven"]. Aristotle, *Physics* VIII, 251b27: ὁ χρόνος πάθος τι κινήσεως ["time being a kind of affection of motion"].

b Bergson's note: Above all, I have almost entirely left to the side those admirable (though somewhat fleeting) intuitions that Plotinus would later catch hold of, develop, and bring into focus.

life, along with everything that comes from its rather rudimentary physics and biology, that is, if we set aside all of the friable materials that enter into the construction of this immense edifice, a solid framework [*charpente*] still remains—and I believe this framework sketches out the broad outlines of the natural metaphysics of human intellect.[128] Indeed, this is precisely the kind of philosophy we reach the moment we follow the cinematographic tendency of perception and thought to the end. Our perception and thought begin by substituting for the continuity of evolutionary change a series of stable forms that would be strung together, one by one, as they pass by, like those rings that children on the merry-go-round attempt to capture on sticks as they pass by. But what does the passing by consist in here, and upon what will the forms be strung together? Since the stable forms have been obtained by extracting from change everything that is found to be definite there, all that remains for characterizing the instability upon which the forms are imposed is a negative attribute, which will be indetermination itself. This is the first step that our thought takes. It dissociates each change into two elements: the first one being stable and definable for each particular case, namely the Form; the second being indefinable and always the same, which would be change in general. And this is also the essential operation of language. The forms are all that language is capable of expressing. It is forced to imply, or it limits itself to merely *suggesting* a mobility that, precisely because it remains unexpressed, is supposed to remain the same in all cases. And so a philosophy emerges that takes as legitimate the dissociation thus executed by thought and language. And what will this philosophy do other than objectify the distinction with even more force, push it to its extreme consequences, and reduce it into a system? As such, this philosophy will construct the real with, on the one hand, definite Forms or immutable elements, and, on the other hand, a principle of mobility that, being the negation of form, will necessarily escape every definition and will be pure indetermination. The more this philosophy directs its attention to the forms that thought delimits and that language expresses, the more it will see them rise up above the sensible and become subtilized into pure concepts, capable of entering into each other and even of ultimately being gathered together into a single concept as the synthesis of all of reality and the culmination of all perfection. To the contrary, the more this philosophy descends toward the invisible source of universal

326

mobility, the more it will sense it slipping away beneath itself, while at the same time emptying itself out and being completely absorbed into what it will call pure nothingness. Ultimately, such a philosophy will have, on the one hand, the system of Ideas logically coordinated amongst themselves or concentrated into a single one, and on the other hand, a quasi-nothingness, namely Platonic "non-being" or Aristotelian "matter." But, after having cut the fabric, it is time to sew.[129] Now it becomes a question of reconstituting the sensible world from the supra-sensible Ideas and an infra-sensible non-being. This will only be possible if we postulate a type of metaphysical necessity in virtue of which the bringing together of this Whole and this Zero is *equivalent* to the positing all of the degrees of reality that make up the interval between the two, in the same way that a whole number—the moment we imagine it as a difference between itself and zero—is revealed as a certain sum of units and, in the same stroke, causes all of the smaller numbers to appear. And this is the natural postulate. It is also the one that we catch sight of at the foundation of Greek philosophy. In order to explain the specific characteristics of each one of these intermediary degrees of reality, all that remains to do is to measure the distance that separates it from integral reality. Each inferior degree consists in a diminution of the higher one, and what we perceive as a sensible novelty would be resolved, from the point of view of the intelligible, into a new quantity of negation that has been added to it. The smallest possible quantity of negation, the quantity that we already find in the highest forms of sensible reality and, consequently, *a fortiori* in the inferior forms, will be the one that expresses the most general attributes of sensible reality: extension and *durée*. Through increasing degradations, we will obtain attributes that are more and more specific. Here the philosopher's imagination takes flight since it is through an arbitrary decree, or at least one that is debatable, that some particular aspect of the sensible world is equated with some particular diminution of being. That is, we do not necessarily arrive, as Aristotle did, at a world of concentric spheres turning in place. But we will be led to an analogous cosmology, by which I mean a construction whose pieces, despite being completely different, will nonetheless have the same relations among themselves. And this cosmology will still be governed by the same principle. The physical will be defined by the logical.[130] From beneath the changing phenomena, a closed system of concepts, subordinated to each other and coordinated together, will be

327

328

shown to be shining through.[131] Science, understood as the system of concepts, will be more real than sensible reality. It will be anterior to human knowledge, which merely spells it out, letter by letter; it will be anterior to the things that clumsily attempt to imitate it. This science need only be distracted from itself for a moment so as to exit from its eternity and, by the same stroke, to coincide with all of human knowledge and with all of these sensible things.[132] Its immutability is thus surely the cause of universal becoming.

Such was the perspective ancient philosophy adopted toward change and durée. Now, it seems incontestable that modern philosophy had, repeatedly, but especially in its beginnings, the vague desire to change this point of view. But an irresistible attraction draws the intellect back to its natural movement, and draws modern metaphysics back to the general conclusions of Greek metaphysics. I will now attempt to clarify this final point, in order to show the invisible threads that connect our mechanistic philosophy to the ancient philosophy of Ideas, as well as how modern metaphysics also responds to the primarily practical demands of our intellect.

*

[BECOMING, ACCORDING TO MODERN SCIENCE. TWO PERSPECTIVES ON TIME][133]

Modern science, like ancient science, proceeds according to the cinematographic method.[134] It cannot do otherwise: every science is subjected to this law. Indeed, the essence of science is to manipulate signs that are substituted for the objects themselves.[135] Of course, these signs differ from those of language by their greater precision and higher efficacy; yet they are no less constrained by the general condition of a sign, which is to indicate a fixed aspect of reality via an unchanging form [forme arrêtée]. In order to think movement, the mind must exert a continuously renewed effort. Signs are designed to free us of this effort by substituting for the moving continuity of things an artificial recomposition that is equal to it in practice and that has the advantage of being easily manipulated.[136] But let us leave aside the procedures and consider only the result. What is the essential aim of science? It is to increase our influence on things. Science might well be speculative in its form and disinterested in its immediate goals; in other words, we can grant it credit for as long as it likes. And yet although the due date of the loan has been pushed back, we must eventually be paid what is owed. Thus, science always ultimately aims at practical utility. Even when it launches into theory, it is required to adapt its procedures to the general configuration of practice. As high as it raises itself up, it must always remain ready to fall back into the field of action and to immediately find its footing there. This would not be possible for it if its rhythm was absolutely different from that of action itself.[137] Now, as I have said, action proceeds by leaps.[138] To act is to readapt oneself. To know—that is, to foresee in order to act—will thus be to go from one situation to another situation, from an arrangement to a rearrangement.[139] Science will be able to consider rearrangements that are closer and closer together; it will thereby increase the number of moments that it isolates, but it will still be isolating moments. As for what takes place in the interval, science pays no more attention to this than do ordinary intellect, the senses, or language: science focuses on the extremities of the interval and not the interval itself. The cinematographic method thus imposes itself on our science, just as it already imposed itself on the science of the ancients.

So where is the difference between these two sciences? I indicated this difference when I suggested that the ancients reduce the physical order to the vital order, the laws to the genera, whereas the moderns want to resolve the genera into laws.[140] But it is important to consider this from another angle, which is nonetheless but a transposition of the first. What does the difference of attitude between these two sciences consist in when it comes to change? I would formulate it as follows: *ancient science believes an object to be sufficiently known when it has observed some of its privileged moments, whereas modern science considers the object at any moment whatsoever.*

Plato's or Aristotle's forms or ideas correspond to the privileged or salient moments in the history of things—the very moments that, in general, were fixed by language. The forms are supposed to characterize a period for which they express what is quintessential, such as the "childhood" or the "old age" of a living being; the rest of the period, filled up by the passage from one form to another form, is of no interest in itself. What about a falling body? They assume the fact to be sufficiently examined once it has been characterized globally: it is a *downward* movement, the tendency toward a *center*, and the *natural* movement of a body that, having been separated from the earth to which it belonged, goes about regaining its proper place. Thus, they note the final term (τέλος) or the highest point (ἀκμή), they reify it into the essential moment, and this moment, which language has selected for expressing the whole of the fact, is also sufficient for science as a way of characterizing that fact.[141] In Aristotle's physics, the movement of a body launched through space or in a free fall is defined via the concepts of high and low, of spontaneous displacement and forced displacement, of proper place and foreign place.[142] But Galileo maintained that there are no essential moments and no privileged instants: to study a falling body is to consider it at any moment whatsoever along its pathway.[143] The true science of gravity will be the one that determines, for any given moment of time, the position of the body in space. This will of course require signs that are precise in a different way than the signs of language.

331   As such, one could say that our physics above all differs from ancient physics by the indefinite decomposition of time that it enacts. For the ancients, time includes as many indivisible periods as our natural perception and our language cuts out of it as successive facts, each presenting a kind of individuality. This is why, from their perspective, each of these facts only admits of a global definition or global description. If the description leads us to distinguish several phases, then we will have several facts rather than a single one, or several undivided periods rather than a single period. But time will always have been divided up into determinate periods, and this mode of division will always have been imposed upon the mind by apparent crises in the real—comparable to that of puberty—or by the apparent releasing of a new form. For scientists like Kepler or Galileo, to the contrary, time is not objectively divided up in one way or another by the matter that fills it. Time has no natural articulations.[144] We can, and indeed must, divide it up however we like.[145] All instants are of value. None of them has the right to set itself up as the representative instant or as the ruler of the others. And, consequently, we only know a change when we know how to determine where it stands at any given moment whatsoever.[146]

The difference is profound. From a certain angle, it is even radical. And yet, from the point of view we are adopting here, it is a difference of degree rather than a difference of nature. The human mind has gone from the first kind of knowledge to the second by gradually perfecting itself and by simply seeking a higher level of precision.[147] Between these two sciences, we find the same relation as between the observations of the phases of a movement by the eye and the much more complete recording of these phases through instantaneous photography. It is the same cinematographic mechanism in both cases, but in the second case it attains a level of precision that it cannot reach in the first. When we see a galloping horse, our eye above all perceives a characteristic, essential, or even schematic attitude, i.e., a form that appears to shine across a whole period and thus to fill up the time of a gallop—and this is the attitude that the art of sculpture captures in the friezes of the Parthenon.[148] But the instantaneous photograph isolates any moment whatsoever; it places them all on the same level, and this is how the horse's gallop is divided up into as many successive attitudes as one likes, rather than being gathered together into a single attitude that would shine in a privileged instant and light up an entire period.

332

All of the other differences flow from this original one. A science that examines indivisible periods of *durée*, one after the other, will only see phases followed by other phases, or forms that replace other forms. Such a science will be satisfied with a *qualitative* description of objects, which it will assimilate to organic beings.[149] But when we look for what happens within one of these periods, within a given moment of time, we aim at something completely different. Here the changes that take place from one moment to the next are, in principle, no longer changes of quality—as a result, they are *quantitative* variations, either of the phenomenon itself, or of its elementary parts. Thus, it was correct to say that modern science differs from ancient science insofar as it focuses on magnitudes and takes its primary task to be measuring them.[150] The ancients were already practicing experimentation, whereas Kepler did not experiment, in the strict sense of the word, and yet discovered a law that is the paradigmatic example of scientific knowledge such as we understand it. What distinguishes our science, then, is not the fact that it experiments, but rather that it only experiments and more generally, only works in order to measure.

This is also why it was correct to say that ancient philosophy focused on *concepts*, whereas modern science seeks *laws*, that is, constant relationships between variable magnitudes.[151] Aristotle was able to define the movement of the celestial bodies with nothing other than the concept of circularity. And yet, even with the more precise concept of the form of the ellipse, Kepler did not believe he could account for the movement of the planets. He needed a law, a constant relationship

333

between the quantitative variations of two or more elements involved in planetary movement.

Nevertheless, these are merely consequences, by which I mean differences that derive from a fundamental difference. The ancients might well have accidentally discovered experimentation with the aim of measuring, just as they might well have discovered a law announcing a constant relationship between magnitudes. Archimedes' principle is a genuine experimental law. It accounts for three variable magnitudes: the volume of the body, the density of the liquid in which it is immersed, and the thrust from low to high to which the liquid is subjected. And indeed, his principle clearly announces that one of these three terms is a function of the other two.

As such, the essential or original difference must be found elsewhere. It is the very difference that I indicated above.[152] Ancient science is static. Either it considers the change it is examining as a block, or, if it does divide it up into periods, it treats each of these periods in turn as a block—this amounts to saying that it does not take time into account. But modern science is constructed around Galileo's and Kepler's discoveries, which immediately provided it with a model.[153] So, what do Kepler's laws say? They establish a relationship between the areas described by the heliocentric radius vector of a planet and how much time it takes to describe them, or a relationship between the major axis of the orbit and the time it takes to cross it.[154] And what was Galileo's main discovery? A law that connected the space traveled by a falling body to the time occupied by the fall. But we can go even further. What was the first of the great transformations of geometry in modern times? It consisted in introducing time (albeit, in a hidden form) and movement directly into the consideration of shapes.[155] For the ancients, geometry was a purely static science. The shapes were given in a single stroke, in their completed state, much like the Platonic Ideas. But the essence of Cartesian geometry (although Descartes himself did not formulate it in this way) was to consider every plane curve as described by the movement of a point along a movable straight line that is displaced, parallel to itself, all along the horizontal axis, with the displacement of the movable line being assumed to be uniform and the horizontal axis thereby becoming representative of time. The curve will be defined, then, if one can formulate the relationship that links the space traveled along the moving straight line to the time employed in traveling it, i.e.,

334

if one is able to indicate the position of the moving point on the line that it travels at any given moment whatsoever of its path. This relationship will be precisely the equation of the curve. In short, to substitute an equation for a shape consists in seeing where one is at on the pathway of a curve at any moment whatsoever, rather than envisioning this pathway all at once, gathered together into a single movement where the curve is in its completed state.[156]

This was surely the driving principle behind the reformation that renewed both natural science and the mathematics that served as its instrument. Modern science is the daughter of astronomy.[157] It is brought down from heaven to earth along Galileo's inclined plane, since it is Galileo who is the link that ties Newton and his successors back to Kepler.[158] Now, how did the astronomical problem present itself to Kepler? Knowing the respective positions of the planets at a given moment, it was a question of calculating their positions at any other moment whatsoever. From then on, the same question was asked of every material system. Each material point became a rudimentary planet, and the question *par excellence*—the ideal problem whose solution was to deliver the key to all the others—was to determine the relative positions of these elements at any moment whatsoever once their positions at a given moment were known.[159] Of course, the problem is only posed in these precise terms in very simple cases, for an oversimplified reality, since we never know the respective positions of the true elements of matter, assuming that there are real elements. And even if we knew them for a given moment, the calculation of their positions at another moment would very often require a mathematical effort that exceeds human capabilities [*les forces humaines*]. But it is enough for us to know that these elements could be known, that their current positions could be found, and that a superhuman intellect could, by submitting these givens to mathematical operations, determine the positions of the elements at any other moment of time whatsoever. This conviction is at the foundation of the questions we ask ourselves regarding nature, as well as of the methods we employ to resolve those questions. This is why every static law appears to us as a provisional down payment, or as a particular point of view, upon the dynamic law that, alone, would provide us with integral and definitive knowledge.

We must conclude that our science is not distinguished from ancient science merely because it seeks laws, nor even because its laws announce

335

relationships between magnitudes. It must be added that time is the magnitude to which we would like to be able to relate all the others, and that *modern science must above all be defined by its aspiration to consider time as an independent variable.* But what time is in question here?

336   I have already said it (and cannot repeat it often enough): the science of matter proceeds like ordinary knowledge.[160] Although it perfects this knowledge and increases its precision and scope, it works in the same direction and puts the same mechanism into action. If ordinary knowledge, as a result of the cinematographic mechanism that it is subjected to, gives up following the movement of becoming, then so too will the science of matter. Of course, this science distinguishes as many moments as we like within the interval of time that is being examined. As small as the intervals might be that this science settles upon, it still authorizes us to divide them again should we have the need to do so. In contrast with ancient science, which settled upon certain supposedly essential moments, modern science deals with any moment whatsoever with indifference. But it still considers moments or virtual stopping points; in short, it still considers immobile parts. This amounts to saying that real time—conceived of as a flux or, in other words, as the very mobility of being—escapes the grasp of scientific knowledge here.[161] I have already attempted to establish this point in a previous work.[162] I also discussed it briefly in the first chapter of this book.[163] But it is important to return to this point one last time to clear up any misunderstandings.

When the positive sciences speak of time, they are referring to the movement of a certain mobile object T along its trajectory. They have chosen this movement as representative of time, and it is uniform by definition. Let us name the points that divide up the trajectory of this mobile object into equal parts, $T_1, T_2, T_3, \ldots$ etc., beginning from its origin, T_0. It will be said that 1, 2, 3, … units of time have gone by when the mobile object is at points $T_1, T_2, T_3, \ldots$ of the line that it travels along. Now, to consider the state of the universe at the end of a certain time t is to examine where it will be when the moving object T will be at point T_t of its trajectory. But the flux of time itself—especially because of its effect on consciousness—is excluded from consideration

337   here; for what enters into the account are the points $T_1, T_2, T_3, \ldots$ taken from along the flux, and never the flux itself. We can shorten the time being considered as much as we like, decompose at will the interval

between two consecutive divisions T_n and T_{n+1}, but we will always be dealing with points and only with points.[164] All that is retained from the movement of the mobile object T are positions taken along its trajectory. All that is retained from the movement of all of the other points of the universe are their positions on their respective trajectories. At each *virtual stopping point* of mobile object T in the points of the division T_1, T_2, T_3, ... a corresponding *virtual stopping point* is posited for all the other mobile objects at the points they pass through. And when we say that a movement or any other change has occupied a time t, this means that we have noted a certain number t of correspondences of this kind. We have thus counted up simultaneities; we have not been dealing with the flux that goes from the one to the other. The proof is that I can, at will, vary the speed of the flux of the universe in relation to a consciousness that would be independent of it and that would perceive the variation through an entirely qualitative *feeling* that it would have of it. Since the movement T would participate in this variation, I would have nothing to change in my equations or in the numbers that figure there.

Let us go even further. Imagine that the flux reaches an infinite speed. Imagine, as I said in the first pages of this book, that the trajectory of the mobile object T is given all at once, and that all of the past, present, and future history of the material universe is instantaneously spread out in space.[165] The same mathematical correspondences will subsist between the moments of the history of the world fanned out, so to speak, and the divisions T_1, T_2, T_3, ... of the line that will be called, by definition, "the course of time."[166] From the scientific point of view, nothing will have changed. But if science has nothing to change in what it tells us despite time being thus spread out in space and succession becoming juxtaposition, this is because, in what it told us previously, it took into   338 account neither *succession* in terms of what was specific to it nor *time* in terms of its way of flowing.[167] Science has no sign for expressing the way our consciousness is struck by succession or *durée*.[168] Science is no more applied to becoming, insofar as becoming is a moving reality, than the bridges thrown up along the river follow the water that flows beneath their arches.

Nevertheless, succession exists; I am conscious of it—it's a fact. When a physical process takes place before my eyes, its acceleration or its slowing down have nothing to do with my perception or my inclination. What matters to the physicist is the *number* of units of *durée* that the process occupies; he need not worry over the units themselves. And this is why the successive states of the world can be all at once spread out in space without the physicist's science changing at all and without him having to stop speaking of time. But when it comes to us, conscious beings that we are, the units matter, for we do not count the extremities of the intervals—we feel and live the intervals themselves.[169] Now, we are conscious of these interval as *determinate* intervals. This again brings me back to my glass of sweetened water: why must I wait for the sugar to dissolve?[a] Even if the *durée* of the phenomenon is relative for the physicist (insofar as it is reduced to a certain number of units of time and the units themselves are whatever one likes), this *durée* is an absolute for my consciousness since it coincides with a certain rigorously determinate degree of impatience. Where does this determination come from? What obliges me to wait, and to wait for a specific length of psychological *durée* that is imposed and against which I can do nothing? If succession, insofar as it is distinct from mere juxtaposition, has no real efficacy, and if time is not some sort of force, then why does the universe unfold its successive states according to a speed that, from the perspective of my consciousness, is a genuine absolute?[170] Why this determinate speed rather than any other speed whatsoever? Why not an infinite speed? In other words, why is it that everything is not given in a single stroke, such as it is on the cinematographic filmstrip?[171] The more I examine this point, the more it seems that if the future is condemned to *succeed* the present rather than to be given alongside it, this is because the future is not completely determined at the present moment. And, it seems that if the time occupied by this succession is something other than a number—if it has, for the consciousness that is situated within it, an absolute value and an absolute reality—this is because it continuously creates something unforeseeable and new, perhaps not in some artificially isolated system, such as my glass of sweetened water, but in the concrete whole with which this system forms a body [*fait corps*]. This *durée* might not be the fact of matter itself, but rather that of the Life that moves back up the pathway of matter; but the two movements are no less interdependent.[172] Thus, *the durée of the universe must be one with the latitude for creation that can find its place therein.*

339

a Bergson's note: See page 11 [TN: See, rather, page 9].

When the child plays at reconstituting an image by assembling the pieces of a puzzle, he succeeds more and more quickly the more he practices. Moreover, the reconstitution was instantaneous, in that the child found the image ready-made when he first opened the box upon leaving the store. As such, the operation does not require a determinate time, and, at least theoretically, it requires no time at all. This is because the result of the operation is given. This is because the image has already been created and because, in order to obtain it, all that is required is a work of recomposition and rearrangement—a work that can be imagined to go faster and faster, and even becoming infinitely fast, to the point of being instantaneous. But, for the artist who creates an image   340
by drawing it forth from the depths of his soul, time is no longer a mere accessory. It is not an interval that can be lengthened or shortened without thereby modifying its content. The durée of his work is an integral part of his work. To contract it or to dilate it would be to modify both the psychological evolution that fills it up and the invention that is its endpoint. The time of invention is one with the invention itself. It is the progression of a thought that changes to the extent that it takes on a body.[a/173] In short, it is a vital process, something like the maturation of an idea.[174]

The painter stands before his canvas, the colors are on the palette, the model poses. We see all of that, and we also know the painter's style—but can we foresee what will appear on the canvas? We possess the elements of the problem, and we know (in an abstract way) how it will be resolved since the portrait will surely resemble the model and will surely resemble the artist as well. And yet, the concrete solution brings with it that unforeseeable nothing that is everything in the work of art.[175] And it is this nothing that takes time. A nothingness of matter, it creates itself as form. The germination and the blossoming of this form stretch across an unshrinkable durée, a durée that forms a body with these processes.[b] The same goes for works of nature. In nature, novelty appears

a TN: ...au fur et à mesure qu'elle prend corps. This could also be translated as "to the extent that it takes shape" or "as it is born." Given that Bergson continues to draw on images of embodiment and the body (corps), I have preserved the allusion to the body in this expression.

b TN: ...qui fait corps avec elles. This could also be translated as "is an integral part of these processes." Again, I have tried to maintain Bergson's use of expressions alluding to the body (corps), and it could be misleading to refer to durée as a "part" here.

to come from an inner thrust that is a progression or a succession. This thrust confers upon succession a particular power or owes to succession all of its power, which, in any case, renders succession—that is, the *continuity of interpenetration in time*—irreducible to a simple instantaneous juxtaposition in space. This is why the idea of reading the future of living forms in a present state of the material universe, and of unfolding their future history in a single stroke, must contain a genuine absurdity. But this absurdity is difficult to bring into relief because our memory is accustomed to lining up in ideal space the terms that it perceives one after the other, and because our memory always imagines succession as already *past* and in the form of a juxtaposition.[176] Indeed, it can do so precisely because the past has already been invented and is now dead; it is no longer creation and life. And so, since the succession to come [*à venir*] will end up being a succession gone by, we persuade ourselves that the *durée* to come admits of the same treatment as the *durée* gone by. We assume that it is already something that could be unrolled, that the future is there, all rolled up, and already painted on the canvas. This may well be an illusion, but it is a natural and ineradicable illusion, and it will endure as long as the human mind!

341

Time is either invention or it is nothing at all.[177] But physics cannot take time-as-invention [*temps-invention*] into account, constrained as it is to follow the cinematographic method. It limits itself to counting the simultaneities between the events that are constitutive of this time and the positions of the mobile object T along its trajectory. Physics detaches these events from the whole that continuously assumes a new form and that communicates to these events something of its novelty. It considers them in their abstract state, such as they would be outside of all that is living, that is, in a time spread out in space. Physics thus only retains the events or the systems of events that can be isolated in this way without causing them to undergo a profound deformation, because these are the only ones that lend themselves to the application of its method. Our physics was born the day we were able to isolate such systems. To summarize, *if modern physics is distinguished from ancient physics insofar as it considers any moment of time whatsoever, then it relies entirely upon a substitution of time-as-length* [*temps-longueur*] *for time-as-invention.*[178]

Parallel to this physics, then, it seems a second type of knowledge should have been constituted that would have retained what physics allowed to escape.[179] Given its attachment to the cinematographic

342

method, science neither wanted to, nor even could, get a hold on the flux of durée itself. This second type of knowledge would have freed us from that method. The mind would have been forced to give up its most prized habits. We would have been transported to the interior of becoming through an effort of sympathy.[180] We would no longer have asked ourselves where a mobile object will be, what configuration a system will assume, or which state a change will pass through at some moment or other. The moments of time, which are nothing but the stopping points of our attention, would have been abolished—it would have been the flowing of time, the very flux of the real itself, that we would have attempted to follow. The first kind of knowledge has the advantage of allowing us to foresee the future and of rendering us, to a certain extent, masters of events.[181] On the other hand, it only retains some possible immobile parts of moving reality, i.e., views or snapshots taken of it by our mind. Rather than expressing the real, this kind of knowledge symbolizes it and transposes it into something human. The other type of knowledge, if it is indeed possible, will be practically useless: it will not extend our empire over nature and will even go against certain natural aspirations of the intellect. But if it were to succeed, it would encompass reality itself in a definitive embrace. In this way, we would not only complete the intellect and its knowledge of matter by training it to situate itself within moving reality; by developing another faculty, complementary to the intellect, we would also open up a perspective upon the other half of the real. From the moment that we find ourselves in the presence of real durée, we see that it signifies creation, and if what is being unmade endures, this can only be through its interdependence with what is being made.[182] As such, the necessity of a continuous growth of the universe would appear, by which I mean a life of the real. And from then on we would conceive of the life that we encounter on the surface of our planet from a new angle, as a life oriented in the same direction as the life of the universe and in the opposite direction of materiality. Intuition would have finally been added to the intellect.

The more that we reflect on this, the more we will find that this conception of metaphysics is the one suggested by modern science.[183]   343

Indeed, time is theoretically insignificant for the ancients because for them the durée of a thing simply manifests the degradation of its essence.[184] Their science focuses on this immobile essence. Change being nothing other than the effort of a Form toward its own realization, the realization is the only thing we need to know. Of course, this realization is probably never complete—this is what ancient philosophy expresses by saying that we do not perceive form without matter.[185] But if we examine the changing object at a certain essential moment—at its apogee—we could say that it *brushes up against* its intelligible form.[186] Our science seizes upon this intelligible, ideal and, so to speak, extreme form. And when it thereby possesses the gold piece, it also eminently holds that spare change that is change itself. Change is thus lesser than being. And the knowledge that would claim to take change as its object, assuming such a knowledge to be possible, would be lesser than science.

But, for a science that places all of the instants of time on the same level, that does not allow for essential moments, a highest point, or an apogee, change will no longer be a diminution of the essence, nor will *durée* be a diluting of eternity. Here the flux of time becomes reality itself, and the things that flow by become the objects of study. True, we are limited to taking mere snapshots of this reality that flows by. But this is precisely why scientific knowledge must call for another one, a knowledge that would complete it. Whereas the ancient conception of scientific knowledge ended up making time into a degradation and change into the diminution of a Form that is itself given for all eternity, had we rather followed the new conception through to its conclusion, we would have ended up seeing in time a progressive growth of the absolute and in the evolution of things a continuous invention of new forms.

True, this would have been to break with ancient metaphysics. The ancients were only aware of one single way of attaining definitive knowledge. Their science consisted in a scattered and fragmentary metaphysics and their metaphysics in a concentrated and systematic science—these were, at most, two species of the same genre.[187] By contrast, according to the hypothesis I am adopting here, science and metaphysics are opposite though complementary ways of knowing: the first only retaining the instants, i.e., that which does not endure; the second focusing on *durée* itself. It was natural that there was some hesitation between such a new conception of metaphysics and the traditional one. The temptation must have even been quite strong to begin again, with the new science, what had been tried with the ancient one, to immediately assume that our scientific knowledge of nature was complete, to unify it completely, and to name this unification, as the Greeks already did, "metaphysics." As such, alongside the new pathway that philosophy could have opened

344

up, the ancient one also remained open. Physics marched along on that very pathway. And when it comes to time, since physics only retained that which could be equally well spread out in space all at once, the metaphysics that set out in this direction necessarily had to proceed as if time neither created nor destroyed anything at all, as if *durée* had no effect at all. Constrained (like modern physics and ancient metaphysics) to the cinematographic method, modern metaphysics arrived at the following conclusion, implicitly assumed from the outset and immanent within the method itself: *Everything is given.*[188]

*

[THE METAPHYSICS OF MODERN SCIENCE. DESCARTES, SPINOZA, LEIBNIZ]

It seems incontestable that metaphysics at first hesitated between these two pathways. The oscillation is visible within Cartesian philosophy.[189] On the one hand, Descartes affirms universal mechanism: from this point of view, movement would be relative, and since time has just as much reality as movement, the past, present, and future must be given for all eternity.[a/190] But on the other hand—and this is why he does not reach the ultimate consequences of his own position—Descartes believes in human free will.[191] He superposes the indetermination of human actions upon the determinism of physical phenomena, and consequently he superposes a *durée* in which there is invention, creation, and true succession upon time-as-length.[192] Descartes supports this *durée* with a God that continuously renews the creative act and who, thereby being tangential to time and to becoming, sustains them and necessarily communicates to them something of his absolute reality.[193] When Descartes adopts this second point of view, he speaks of movement, even movement in space, as an absolute.[b/194]

Having decided to follow neither pathway to the end, Descartes thus attempted to follow them both in turn. The first pathway would have led him to the negation of free will in man and any genuine will in God.[195] This was the suppression of all effective *durée* and the assimilation of

345

---

a Bergson's note: Descartes, *Principles*, Part II, §29.
b Bergson's note: Descartes, *Principles*, Part II, §36ff.

the universe to a *given* thing that a superhuman intellect could encompass all at once, either instantaneously or eternally. To the contrary, by following the second pathway, we arrived at all of the consequences that the intuition of true *durée* implies. Creation no longer appeared merely as *continued*, but rather as *continuous*. The universe, conceived of as a whole, truly evolved. The future could no longer be determined in relation to the present; the most that could be said was that, once the future had been realized, it could be found in its antecedents, just as the sounds of a new language can be expressed with the letters of an ancient alphabet—

346   here the value of the letters is dilated, and the sonorities that no combination of ancient sounds could have made foreseeable are retroactively attributed to them.[196] Ultimately, the mechanistic explanation was able to remain universal insofar as it was extended to as many systems as one wanted to cut out of the continuity of the universe. But mechanism thereby became a *method* rather than a *doctrine*. It claimed that science must follow the cinematographic approach, and that the role of science is to articulate the rhythm of the flowing by of things—not to insert itself into it. These were the two opposing conceptions of metaphysics available to philosophy.

It set out on the first pathway. The reason for this choice is surely to be found in the tendency of the mind to proceed according to the cinematographic method, a method so natural to our intellect and so well-adjusted to the demands of our science that it would be necessary to be doubly certain of its speculative impotence for us to renounce it in metaphysics. But here the influence of ancient philosophy also played a role. Forever the admirable artists, the Greeks created a type of supersensible truth, like sensible beauty, whose attraction is difficult to escape. From the moment one leans toward making metaphysics into a systematization of science, one slides in the direction of Plato and Aristotle. And, once one has entered into the zone of attraction in which the Greek philosophers move, one is pulled into their orbit.

This is how the doctrines of Leibniz and Spinoza were constituted. Now, I am not unaware of the treasures of originality that these doctrines contain. Spinoza and Leibniz poured the contents of their souls into their doctrines, which are rich with the inventions of their genius and the acquisitions of the modern spirit. For both, and above all for Spinoza, there are thrusts of intuition that cause the system to crack.[197] And yet, if we eliminate from these two doctrines what animates them and what gives them life, if we retain of them only their skeletal structure, we will have before us the very image that we would obtain by looking at Platonism and Aristotelianism through the lens of Cartesian mechanism. We are in the presence of a systematization of the new physics, a systematization constructed according to the ancient model of metaphysics.

What might the unification of physics actually be? The inspiration for this science was the idea of isolating systems of material points within the universe such that, if the position of each one of these points was known for a given moment, then its position at any other moment whatsoever could be calculated. Moreover, since the systems thus defined were the only ones that the new science was able to grasp, and since it could not be said, *a priori*, whether a given system satisfied or did not satisfy the desired condition, it was useful to proceed always and everywhere *as if* that condition obtained.[a] There was in this a wholly adequate methodological rule, one so obvious that it was not even necessary to formulate. Indeed, simple good sense tells us that when we are in possession of an effective research instrument, and when we do not know the limits of its applicability, then we should act as if its applicability were limitless— there will always be time to lower our expectations later. But for the philosopher, it must have been very tempting to hypostatize this hope, or rather this *élan*, of the new science and to thereby convert a general rule of method into a fundamental law of things. They thereby transported themselves to the limit: they assumed physics to be completed and to embrace the totality of the sensible world. The universe became a system of points whose positions were rigorously determined at each instant in relation to the preceding instant, and theoretically calculable for any moment whatsoever. In short, they arrived at universal mechanism. But it was not enough to merely formulate this mechanism; it was necessary to ground it, that is, to prove its necessity and to justify it. And given that the essential claim of mechanism is that there is a mathematical interdependency among all of the points of the universe and among all of the moments of the universe, the justification of mechanism must be found in the unity of a principle that contains everything that is juxtaposed in space and everything that is successive in time. That being the case, the totality of the real was taken as given in a single stroke. The reciprocal

347

348

---

a TN: The "condition" in question here is the calculability of isolated material systems described in the previous sentence.

determination of juxtaposed appearances in space came from the indivisibility of true being. And the rigorous determinism of successive phenomena in time simply expressed that the whole of being is given for
all of eternity.[198]

The new philosophy would thus be a renewal, or rather a transposition, of ancient philosophy. Ancient philosophy focused on the concepts
in which a becoming was concentrated or that marked its apogee.
It then assumed all of the concepts to be known and gathered them
together into a single concept: the form of forms, the idea of ideas, such
as Aristotle's God.[199] The new philosophy would focus on the laws that
condition one becoming in relation to others and that are like the permanent substrate of phenomena. It then assumed all of these laws to be
known and, as above, gathered them together into a unity that expressed
them eminently, but that—like Aristotle's God (and indeed for the same
reasons)—had to remain immutably enclosed within itself.[200]

True, this return to ancient philosophy did not occur without facing
some significant difficulties. When a thinker like Plato, Aristotle, or
Plotinus grounds all of the concepts of their science in a single one, they
thereby embrace the totality of the real, since the concepts represent
the things themselves and possess at least as much positive content as
them. But, in general, a law merely expresses a relation and, in particular, physical laws merely translate quantitative relationships between
concrete things. As a result, if a modern philosopher operates on the
laws of the new science in the manner that ancient philosophy did on
the concepts of ancient science, if he brings all of the conclusions of a
supposedly omniscient physics together into a single point, he leaves
to the side what is concrete in the phenomena: the perceived qualities
and the perceptions themselves. The modern philosopher's synthesis
only includes, or so it seems, a fraction of reality.[201] In fact, the first
outcome of the new science was to cut the real into two halves: quantity and quality, one being attributed to bodies and the other to souls. The
ancients did not erect any such barriers—neither between quality and
quantity, nor between soul and body. For them, mathematical concepts
were concepts like any other; they were akin to the others and could be
inserted quite naturally into the hierarchy of ideas. The body was not
defined by geometrical extension, and the soul was not defined by consciousness. If Aristotle's ψυχή [soul], the entelechy of a living body, is
less spiritual than is what we call "soul," this is because his σῶμα [body],

which is already saturated with ideas, is less corporeal than what we call "body."[202] The schism between the two terms was not yet irremediable. It became irremediable, and from then on, any metaphysics that aimed at an abstract unity had to resign itself either to including only one half of the real in its synthesis, or to making use of the absolute irreducibility of the two halves between themselves so as to consider one to be a *translation* of the other.[203] If two different phrases belong to the same language, that is, if they have a certain shared kinship of sounds, then they will necessarily mean different things. To the contrary, if they belong to two different languages, they might well express the same thing precisely because of their radical difference in sounds. The same goes for quality and quantity, and for soul and body. Having cut all ties between the two terms, philosophers were led to establish a rigorous parallelism between them (of which the ancients never dreamed), to consider them translations, and not inversions, of each other, and finally, to posit a fundamental identity as the substrate of their duality. As such, the synthesis that they were able to reach became capable of embracing everything. Some divine mechanism established a one-to-one correspondence between the phenomena of thought and those of extension, between qualities and quantities, and between souls and bodies.

350

This is the parallelism we find in both Leibniz and Spinoza, though, it is true, in different forms, given that they do not attach the same importance to extension. For Spinoza, the terms Thought and Extension are, at least in principle, placed on the same level. They are thus two translations of the same original, or, as Spinoza puts it, two attributes of a single substance that must be named God.[204] And these two translations, along with an infinity of other ones in languages that we do not know, are called forth and even required by the original, just as the essence of a circle is automatically translated both by a shape and by an equation.[205] For Leibniz, on the contrary, although extension is surely still a translation, thought is the original, and thus it might exist without being translated, since the translation is only made for us. By positing God, we necessarily also posit all possible views of God, i.e., the monads.[206] But we can always imagine that a view had been taken from a point of view, and it is natural for an imperfect mind like our own to classify the views (which are qualitatively different) according to the order and the position of points of view from where they were taken (which are qualitatively identical). In reality, the points of view do not exist since all that exists are views, each one given in an indivisible block and representing, in its manner, the whole of reality, which is God. But we need to translate the plurality of views that are dissimilar among themselves by the multiplicity of these points of view, which are external to each other, just as we need to symbolize the more or less close kinship among the views by the relative situation of these points of view among themselves, that is, by their close relations or their distance [écart] from each other, by a magnitude. This is what Leibniz expresses when he says that space is the order of coexistences, that the perception of extension is a confused perception (relative to an imperfect mind), and that all that exists are monads, meaning by this that the real Whole has no parts, but that it is repeated infinitely, and each time integrally (though differently) within itself, and that all of these repetitions are mutually complementary.[207] This is why the visible depth of an object is equal to the set of stereoscopic views that could be taken of it from all points. This is also why, rather than seeing depth as a juxtaposition of solid parts, it could equally be considered to be made of the *reciprocal complementarity* of these integral views, each given as a block, each indivisible, each different from the others, and, nevertheless, each representative of the same thing. For Leibniz, the Whole, which is to say God, is this very depth, and the monads are these flat views that are complementary of each other. This is why he defines God as "the substance that has no point of view," or again as "the universal harmony," that is, the reciprocal complementarity of monads.[208] In short, Leibniz differs here from Spinoza insofar as he considers universal mechanism as an aspect that reality takes on for us, whereas Spinoza turns it into an aspect that reality takes on for itself.

True, after having concentrated the totality of the real into God, it becomes difficult for them to go from God to the things, from eternity to time. The difficulty was even greater for these philosophers than it was for philosophers such as Aristotle or Plotinus. Indeed, Aristotle's God was obtained through the compression and the reciprocal compenetration of the Ideas that represent the changing things in the world in their completed state or at their highest point. As such, this God was transcendent to the world, and the durée of the things was juxtaposed to his eternity, durée being a degradation of eternity. But the principle we reach through the consideration of universal mechanism, the principle that must serve as its substrate, no longer condenses concepts or *things* into itself, but rather laws or *relationships*. Now, a relationship does not exist independently. A law ties together terms that change; it is immanent to what it governs.[209] The principle into which all of these relationships are condensed, and that grounds the unity of nature, can thus no longer be transcendent to sensible reality; it is immanent to it, and it is necessary to assume simultaneously that it is in time and outside of time, gathered up into the unity of its substance and yet condemned to unroll that substance in a chain that has neither a beginning nor an end. Rather than formulate such a shocking contradiction, these philosophers were led to sacrifice the weaker of the two terms and so to take the temporal aspect of things for pure illusion. Leibniz says so explicitly, since he makes time, like space, into a confused perception.[210] If the multiplicity of his monads expresses merely the diversity of the views taken of the whole, then for Leibniz the history of an isolated monad could hardly appear to be anything other than the plurality of the views that one monad can take of its own substance in such a way that time would consist in the collection of the points of view of each monad upon itself, just as space consists in the collection of points of view of all of the monads upon God. But Spinoza's thought is much less clear, and it seems that this philosopher wanted to establish between eternity and the things that endure the same difference as the one that Aristotle posited between the essence and the accidents. This was indeed one of the most difficult tasks, since Aristotle's ὕλη [matter] was no longer available to fill in the gap and explain the passage from the essential to the accidental (given that Descartes had eliminated that idea once and for all). Whatever the case may be, the more we examine the Spinozist conception of the "inadequate" in its relations with the "adequate," the more we sense ourselves moving in the direction of Aristotelianism.[211] The same goes for Leibniz's monads: to the extent that they come more clearly into focus, they tend to become closer and closer to Plotinus's Intelligibles.[a/212] The natural incline of these two philosophies draws them back to the conclusions of ancient philosophy.

In sum, the similarities between this new metaphysics and that of the ancients come from the fact that both take as ready-made a unified

353

a Bergson's note: I attempted to bring out these resemblances in a course on Plotinus that I gave at the Collège de France during the 1897–1898 academic year. The similarities are numerous and quite striking. The analogy continues even into the very formulations used by the two authors.

and completed Science that would coincide with everything real that the sensible contains—the new metaphysics placing this Science above the sensible, the ancient one placing it within the sensible itself.[213] For *both of them, reality, along with truth, would be fully given for all eternity.*[214] Both are repulsed by the idea of a reality that would create itself as it advances, or in other words, the idea of an absolute *durée.*[215]

Moreover, it is easy to demonstrate that the conclusions of this metaphysics, which emerge from science, must have reappeared at the very heart of science through a sort of ricochet. This metaphysics continues to penetrate all of our so-called empiricism.[216] Physics and chemistry study nothing but inert matter; when biology examines the living being physically and chemically, it only considers it from the perspective of inertia.[217] As such, and no matter how much they develop, mechanical explanations encompass but a small part of reality. To assume, *a priori*, that the totality of the real can be resolved into elements of this kind, or at least that mechanism could give an integral translation of everything that takes place in the world, is to opt for a certain metaphysics, precisely the one whose principles and consequences were set out by thinkers like Spinoza or Leibniz.[218] Of course, a psycho-physiologist who claims that there is an exact equivalence between the brain state and the psychological state, who imagines the possibility for some superhuman intellect of reading in the brain what is happening in consciousness, believes himself to be quite far from the seventeenth-century metaphysicians and much closer to experience.[219] Nevertheless, pure and simple experience tells us nothing of the sort.[220] Rather, experience shows the interdependency [l'interdépendance] between the physical and the spiritual [moral], along with the necessity for a certain cerebral substratum for the psychological state, but nothing more. From the fact that one term is interdependent [solidaire] with another, it does not follow that there is an equivalence between them. From the fact that a certain nut is necessary for a certain machine (i.e., the machine works when we leave the nut in place and stops when we remove it), we cannot say that the nut is the equivalent of the machine.[221] In order for the correspondence to be an equivalence, it would be necessary for some particular part of the nut to correspond to some part of the machine—such as in a literal translation where each chapter renders a chapter, each phrase a phrase, each word a word. But the relationship between the brain and consciousness appears to be something completely different. Not only does

354

the hypothesis of an equivalence between the psychological state and the brain state imply a genuine absurdity (as I have attempted to show in a previous work), but the facts, when examined without prejudice, also seem to indicate clearly that the relationship between them is precisely the same as the one between the machine and the nut.[222] To speak of an equivalence between them is simply to truncate—to the point of rendering it more or less unintelligible—the metaphysics we find in Spinoza and Leibniz. It is to adopt this philosophy at face value on the side of Extension but to disfigure it on the side of Thought. With Spinoza and with Leibniz, the unifying synthesis of the phenomena of matter is assumed to be completed: everything in matter could be explained mechanically. But when it comes to conscious facts, the synthesis is no longer pursued to its conclusion. The psycho-physiologist stops midway. They assume consciousness to be coextensive with some part of nature and no longer with nature as a whole. In this way, they arrive either at an "epiphenomenalism" that attaches consciousness to certain particular vibrations and places it here and there in the world, in a sporadic state; or, they end up with a "monism" that scatters consciousness into as many small grains as there are atoms.[223] But, in both cases, one comes back to an incomplete Spinozism or an incomplete Leibnizian philosophy. Moreover, between this conception of nature and Cartesian philosophy, many historical intermediaries can be found. The philosophical physicians [les médecins philosophes] of the eighteenth century, with their narrow Cartesian philosophy, greatly contributed to the genesis of contemporary versions of "epiphenomenalism" and "monism."[224]

355

*

[KANT'S CRITICAL PHILOSOPHY]

As such, these doctrines are not as advanced as Kant's critical philosophy. True, Kant's philosophy is also imbued with the belief in a unified and integral science that embraces the totality of the real. And indeed, from a certain angle, Kant's philosophy is merely an extension of modern metaphysics and a transposition of ancient metaphysics. Spinoza and Leibniz, following the example of Aristotle, had already hypostasized the unity of knowledge in God. Kant's critical philosophy, at least from one angle, consisted in asking whether the totality of this hypothesis was necessary for modern science, as it had been for ancient science, or whether a part of the hypothesis might not suffice. For the ancients, science focused on concepts, which is to say on kinds of things. By compressing all concepts into a single one, they necessarily arrived at a *being*, which of course they could call Thought, but which was thought-as-object rather than thought-as-subject.[225] When Aristotle defined his God as the νοήσεως νόησις [thinking of thinking], he probably placed more emphasis on the νοήσεως than on the νόησις.[226] God was, in this case, the synthesis of all concepts, the idea of ideas. But modern science is focused on laws, which is to say on relationships. Now, a relationship is a link established by a mind between two or more terms.[227] A relation is nothing outside of the intellect that does the relating. As such, the universe can only be a system of laws if the phenomena pass through the filter of an intellect. Of course, this intellect could be that of a being infinitely superior to man, one that would ground the materiality of all things at the same time as he would relate them together—this was indeed the hypothesis proposed by Leibniz and by Spinoza.[228] But, it is not necessary to go that far, and human intellect is sufficient to obtain the outcome we are hoping to obtain here. And this is precisely the Kantian solution. There is the same distance between Spinoza's or Leibniz's dogmatism and Kant's critique as there is between the "it is necessary that" and the "it is sufficient that."[229] Kant brings this dogmatism to a halt along the incline that caused it to slide too far toward Greek metaphysics; he was able to reduce to a strict minimum the hypothesis that must be made in order to assume Galileo's physics to be infinitely extensible. True, when Kant speaks of human intellect, it is not a question of either yours or mine. The unity of nature would surely come from the human understanding that unifies, but the unifying function at work here is impersonal. This function enters into our individual consciousnesses, but it also exceeds them. It is much less than a substantial God; it is a bit more, however, than one man's isolated work or even than humanity's collective work. It is specifically not a part of man. It is rather man who is within it, as if within an atmosphere of intellectuality that his consciousness would breathe. It is, if you will, a *formal God*—something that is not yet divine in Kant, but that tends toward becoming so. This can be seen in Fichte.[230] In any case, its principal role for Kant is to give to the whole of our science a relative and *human* character, even if this character is already that of a somewhat deified humanity. Kant's critical philosophy, seen from this point of view, consisted above all in limiting the dogmatism of his predecessors by accepting their definition of science while reducing to a minimum what that conception implied for metaphysics.

The same is not true of Kant's distinction between the matter of knowledge and its form. By seeing the intellect as, above all else, a faculty for establishing relations, Kant attributed an extra-intellectual origin to the terms between which those relations were established. He thereby claimed, contrary to his immediate predecessors, that knowledge cannot be entirely resolved into intellectual terms.[231] He reintegrated into philosophy—though modifying it and transporting it to another level—that essential element of Descartes' philosophy that had been abandoned by the Cartesians.[232]

In this way, Kant opened the pathway to a new philosophy, one that would place itself within the extra-intellectual matter of knowledge through a superior effort of intuition. By coinciding with this matter of knowledge, by adopting its very rhythm and movement, might not consciousness itself—through two efforts moving in opposite directions, in turn raising itself up and lowering itself down—grasp from within (and no longer perceive from the outside) the two forms of reality, namely body and mind?[233] Might not this double effort allow us, at least as far as possible, to relive the absolute? And moreover, in the course of this operation, we would see the intellect surging forth on its own accord, cutting itself out of the whole of the mind [esprit]; intellectual knowledge would thus appear such as it is—limited, but no longer relative.[234]

This was the direction that Kantian philosophy might have indicated to a revivified Cartesian philosophy. But Kant himself did not take up this direction.[235]

Kant did not want to take up this direction because, although he   358
assigned knowledge an extra-intellectual matter, he believed this matter to be either coextensive with the intellect or narrower than the intellect. This being the case, he could no longer dream of cutting the intellect out of this matter, nor consequently of retracing the genesis of the understanding and its categories.[236] The frameworks of the understanding and the understanding itself had to be accepted at face value and as ready-made. There could be no kinship between the matter of knowledge presented to our intellect and this intellect itself. The harmony between the two came from the fact that the intellect imposed its form on the matter. Such that not only was it necessary to posit the

intellectual form of knowledge as a kind of absolute and to give up looking for its genesis, but the matter itself of this knowledge seemed too tainted by the intellect for us to be able to hope to reach it in its original purity. It was not the "thing in itself," it was only the refraction of it through our atmosphere.

If we now ask why Kant did not believe that the matter of our knowledge overflowed its form, here is what we find. The critique that Kant instituted of our knowledge of nature consisted in untangling what our mind and what nature must be if the claims of our science are justified. But Kant did not critique those claims themselves.[237] By this I mean that Kant accepted, without discussion, the idea of a unified science capable of embracing, with the same force, all of the parts of the given and of coordinating them into a system that was equally solid from all sides. In his Critique of Pure Reason, he did not think that science becomes less and less objective, and more and more symbolic to the extent that it goes from the physical to the vital and from the vital to the psychical [psychique].[238] Kant does not see experience as moving in two different and perhaps opposite directions, the one in the direction of the intellect and the other its opposite. According to him, there is but one experience, and the intellect covers its entire extension.[239] This is what Kant means when he says that all of our intuitions are sensible, or, in other words, infra-intellectual.[240] And indeed, we would have to accept this conclusion if our science presented an equal objectivity in all of its parts. But let us imagine that, to the contrary, science is less and less objective and more and more symbolic to the extent that it goes from the physical to the psychical, passing through the vital. Now, since it is surely necessary to perceive a thing in some way in order to actually symbolize it, there would have to be an intuition of the psychical, and more generally of the vital, that the intellect would transpose and translate, but that would no less go beyond the intellect. In other words, there would have to be a supra-intellectual intuition.[241] If this intuition exists, then the mind can take possession of itself and is no longer restricted to merely external and phenomenal knowledge.[242] And what is more: if we have an intuition of this kind, by which I mean one that is ultra-intellectual, then sensible intuition is surely in continuity with it by means of certain intermediaries, the way that infrared is continuous with ultraviolet. Sensible intuition will thus be elevated above itself. It will no longer merely attain the phantom of an ungraspable thing in itself. This intuition (so long as we apply certain indispensable corrections to it) will again introduce us into the absolute. So long as sensible intuition was taken to be the unique matter of our science, something of the relativity that strikes any scientific knowledge of the mind was reflected back upon all science; and, that being the case, the perception of bodies, which is the beginning of the science of bodies, itself appeared to be relative. Sensible intuition thus seemed to be relative. But this is no longer the case if we distinguish between the various sciences and if we see in the scientific knowledge of the mind (as well as, consequently, the scientific knowledge of life) the more or less artificial extension of a certain way of knowing that, applied to bodies, was in no way symbolic.[243] We can go even further: if there are thus two intuitions of different orders (the second being obtained, moreover, by a reversal of the direction of the first), and if the intellect naturally carries itself to the side of the second one, then there is no essential difference between the intellect and this intuition itself.[244] The barriers between the matter and the form of sensible knowledge are knocked down, along with the ones between the "pure forms" of sensibility and the categories of the understanding. We see the matter and the form of intellectual knowledge (restricted to its proper object) engendering each other through a reciprocal adaptation—the intellect modeling itself upon corporeity and corporeity upon the intellect.[245]

360

But Kant neither wanted to, nor could, accept this duality of intuition. In order to accept it, it would have been necessary to see durée as the very fabric of reality, and consequently to distinguish between the substantial durée of things and time as scattered out into space.[246] It would have been necessary to see space itself, along with the geometry that is immanent within it, as an ideal term toward which material things are developing, but where they are never fully developed.[247] Nothing could be more contrary to the letter, and perhaps also to the spirit, of the Critique of Pure Reason. Of course, Kant presents knowledge as a list that remains forever open and experience as a thrust of facts that goes on indefinitely.[248] But according to Kant, as these facts develop, they are scattered out on a plane; they are exterior to each other and exterior to the mind. Kant never considers a knowledge from within, a knowledge that would grasp these facts in their springing forth itself rather than taking them as already sprung, a knowledge that would thereby dig beneath space and spatialized time. And yet our consciousness surely places us beneath this plane—that is where true durée is found.

361   From this angle, Kant is again fairly close to his predecessors. He does not admit any middle ground between the non-temporal and time as scattered out into distinct moments. And since there is no intuition that transports us into the non-temporal, every intuition would therefore be, by definition, sensible.[249] But between physical existence, which is scattered out in space, and a non-temporal existence, which could be nothing other than a conceptual and logical existence, such as the one discussed by dogmatic metaphysics, is there not a place for consciousness and for life?[250] Yes, incontestably. We catch sight of it the moment we place ourselves within durée in order to go from there to the moments, rather than beginning from the moments in order to connect them together in durée.[251]

        In order to escape Kantian relativism, however, Kant's immediate successors moved toward the side of a non-temporal intuition. Of course, the ideas of becoming, progression, and evolution seem to occupy a significant place in their philosophy. But does durée truly play a role there? In real durée, each form derives from previous forms while also always adding something, and each form (to the extent that it can be explained) is explained by the previous forms. But to deduce this form, directly, from the overall Being that it is assumed to manifest is to go back to Spinozism.[252] It would be to deny durée any effective action, as both Leibniz and Spinoza did. Post-Kantian philosophy, as severe as it could be toward mechanistic theories, adopts from mechanism the idea of a unified science that would be identical for every type of reality.[253] And it is closer to mechanism that it imagines. For even if, when considering matter, life, and thought, it replaces the successive degrees of complexity (that mechanism posited) by the degrees of the realization of an Idea or by the degrees of objectivation of a Will, post-Kantian philosophy still speaks of degrees, and these degrees are the degrees of a scale that being moves through in a single direction.[254] In short, it distinguishes in nature the very same articulations that mech-

362   anism did. It retains the entire outline [dessin] of mechanism and merely fills it in with different colors. But it is the outline itself, or at least half of the outline, that must be redone.[255]

        True, to accomplish this it would have been necessary to renounce the method of construction, which was the method adopted by Kant's successors. It would have been necessary to appeal to experience—to a purified experience, by which I mean one that is freed, where necessary, from the frameworks that our intellect had constituted during the progression of our action on things. This kind of experience is a non-temporal experience. Beyond the spatialized time in which we believe we perceive continual rearrangements among the parts, it seeks nothing but concrete *durée* in which a radical recasting of the whole is continuously taking place. This experience follows the real in all of its sinuosities. It does not lead us, like the method of construction, to higher and higher generalities, to some new floors superposed atop some magnificent edifice. At least it does not leave any play between the explications that it suggests and the objects that it is supposed to explain.[256] This new method claims to illuminate the detail of the real and no longer merely the overall whole.[257]

*

[SPENCER'S EVOLUTIONISM]

That nineteenth-century thought called for a philosophy of this kind, shielded from what is arbitrary and capable of descending into the details of particular facts, is beyond doubt. And it indisputably sensed that such a philosophy had to place itself within what I have called concrete *durée*. The advent of the moral sciences, the progress in psychology, the growing importance of embryology in the biological sciences—all of this should have indeed suggested the idea of a reality that inwardly endures, a reality that is *durée* itself.[a/258] And so when a thinker appeared who proclaimed a doctrine of evolution in which the progression from matter to perceptibility would be traced out at the same time as the march of the mind toward rationality, in which the complexity of correspondences between the external and the internal would be followed degree by degree, and in which change would ultimately become the very substance of things, all eyes turned toward him.[259] Hence the powerful attraction that Spencer's evolutionism exerted on contemporary thought. As far from Kant as Spencer might seem to be, and as ignorant of Kant's critical philosophy as he indeed was, he no less sensed, from his earliest contact with

363

---

a TN: For the translation of Bergson's term *les sciences morales* (which, being broader than ethical or moral questions, grouped together research in human, spiritual, and psychological questions), see translator's note to page 213.

the biological sciences, in what direction philosophy could continue to march by taking Kant's critique into account.

But having barely set out on this path, he took a shortcut.[260] He had promised to retrace a genesis, but in fact he did something completely different. Sure, his doctrine was called "evolutionism," and it claimed to go back up and to come back down the course of universal becoming. In reality, it had nothing to do with becoming or with evolution

It is not necessary to enter into a deep examination of this philosophy. Let me simply say that *the standard trick of Spencer's method consists in reconstituting evolution from the fragments of the evolved.* If I glue an image onto cardboard and then cut the cardboard into pieces, I will be able to reproduce the image by grouping the small pieces of cardboard together as necessary. The child who works on a puzzle in this way, who juxtaposes the formless image fragments and who ends up with a beautifully colored drawing, probably imagines that they have *produced* the drawing and the colors.[261] And yet, the act of drawing or of painting has no relation to the act of assembling the fragments of an image that has already been drawn or painted. In the same way, when it comes to arranging among themselves the simplest results of evolution, you will imitate as best you can the most complex effects. But you will have retraced the genesis of neither the simple results nor of the complex effects, and this adding of the evolved to the evolved will not at all resemble the movement of evolution itself.

364   Nevertheless, this is Spencer's illusion. He begins from reality in its current form; he then breaks it apart, scatters it into fragments, and throws them to the wind; finally, he "integrates" these fragments and "dissipates" the motion.[262] Having imitated the Whole through a work of mosaic, he imagines himself to have retraced the drawing and to have brought about its genesis.

Are we dealing with matter here? The diffuse elements that Spencer integrates into visible and tangible bodies have the air of being the very particles of the simple bodies, which he first assumes to be already disseminated throughout space. In any case, they are "material points," and consequently invariable points, that is, veritable little solids—as if solidity, being that which is closest to us and the most manipulable by us, could be at the very origin of materiality! The more physics progresses, the more it shows the impossibility of representing the properties of the ether or of electricity (the probable foundation of all bodies) according to the model of the properties of the matter that we perceive.[263] But philosophy goes still higher than the ether, that simple schematic representation of the relations among phenomena as grasped by our senses. Indeed, philosophy knows that whatever is visible or tangible in things represents our possible action upon them.[264] We will never attain the principle of what evolves by dividing up the evolved. We will never reproduce evolution by recomposing the evolved with itself, which is rather the endpoint of that evolution.

Are we dealing with mind? By combining reflexes together, Spencer believes he can engender, one after the other, instinct and the rational will.[265] He does not realize that the specialized reflex—being as much an endpoint of evolution as is the consolidated will—cannot be taken for granted at the outset. Of course, it is quite likely that the first of these two attained its definitive form more quickly; but both of them are deposits of the evolutionary movement, and the evolutionary movement itself can no more be expressed solely as a function of the first than merely of the second. We would have to begin by mixing reflex and will together. Then we would have to go looking for the fluid reality that rushes forward beneath this double form and that surely participates in both without thereby solely being either of them.[266] At the lowest degree on the scale of animals, among living beings that are limited to an undifferentiated protoplasmic mass, the reaction to a stimulus does not yet put into action a determinate mechanism in the manner of a reflex; the reaction does not yet have a choice among several determinate mechanisms in the manner of a voluntary action.[267] Thus, it is neither voluntary nor a reflex, and nevertheless it foreshadows both. We experience in ourselves something of this genuine originary activity when we execute semi-voluntary and semi-automatic movements in order to escape an urgent danger. But then again, this is but an imperfect imitation of the primitive process since here we are dealing with a mixture of two already constituted activities that are already localized in a brain and in a spinal cord, whereas the primordial activity is something simple that is diversified by the very production of mechanisms such as those of the spinal cord and of the brain.[268] But Spencer closes his eyes to all of that since the essence of his method is to recompose the consolidated with the consolidated, rather than to rediscover the gradual work of consolidation that is evolution itself.

365

Finally, are we dealing with the correspondence between mind and matter?[269] Spencer was right to define the intellect by this very correspondence. He was right to see it as the endpoint of an evolution. But when he tries to retrace this evolution, he again integrates the evolved with the evolved, not noticing that his efforts are useless. By taking as given the slightest fragment of what has actually evolved, he posits the whole of what is currently evolved, and it is in vain that he would thus claim to have explained its genesis.

366   Indeed, Spencer holds that the phenomena that succeed one another in nature project images that represent them into the human mind.[270] As such, the relationships among the representations systematically correspond to the relationships among the phenomena. And the most general laws of nature, which condense together the relationships among phenomena, are thereby found to have engendered the guiding principles of thought, which integrate the relationships among representations. Nature is thus reflected in the mind. The intimate structure of our thought corresponds, part by part, to the very skeletal structure of things.[271] Now, I would love if this were the case. But in order for the human mind to be able to imagine the relationships between phenomena, there must first be phenomena, that is, distinct facts cut out from the continuity of becoming. And as soon as we take this particular mode of decomposition (as we perceive it today) as given, we also take the intellect (such as it is today) as given, since the real is only decomposed in this particular manner in relation to the intellect and in relation to it alone. Does anyone really think that the Mammal and the Insect notice the same aspects of nature, trace out the same divisions in it, and disarticulate the whole in the same manner?[272] And yet, insofar as the Insect is intelligent, it already possesses something of our intellect. Each being decomposes the material world along the very lines that its actions will follow in the world. These lines of *possible action*, by interweaving, sketch out the network of experience in which each stitch is a fact. A town is surely composed exclusively of houses, and the town's streets are merely the intervals between the houses. In the same way, we could say that nature contains nothing but facts and that, once the facts have been posited, the relations are simply the lines that run between them. In a town, however, the gradual zoning of the land simultaneously determined the placement of the houses, their configuration, and the

367   direction of the streets. It is necessary to look back to this zoning to

understand the particular mode of subdivision that led to each house being where it is and to each street going in the direction that it goes. Now, Spencer's fundamental error is to take for granted experience as having already been zoned, whereas the real problem is to know how this zoning itself came about.[273] I admit that the laws of thought are nothing other than an integration of the relations between the facts. And yet, the moment I posit the facts along with the configuration they have for me today, I also presuppose my faculties of perception and intellection such as they are in me today. After all, these faculties are responsible for the zoning of the real and for cutting out the facts from the whole of reality. This being the case, rather than saying that the relationships among the facts have engendered the laws of thought, I can just as well claim that it is the form of thought that has determined the configuration of the perceived facts, and hence the relationships among them. The two ways of expressing it are equivalent.[274] They basically say the same thing. True, with the second we give up speaking of evolution altogether. But with the first, we limit ourselves to merely speaking *about* evolution and again to not *thinking* evolution. For a true evolutionism would aim to discover the gradually obtained *modus vivendi* by which the intellect adopted its blueprint [*plan de structure*] and matter its mode of subdivision.[275] This structure and this subdivision gear into each other. They are mutually complementary. They must have developed together. And whether the current structure of the mind is posited or whether the current subdivision of matter is taken as given, in both cases we remain within what has evolved—this tells us nothing about what evolves and nothing about evolution itself.

And yet, it was this evolution that was to be found. In the domain of physics itself, the scientists who push the examination of their science the furthest are already inclined to believe that we cannot reason in relation to the parts in the same way we reason in terms of the whole, that the same principles are not applicable at the origin and at the endpoint of a progression, and that neither creation nor annihilation, for example, are necessarily excluded when it comes to the question of the constitutive corpuscles of the atom.[276] As such, they tend toward placing themselves within concrete *durée*, the only form of *durée* where there could be generation and not merely the composition of parts. True, the creation and the annihilation they are talking about concerns movement or energy, and not that imponderable milieu through which energy and movement circulate. But what can remain of matter when everything that determines it is taken away, i.e., precisely energy and movement?[277] The philosopher must go further than the scientist. Making a clean sweep of everything that is merely an imaginary symbol, the philosopher will see the material world resolve into a simple flux, a continuity of flowing by, a becoming.[278] And he will thereby prepare himself for finding real durée where it is even more useful still to find it: in the domain of life and consciousness. For so long as it is a question of brute matter, the flowing by can be neglected without committing any serious errors. Matter, as I have said, is weighed down with geometry; this reality that descends does not endure, other than through its interdependence with what ascends.[279] But life and consciousness are this very ascent itself. When, by adopting their movement, we have grasped them in their essence, we will understand how the rest of reality derives from them.[280] Evolution appears, and at the heart of this evolution, so too does the progressive determination of materiality and intellectuality through the gradual consolidation of the two. But as such, we place ourselves within the evolutionary movement in order to follow it to its current results, rather than artificially piecing them together out of some fragments of themselves. This, it seems to me, is the proper function of philosophy.[281] Understood in this way, philosophy is not merely the return of the mind to itself, nor the coinciding of human consciousness with the principle from which it emanates, nor even a coming into contact with the creative effort. Rather, philosophy is the profound examination of becoming in general, the true evolutionism, and consequently, the true extension of science—so long as we understand science to be a collection of observed and demonstrated truths, and not a certain new scholasticism that germinated during the second half of the nineteenth century alongside Galileo's physics, just as the ancient science did alongside Aristotle's physics.

369

CORRESPONDENCE, RECEPTION, AND COMMENTARIES

INTRODUCTION

This "Correspondence, Reception, and Commentaries" section begins from the meticulous research and extensive apparatus created by Arnaud François for the French scholarly edition of *L'évolution créatrice*, which appeared in 2013 as part of a series of critical editions of Bergson's works under the direction of Frédéric Worms and published by Presses Universitaires de France.[1] For a discussion of the creation of and motivation behind the original French apparatus, I refer the reader to the French publication.[2]

For this English edition, the selections have been rearranged and some additional materials have been added. As above with the editorial endnotes, the introductions to the following selections are generally based upon Arnaud François's introductions from the French edition. I have also included additional information with the English reader in mind, and the introductions to the selections from Znaniecki, Ruyer, Canguilhem, Merleau-Ponty, and Deleuze are my own.

The notes included in this section are identified as follows: all notes that begin "TN" are my own; all substantial discursive endnotes that have been added by the editors or translators of these selections include

DOI: 10.4324/9781315537818-7

their name; and, finally, all unattributed endnotes are the original notes from the author of the selection.

I have generally not adjusted the translations included here. Nevertheless, I have updated all citations of Bergson's *L'évolution créatrice* to reflect this new translation. Moreover, I have occasionally changed the translation of certain terms (such as "the intellect") to reflect my translation decisions and corrected errors in the translations where necessary (erroneous citations, lost capitalizations, etc.). Note that several of the authors originally cited the 1907 French original of *L'évolution créatrice*; their page references have thus also been updated to reflect the new standard French edition whose pagination was conserved in the 2013 critical edition as well as in all of the critical editions in Worms' series.

It is with profound appreciation that I acknowledge the incredible research by Arnaud François and the editorial oversight of Frédéric Worms, and thank them for their permission to adjust this apparatus as necessary for the English reader.

Donald A. Landes

I

CORRESPONDENCE

JAMES–BERGSON CORRESPONDENCE (1907)

American philosopher and psychologist William James (1842–1910) and Henri Bergson started corresponding in 1902, and the letters published here show the importance of their exchange particularly in relation to *L'évolution créatrice*. The first letter reveals James's enthusiastic initial reaction to the book, as well as his interest in having Bergson's thoughts on his own book *Pragmatism*.[1] Bergson's response bears witness to his genuine respect for the American pragmatist and a deep appreciation for James's praise. Moreover, their exchange points to the emerging questions that remained between them regarding the interpretation of the intellect, the nature of reality, and the nature of truth. The French translation of *Pragmatism* would be published in 1911, after James's death, and it included an introduction from Bergson that addressed precisely these questions.[2]

* * *

DOI: 10.4324/9781315537818-8

CHOCORUA [New Hampshire], June 13, 1907[3]

To Henri Bergson.

Oh, my Bergson, you are a magician, and your book is a marvel, a real wonder in the history of philosophy, making, if I mistake not, an entirely new era in respect of matter, but unlike the works of genius of the "transcendentalist" movement (which are so obscurely and abominably and inaccessibly written), a pure classic in point of form. You may be amused at the comparison, but in finishing it I found the same after-taste remaining as after finishing "Madame Bovary," such a flavor of persistent *euphony*, as of a rich river that never foamed or ran thin, but steadily and firmly proceeded with its banks full to the brim. Then the aptness of your illustrations, that never scratch or stand out at right angles, but invariably simplify the thought and help to pour it along! Oh, indeed you are a magician! And if your next book proves to be as great an advance on this one as this is on its two predecessors, your name will surely go down as one of the great creative names in philosophy.

There! have I praised you enough? What every genuine philosopher (every genuine man, in fact) craves most is *praise*—although, the philosophers generally call it "recognition"! If you want still more praise, let me know, and I will send it, for my features have been on a broad smile from the first page to the last, at the chain of felicities that never stopped. I feel rejuvenated.

As to the content of it, I am not in a mood at present to make any definite reaction. There is so much that is absolutely new that it will take a long time for your contemporaries to assimilate it, and I imagine that much of the development of detail will have to be performed by younger men whom your ideas will stimulate to coruscate in manners unexpected by yourself. To me at present the vital achievement of the book is that it inflicts an irrecoverable death-wound upon Intellectualism. It can never resuscitate! But it will die hard, for all the inertia of the past is in it, and the spirit of professionalism and pedantry as well as the aesthetic-intellectual delight of dealing with categories logically distinct yet logically connected, will rally for a desperate defense. The *élan vital*, all contentless and vague as you are obliged to leave it, will be an easy substitute to make fun of. But the beast *has* its death-wound now, and the manner in which you have inflicted it (interval versus *temps d'arrêt*, etc.) is masterly in the extreme. I don't know why this later *rédaction* of your critique of the mathematics of movement has seemed to me so much more telling than the early statement—I suppose it is because of the wider use made of the principle in the book. You will be receiving my own little "pragmatism" book simultaneously with this letter. How jejune and inconsiderable it seems in comparison with your great system! But it is so congruent with parts of your system, fits so well into interstices thereof, that you will easily understand why I am so enthusiastic. I feel that at bottom we are fighting the same fight, you a commander, I in the ranks. The position we are rescuing is "Tychism" and a really growing world. But whereas I have hitherto found no better way of defending Tychism than by affirming the spontaneous addition of discrete elements of being (or their subtraction), thereby playing the game with intellectualist weapons, you set things straight at a single stroke by your fundamental conception of the continuously creative nature of reality. I think that one of your happiest strokes is your reduction of "finality," as usually taken, to its status alongside of efficient causality, as the twin-daughters of intellectualism. But this vaguer and truer finality restored to its rights will be a difficult thing to give content to. Altogether your reality lurks so in the background, in this book, that I am wondering whether you *couldn't* give it any more development *in concreto* here, or whether you perhaps were holding back developments, already in your possession, for a future volume. They are sure to come to you later anyhow, and to make a new volume; and altogether, the clash of these ideas of yours with the traditional ones will be sure to make sparks fly that will illuminate all sorts of dark places and bring innumerable new considerations into view. But the process may be slow, for the ideas are so revolutionary. Were it not for your style, your book might last 100 years unnoticed; but your way of writing is so absolutely commanding that your theories have to be attended to immediately. I feel very much in the dark still about the relations of the progressive to the regressive movement, and this great precipitate of nature subject to static categories. With a frank pluralism of *beings* endowed with vital impulses you can get oppositions and compromises easily enough, and a stagnant deposit; but after my one reading I don't exactly "catch on" to the way in which the continuum of reality resists itself so as to have to act, etc., etc.

The only part of the work which I felt like positively criticizing was the discussion of the idea of nonentity, which seemed to me somewhat overelaborated, and yet didn't leave me with a sense that the last word had been said on the subject. But all these things must be very slowly digested by me. I can see that, when the tide turns in your favor, many previous tendencies in philosophy will start up, crying "This is nothing but what we have contended for all along." Schopenhauer's blind will, Hartmann's unconscious, Fichte's aboriginal freedom (reedited at Harvard in the most "unreal" possible way by Münsterberg) will all be claimants for priority. But no matter—all the better if you are in some ancient lines of tendency. Mysticism also must make claims and doubtless just ones. I say nothing more now—this is just my first reaction; but I am so enthusiastic as to have said only two days ago, "I thank heaven that I have lived to this date—that I have witnessed the Russo-Japanese war, and seen Bergson's new book appear—the two great modern turning-points of history and of thought!"

Best congratulations and cordialest regards!

Wm. JAMES.

* * *

Villa Montmorency, June 27, 1907.[4]

Dear Professor James,

Your letter has brought me great joy, and I must thank you without delay. You are certainly right to say that the philosopher loves praise and that, in this, he resembles the most common of mortals—but you must allow me to add that the approbation I desired above all was the one that comes from the thinker who has done so much to reshape the soul of the new generation, and whose work I have always admired so deeply. And so, nothing could be more touching for me than this letter in which you declare your readiness to enter into the essential ideas of my work, and in which you even defend them in advance against the attacks that they will undoubtedly elicit. I will treasure your letter as a sufficient payment for the ten years of effort that this book has cost me.

I started reading your book *Pragmatism* the very moment it arrived in the mail, and I could not put it down until I had finished it. It is the program, admirably sketched out, for the philosophy of the future.

Through such a diverse series of considerations, which you have always been able to bring back to the same central point (through suggestions as much as through explicit reasons), you provide us with the idea— and above all the feeling—of the supple and flexible philosophy that is destined to replace intellectualism. Never had I been more aware of how analogous our two points of view really are than while reading your chapter "Pragmatism and Humanism." When you say that "*for rationalism reality is ready-made and complete from all eternity, while for pragmatism it is still in the making*,"[5] you give the very formula of the metaphysics that I am convinced we are heading toward, a metaphysics we would have already reached long ago had we not remained under the spell of Platonic idealism. Would I go as far as adopting your claim that "truth [itself] is mutable"?[6] I believe in the mutability of *reality* rather than that of truth. If we could tune our faculty of intuition to the moving nature of the real, would not this tuning-in itself be something stable? And truth—which can be nothing other than this tuning-in itself—would it not also have something of the nature of this stability? But before reaching that point, we would first have to make many tentative steps. Once again, thank you, dear Professor James, and please accept my sincere compliments for your new work, destined to exert a considerable influence.

LETTER TO H. WILDON CARR (1908)

The next piece by Bergson is part of a letter he wrote to British philosopher H. Wildon Carr (1857–1931). Carr had sent Bergson a draft of an article in which he critically engages with *L'évolution créatrice*.[1] In general, Carr praises Bergson's thought, but he also includes some criticisms and acknowledges the difficulties that he has with certain arguments in *L'évolution créatrice*. Bergson's response was included as an appendix to Carr's article upon its publication. The following brief summary of Carr's criticisms, which paraphrases Arnaud François's introduction in the French edition, will help to contextualize Bergson's response.[2]

After discussing Bergson's account of the relationship between the theory of knowledge and the theory of life, Carr compares Bergson's thought to pragmatism and criticizes Bergson's discussions of Zeno's paradoxes. Carr thinks it unlikely that we can conceive of becoming in itself without positing some underlying thing that becomes, and he goes on to question Bergson's radical distinction between space and *durée*.[3]

Carr questions Bergson's solutions to Zeno's paradoxes by claiming that they can in fact be presented solely in terms of space. Second, Carr suggests that Bergson's account of radical or absolute becoming risks plunging us into a "pure absolute becoming" in which there could be no degrees of motion, thus rendering impossible any genuine knowledge of the durational realities that Bergson mentions.[4] Third, if we admit that movement is an absolute and that nothingness is genuinely nothing, then, asks Carr, how can we understand the status of any true idea?[5] Is not the correct idea itself drown in a process of constant change? And does not knowledge thereby again become impossible?[6] Fourth, Carr confides that he does not understand why life must, according to Bergson, be partially inaccessible to the intellect.[7] In general, Carr worries that Bergson, like the pragmatists, is bent on "discrediting the intellect."[8] After addressing the question of Zeno's paradoxes and space, Bergson primarily responds to Carr's third and fourth criticisms. He is most compelled to address the charge of anti-intellectualism and to clarify the relationship between intellect, intuition, and instinct.

* * *

[Dear Professor Carr,][9]

Regarding Zeno's arguments, I do not think that any of them consider space alone, nor that "the antinomy of movement may be stated in terms of space [alone]."[10] It will always be necessary to introduce movement, since movement is what is in question, and consequently it will always be necessary to introduce time as well. True, you claim that the continuity of time is the same as that of space. But that is precisely what I cannot accept. In a work that predates L'évolution créatrice (namely, Essai sur les données immédiates de la conscience), I explained that real time radically differs from space in terms of all of its attributes, and that if we place time and space on the same line, if we assume that they have the same attributes, this is because (for the convenience of language and science) we substitute spatialized time for real time, i.e., a space that has become symbolic of true durée.

You add that, on my theory, one cannot see how a rapid movement could be distinguished from a slow movement. In L'évolution créatrice,

I respond to this question only in passing. I think that between these two movements there is a difference as radical—or even more radical—than is ordinarily assumed. The speed of a movement is, for me, a *quality* of the movement much more than a *quantity*. In other words, something that is (at least in certain cases) an absolute rather than being a relationship.

I am worried now that there might be a misunderstanding between us in terms of the relation of Life to Science and the Intellect. When I say that Life cannot be embraced by science, I mean to speak of science such as it has been understood up until now, such as we understand it today, namely a science whose goal is to analyze each whole into its elements. The *élan vital* escapes all studies of this kind, both because of its *indefinite richness* and because of its *indivisibility*. Science, with its current methods, can take views or snapshots of the *élan vital* by circling around it, but these views will always be external ones. Philosophy, such as I conceive of it, aims rather at placing itself within it [*s'installer dans*] through sympathy and intuition. This is why I say that life, *in itself*, is a question for philosophy and not science. Now, it is possible that Science might one day decide to take the pathway along which philosophy is already moving. On that day, we will be able to call science what scholars today call philosophy. But I worry that that day is still far off since the tendency of positive science appears to me to be moving in the opposite direction. More and more it proposes to *analyze* life, that is, to resolve the vital into the physico-chemical.

On the other hand, when I say that life escapes the intellect, I again take the word "intellect" in its usual and current sense: the faculty of attributing a predicate to a subject, the faculty of placing things into categories or into pre-formed frameworks, or the faculty of induction and deduction. Life overflows our frameworks and categories, our deductions and inductions. This is how life transcends the intellect, in my opinion. But then how, you ask, is it possible to know life? For me it is possible because our intellect it surrounded by a fringe of intuition that allows us to sympathize with what is properly vital in life. If one would like to name this fringe "intellect," one is free to do so. But one thereby extends the meaning of the word considerably. And, in fact, this fringe of intuition appears to me to resemble less the intellect than instinct, which is just about the opposite of the intellect.

[H. Bergson]

LETTER TO FLORIAN ZNANIECKI (1911)

This letter is from Bergson to sociologist Florian Znaniecki (1882–1958), the translator who was working on the annotated Polish edition of *L'évolution créatrice* that would be published in 1913.[1] Bergson's letter was included in the first edition of the Polish translation upon its publication. The letter shows Bergson's keen interest in the translation of his works, along with a certain reserve as to the possible success of such a venture! Bergson's note to Znaniecki was written the same year as the English publication of *Creative Evolution*, and perhaps Bergson had in mind some of the criticisms that the English version elicited.[2]

* * *

[December 3, 1911][3]

[Dear Professor Znaniecki,]

I would be delighted if, thanks to a good translation, the ideas that I cannot help but take to be at least the approximation of the truth were to find an audience in Poland. You are absolutely right to start with *L'évolution créatrice*. It is a book that can help to clarify my two previous ones. Nevertheless, if a reading of this book is (or seems) relatively easy in certain respects, I am persuaded that it is a particularly difficult book to translate given the different translations that have been already attempted. So difficult, in fact, that—in the unanimous opinion of all of those working on it—it is impossible for a sole translator to accomplish it. At the very least, he will be obliged to turn to specialists and to present the whole work to one or several professional philosophers who know the foundations of the history of philosophy. The Swedish translator had to secure the collaboration of a biologist, a mathematician, and two writers or philosophers. Just as many people participated in the English translation. The German translation, which has been underway now for several years and is still not completed, revealed obstacles that seemed insurmountable for a while. The difficulties come first come from the fact that this work is the result of research I accomplished across many years and in several completely different domains. But also (and above all) from the fact that, in this book, I aim to lead my reader to a certain manner of thinking that goes beyond "concepts" and abstract ideas, and that can only be expressed through images: the image being here not a simple ornament, but rather the only means of expression that adapts itself to the thought.

[H. Bergson]

II

CRITICAL RECEPTION
IN BIOLOGY

BERGSON AND LE DANTEC IN DIALOGUE

The first important critique of *L'évolution créatrice* came from biologist Félix Le Dantec (1869–1917), Chair of General Embryology at the Sorbonne. The review appeared in the journal *Revue du mois*, which was edited by Émile Borel (whose own response to Bergson's reply also appears below). According to Le Dantec, the future belonged to mechanism, which he understood in a neo-Lamarckian way. On Le Dantec's reading, Bergson—from the lofty heights of his Chair of Philosophy at the Collège de France—had done little more than repackage several more or less finalist doctrines into an equivocal idiom. The review is intentionally irreverent in tone and deploys humor and irony at several moments. Le Dantec repeatedly focuses on the need for Bergson to be "translated," a theme that Bergson himself picks up on in his response. Despite acknowledging some convergence between their ideas, Le Dantec's main point is that Bergson's method of intuition or "sympathy" (understood by Le Dantec as a return to merely subjective experience) holds no value

DOI: 10.4324/9781315537818-9

for science. Contra Bergson, Le Dantec advocates for an understanding of life as "struggle," a notion that he believes is completely foreign to the principles of *L'évolution créatrice*.

Bergson's response to Le Dantec came in the form of a letter to the director of *Revue du mois*, Émile Borel. Bergson returns to the theory of absolute movement that he developed in "Introduction à la métaphysique" and that was already present in *Matière et mémoire*. This is not only because Le Dantec focuses on this theory, but also because it plays a central role in *L'évolution créatrice*. Bergson betrays some impatience with regard to his critic, a gesture that is relatively rare for him, and he takes up Le Dantec's image of translation in order to emphasize the incompatibility between Le Dantec's point of view and his own. Although the two thinkers are clearly comfortable simply remaining within their own incompatible positions, Bergson does take seriously the need to reject Le Dantec's "mathematical" point of view, which he believes would lead to the negation of the contingency that is at the core of life and reality in general.

* * *

"BERGSON'S BIOLOGY" (1907)[1]

By Félix Le Dantec
Translated by Kathleen Hulley

Some friends strongly encouraged me to read Monsieur Bergson's new work, *L'évolution créatrice*.[2] Up until now, I had but a vague knowledge of the writings of this eminent philosopher. The language he writes in is so different from the one I am accustomed to that I would need a translator to readily follow it, and I have not yet encountered such a translator. So I limited myself to reading his article from *Revue de métaphysique* (January 1903), since I was assured that I would find there a good summary of his way of thinking.[3] Having read that article, I was convinced that this famous professor's method only differed from that of the experimental sciences in terms of its point of view, or in the form of its language. After all, he says as much at the beginning of that article:

> If we compare the various ways of defining metaphysics and of conceiving the absolute, we shall find, *despite apparent differences*, that

philosophers agree in making a profound distinction between two ways of knowing a thing. The first implies going all around it, the second entering into it. The first depends on the viewpoint chosen and the symbols employed, while the second is taken from no viewpoint and rests on no symbol. Of the first kind of knowledge we shall say that it stops at the *relative*; of the second that, wherever possible, it attains the *absolute*.

Take, for example, the movement of an object in space. I perceive it differently according to the point of view from which I look at it, whether from that of mobility or of immobility. I express it differently, furthermore, as I relate it to the system of axes or reference points, that is to say, according to the symbols by which I translate it. And I call it *relative* for this double reason: in either case, I place myself outside the object itself. When I speak of an absolute movement, it means that I attribute to the mobile an inner being and, as it were, states of soul; it also means that I am in harmony with these states and enter into them by *an effort of imagination*. Therefore, according to whether the object is mobile or immobile,[4] whether it adopts one movement or another,[5] I shall not have the same feeling about it. And what I feel will depend neither on the point of view I adopt toward the object, since I am in the object itself, nor on the symbols by which I translate it, since I have renounced all translation in order to possess the original. In short, the movement will not be grasped from without and, as it were, from where I am, but from within, inside it, in what it is in itself. I shall have hold of an absolute.[6]

The clarity of this passage from Monsieur Bergson is extremely valuable for those, like me, who have so much difficulty following the arguments of the metaphysicians. First, we see that the author's language differs from pure objective language since he adds an "effort of imagination" to it. This "effort of imagination" hardly raises the ire of the majority of philosophers, if I am to believe the high esteem in which they hold Bergson's works. But I do not see the value of this effort when it comes to the knowledge of the movement being studied. And I am convinced that, in order to give an account of a movement, the eminent professor begins (as I do) with a purely objective account of it, prior to giving himself over to his dreams and imagining himself to be the moving thing itself.

I, for one, mistrust the sensation of movement that I experience in myself when I am moving, that is, in the case where I myself am indeed the moving object that I observe. On a train, in the darkness of night, I often wonder which direction I am going because I only see the walls of the car in relation to which I am immobile. When the train passes the light from a railway signal, I immediately know the direction in which I am moving along the tracks, unless I have unknowingly seen the light in a mirror, in which case I seem to be moving in the opposite direction. In that case I even experience the sensation of moving in the opposite direction—and this sensation is so deceptive that if I pass the lights of a railway station a moment later, I immediately experience (prior to regaining my bearings) a profound astonishment at seeing this station move in the same direction as the train. This observation, which I have made a hundred times, has removed all confidence I might have had in the sensation of movement, and I do not see any advantage in imagining myself to be in the moving object that I am observing. I admit, quite humbly, that by objectively observing the movement, I only have a relative knowledge. But this is hardly surprising given that all of the movements I know differ in relation to the bodies by which we observe them. So if I grasp, through an effort of imagination, the absolute of the movement, then I have dreamed up an obvious error. Seated at my desk, I am immobile in relation to the objects that surround me, but I am in fact turning very quickly with the earth and around the sun.

In which one of these relative movements will I sense the absolute in myself? I do not see what we stand to gain by introducing an "effort of imagination" into measurable phenomena, an effort that might well alter the nature of these phenomena. Nor will I find an advantage in this, unless Monsieur Bergson is able to teach us how we might directly examine the movement *in itself*. Thus far, it seems that this brilliant metaphysician is limited to employing the same measures that I do, i.e., to first studying and measuring the relative by means of the objective method. If, having measured the relative, we then find invariables that are independent of our system of measurement, we will have made a step toward the knowledge of the absolute by setting forth the principle of the conservation of energy. But this would be an *a posteriori* result, and not something imagined *a priori*, such as the one that would consist in "sympathizing with the movement of a mobile object into which we transport ourselves in thought."[7]

Therefore, I do not see the scientific value in the effort of imagination by which Monsieur Bergson goes from objective language to metaphysical language, but I do quite clearly see the poetic value in this method, which panders to our inveterate taste for mysticism. When it comes to understanding (?) life, nothing will have been more advantageous than having often employed the metaphysical method.[8] If we were accustomed to thinking of any molecule whatsoever as if we were inside of it, then we would have no difficulty explaining how we find ourselves to be in an assemblage of molecules. In this way, we would basically fall back into the theory of epiphenomenal consciousness, of human consciousness considered to be a whole made up of elementary consciousnesses. But this distressing theory will be poeticized, sublimated, and, if I dare say, surrounded by melodious phrases, just as certain candies of quite average quality are sold at much higher prices when confectioners place each one into an elegant paper wrapping. From what I could understand of it, Bergson's theory does not essentially differ from the system we find in mechanism—it only differs by the paper wrapping that hides each candy. This eminent professor from the Collège de France is an incomparable artist who excels at presenting, in a seductive way, theories that would frighten off his readers if they were presented in an unvarnished way. I do not believe I go too far in saying that a good translator might well offer an interpretation of Monsieur Bergson's latest book in objective and ordinary language, a version for those who, like me, get hopelessly bogged down in the depths of measurement-based science, stripped as it is of the ideal. I make no claim to be that translator. But I do find it interesting to note that two researchers—as opposite in their tendencies as are Monsieur Bergson and myself—are quite naturally led by the current state of science to ask the same questions and even sometimes to adopt the same formulas. To me, such an observation appears just as important in terms of scientific value as the observation that there is an equivalence between quantities measured by our nose or sense of temperature and other quantities measured by the eyes. A metaphysician and a monist coming together on the interpretation of life, despite having set out from the extreme opposite ends of the human horizon, now that makes me think of the encounter between heat and movement in the domain of energy. I have no intention of determining who, Monsieur Bergson or myself, employs the most advantageous method. This would have no value—we are both but irrelevant and fleeting vortexes, and I believe that neither of these vortexes has influenced the movement of the other. But despite all that, we have sometimes come together. I accord much more importance to the points on which we agree than to those on which we cannot understand each other. True, perhaps where I do not understand Monsieur Bergson, this is only for lack of knowing how to translate him.

Now, translation is useless when the formulas are the same in the two languages. I thus begin by pointing out, with much satisfaction, the following claim found on the first page of *L'évolution créatrice*:

> We will see that human intellect feels at home so long as we allow it to remain among inert objects, particularly among solids, where our action finds its footing and where our industriousness finds its tools. We will see that our concepts have been formed in the image of solids and that *our logic is, above all else, the logic of solids...* etc.[9]

I expressed this very idea four years ago; it was important in my *Les lois naturelles* as well as in *Influences ancestrales*, where I specifically wrote:

> The role of solid bodies in our ancestral education was such that we can say today, without over exaggeration, *that our logic*, which is the hereditary summary of the experience of our ancestors, *is above all a logic of solid bodies*. I had previously attempted to demonstrate how the experimental concepts that came out of the encounters between our ancestors and solid bodies have contributed to the creation of arithmetic and geometry...[10]

The striking similarity between Bergson's views and my own appeared even more interesting to me given that our starting points are so distant from each other. In *L'évolution créatrice*, the eminent metaphysician devotes a chapter to the "cinematographic mechanism of thought," a topic I addressed at least seven or eight years ago in several books, in particular in *Le conflit*.[11] As such, we were not only led to the same formulations, but we also independently encountered the same comparisons. And we were inevitably led to ask the same questions. This explains why Bergson asks, on the second page of his introduction [to *L'évolution créatrice*], the very same question I tried to answer in a recent article in *Revue philosophique*.[12] Here is the passage in which it *seems* that Bergson must be led to conclusions altogether in opposition to my own:

But it also follows from this that our thought, in its purely logical form, is incapable of conceiving of the true nature of life and the deep meaning of the evolutionary movement. Our thought was created by life, in specific circumstances, so as to act upon specific things. How could it encompass life, of which it is merely an emanation or an aspect? Deposited along the way by the evolutionary movement,[13] how could it be applied to the entire evolutionary movement itself? One might as well claim that the part is equal to the whole, that the effect can reabsorb its own cause, or that the pebble left behind on the beach sketches out the form of the wave that carried it there.[14] In fact, we clearly sense that none of the categories of our thought (unity, multiplicity, mechanical causality, intelligent finality, etc.) can be unequivocally applied to the things of life[15]—who can say where individuality begins or ends, whether the living being is one or many, or whether the cells come together to form the organism or the organism divides up into cells? It is in vain that we force the living being into one or another of our frameworks. All of the frameworks crack. They are too narrow and, above all, too rigid for what we would like to fit into them. Moreover, our reasoning, so sure of itself when it circulates among inert things,[16] feels ill at ease on this new terrain. It would be very difficult to cite a single discovery in biology owing to pure reason alone. And, more often than not, when experience finally shows us how life goes about obtaining a certain result, we find that its way of proceeding is precisely the one that we would never have thought of.[17]

None of this prevents Bergson from courageously endeavoring to establish a theory of life. At the end of this introduction, he writes: "The fourth and final chapter is designed to show how our understanding itself, by disciplining itself in a certain manner, can prepare the way for a philosophy that goes beyond it."[18]

This is precisely the result I believe we reach by discovering—thanks to mathematical language—the invariants of the world. Unfortunately, the end of L'évolution créatrice is written in a language that I do not yet know how to read. Thus, I cannot confirm that there is a parallelism in our conclusions, despite having certainly understood the identical nature of the problems we have each asked at the outset.

Insofar as I could follow it, Monsieur Bergson's idea seems to be that mechanism and finalism are incomplete systems. He thus proposes another system that goes beyond them and that reaches reality itself. To explain his system, he again returns to his example of the movement in which he earlier taught us how we might grasp an absolute:

> If I raise my hand from point A to point B, this movement simultaneously appears to me in two different ways. Sensed from within, it is a simple and indivisible act. Viewed from the outside, it is the pathway of a particular curve, AB. I will be able to identify in this line as many positions as I like, and the line itself can be defined as a particular coordination of these positions among themselves. But the infinite number of positions and the order that ties them together sprang forth automatically from the indivisible act by which my hand moved from A to B. Here, mechanism would consist in seeing only the positions. Finalism would only take their order into account. But both mechanism and finalism would miss the movement, which is the reality itself.[19]

I think that it is this "reality itself" that Monsieur Bergson would like to grasp through a method that, if I understand correctly, attempts to analyze and synthesize things at the same time, to study them from both the objective and the subjective points of view. At least this is what seems to result from what he said regarding his manner of grasping an absolute movement. I wonder whether Monsieur Bergson's philosophy does not consist in recounting a mechanism in finalist language, and in recounting what is objective in a subjective language. For me, this creates a rather large difficulty, and I get a terrible headache when I try to follow it for too long. It produces in me the same effect as when one tries to look in two different directions at the same time. But it must be that I am ill-formed, since many people in whom I have much confidence have assured me that they easily follow, and with boundless pleasure, the elegant lectures offered by this eminent professor at the Collège de France. However artificial my methods of analysis might be,[20] I believe that it is useful to employ them SUCCESSIVELY and SEPARATELY, rather than giving myself the illusion of following two different paths at the same time. For example, I am fully convinced, as Bergson writes in his introduction [to *L'évolution créatrice*], that: "[T]he *theory of knowledge* and the *theory of life* appear [...] inseparable."[21] Nevertheless, I believed it necessary to first study the objective phenomena of life via an objective method. I have tried, only *afterwards*, to give an account of the facts of consciousness that clearly accompany observable phenomena, both in others as much as they do in myself. Monsieur Bergson wants to do both at once, but in acting this way, he merely does the work of observation and the work of imagination simultaneously, as he himself says regarding absolute movement. He observes a movement and studies it without knowing whether the moving object is conscious of its own movement; but, at the same time, he imagines himself being the moving object and experiencing the sensations that are perhaps misleading. And yet, he is still obliged to be content with the givens of the objective method—these are the very givens that allow his imagination to follow a certain course, and even if he draws a certain satisfaction from his effort of imagination, this changes nothing in the indisputable results of the objective measurements.

When Monsieur Bergson studies life, he does the same thing as when he studies movement. But the imagination that allows him to place himself within the living body under examination is even more dangerous here given the resemblance that exists between observer and observed. They are both living, and they resemble each other insofar as they are living beings. But they also differ, and I think that their subjective differences are of the same order as their objective differences, such that if the observer "puts himself in the shoes of the observed," he commits a gratuitous error—unless he makes the effort to adapt his mindset to what he assumes (according to objective observations) to be the mindset of the observed. It is always necessary to return to observation. The work of imagination, which is added on to observation, is misleading if it does not consider all of the observed facts and those facts alone. And even when the work of imagination is not misleading, it does not add anything to the direct results of the observation. The study of knowledge and the study of life thus appear to me as needing to be separated, despite the fact that knowledge and life have evolved in parallel, and despite the fact that they are the same thing considered from two different points of view. We would have to consider, at each moment, the realization of absolute beginnings only in cases where a complete objective study of life would be impossible. Observation does not lead toward anything like this at all; a complete objective study of life is possible, and the fact that vital phenomena are conscious has no importance in this objective study. As far as I could understand him, Monsieur Bergson does not seem to deny this finding. He limits himself to presenting objective things in a subjective language and to speaking of living bodies as if he were inside them. But he does not claim that the animal initiates absolute beginnings and would thereby be in violation of the law of the conservation of energy. And nevertheless, he asks that we examine the theory of knowledge and the theory of life simultaneously:

> Together they will be able to resolve philosophy's most difficult questions through a method that is more reliable and that remains closer to experience. For, were they to succeed in their common project, they would allow us to witness the formation of the intellect [...] They would substitute for Spencer's false evolutionism—which consists first in cutting up an already evolved reality into smaller bits, themselves no less evolved, then recomposing that reality out of these fragments and thereby taking for granted everything that it was supposed to explain—a true evolutionism that would follow reality in its genesis and growth.[22]

Now, I subscribe wholeheartedly to what Bergson says here about false evolutionism. I have been criticizing this false evolutionism for more than ten years now—and not only in Spencer, but also in Darwin with his gemmules, in Weismann with his determinants, in short, in all of the supporters of representative particles. And yet, from my point of view, I believe it is useless—and even detrimental—to simultaneously follow two theories that represent the same thing from different points of view. Of course, the objective study of evolution, easy when it comes to organs and other external appendages of the body, is very difficult when it comes to the brain. And of course, the variations in the brain that correspond to very important mental facts are difficult to measure and to study—but, for all that, they no less exist, and they must be discussed in an objective language, forgetting the fact that they are above all known when it comes to ourselves and through the method of inner observation. Objective or mechanical language must be separated from subjective or psychological language. The principal accomplishment of modern biology even resides in the possibility of this very separation. Monsieur Bergson prefers to use the two languages at the same time, and in doing so he manages to bring pleasure to many. To me, this practice results in genuine torture, and I cannot follow it without a painful effort. And this is why it was so painful to discover, in the form of the *élan vital* of *L'évolution créatrice*, more or less my own familiar concept of a "universal struggle" and a victory of the living beings who do not perish.[23]

Bergson recounts the story of the struggle by putting himself in the shoes of the one who struggles, whereas I watch the battle as a disinterested observer. And the consequence of our different attitudes with regard to the phenomenon is that I always envision the possibility of the death of the object I am studying, whereas Bergson, identifying himself with that object and yet not dying, believes that there is a necessity in that object's victory: "The *élan of life* that I am discussing consists, in short, in a need for creation. This *élan* is not capable of absolute creation since it encounters matter, i.e., the inverse movement of its own movement."[24] This is exactly how I understand the struggle of the living being against the food that it assimilates, unless, however, it is not assimilated or at least not digested by him (i.e., toxins). Bergson's *élan vital* appears to me to be the translation, into subjective language, of the objective observation that the life of a being is defined by his victory in the struggle. I only observe the results; but through an "effort of imagination," Monsieur Bergson embodies one of the combatants and sympathizes with him in order to then share in his victory. This only changes the procedure of the narration. The result remains the same. But the author oversteps his rights when he claims to show the incomplete side of the purely mechanical narration of the facts. He only gets there by neglecting the part of this mechanical narration that concerns phenomena that are difficult to measure. For example, here is his study of the formation of the eye in mollusks and in vertebrates:

> A mechanistic theory will be any theory that presents to us the gradual construction of the machine through the influence of external circumstances, whether they intervene directly by acting upon the tissues, or indirectly by selecting the individuals that are best adapted. And yet, no matter what form this theory takes, and even if we grant that it is somewhat valuable for understanding the details of the parts, it casts no light upon their correlation.
>
> And so the doctrine of finality arises. This theory claims that the parts have been assembled according to a preconceived plan and to reach a goal. In this way, it likens nature's labor to that of the worker who also proceeds by assembling parts in order to realize an idea or to

imitate a model. Mechanism will thus be correct to criticize finalism for its anthropomorphic character. But mechanism does not notice that it too proceeds according to the same method, simply truncating it. Sure, it has done away with the end that is pursued or the ideal model, but it too implies that nature works like the human worker, namely by assembling parts.

Thus, we must move beyond both points of view, both mechanism and finalism, which are ultimately nothing more than points of view to which the human mind had been led by the spectacle of human work. But in what direction should we move beyond them? I said that even though the functioning of the whole is something simple, when the structure of an organ is analyzed, from one decomposition to the next, this process goes on to infinity. This contrast between the infinite complication of the organ and the extreme simplicity of the function is precisely what should open our eyes to the right direction.[25]

Bergson's remark does not only apply to living beings, but rather to all of the objects that we study. For example, consider a circle. We can define it by saying that it is a circle, i.e., by a simple word. And so, in terms of our language, it is a *simple thing*, like the function of sneezing or digestion. But if we analyze this circle, taken as a whole, into arbitrarily chosen differential elements, we know that this analysis can be pursued to infinity. We thus have two manners of speaking of the whole: a synthetic manner and an analytic one. But I would not say that the analytic manner fails to provide any detail regarding the "the correlation of the parts to the whole." In fact, I can only use this analytic manner by examining the connections among the parts of the circle, that is, by introducing the equation of the circle into my considerations. In the same way, in order to adequately analyze a living being, I must never forget the reciprocal relations among its constitutive parts, otherwise my analysis will be incomplete. None of this will require that I make an "effort of imagination" to place myself (in thought) into the living body being studied—I will be able to hold myself to an objective language by employing synthetic expressions. This is what I did in my book *La lutte universelle*, where I considered each function as a struggle between the overall heredity of the individual (what the individual carries along with him) and his general education or training (the collection of circumstances that he lives through). It would not take much for me to be convinced that the original *élan* of life, which Monsieur Bergson discusses, amounts to a translation of the notion of heredity into a metaphysical language. And yet, this interpretation does not seem to fit with the following claim:

> This *élan*, preserving itself across the lines of evolution among which it is divided and shared, is the deep cause of the variations, or at least of those variations that are consistently transmitted, that accumulate, and that create new species.[26]

Heredity is, rather, the conservator of the initial directions. On pain of death, heredity is opposed to the introduction of any variations that might go beyond a certain range into the lines of evolution. And I am altogether convinced that the similarities that we observe between the eyes constructed by octopuses and humans must be attributed to something common in their heredities. The entire formation of species can be recounted in objective language, and, if the story is properly recounted, then none of the correlations or coordinations will be neglected. Only, language is more difficult to the extent that we, at each moment and for the purposes at hand, intermingle the *incompatible* languages of the subjective and the objective. These two languages are mixed together in everyday language, but not in mathematical language, and so Monsieur Bergson could not be translated into a *universal mechanism*. In short, it seems to me that if Monsieur Bergson seems to reconcile mechanism and determinism, this is because he speaks *a priori* of what is only known *après coup*. He neglects the clause "on pain of death," which is the major difference between biology and the science of what is dead. It should not be said that "some particular thing will be *produced* in virtue of hereditary tendencies," but rather that "a characteristic that is manifested by the descendants along an evolutionary line, *when they have survived*, is the consequence of the collection of those transportable properties that we call *heredity* and of the circumstances it has passed through, its *education or training*." In his own distinctive language, Bergson himself says as much when he defends himself against the accusation of being a finalist:

> [B]y speaking of a march toward vision, am I not simply going back to the ancient notion of finality? This would certainly be the case if

this march required the conscious or unconscious representation[27] of a goal to be reached. But the truth is that this march takes place in virtue of the original *élan* of life and is implied in this movement itself,[28] which is precisely why we find it along independent lines of evolution. If someone now asked me why and how this march toward vision is implied in this movement, I would reply that life is, above all else, a tendency to act on brute matter. The direction of this action is surely not predetermined, and this is the reason for the unpredictable variety of forms that life sows along its pathway as it evolves.[29]

If we consider the universal struggle in its broadest sense, and without any sentimental preference for living bodies, then we have just as much reason to say that the existence of a brute body is a tendency to act upon other living or brute bodies. Only, *to live is to vanquish* and, consequently, the *tendency to act upon brute matter* is preserved in the lineage that avoids death, except for the acquisition of new characteristics that represent a partial victory of the surroundings on the living bodies.

To summarize, I do not believe that Bergson's story presents the slightest advantage over an objective story, other than to pander to our taste for mysticism. And of course, this means it has a good chance of being adopted by many. Darwin himself owed the immense success of his book to the fact that he gave an appearance of finalism to an *a posteriori* story. I admit that I prefer Lamarck's system, which teaches factual truths without hiding their distressing humility.

In a work that will soon appear, I have tried to show that we can find the starting point of all vital manifestations in the phenomena of resonance. These phenomena of resonance are known to exist in brute matter, and we know how to describe them in objective language. Monsieur Bergson, who likes to exert an "effort of imagination" so as to place himself into the moving object whose movement he is studying, will also be able to place himself (in thought) into the tuning fork or into its resonator, and then into the colloid and into the diastase, and then into the living being that we examine under the microscope. All of this will surely be agreeable to him, but I do not believe that this would change anything in terms of the philosophy of life. In the most recent issue of *Revue philosophique*, Bergson's student and admirer extolled the merits of *L'évolution créatrice*, writing:

> Bergsonism, by its very nature, escapes ordinary discussion. Just like the reality whose inexhaustible creative mobility it aims to reveal, *Bergsonism itself is ungraspable* [...] Ultimately, it opens out into poetry, and the history of the world that it recounts is more of an evocation. Being indemonstrable, it does not seek to provide demonstrations [...] By necessity and by destiny, this metaphysician can only proceed like an artist. Bergson is such an artist; he knows it and he embraces it. This is the source of the richness, the abundance, and the variety of metaphors in his writing. It would be a terrible mistake to take these brilliant images to be mere ornaments—*they are arguments.*[30]

I end my citation with these words. Had I wanted to warn the new generation of philosophers against Monsieur Bergson's seductions, I would not have dared to say any more.

FÉLIX LE DANTEC

* * *

"LETTER TO THE EDITOR OF *REVUE DU MOIS*" (1907)[1]

Saint-Cergues (Suisse), August 20, 1907

Monsieur le Directeur [Émile Borel],

The article that Monsieur Le Dantec clearly intended to devote to my book *L'évolution créatrice* has plunged me into the greatest confusion. I simply cannot, despite all my efforts, find even the slightest relation between what I actually said and what Monsieur Le Dantec has me saying.

Monsieur Le Dantec takes as his starting point a much earlier article that I wrote on the function of science and the function of metaphysics. There I was seeking a definition of the terms *relative* and *absolute*. While examining movement, for instance, I said that we perceive the movement of an object differently depending on the point of view (either mobile or immobile) from which we observe it—that is, we express the movement differently depending on which system of coordinates we relate it to—and that this is why we call the movement "relative." In both cases, I said, we place ourselves outside of the object itself. I added that when we speak of an "absolute movement," we "attribute to the mobile an inner being and, as it were, states of soul, that we sympathize with these states, that we enter into them through an effort of imagination."[2] On this topic, Monsieur Le Dantec warns me that the sensation of movement is prone to illusions. He reminds me that when we are on a train, we often make errors in terms of knowing the direction in which we are moving and even the direction in which we see the station moving. And then he adds, "we only objectively observe relative movements."[3] But who ever contested that? For my part, I limited myself to saying that *when we speak* in everyday language of an absolute movement, this is because we attribute an interiority to the moving object, and that we refer back to the consciousness of the movements that we ourselves execute voluntarily—a consciousness that, itself, does not depend on any symbol or any point of view. Of course, we could define the idea of an absolute movement in more scholarly terms. But in that case, the definition will be a *negative* one: it will still consist in first evoking, either explicitly or implicitly, the image of relative movement and then eliminating something from that image. From the moment we want to give a *positive* content to the idea of absolute movement, we necessarily come back to a psychological image. Descartes clearly illuminated the mathematical character of the idea of relative movement when he explained that all movement is "reciprocal," and that if A moves toward B, then we just as well say that B moves toward A. And Henry More clearly illuminated the psychological character of the idea of absolute movement when he responded to Descartes, taking the interior sensation of the work of the muscles into account: "If I am quietly seated, and another, going a thousand paces away, is flushed with fatigue, it is certainly he that moves and I who am at rest."[4] In the passage cited by Monsieur Le Dantec, I was doing nothing more than defining first Descartes' point of view and then Henry More's.

Now, when it is a question of resolving the enormous difficulties raised by philosophers around the question of movement, I am on Henry More's side. Of course, not that I see some sort of advantage in imagining "that we are the moving object itself" in order to "give an account of a movement."[5] Who, prior to Monsieur Le Dantec himself, had ever thought of such an extraordinary method? In order to study a determinate movement, it is always necessary—as Monsieur Le Dantec says, as I myself have said more than once, and as everyone in the world already knows—to rely on objective procedures of measurement. But it is still quite acceptable to hold that the interior sensation of muscle movement allows us to penetrate even further into the inner nature of movement than the visual perception of external displacement does. The former is immanent to the action itself, whereas the latter shows us but the relation between the moving object and the surrounding objects. And I never claimed otherwise.

But Monsieur Le Dantec would have it that every time I see a moving object moving along, I amuse myself by placing myself within it in thought and by moving along with it. And from there he reconstructs my book. I supposedly have the habit of thinking about molecules by placing myself inside them. Moreover, I will thereby necessarily fall back "into the theory of epiphenomenal consciousness, of human consciousness considered to be a whole made up of elementary consciousnesses," with the only difference being that "this distressing theory will be poeticized."[6] I am entirely indifferent as to whether the theory of epiphenomenal consciousness is distressing or consoling. For me, the only question worth asking is whether or not it is true. In fact, I believe that such a theory is completely false. All of my works, from the very first ones right up to the one that Monsieur Le Dantec has just read, or at least the one that he has just discussed, tend toward establishing that consciousness is effective and genuinely creative. In a previous work, I tried to determine—and not through a priori deductions, but through the study of the normal and the pathological phenomena of memory— the sui generis relationship that links the psychological state to the cerebral fact.[7] This is an extremely complex relationship that has nothing in common with the concept of epiphenomenal consciousness.

This first misunderstanding by Monsieur Le Dantec has led to the others. Convinced as he is that I must believe, like him, in a mathematics of life and of consciousness, he concludes that my last book could be nothing but a "poetic" transposition of that mathematical mechanism. But the truth is that my work, from beginning to end, is the very negation of a mathematics of this genre. I will not attempt to define here the point of view that philosophy must adopt, in my opinion, with regard to the phenomena of life. Were I able to explain myself on this topic in just a couple of lines, I would be entirely unforgivable for having written a 400-page tome on it. Let it suffice to say that, beginning from the data of current biology, we can attempt to tie them together either through mathematical schemas (which, I believe, is what Monsieur Le Dantec does) or through psychological schemas (which is what I attempted to do). These are not, as Monsieur Le Dantec seems to believe, two different ways of saying the same things. These are two opposite points of view on the evolution of life. The first method eliminates every kind of contingency from the evolution of life. The second allows contingency to play a role and then aims precisely at defining that role.

This is why the conclusions that I reach cannot meet up with those reached by Monsieur Le Dantec. To mention only what is essential, in terms of what I call "*élan vital*," I do not see either how it is a "universal struggle" nor how one could confuse it with "heredity." As Monsieur Le Dantec himself notes, it is a much more a principle of change than one of conservation. But above all, it is a principle whose approximation we will never obtain other than through schemas that are psychological in nature.

I can at least congratulate myself on having come together with Monsieur Le Dantec in the choice of certain formulations. But even there, where we find ourselves speaking the same language, I worry that we might still be very far from coming to a mutual understanding. When I said, in my most recent book, that our logic is above all the logic of solids, I was simply recalling a theory that I have held for nearly twenty years, namely that the essential function of our intellect is to "spatialize" and to "solidify." This was already one of the guiding ideas of my book *Essai sur les données immédiates de la conscience*. This theory has always allowed me to show the illusory side of radical determinism, which consists in reifying mechanical symbols (those habitually used by our intellect) into absolute realities, symbols that, like the intellect, are oriented toward space and above all absorbed in the consideration of solid objects. But even in *Essai sur les données immédiates de la conscience*, I already insisted upon the fact that the intellect can only envision time as a series of moments, becoming as a series of states, and movement as a series of positions, and thereby can only artificially reconstitute mobility by combining immobile parts with other immobile parts. True, I did not characterize that procedure at the time as "cinematographic." But then again, the cinematograph had not yet been invented. In any case, and by whatever name we call it, this mechanism that is inherent to our intellect is, in my opinion, the true cause of our tendency to eliminate *concrete duration* from the real, to only take mathematical time into account, to only see arrangements, disarrangements, and rearrangements of parts, even in cases where an indivisible and irreversible becoming exists. This is to say that the second remark, like the first, has been useful in showing the artificial character that the mechanical schemas take on when we use them to represent the evolution of consciousness and of life.

Recevez, je vous prie, Monsieur le Directeur, l'assurance de mes sentiments les plus distingués.

H. Bergson

RUYER AS READER OF BERGSON

Raymond Ruyer (1902–1987) was a French philosopher of science who, although often critical of Bergson, nevertheless always demonstrated a high level of respect toward him. In these pages, Ruyer argues that Bergson had been unfairly dismissed in biology, and he emphasizes the importance of Bergson's comparison of the diversity of instincts with so many "variations" on a given theme.

In "Instinct, Consciousness, Life: Ruyer Contra Bergson," which includes the full English translation of this article, Tano S. Posteraro and Jon Roffe provide the following useful summary of Ruyer's article, which I quote here by way of introduction:

> "Bergson et le Sphex ammophile" does three things. First, it provides a close and charitable reading of an aspect of Bergson's philosophy of biology that has been badly understood and both overblown and unduly neglected for that reason. This is Bergson's theory of biological sympathy as it is evidenced by the parasitic instincts of solitary wasps. Against its detractors, Ruyer attempts to demonstrate the viability of the theory, at least in outline. Second, Ruyer provides a helpful and clarifying supplement to Bergson's dense and difficult argument that the wasp appears to "know" the anatomy of its prey, the cater-pillar, when in fact the wasp acts out of an intuitive sympathy for it. Ruyer introduces an analogy between parasitism and fertilization, and uses the cases of sexual dimorphism and the differentiation of sexual organs in hermaphroditic organisms to arrive at Bergson's conclusion via what is most likely a clearer line of reasoning. Finally, third, Ruyer recalls his criticism of Bergson's theory of perception and situates it in the broader context of the article's topic. We are thus given a fresh sense for the significance of that criticism to Ruyer's reception of Bergson's philosophy of biology more generally.

Ruyer's "Bergson et le Sphex ammophile" offers a complex defence of one of the most eccentric passages of Bergson's *Creative Evolution* from the critics that he claims misunderstood it. This is the infamous passage in which Bergson invokes Fabre's reports on the instincts of parasitizing wasps in order to connect a theory about the "thematic" nature of instinct to a theory about the "sympathetic" nature of instinctual knowledge. [...] Bertrand Russell famously remarked that the "love of the marvellous may mislead even so careful an observer as Fabre and so eminent a philosopher as Bergson."[1] Most others agreed: the story of the wasp and the theory of instinct derived from it constituted a series of mistakes better forgotten. It was no accident that contemporaneous with Russell's disparagement was the growing success of behaviourism's endeavour to reduce all instincts to mechanical reflexes.

[Ruyer's] defence of Bergson's theory is threefold. (1) Ruyer correctly observes that the thematic conception of instinct is in principle distinct from Fabre's account [...] (2) Ruyer insists that the story of the wasp is [...] perfectly appropriate as evidence in support of Bergson's position [...] (3) Ruyer argues, finally, that Bergson's theory of instinct-sympathy is best understood as the attempt to outline a positive solution to the problem that the case of solitary wasps poses for Neo-Darwinian theories of instinct.[2]

* * *

"BERGSON AND THE AMMOPHILA SPHEX" (1959)[3]

By Raymond Ruyer

Translated by Tano S. Posteraro

On the question of instinct, Bergson was victim of a veritable prank played by the memory of readers, then aggravated by the writers of textbooks. All his concepts have clung, at least in the minds of distracted readers, to Fabre's story of the Ammophila Sphex that "inflicts nine successive stinger strikes to nine nervous centers of its Caterpillar, and finally grabs the head and bites it just enough so as to ensure paralysis without death."[4] Bergson may well have immediately added: "Of course, it is hardly the case that the operation is always executed to perfection," and quoted the Peckhams to correct Fabre,[5] but the impression remained that Bergson supports his whole theory of instinct on Fabre's embellished stories, and that the Peckhams later demolished the observations on which Bergson relied. It is a little like the way children read the simplified story of the beginning of the war of 1914 in their small History of France. The conspicuous uniforms of the French soldiers are mentioned there, and the children retain the impression that the first of France's defeats is due only to red trousers.

It would already be unfair to condemn the entire Bergsonian theory of instinct even if the story of the Sphex was false because Bergson cites and uses many other very accurately reported facts. But most curious is that the story of the Sphex is accurate—whatever its critics say—and is also perfectly suitable as an argument in support of the thesis for which Bergson uses it. Bergson cites it in the course of a critical discussion of Neo-Darwinian theories, according to which "some useful procedure, accomplished naturally by the individual in virtue of an accidental predisposition of the germ, would be transmitted from one germ to the next until such time as chance might come to add new improvements to it through the same procedure"—according to which, in short, the instinct is constituted by the progressive addition of new elements.[6]

Bergson demonstrates the improbability of this theory by logical arguments on the one hand, and by the examination of facts on the other. When we compare the different forms of the same instinct, of the same instinctive theme in various species of Hymenoptera, we cannot arrange these forms in linear series in order of increasing complexity. The instinctive theme is really a central theme "which transposes itself as a whole in different directions and on which, also as a whole, is performed in every species different variations," the differences of complexity not always corresponding to an order of filiation, as they should on the Neo-Darwinian hypothesis.[7] One could cite innumerable examples in favor of the thesis of "thematism": various uses of thread by the various species of spiders; types of display among the birds; types of camouflage or parasitism among the crabs; the theme of "brandishing the pincers" among the twenty-seven species of Uca crabs in Panama; the "courtship dance" theme among the various Sticklebacks. The proponents of comparative ethology have been able to make veritable etymologies of instinctive behaviors. A certain "root" (in the linguistic sense) of behavior undergoes displacements, variations, complications, simplifications, changes of sense in the related neighboring species. Bergson limits himself to briefly citing the social instinct among the various Apidae (where the differences in complexity do not correspond to an order of filiation) and finally the paralyzing instinct of various species of Hymenoptera, where the theme—the etymological root— "to paralyze a prey and to deposit its eggs there" is varied or weakened according to the different species of prey: spider, beetle, caterpillar, cricket.

The fact that one can so easily combine Bergson's thesis with the modern conceptions of comparative ethology proves that far from being behind on the science Bergson anticipated it. He was perfectly right to challenge the pseudoscientific thesis, fashionable at the time when he wrote *Creative Evolution*, which considered instinct as a mosaic of reflexes placed alongside [*juxtaposés*] genetic variation and natural selection. The variations of an instinctive theme resemble linguistic variations, and it is almost as absurd to explain the variations of the theme "paralyze" in the various species of Hymenoptera by accumulation of genetic mutations as it would be to explain the variations of the unknown Indo-European root signifying "girl," which makes *duhitar* in Sanskrit, *thugatêr* in Greek, *Tochter* in German, by genetic mutations of various Indo-European groups.

The contemporary ethologists do not reject natural selection, or even the possibility in certain cases—never directly verified, to tell the truth— of a genetic origin of behavioral variations. But they more often invoke the thematic variation of behavior as the primary material on which selection then works, and as the first origin of specification by isolation and segregation.[8] As if, to continue our example, a linguistic difference in a group led secondarily to a separation of two or more subgroups, or reinforced an already initiated [*amorcée*] segregation.

To put it more simply, Bergson's discussion bears on both the Neo-Darwinism and the Neo-Lamarckism of the time. Our comparison, in Lamarckian spirit, of the variations of the instinctive theme with those of a linguistic theme, is limited in that the linguistic transmission is socio-cultural, and not biological like the transmission of instinct. Instinct, according to Bergson, is no more an intelligent discovery of some individuals of the species, degraded into habit and transmitted by heredity, than a mosaic of fortuitous mutations.

The linguistic theme is unconscious or subconscious in individual speakers. That is why it seems to have a life of its own, independent of its bearers and transmitters—in which it resembles an instinct. But it is evidently not related to the organic form of the species and in this it differs from instinct. Instinct is inseparable from the organic formation and has its origin in a super-individual [sur-individuel] and even super-specific [sur-spécifique] "domain of consciousness," which is to say in the cosmic consciousness of the élan vital. This consciousness "gives us the key to all vital operations," and is the common origin of all species, of all formative instincts, and of all instincts of behavior. We see that if Bergson rejects Neo-Darwinism, he limits himself to correcting the Neo-Lamarckism in a way that is truly radical, by transporting it from the psychological or psychobiological level to a metaphysical level. The vital consciousness is a consciousness, but it is no longer an intelligent, calculating consciousness, as is human consciousness, which is based on the perception of external objects and bears on these objects from the outside.

This yields the key to a Bergsonian theory that has until now appeared entirely verbal, literary, metaphorical: the theory of instinct-sympathy. After having made critical usage of the example of the paralyzing Sphex, Bergson proposes a positive solution, in rather vague terms, we must admit, and with a discomfort emphasized by the use of conditional verbs. We are wrong, says Bergson, in wanting to translate the "science" of the Hymenoptera in terms of intelligence and consequently assimilating the Sphex with the biologist who knows the caterpillar from the outside:

> [T]he Sphex would have to learn, one by one, like the entomologist, the positions of the Caterpillar's nervous centers. [...] But none of this would be required if we assumed rather a *sympathy* (in the etymological sense of the word) between the Sphex and its victim, a sympathy that would inform it from within [du dedans], so to speak, of the Caterpillar's vulnerability. This feeling for its vulnerability might well owe nothing to external perception and might result from a simple present encounter between the Sphex and the Caterpillar, no longer considered as two organisms but rather as two activities. The feeling would express in a concrete form the relation of the one to the other. True, a scientific theory cannot appeal to these types of considerations. Such a theory

> must not place action prior to organic development [*organisation*], nor
> sympathy prior to perception and knowledge.[9]

Reading such passages, the biologist shrugs his shoulders and taking advantage of the confession that the conception of instinct-sympathy has nothing in common with a scientific theory he returns to his experiments on the role of hormones or on the "innate trigger mechanisms."

However, the biologist would do well to remember the instinctive situations where he found himself as representative of the human species. For example, married very young and naive, before having completed his studies and having very precise scientific notions about procreation, embryology, sexual hormones, he knew how to behave in the presence of his wife in such a way that he had a child—though not without a lot of awkwardnesses and clumsinesses analogous to what the Peckhams detect in the behavior of the Sphex.

Everything happened as if he had known—before having consulted any treatise—the anatomy, or the physiology, of the reproductive organs of man and woman. He would have done entirely as well if he had been, like the biologists of the eighteenth century, "ovist" or "animalculist," or if he had been a convinced Aristotelian and if he had believed that man brings form to feminine matter. He would have done just as well – probably better—if he had not been an intellectual, and even if he had not been a civilized man, and had ignored the role of the father in procreation and had believed to be making [*frayer*] only "a way to the spirits" [un *chemin aux esprits*], like the Melanesians or Kanakas.

By "sympathy," Bergson means a "non-intellectual knowledge" [*savoir*] or, more exactly, a "non-perceptual knowledge," a knowledge that has no need of information about the external object conveyed step by step along transmission channels. What to say? The problem of instinctive behavior becomes much clearer if it is posed in the form: "Is it a matter of knowledge [*savoir*]? or of ability [*pouvoir*]?" In general, if a being does what he does, it is first that he knows and second that he can do it. I swim, because I can swim (I do not have muscular cramps), and because I know how to swim (I learned to swim, and I do not have neurological disorders, no apraxia). In many languages, knowledge [*savoir*] and ability [*pouvoir*] are poorly distinguished when it comes to a learned action and not an acquired notion (for which all the languages agree to use the word "knowing" [*savoir* and *connaître*], and not the word "the ability to know" [*pouvoir*]). We say in English and in German: "I can swim," whereas we say in French "I know how to swim." Is it English or French that is right? English or German would certainly be wrong to talk about "ability" [*pouvoir*] if these languages meant by this—what they do not—a purely physical power [*pouvoir*], a "power to function." In fact, they mean a capacity, a psycho-physiological "competence," an "effective knowledge" [*savoir*]. The danger of confusion is between physical power [*pouvoir*] and effective mental knowledge [*savoir psychique*]. And this possible confusion is extremely important in the problem of instinct.

If the yellow-winged Sphex paralyses the cricket and its three pairs of legs, if the Ammophila paralyses the nine nervous centers of the caterpillar, Bergson says it is because it knows how to do it. The positivist biologists would object to this answer: it is simply that it can do it mechanically. Given the structure of its body with the stinger, and the structure of the body of the victim, it stings as it can. But, the Bergsonian could respond, even admitting all possible clumsinesses, whence does the Sphex know that his stinger is at least made for stinging and that he uses it for that in a more or less adroit fashion instead of remaining unresponsive in front of the cricket or the caterpillar? This "knowing how to use" [*savoir employer*] an organ, retorts the positivist biologist, is itself only an "ability" [*pouvoir*] or a "microability" of nervous systems of the Sphex. These systems are arranged so that, stimulated or triggered by the image of the caterpillar, they function so as to control in turn the operation of the stinger. Nowhere is there knowledge, only abilities. And abilities in the strictly positivist sense of the term: it is not a matter of potentiality beyond the actual [*potentialité sur-actuelle*], which would revert to a competence, to a "virtuality of ability," that is to say again to a knowledge. It is a matter only of an "ability to function," as when it is said that a gearshift in a car can take four positions or that a ratchet can rotate in one direction.

Thus defined, today the debate is evidently decided in Bergson's favor, at least in favor of instinct-knowledge, and against the notion of an instinct-ability, an instinct-function. Instinct is thematic in each of its phases. In each of its phases it includes an element of "sense," of abstract and unspecified knowledge [*savoir abstrait et non précisé*], cognition [*connaissance*] beyond mechanical actuality [*l'actuel mécanique*]. The internal hormonal sensitization evokes a half-oriented search behavior, an observable counterpart of a concern, of an anxious presentiment. The stimulus, or rather the external object putting an end [*mettant fin*] to the search behavior, is not a mechanical trigger, a key of precise form that would correspond to an automaton lock [*un automate à serrure*], it is an object or a situation corresponding to a certain instinctive "gnosis"— according to the very exact technical term—that is to say to a cognition [*connaissance*] or to a capacity for re-cognition [*re-connaissance*].

Experimental studies—especially those of Buytendijk, Tinbergen and the Dutch school—have used lures and simulacra to show that only a part of the stimulant is really effective as stimulus-signal. But this "part" is not a part in the geometrical sense of the word, it is a "schema": *angeborene auslösende schéma* [innate triggering schema], a kind of particularly expressive extract. The animal perceives this attractive or dangerous "expressive extract" as the man perceives a very familiar situation, by a trait, a detail, or a characteristic aspect, as when he "runs after a petticoat" or "avoids a military cap." The stimulus-signal always seems like a simplification, a secondary reduction of a "gnosic" situation; it is not the primary and constituent element. Also, it is often impossible to use the method of simulacra to extract a specific stimulus-signal to which the animal is sensitive in the general situation, as E.S. Russell pointed out.

Last, instinct is also thematic in the movements, or rather, the actions of consumption. They might sometimes take on the appearance of a motor melody that can even unfold itself meaninglessly [*se dérouler à vide*]. This might make one think, according to Tinbergen, of well-localized and hierarchized nervous mechanisms that function once triggered in an autonomous way. Yet the actions of effection or consumption—that is, precisely the case of the Sphex stinging the caterpillar—are more often very flexible, regulated by [*régulés à mesure sur*] indication stimuli or orientation stimuli, or again, by internal irritations that the actions seek to relax in one way or another. In all these cases, both those described by Lorenz and Tinbergen and those described by the American school hostile to "autonomous melodies," the observable movements of the animal always constitute actions. They are not really stereotypes like the movements of an automaton of cams and spikes, they are significant, they seek a realization, even if they do not imagine it in advance. They are oriented by a knowledge [savoir], and not pushed according to the power of a fully assembled mechanism.

But the decisive argument against pure instinct-ability, actual mechanical ability, is provided by the basic complementarity of equivalence between instinct and organization, as Bergson saw perfectly. Let us admit, in fact, that the Sphex uses its stinger as it can, according to the structure of this organ, and according to the actual structure of nervous mechanisms which, once triggered by the sight of the caterpillar, control the use of the stinger. In any case, it was necessary over the course of the ontogenesis of the Sphex that every part of its hypothetical nervous mechanisms, as well as the stinger with its muscles and accessory glands, be constructed from scratch from the fertilized egg. This construction was epigenetic for the one as for the other. It is not explicable by an "actual mechanical power" of the egg, but by a "capacity," a "competence," a "virtual power," a "potentiality." Either all these words mean nothing, or else they designate more or less well what must be considered as a "knowledge" [savoir]. But then, against the well-established facts demonstrating the thematic character of instinct, why connect the instinctive actions to fully assembled nervous mechanisms? For in any case these mechanisms themselves have been constructed not by the functioning of other mechanisms, by hypothesis not yet given, but by what must be called a knowledge or a morphogenetic instinct "behind" the egg or embryo observable in actual space. Unless we adopt a long abandoned and refuted conception of the nature of genes or fall back on the old preformationist errors about DNA molecules there is no way to escape the consequence.

Rejecting the epigenetic character of the ontogenesis of instinct is useless since we must in any case accept the epigenetic character of the ontogenesis of the organs that instinct uses and that nervous mechanisms are supposed to control. The bird makes its nest—in the strong sense of "make" [faire], with know-how [savoir-faire], if not with "know-why" [savoir pourquoi]. It is not a simple mechanical intermediary between the mechanisms of its brain and the materials of the nest, because the body of the bird, brain included, is made precisely like the nest, with know-how, and without us able to find any part of effector mechanisms already assembled.

Why would epigenesis, that is to say the emergence of formative action, be realized in one go? Why should it, in one go, assemble mechanisms that would henceforth only have to function, instead of continuing to improvise assemblages [montages] and formative behaviors, to complete the assemblages already acquired and integrate their functioning with a meaningful global action? Instincts are relatively complementary with organic forms precisely because organic forms are themselves the result of embryonic instincts, of the embryo's "knowing how to act" [savoir agir].

In certain cases by the mere sight of organic forms one can guess which instincts use them: a stinger is made to sting, a wing to fly, the hard beak of the unhatched chick to break the shell. But from the examination of forms alone it is not always possible to guess the instinct that uses them. All spiders possess just about the same seric glands, and yet some of them make a regular web, others an irregular web, others emit threads to be lifted away by the wind, others simply line their nest. Lloyd Morgan argues that no one could have deduced the remarkable migrations of the eel by examining the structure of its body—and even less, we will add, by examining the structure of its genes and of DNA molecules, even with an ideal precision. From the anatomy of the antlion, one can conclude that it is carnivorous, not that it digs a hole in the sand and waits for an ant to fall in. For that matter, it happens that similar instincts are manifested by animals with very different corporeal structures. Ants and termites have very similar social behaviors, yet they belong to quite distant families.

Forms and instincts—or, more exactly, formative instincts and behavioral instincts—are relatively independent while complementing each other precisely because they are of the same nature. Instinct is not only a secondary function—and functioning—of forms already constructed by formative instincts. Instinct uses acquired knowledge [l'acquis], but it overflows it [le déborde]. It is the form (of the organism as of the behavior) which is a function of the activity. It is necessary not to get things backwards. So, when we say that the bird makes its nest, that the nest is its work [œuvre] and not an automatic result, this thesis is true contra the mechanical conception of instinct; but it should not be used contra the conception of an ontogenesis of instinct. The bird that makes its nest is at once the individual bird (already "made" by embryogenesis), and, through it, the specific bird-type. The individual bird is at once work and worker; by working at its nest it still continues to be worked [œuvré]. The knowledge is not entirely its own knowledge. Above all, by its already constructed nervous system, it brings orienting or controlling "gnoses" to an instinctive theme that "arrives" in its nervous system in the same way that formative themes arrive in embryonic primordia over the course of their organic formation.

Tinbergen's experiments on the instinctive reactions of young blackbirds in the nest to the arrival of the nourishing parent have shown this mixed character of instinctive action. When the young chick extends its open beak towards the nourisher, the action seems unified. Yet it is composed, with a certain observable lag [décalage], of an autonomous motor theme (open the break upward), combined, when the chick has ceased to be blind for two or three days, with an orientation towards the nourisher. This orientation represents the part played by the individual nervous system that has become functional. In short, to use Weismann's scheme, the bird does not "make" the egg, the egg is made in it. On the contrary, it "makes" its nest, but the nest is in part also made in it or through it, like the egg, by the same specific knowledge that follows the continuity of the germ.

Embryology and the psychobiology of instinct are increasingly revealing themselves to be two chapters of the same science. The same scientific concepts are used: primordia, territories, formative themes, autonomous development, stimuli-signals, stimuli-orientors, induction, regulation, competence, importance of timing. The "chemistry" of instinct is the same as the chemistry of embryogenesis; the hormones of the sexual instinct are the same as the hormones that realize the genetic sex or potentially make it change in laboratory experiments. Organic teratology, by arresting development, fusion and condensation, displacement and translocation, is exactly analogous to the teratology of instinct, as we can verify by comparing E. Wolff and Freud.

We seem to have lost sight of Bergson. Instead, we have followed a parallel path that leads to a very similar point. If "action is before organization" as Bergson suspected, then instinct is a "knowledge," a consciousness, not a functional ability but a knowledge and a primary consciousness more fundamental than individual consciousness and knowledge, which are mere developers of primary knowledge. This is the same definition that Bergson gives of "sympathy," and the life-consciousness equivalence.

However, it must be recognized that Bergson's own considerations, and the way in which he represents this "primary consciousness" of life, are vitiated by a grave error due to their recapitulation of *Matter and Memory*'s misguided theory of perception. According to this theory, the sensory apparatuses do not produce the image, the sensory representation, because the universe is in principle already only an ensemble of images. Here Bergson has given in to Berkeleyan influences and generated elegant solutions by systematic reversal of the terms of the problem and by return to the immediate. The sensory apparatuses are, like the rest of the body, apparatuses for action, not for the generation of representations. Perception is a selection. The image (in principle) in its entirety is reduced to what interests my action. The image of the object (which interests me) is not formed in my visual center, then projected in the place where the real object would be: the object, "the rays which it emits, the retina and the nervous elements affected, form a single whole; and that is really where the object is, and not elsewhere, that the image is formed and perceived."[10]

If we are not mistaken, to try to understand how the Sphex possesses a non-intellectual, extra-sensory "knowledge" [*savoir*] of the anatomy of the caterpillar, Bergson uses two different theories without being clearly aware of the duality:

(1) The thesis that life is unitary, that it forms a "whole that is sympathetic with itself,"[11] that this unity pre-exists the super-imposed secondary dissociations, the divisions of labor, of the cells of the organism, individuals in a colony or society, or different species in a fauna or a flora, and that this unity reappears when necessary: "The instinctive knowledge that a species possesses of another species on a certain specific point thus has its roots in the very unity of life."[12]

(2) The thesis (recalling *Matter and Memory*) that perception (which bears on another living being, or on a distant star) is in principle a knowledge [*connaissance*] at a distance, without an intermediate progression of information:

A person blind from birth, living among others who were also blind from birth, would not admit that it is possible to perceive a distant object without first having passed through the perception of all the intermediary objects. Nevertheless, vision performs this miracle. True, we could say that the blind person is right insofar as vision—having its beginnings in the stimulation of the retina caused by the vibrations of the light—is in fact nothing other than a touching with the retina [...] But I have shown elsewhere that the philosophical explanation of perception must be of an entirely different nature [...] But instinct too is a knowledge from a distance.[13]

This second thesis is very different from the first (as stated previously). It relates not to life but to perception in general, to which is attributed a quasi-magical character of influence at a distance. Apart from Bergson's very artificial reasoning here, it would be proven if Rhine's contestable and suspicious experiments on extra-sensory perception could be verified and confirmed. On the contrary, the first thesis (on the primary unity of life, on the unity of spatially distinct organisms, and even of organisms of different species) is much more solid and today relies on a large number of biological facts.

An organism has a primary unity underlying the spatial distinction of its parts. The experimental study of ontogenesis, regulation, and regeneration has shown that the living individual is primarily an equipotential domain, not an assemblage [mosaïque]. Even after a certain spatial distribution of formative themes, each region still retains equipotentiality in relation to the theme it has received. It differentiates itself first as a whole and according to this general theme before distributing it in its turn. From one equipotential region to another the synchronizations and harmonizations are operated by stimuli (chemical inducers); but in the interior of each region the stimuli pathways play only a secondary role. If one cuts one of these regions it forms two complete organizations in the place of one; if one joins two of these symmetrical regions (before any internal "distribution") they form only one organization (whence symmelia, cyclopia, and all the monstrosities by fusion). In the adult organism a whole network of internal communications, nervous or chemical, is indispensable to harmonize the life of the whole. But in the egg or young embryo it is not the same; the unity depends on a primary equipotentiality, as if the organism in formation were analogous to a field of consciousness, to an acquired visual sensation, where the details are at once distinct and yet form an immediate conscious unity without needing to be seen again by an internal eye. Or, better understood, the analogy or dependence is in the opposite direction, and it is the visual cortical area that is clearly conscious, *unitas multiplex*, of a field only because it retains the equipotential property of a young embryonic organism or area in the adult.

From the primary unity of an organism one passes easily to the primary unity of two organisms. Two univitellin twins [i.e., of a single egg] differ little more than the right hand and the left hand—especially if like Marie and Émilie Dionne they manifest a mirror symmetry. The male and female of a dioecious species [having male and female reproductive organisms in separate individuals] are two distinct organisms. The male and female organs of a monoecious or hermaphroditic species are formed in two regions of the same embryo. There is nothing unwarranted in admitting, with Bergson, or in Bergson's style, that when a male and female of the same species are brought together, like the Sphex and the caterpillar, the stimuli-signals evoking instinctive sexual behaviors represent only an accessory guidance; that the male "does not have to learn one by one, like the biologist, the positions [of the reproductive organs in the female], nor to acquire the practical knowledge [*connaissance*] of these positions by experimenting [with the effects of copulation]."[14] One can even add, with the text of *Creative Evolution*, that "everything happens as if there were, [between the male and the female], a sympathy which informs [the male] from the inside about [the sexual characteristics of the female]."[15]

It is better to stop here, and not continue with Bergson, who adds: "This feeling [of the sexuality of the female] may owe nothing to the external perception and result from the mere presence together [of the male and the female] considered no longer as two organisms, but as two activities."[16] In fact, it is highly doubtful that orienting stimuli-signals, guiding external perceptions, could be completely absent in the most surprising instinctive performances, such as eel migrations, for example. There is nothing magical about the evocation of instinct; like development, it always has inductors, most often chemical. But it is true that once evoked, instinctive actions overflow the evocator, the inductor, the stimulating perception by their complexity. The animal is guided by light, or smell, it is not guided by it like an automaton. The external stimulating perception does not bring it [the animal] complete information. The instinct evoked is informed by itself: once again, it is "knowledge" [*savoir*]. This knowledge is of the same nature as "competence" (in the sense that embryologists use this word), as the competence, for example, of an embryonic area to unitarily differentiate according to the theme that was distributed to it when it is induced to do so by a chemical inducer emanating from a neighboring area.

The obligation for the male to be sensitized by hormones on the one hand, and to perceive the female according to a certain *auslösende schéma* [triggering schema] on the other, does not exempt him from having a "competence" of the same nature as the competence of the primordium of the male organs to differentiate in the hermaphroditic organism.

Bergson does not give enough credit to science when he adds: "True, a scientific theory cannot appeal to these types of considerations. Such a theory must not place action prior to organic development [*organisation*], nor sympathy prior to perception and knowledge."[17] It is the experimental embryology itself which furnished the theory of instinct with these notions of "competence," "equipotentiality," "evocation," which Bergson anticipates with a vocabulary obviously less precise, but less inadequate than it seems at first sight.

Can we finally pass, then, from the male-female situation (of the same species), to the Sphex-caterpillar situation? The step is important, and the jump is difficult, especially if one renounces the "magical" theory of perception at a distance of any object as one certainly must, to contaminate or to "complicate" [*corser*] the thesis of the primary unity of the organism, of organisms and even species. Difficult, but not absolutely impossible, because after all, living species, even the most distant from each other today, are born from a bifurcation in an organic domain which does not differ essentially from a twin bifurcation [*une bifurcation gémellaire*]. It is not even absolutely excluded that there is a "finality in the service of others," or, to use Bergson's expression, "primary sympathy" from plant to animal, as we have believed to be the case with galls where the plant seems to accommodate a dwelling for the parasitic insect, or again with entomophilic plants.

To tell the truth, we are personally very skeptical on this point, and this is why we preferred to comment and to defend Bergson's general theses on instinct-knowledge [*savoir*] and instinct recovering primary consciousness, which is the very texture of life, substituting the example of the male-female situation for the Sphex-caterpillar situation.

The very possibility of this hesitation and substitution is strange only at first glance. If we have to dissociate the two types of instinctive situation, if the male and female organisms present together still benefit from a primary organic unity in the one case and they no longer benefit in the other—where it is a matter of predator-prey or host-parasite—then from where does their resemblance come? And, finally, how to explain the Sphex's "knowledge" [savoir] of the caterpillar? Must we return decidedly, at least in cases of this kind, to Neo-Darwinism, and to the accumulation of accidental mutations? But no Neo-Darwinian can argue that in the male-female case and in the often complex courtship instincts, the instincts of the male and the instincts of the female are formed separately by accumulation of mutations. The two instincts are often coordinated in a very rigorous way, like the male and female forms of a dioecious plant. Selection, if it intervened, had to intervene on the male-female organic unity with which we therefore cannot, in any case, dispense. Orthodox Neo-Darwinism therefore compounds the incongruity rather than elucidating it. It unduly accentuates the contrast between the two situations.

The solution to this problem has been given by studies on disjunctions, displacements, transpositions, and readjustments of instinctive themes in comparative animal ethology,[18] on the role of selection in these displacements and readjustments on the one hand, and of these displacements on the selective action on the other. Once distributed, the initially unitary theme—that is, capable of inaugurating the formation of an entire individual or a global behavior—becomes a multiplicity of themes, coordinated but nevertheless obliged from now on to perfect that coordination. Selection bears on the adjustment from theme to theme and also on the displacements used by these dissociated themes much more than on the fortuitous mutations of the genes or molecules of DNA.

Selection does not resemble a mechanical sorting, but an organic learning by trial and error at once directed and fortuitous. Even in sexual formations and behaviors, there are, despite the primary organic unity, many thematic dissociations and consequently many secondary adjustments obtained at first haphazardly. The distance is smaller than it appears between sexuality and parasitism or predation. Fertilization is a break-in; in mammals, the embryo is a parasite on the mother, and is implanted in the uterus like mistletoe on an apple tree. Many animals are killed by sexual *rapprochement*. Laying eggs on a parasitized animal is a behavior that differs little from that of some species of male whose gametes occasionally migrate throughout the whole body of the female. It is very probable that in many symbioses, parasitisms, or predations, whole stretches of sexual behavior have been introduced after displacement and modification, just as one can still recognize themes of attack, combat, flight, as well as the offer of nourishment or infantile behaviors in many courtship behaviors.

The resemblance between the male-female situation and the Sphex-caterpillar situation is thus explained in this way. In the abstract, nothing is more opposed than natural selection and the unity of an organic domain. But in fact, one penetrates the other to various degrees. The unitary organic themes are still at work, but more or less dissociated and readjusted to the potential of the species. Chance and selection count little in the interplay of the sexes. They count more in symbioses and parasitisms. The evolutionary history of paralyzing Hymenoptera, entomophilic flowers, mimetic forms—especially of those that even imitate physical accidents—must look a little less than Bergson believes like an invention in orthogenesis, or like the linear evolution of a technical or artistic form in a homogeneous human culture, and a little more like a "history" in the strong sense of the word, like a political or linguistic history, complex and capricious. The history of the evolution of instincts, like organisms, is a mixture of organization and chance, of sense and of fortuitous displacements of sense, of improvisations to suit the circumstances, and corrections [*rattrapages*] by means of fortune. Bergson is certainly wrong to speak of a theme "which transposes itself as a whole and on which, also as a whole, variations can be performed."[19] Biological themes, like words, are composites. But, overall, Bergson's mistake is minor.

* * *

The Bergsonian conception of instinct is therefore generally correct. Its weaknesses are due rather to its less auspicious connections to neighboring theories.

The instinct-intelligence opposition, and especially the bifurcation between animal instinct and human intelligence, is certainly forced. This is incompatible with the incontestable thesis that does not dissociate instinct and organization. The body of man is formed like that of animals. His psyche is fundamentally the same. Comparative ethology finds in man the same instinctive bases as the mammals: internal sensitization, stimuli-signals and gnoses, phenomena of displacement, etc. Autonomous motor melodies are rarer, but they are not lacking, as A. Gesell and Piaget have shown. And, moreover, they are also rare in other mammals.

As Baerend's observations have demonstrated, on the one hand, even the instincts of insects include variable phases of rigidity and flexibility [*souplesse*], the phases of flexibility and "broad" adaptation not being due to a mixture of intelligence but to a sort of thematic breadth of instinct itself, "foreseeing" a collaboration with learning, if one can so speak. The instincts in man and the higher animals are exceptionally "broad," but they are authentic instincts.

On the other hand, the characterization of instinct as the capacity to form organic tools, and intelligence as the ability to form inorganic tools, without being false does not help us grasp—at least as Bergson presents it—the modality of the passage from the organic tool to the fabricated tool. The use and fabrication of a tool, by man, and exceptionally by the monkey, is still an "organic formation" if one thinks about it, like any intelligent activity more generally. In fact, the use and fabrication of the tool are directed by the brain and imply spontaneous changes in the brain tissue of the motor and sensory areas that direct the movements of use and fabrication. Fabrication, technical invention, is a provisional and erasable cortical morphogenesis. It thus relies on the primary organic self-structuring property applied to these objects instead of being applied to organic tissues as tissues, thanks to the fact that the brain is at once organic tissue and a screen where the forms of external objects are projected.

Why do all animals with a cortex "informed" by sensory organs— that is to say a good part of vertebrates, and even metazoans—not fabricate tools if they have organs, beak, mandibles, hands, etc., capable of holding a material object? Why do only some insects, some birds, and the anthropoid apes sporadically use thorns, pebbles, sticks? It is because the crucial transition [*le passage capital*] is less the use of tools than the constitution of symbolism. The monkey intelligently uses a stick, but for lack of an idea, of a cerebral assembly [*montage*] corresponding to the stick, it does not put it in reserve or set it aside as a "permanent tool." The constitution of a symbolic activity was the decisive fact, much more than the use of a tool for a *present* problem. It is this fact that has allowed for both social culture and external technics.

Bergson has tried to escape from the common error for which all panpsychists fall, which consists in attributing to instinct, formative or not, an evanescent or asymptotically null consciousness. Consciousness of the instinctive act, says Bergson, is not null, but annulled [*annulée*]: the representation is stopped up [*bouchée*] by action. Consciousness is there, but neutralized. As the action hesitates, the representation appears.[20] But this theory of Bergson's is very bizarre in addition to the fact that it agrees badly with the conception of instinct-intuition in contrast with pragmatic intelligence. It is applied only to certain actions of man or animals with advanced brains. The obstacle to the action in progress requires "mental experiments" based on imagination. But let us apply it to the formative instinct or the instinct of animals without developed brains: if we prevent an embryo from gastrulating, or if we disturb a Stentor with grains of carmine, it is very doubtful that a representational consciousness appears. The mistake is to make "consciousness" the synonym for "representation." Representational consciousness is only a secondary consciousness, involving a cerebral tissue modulated by external objects. Primary consciousness is the very form of organization and of its behaviors, as it is in itself, without having to reflect the image of another thing. It is presentation (of itself), and not re-presentation (of another thing). This form for itself is perfectly distinct and precise, as precise and distinct as a visual field. But it is a visual field in which it would see only itself and its own possibilities. Existing as an absolute domain of space and time, it is present unitarily to its own parts and to its own thematic development as well.

As for the fabrication of tools, Bergson does not realize that representational, cerebral consciousness is itself possible only through the primary consciousness of the cerebral tissue as organic tissue. We only see things, and we only fabricate them, because our protoplasmic tissue "sees" and "fabricates" itself.

All individualized forms, cerebral, protoplasmic, or even chemical, are indissolubly "conscious." Consciousness is not a "current launched through matter," as Bergson says in an unfortunate metaphor.[21] Matter as such, apart from religious or philosophical myths, is only a secondary mode of primary individualized forms when they constitute aggregates edge to edge [bord à bord], without their own unity.

Bergson's real philosophical misfortune is to have been born at a very bad time. At the end of the nineteenth century, one had doubts about the artificial character of mechanistic and deterministic science but one saw no other solution than to seek the truth outside of science. One could not anticipate the scientific revolution that micro-physics was about to bring by discovering the domains of individuality beneath the secondary and statistical laws. Micro-physics also permits one to follow the progressive complication of lineages of individualities, from the atom to macro-molecules, to viruses, to cells, to multicellular organisms on the one hand, and the appearance of classical secondary laws when multitudes of individuals are considered in their assembly [ensemble] and in their statistical effects on the other.

Bergson works true wonders to catch a glimpse of what would come clearly to the scientists themselves a few years after *Creative Evolution*. His conception of the "ideal genesis of matter," relaxing in geometry and numbers; his conception of movement as primary relative to substantial mobiles;[22] his conception of the appearance of quality, are so many presentiments: in effect, two physics were to be constituted, or rather two stages of physics and of science in general. Science itself would soon realize that mechanisms at our scale and the principles of mechanics, far from being fundamental, represent, if not a "relaxation," as Bergson said, at least a statistical appearance obtained from leaving out a very large number of "actions." These actions, individualized, irreducible to mechanical operations—since they appear only as the result of these actions—already have the essential characteristics of actions attributable to the most complex individualities. They are expressed by "an energy multiplied by time," which is not very far from Bergson's fundamental thesis. Time is inherent in their structure, time is structuration of action, emerging and creating, and is not the simple unfolding of an already given spatial structure.

Bergson was born relative to the scientific revolution of the twentieth century as if Descartes had been born before Galileo and had written almost all the same things, the *Meditations* and the *Principles*. We will appreciate Bergson's merit all the more if we consider that even today so many philosophers, so many scientists lost in their specialty, and especially so many biologists—such as Fontenelle, too faithful to the Cartesianism of his youth—still imagine that science has remained with the principles of nineteenth-century physics and completely misunderstands the extent of the revolution brought about by micro-physics. This revolution is as important as the Galilean revolution, which it corrects—otherwise than Bergson had anticipated, but in the sense of his presentiments.

III

CRITICAL RECEPTION
IN MATHEMATICS

BERGSON AND BOREL IN DIALOGUE

Émile Borel (1871–1956) was one of the most important French mathematicians of the twentieth century, and his work on the theory of measurement (which he helped establish), on functions, and on divergent series was influential for subsequent developments in the field. Borel criticizes Bergson for failing to recognize an "evolution" in the geometrical intellect between the Greeks and modern geometry. It is worth noting, however, that Borel's critique of Bergson is interwoven within a discussion of his own recently published article, "L'intuition et la logique en mathématiques"—and the appearance of "intuition" in the title is intriguing given Bergson's own interest in the term.[1] "L'intuition et la logique en mathématiques" was situated within a complex debate between Borel, philosopher and mathematician Louis Couturat, and Édouard Le Roy (Bergson's disciple) over the foundations of mathematics. Le Roy emphasized the role of "invention" and "intuition" in mathematics, whereas Couturat had sided with

DOI: 10.4324/9781315537818-10

Bertrand Russell in insisting that mathematics be reduced to pure logic, thereby dismissing Le Roy's considerations as merely "psychological." Borel entered the debate by arguing that "intuition," far from being simply a "psychological" consideration, plays an important role in the development of mathematical theorems. In a formula that is indeed quite Bergsonian, Borel suggested that it is worth distinguishing mathematical theories that are "ready-made" [toutes faites] from those that are "being formulated" [en voie de formation].[2] Borel thus insisted on the positive contribution of "intuition" to recent mathematics. Given this context, "L'évolution de l'intelligence géométrique" can be read as less concerned with Bergson's ignorance of the history of mathematics, and more with demonstrating that Bergson, understood as something of a natural ally through the concept of "intuition," has in fact unwittingly abandoned his own team by equating all mathematical cognition with an empty and dry, logical form of spatialized thinking. This context helps to explain the respect and praise Borel shows for Bergson for having recognized (or "rediscovered") the introduction of movement and transformation in modern mathematics. Ultimately Bergson will resist this attempt to efface the difference of nature between intellect and intuition. The distance between Borel and Bergson is apparent in their differing opinions regarding the contemporary value of studying Zeno's paradoxes.

In his response, Bergson defends himself on two fronts. On the one hand, he wants to show (and not without some humor) that he indeed knows the history of developments in mathematics. On the other hand, he wants to show that he is precisely the thinker who makes intelligible the very possibility of recognizing an evolution in geometrical intellect, insofar as such an evolution would require a simultaneous evolution of materiality—though geometry will always remain for Bergson the "natural inclination" of intellect. Moreover, it is precisely because the intellect is evolving that one can (against Borel's claims) continue to confer a value on Greek geometry, since the relation between Greek geometry and modern geometry is not an external relation between two fragments, but an internal relation in the manner of a continuity. This results in a theory by which the evolution of the individual mind would recapitulate, so to speak, the evolution of humanity. We might even say that a central thesis defended in L'évolution créatrice, according to which the intellect can touch upon an absolute, finds here one of its key applications. Despite being at times ironic or biting, Bergson maintains a relatively neutral tone. Perhaps this is because he saw an opportunity here to focus on correcting a certain misinterpretation of his position regarding the relationship between instinct, intellect, and intuition.

Borel had the final word in this exchange, which came in the form of a letter to the editor of *Revue de métaphysique et de morale*, Xavier Léon. Borel seems genuinely offended at Bergson's reproach for having undersold Greek geometry. The letter also returns to the question of "anti-intellectualism" and concludes with the lingering tension in their understandings of the relationship between "intellect" and "intuition."

* * *

"THE EVOLUTION OF GEOMETRICAL INTELLECT" (1907)[3]

By Émile Borel

Translated by Kathleen Hulley

I would like to offer some brief reflections that are connected to my previous article[4] and to Henri Bergson's recent book, *L'évolution créatrice*.[5] Of course, I do not claim to discuss here more than a few pages of this important and suggestive work. In fact, I will only discuss one particular point: the notion of the geometrical intellect. I will attempt to show that Monsieur Bergson's idea of the geometrical intellect, which seems to play an important role in his system, is adequate in terms of the geometrical intellect of the Greeks, but that geometrical intellect has evolved and currently is much less rigid and much more lively.

Let me begin by noting that if Monsieur Bergson did not discover this evolution in the geometrical intellect, this is because he was not looking for it—he assumed *a priori* that such an evolution does not exist. Wanting to discuss biological questions, he did not recoil before the immense effort required for studying the most recent works in biology; he did not judge that an analogous labor was necessary in geometry. The dogma of the unity of mathematical thought rendered this labor useless—from the moment the mechanism of mathematical thought is understood on some particular points, it is supposedly known in its entirety, given that it is always essentially similar to itself.

Nevertheless, when he is attempting to define modern science in relation to ancient science, Monsieur Bergson offers some highly pertinent comments on the beginnings of the transformation of geometry.

> What was the first of the great transformations of geometry in modern times? It consisted in introducing time (albeit, in a hidden form) and movement directly into the consideration of shapes. For the ancients, geometry was a purely static science. The shapes were given in a single stroke, in their completed state, much like the Platonic Ideas. But the essence of Cartesian geometry (although Descartes himself did not formulate it in this way) was to consider every plane curve as described by the movement of a point [...] [T]o substitute an equation for a shape consists in seeing where one is at on the pathway of a curve at any moment whatsoever, rather than envisioning this pathway all at once, gathered together into a single movement where the curve is in its completed state.[6]

Only, rather than developing these indications by researching how these transformations (whose principle he defined so well) were pursued after Descartes, Monsieur Bergson turns to a metaphysical discussion of the notion of time (of which I will have nothing to say here). But he does not seem to have really believed in the possibility of an evolution of geometrical intellect after Euclid—otherwise, how could he have written a passage such as the following, which I must cite in full, since it is characteristic:

> Let us begin with deduction. The same movement by which I trace a figure in space engenders its properties. These properties are visible and tangible in this very movement, and it is in space that I sense and live the relation between the definition and its consequences, between the premises and the conclusion. For all the other concepts whose ideas are suggested to me by experience, they can only be partially reconstituted *a priori*; the definition of those concepts will thereby be imperfect, and the deductions that include them—as rigorously as we link the conclusion to the premises—will participate in this imperfection. But when I trace out a rough sketch of a triangle in the sand, and when I begin to form the two angles at the base, I know with certainty

and I understand absolutely that if these two angles are equal, then the sides will be as well, and that the figure will thus be able to be flipped over on itself without anything therein being changed. This is something I know long before having learned geometry. As such, prior to scientific geometry, there is a natural geometry whose clarity and whose self-evidence go beyond those of other deductions.[7]

There are many ideas in this passage that are worth pausing over. I will, however, only discuss his use of the process of demonstration by a flip: "[T]he figure will thus be able to be flipped over on itself without anything therein being changed." A moment ago, I was surprised that Monsieur Bergson did not believe it necessary to study the new methods of geometers. But I was mistaken. He did something better than studying them—he discovered them anew for his own personal use. In effect, the "demonstration" that he gives is the modern one: simpler, more intuitive than the Euclidean demonstration, and just as "rigorous," i.e., just as satisfying for the mind of the geometer (even more satisfying because it is simpler).

The geometry in which this demonstration and analogous demonstrations are used is just as "scholarly" as Euclidean geometry. I do not know if it is anterior psychologically, but I believe that I can confirm it to be posterior historically. Of course, prior to Euclid, it is quite likely that people were able to have an intuition of the properties of an isosceles triangle by roughly tracing the lines in the sand. But these observations would have remained sporadic, and the Euclidean method alone allowed for progress in geometry for more than twenty centuries. It has only been for one or two centuries that mathematicians, under the influence of the greatest among them, have been employing different procedures in their research into the new parts of the science. And it has only been a mere few years[8] since they have attempted to bring into teaching at the elementary levels the ideas and methods that have been regularly used for higher mathematics for a century. So it is natural that the geometrical intellect, for all those who are not professional mathematicians, would today be the same as it was in Euclid's day. But the fact that its advances are little known does not make them any less real. And if this evolution brings it closer to a certain instinctive intuition that Monsieur Bergson seems to oppose it to, would it not be better to hope that the intellect thereby continues to perfect itself by evolving, without changing its nature, rather than to condemn it, claiming to replace it by something different?

I worry that I am leaving behind the purely mathematical terrain upon which I had wanted to remain and thereby getting myself lost on the terrain of psychology or metaphysics. And yet, I must say a few words regarding an objection that some of Monsieur Bergson's disciples will surely raise. They will say: "All that precedes has not even the least connection to Monsieur Bergson's ideas. What interest could the evolution of the intellect be to him, given that he is not an intellectualist? He is a psychologist, and you understand nothing of his theories by believing that you could reconcile him with the intellect by modifying it a bit. It is the intellect, in all of its forms, that he has condemned without appeal, and through arguments whose significance singularly go beyond all of these details." Indeed, I do not understand—and I will always refuse to understand—how a man who writes books filled with thought could be anti-intellectual, or that he could oppose instinct to the intellect. There is an internal contradiction here, from which a book such as L'évolution créatrice cannot escape (insofar as it is a negative work, since its positive value is in no way diminished). That someone like Napoleon, Marat, or Saint Vincent of Paul can be anti-intellectual, I'll admit this right away; and I'll also admit that we judge them to be "better" than someone like Montesquieu, Descartes, or Pasteur in the sense in which they are seen as representing higher social forces. This is a question of opinion. But if someone claims to reach anti-intellectualism through arguments, this will forever appear to me to be a meaningless absurdity. I can understand that an anti-intellectual might perform some vocation, that they might be a shopkeeper, a soldier, a nun, a politician, etc. Only one vocation is forbidden to them: writing books. These somewhat absolute ideas probably come from some structure in my mind that does not allow it to consciously think two contradictory ideas simultaneously. But I obviously cannot stop others from thinking them—indeed, that ability must be often quite useful for them.

Of course, these observations do not have the ridiculous pretention of "killing off," in just a few lines, a philosophy as rich as Monsieur Bergson's. They are solely addressed to those who only see but one side of this philosophy, its negative side, and who exaggerate that side as they please. I admire as much as anyone the positive, new insights into foundations and form that Monsieur Bergson has proposed. Even for those who do not always share his manner of seeing, he gives much to think about, which is the philosopher's essential role. Systems come and go, ideas remain and enrich the common foundation of humanity.[9]

[…]

Someone accustomed to [the new] forms of geometrical thinking will experience, when encountering the sophisms of Zeno of Elea, the same indulgent wonder that we all feel when a four-year-old child asks us to take the stars down out of the sky. When the child is just a few years older, we will be able to explain why that is impossible—but then the child will not request that explanation. There is no reason to be surprised that humanity matures slower than a child; humanity will continue to amuse itself with the games of the Greek sophists for a long time to come. Men of science are naturally led to wonder how philosophers attach any value to cosmological (or metaphysical) ideas from men as nourished by inaccurate ideas regarding the world as were thinkers such as Plato or Aristotle. When scholars become indignant toward this attitude among philosophers, it is enough, to render them more modest, to remind them how much they themselves find it difficult to extract themselves from the immense force of tradition. It is enough to remind them that, at the threshold of the twentieth century, their sons have still learned the beginnings of geometry by stammering through the succession of theorems from Euclid, as if the entire effort of the human spirit since that time forward had not existed. There is no doubt that traditionalism is less excusable among geometers than it is among philosophers since the knowledge of the material universe has progressed differently than the knowledge of the moral man. It is thus legitimate to read Aristotle's "Moral Theory," and the well-known love of those bibliophiles for "Complete Works" explains why we continue to publish it alongside his "Physics." But Euclid did not write a "Moral Theory," and so nothing, not even in the eyes of a librarian, justifies the reprinting of his "Elements." His works are only useful for the history of the sciences, the study of which I doubt draws much interest. But historical importance must not be confused with current influence—we have museums for armor from the Middle Ages, but our Minister of War does not go there to choose the models of armor for equipping today's soldiers.

Émile Borel

"IN RESPONSE TO 'THE EVOLUTION OF GEOMETRICAL INTELLECT'" (1908)[1]

By Henri Bergson

Monsieur Borel's interesting article, "The Evolution of Geometrical Intellect," contains certain errors of interpretation regarding L'évolution créatrice that I believe need to be corrected. First, Monsieur Borel assumes that I take geometrical intellect to be something rigid and incapable of evolving, and that it would thus today be the same as it was during the time of the Greeks. This assertion might well surprise my audience at the Collège de France, who will recall that I devoted two entire academic years[2] to showing (as much as one can when one is not a "professional mathematician") the radical transformations that abstract scientific thought has undergone since antiquity and right up to the last century, and how geometry itself was brought back to life by conscious or unconscious (i.e., explicit or implicit) appeals to considerations of movement. Perhaps too readers of this very Revue will be a bit surprised by Monsieur Borel's claims, if they happen to remember an article that I published in these pages, just a few years ago, on the relation between intuition and the intellect.[3] If Monsieur Borel would be kind enough to glance back at that work, he will see if I adhere to the "dogma of the unity of mathematical thought" and if I deny the scientist in general, and the mathematician in particular, the gift of imaginative vision. On this last point, I support the comments from Monsieur Mittag-Leffler, and I see nothing strange—and even find it quite beautiful—in Weierstrass's phrase, "the true mathematician is a poet."[4]

But Monsieur Borel bases his interpretation of me on a couple of lines from L'évolution créatrice. Here is the passage in question:

> But when I trace out a rough sketch of a triangle in the sand, and when I begin to form the two angles at the base, I know with certainty and I understand absolutely that if these two angles are equal, then the sides will be as well, and that the figure will thus be able to be flipped over on itself without anything therein being changed. This is something I know long before having learned geometry. As such, prior to scientific geometry, there is a natural geometry whose clarity and whose self-evidence go beyond those of other deductions.[5]

According to Monsieur Borel, these lines reveal that I do not believe in the possibility of an evolution in geometrical intellect after Euclid. They also apparently imply that I consider the method of demonstration by flipping to be less rigorous than Euclidean demonstration. And finally, he thinks they show that I do not believe it necessary to study the new methods being introduced into the teaching of elementary geometry. Nevertheless, they also show that I have the merit, according to Monsieur Borel, of having "rediscovered" one of those methods on my own.

To begin with the last of these points, I believe that I merit neither the reproach that Monsieur Borel makes against me nor the compliment that he attaches to it. It would hardly be appropriate here for me to go into certain personal memories. I will thus limit myself to telling Monsieur Borel that, as a schoolboy, I had the exceptionally good fortune of being initiated into elementary geometry through a method that was not dissimilar to Monsieur Méray's geometry and that must have indeed been somewhat inspired by it (while, of course, making it accessible for beginners). The "new" procedures Monsieur Borel describes are thus quite old for me (already too old, alas!), and I must decline the honor, as slight as it might be, of having reinvented any of it on my own.

But that is not the most important point. For when he suggests that I oppose "natural" geometry to Euclid's geometry, and when he says that I hold a certain kind of demonstration to be less rigorous than another kind, Monsieur Borel is completely misinterpreting the meaning of the passage that he cites. The "scholarly" geometry that I am speaking of is no more Euclid's than it is Descartes' or any other mathematician's for that matter. It is, rather, and in a general way (as I said in the passage immediately preceding the one he cites), every geometry that has to be learned. No human intellect, not even that of the most powerful geometer, will all by itself reconstruct geometry as a whole, such as the centuries have constructed it— at some point even he will have to seek out some instruction. But every human intellect, as humble as it might be, is on the way toward the elementary truths of geometry and need only make some effort of attention to arrive at them. Now, I distinguish this second geometry from the first, just as we distinguish the natural from the acquired. But in the very lines that Monsieur Borel quotes, I declare that the "clarity and the self-evidence of this natural geometry go beyond those of all other deductions."[6] How can Monsieur Borel conclude from this that I believe one of them to be less rigorous or less satisfying than another? Moreover, even if I establish a distinction between these two geometries, I am far from opposing them to each other. To the contrary, I suggest—and this is one of the essential theories of *L'évolution créatrice*—that every effort of the mathematical mind, as personal and as genius as it might be, and as high as it might seem to lift itself up above material reality, sooner or later comes back to following the natural inclination of our intellect. I set myself the task of marking out the precise direction in which the intellect progresses—so how could I have possibly believed the intellect incapable of progress? In order to attribute such an opinion to me, Monsieur Borel must have read me through the lens of an idea expressed in his most recent article, an idea already mentioned by him in a previous work, that "for all of those who are not professional mathematicians, the mathematical intellect is today still the same thing that it was in the time of Euclid."[7]

Nowhere did I claim it necessary to "replace the intellect by something different," nor that I prefer instinct to the intellect. I simply attempted to show that when we leave the domain of mathematical and physical objects in order to enter into the domain of life and consciousness, we must appeal to a certain *direction [sens] of life* that contrasts sharply with pure understanding, and that has its origin in the same vital thrust as instinct—even though instinct, properly so-called, would be something else. This direction [*sens*] of life is nothing other than consciousness deepening itself more and more, and seeking, through a kind of twisting back upon itself, to place itself back in the direction of nature. It is a certain kind of experience that is as old as humanity, but philosophy is still far from having drawn out all that can be drawn from it. To describe this particular experience, to determine the precise limits of its competence, to show how it superposes itself upon sensible experience that is itself oriented in the same direction as the intellect—does all of that amount to adopting an "anti-intellectual" attitude? Sensible experience, so dear these days to the positive sciences, was itself practiced by humanity, however crudely, for centuries without anyone seeking to purify it. And when the creators of our modern science protested on behalf of this sensible experience against the superbly intellectual constructions that science was made up of at the time, when they insisted that no argument can prevail over an experience, no principle over a fact, they would have surely passed for "anti-intellectuals." If the word is meant in that sense, then I would happily accept being placed among the anti-intellectuals. I will be in good company.

But the true anti-intellectual is surely rather the one who, by persuading philosophy to be nothing but a systematization of the sciences (that is, at root, to simply fill in the gaps of what is actually known by adding some arbitrary hypotheses), gently guides philosophy toward a point at which it will no longer have any choice other than between an indefensible dogmatism and a resigned agnosticism, two manners of falling into complete bankruptcy. The true anti-intellectual is the one who—for having refused to distinguish between cases in which the intellect reaches reality and cases in which it no longer does anything other than manipulate symbols—comes to hold every form of knowledge to be symbolic and every science to be relative to our intellect. If there is one conclusion that emerges from L'évolution créatrice, it is rather that human intellect and the positive sciences, wherever they are working on their proper object, are surely in contact with the real and penetrate more and more profoundly into the absolute.

Monsieur Borel offers the following examples of anti-intellectuals: Marat, Saint Vincent de Paul, and Napoleon. But he should be careful! One can always find someone more intellectualist than oneself, and when he throws overboard Greek philosophy, that perfected form of intellectualism, he runs the risk of finding himself tossed overboard in turn, along with Napoléon, Marat, and Saint Vincent de Paul, who will surely be a bit surprised at his visit, but even more shocked to find themselves lumped together.

For it is what is essential to philosophy, it is Aristotle's general philosophy (and not some random cosmogony) that we find filling the eight books of his Physics, precisely where Monsieur Borel sees nothing more than some things to pique the curiosity of bibliophiles. Now, I am far from believing that the Greek philosophers bequeathed the definitive truth to us. I even dedicated the final hundred pages or so of L'évolution créatrice to showing that the principal theoretical difficulties against which we struggle today come from the fact that philosophers and scientists often still go back, without fully realizing it, to the Greek point of view. But we cannot do without studying Hellenistic philosophy precisely because our mind is still impregnated with Hellenism. This is already necessary when one is content to philosophize in the manner of the Greeks. It is even more necessary when one wants to philosophize differently.

Thought, deepening itself more and more since the first moments of philosophical reflection, encountered, one by one, these large strata of ideas that correspond to the great periods of the history of philosophy. Each one of us can repeat the journey, but it is in vain to want to skip the steps. To go beyond Greek philosophy, we must first pass through it, just as we must study Euclid's geometry (or some other form of elementary geometry) in order to arrive at Descartes' analytical geometry. Someone who opens up a treatise on differential calculus without knowing the basics of algebra just as quickly closes the volume—he is immediately alerted to his imprudence by the very sight of certain signs that he simply cannot understand. But philosophy, in its most profound analyses and in its highest syntheses, is obliged to speak the language that everyone speaks. From this, a fairly widespread illusion emerges that consists in believing that anyone can immediately approach the work of a contemporary philosopher, enter into it at full stride, and refute it immediately, coming to a decision on the problems that it poses or setting them aside as so many futilities, without having taken into account the twenty-five centuries of meditation, anxiety, and effort that are, so to speak, condensed into the current form of the problems, and even into the terms themselves that we use to articulate them. In reality, it is just as difficult to understand Renouvier, for example, if one has not read Kant, or Kant, if one does not know Hume and Berkeley, or Hume and Berkeley if one knows nothing of Locke and Descartes, or the modern philosophers in general if one is ignorant of Ancient philosophy, as it is to read Monsieur Borel's *Leçons sur les séries divergentes* having never heard tell of "integration."[8]

So, let us not look down on Greek philosophy, not even on the arguments of Zeno of Elea. Of course, the Dichotomy, Achilles, the Arrow, and the Stadium would be mere sophisms if we claimed to make use of them to prove the impossibility of real movement. But these arguments acquire a high value when we draw from them what is actually found in them, namely the impossibility for our understanding to reconstruct movement *a priori*, since movement is a fact of experience. Moreover, it seems to me that the difficulties and the contradictions raised around the question of movement fall away by themselves when we consider movement to be a simple thing (in short, when we give up the idea of reconstructing it). But it took time for us to reach that point, and during this time, Zeno's arguments were studied, discussed, and refuted in very diverse ways by men named Descartes, Leibniz, Bayle, Hamilton, Stuart Mill, and Renouvier.[9] All of these men were thinkers of some merit. Two of them were great mathematicians. And, when encountering Zeno's arguments, not one of them experienced "the same indulgent wonder that we all feel when a four-year-old child asks us to take the stars down out of the sky."[10]

H. Bergson

* * *

"LETTER TO THE EDITOR OF *REVUE DE MÉTAPHYSIQUE ET DE MORALE*" (1908)[1]

Dear Sir [M. Xavier Léon],

I do not wish to indefinitely prolong the polemic that has been unleashed by my article "L'évolution de l'intelligence géométrique," but I would ask of you the permission to briefly respond to Monsieur Bergson.

Monsieur Bergson reproaches me for not having fully understood his thought, given that I have not read all that he has written. This proves once again just how difficult it is to penetrate into the minds of others, and I would not be at all surprised that he had interpreted certain passages of the article to which he is responding otherwise than I myself interpret them. I would like to try to clarify the thoughts that I had tried to express in those passages, and this is the primary goal of this brief response.

Monsieur Bergson would like to remind me that philosophy is a specialized discipline and that a mathematician can be ignorant in philosophy just as much, and even more than, a philosopher in mathematics. I never doubted this, and anyone would see in my article, were they so kind as to reread it, that I made every effort to remain strictly on the terrain of science, as much as possible. But philosophers have a habit of speaking about scientific things. Will it be forbidden for those whose vocation is scientific research to give their opinion on the claims philosophers thus make? And indeed, how, when discussing something with a philosopher, can we avoid being at least partially drawn onto his terrain, almost in spite of ourselves?

Nevertheless, there is one point on which I made so many qualifications that Monsieur Bergson's interpretation of it seems hardly possible. I explicitly distinguished between Greek science and Greek philosophy, and I only claimed that the former seemed to lack any value, *other than from an historical point of view.* After all, in science, once the house is built, we take down the scaffolding, and the mind of a student today is not obliged to follow the entire journey of the human spirit, nor would following that journey be sufficient. That is why no genuine mathematician has taken an interest in Zeno's *Achilles* since Leibniz, for Leibniz already solved the problem for us. In the same way, a student in grade three today, through negative proofs of equations, easily learns concepts whose discovery cost humanity so much effort. I would be tempted to believe, despite what we might infer from Monsieur Bergson's text, that the same it true in philosophy and that, there too certain results stand out in an easier and clearer way with time. If it were necessary to follow the thought of all of the great philosophers throughout all of their many detours before beginning to think for ourselves, we would never begin—life is too short.

I will not dwell on the question of anti-intellectualism. I am happy to have provided Monsieur Bergson the opportunity to clarify some of the aspects of his thought that are often disfigured by disciples who are more orthodox than the master. I must, however, make one remark. Monsieur Bergson claims that consciousness, "by a sort of twisting back upon itself, seeks to place itself back in the direction of nature." I am not sure that I understand the meaning of this sentence, but it seems to me that this "twisting of consciousness" might show itself in two principal ways: in the ascetic and contemplative life, which is not something intellectual; or in a metaphysical dissertation on this twisting itself, translating itself into philosophical speculations—and I would call this an intellectual attitude because the faculty that is commonly called "intellect" comes into play. Moreover, for example, Monsieur Bergson would not have been able to write his books if he possessed that faculty to a less eminent degree. I will not belabor this debate over terminology; I will simply recall that I intentionally employed (except for one time, in quotation marks) the adjective *intellectual*, believing myself to be authorized to give to it the habitual meaning it has in French, and not the adjective *intellectualist*, which belongs to philosophical language, a language that I prefer not to write in for precisely the reasons that Monsieur Bergson has aptly outlined.

To summarize, I believe that I am, at root, more in accord with Monsieur Bergson than it might appear. I agree with him that the intellect must be accompanied by intuition; or, if you like, by a certain "direction [*sens*] of life." Only I believe I have observed that the mathematical intellect (to which I had after all limited myself) naturally evolves in this direction, and I continue to reserve for it the same name of "intellect" despite this evolution.

Émile Borel

IV

CRITICAL RECEPTION IN THEOLOGY

BERGSON AND TONQUÉDEC IN DIALOGUE

It is hardly surprising that *L'évolution créatrice* invited commentaries from theology given the metaphysical infrastructure of Bergson's argument (particularly in Chapter III) as well as his sometimes-fleeting claims regarding Consciousness (with a capital C), the Absolute, and God. An important example was Bergson's exchange with Joseph de Tonquédec (1868–1962), a Jesuit priest specializing in the philosophy and history of science. What is most often recalled from this exchange is simply that Tonquédec presents Bergson as a "monist." But Tonquédec's interpretation of Bergson is in fact quite nuanced. He is primarily concerned with how Bergson leaves the ultimate questions indeterminate—particularly whether God is immanent or transcendent to creation.

Bergson twice responded to the charge of "monism" in his personal correspondence with Tonquédec. For Bergson, "monism" meant primarily two things: (1) the Spinozist doctrine of panpsychism (which he explicitly rejected in *L'évolution créatrice*); or (2) an account of reality

DOI: 10.4324/9781315537818-11

without *differences* (a position Bergson found wholly incompatible with his account of becoming and the virtual). By rejecting the label of "monism," Bergson is clearly intent on emphasizing the originality and novelty of his own metaphysics. But despite Bergson's resistance to Tonquédec's interpretation, it should be noted that Tonquédec's position is clear and fair in terms of the obscure passages that he cites from *L'évolution créatrice*. Nevertheless, for Bergson, the rejection of "monism and pantheism more generally"[1] was an essential part of an attempt to rework the very categories that Tonquédec wanted to impose upon his thought.

This section includes a preface by Tonquédec introducing his exchange with Bergson, the two articles by Tonquédec, and Bergson's two responses.[2]

* * *

PREFACE (1912)[3]

By Joseph de Tonquédec

The following two articles were originally published in *Études* [*par des Pères de la Compagnie de Jésus*].[4] The first one had already appeared as a pamphlet. Two questions are addressed here. The first is of general interest and it concerns the argument for the existence of God that relies upon the ideas of order and finality. By examining these notions, I attempt to free them from their misrepresentations and to show that, considered in their purity, they are not at all vulnerable to the objections formulated in *L'évolution créatrice*.

The second question, which is more specific and current, has to do with Bergson's own opinions regarding the question of God. For the moment, this question does not have a precise and complete response. That is why I have decided to reproduce here, just like the various parts of a case file, the articles written about this question and my correspondence with Monsieur Bergson himself. Today I would only add one new piece of evidence to this case file, namely the approval given by the author of *L'évolution créatrice* to the second of the two articles. Regarding the second article, he was kind enough to write: "It seems to me to define, with much precision and clarity, the position that I take vis-à-vis the problem of the nature of God."[5] Thus, the question mark that I placed in the title of the second article ["Is Bergson a Monist?"]—and this article is not a correction of the first article, but rather a return to the question—also exists in the philosopher's thought itself. As of yet, no definitive statements have been made that would show that Monsieur Bergson is not a monist in at least some sense of the term.[6]

In this way, and at least for the time being, those hasty commentators who casually speak of a "Bergsonian monism" are merely presenting a caricatured system. But no less hasty are those who, responding to the first ones, claim that L'évolution créatrice is incompatible with monism in all of its forms. To settle this question of monism, we need not choose (as, for example, Monsieur Le Roy thinks[7]) between an immobile God and a God-in-action, if this option even has any meaning. Nor must we choose between a static conception of the real or a dynamic one. But it is necessary to take a position on the distinction between God and the world. Monism opposes dualism, and not dynamism or evolutionism. My apologies for having to repeat this once again.

* * *

"HOW SHOULD WE INTERPRET THE ORDER OF THE WORLD?" (1908)[1]

By Joseph de Tonquédec

Monsieur Bergson's previous books were primarily studies in psychology, even though this powerfully synthetic mind had already sketched out in them some features of a system of the world. L'évolution créatrice contains the full development of this system, if we can still speak of a "system" in relation to a philosophy that declares itself to be, like all things, in a state of becoming. We find in this book a philosophy that is mature and full, rich in positive knowledge, fertile in viewpoints, and unfolding itself freely into varied and novel images.[2] Perhaps his thought is even clearer here than it ever was before (and this is surely a result of its more complete point of view). When faced with so many treasures, we are forced to make a choice. I will go directly to the ideas that interest me the most—those that have to do with the ultimate explanation of things, the Absolute, and God.

What is the meaning of the world and, in particular, of the organic world? Must we see the order that it presents to us as a series of lucky accidents or rather as the result of a conscious plan [*dessein*]? Ought we to opt for blind *mechanism* or for intelligent *finality*? Monsieur Bergson declares himself to be strongly opposed to the first doctrine. The part of the book where he demolishes that mechanistic evolutionary theory— that, despite complicating itself through ingenious hypotheses and serving as a framework for many scientific observations, is basically nothing more than the very naïve explanation of things via chance—is brilliant and truly magisterial. Monsieur Bergson counters it with the "strongest argument."[3] Evolution does not take place in a single direction. Certain living beings are related, not in a direct line, but in an indirect line [*ligne collatérale*], that is, they do not represent the more or less developed achievement of a single type, but parallel types, diverse and equally fortuitous adaptations of the same plan. And these adaptations appear in very different environments and among the most varied of influences. We observe, for example, "the identical structures of extraordinarily complex organs along diverging lines of evolution."[4] Faced with these facts, mechanism declares that the primitive being, capable of very diverse variations, evolved in some determinate and identical directions under the pressure of several parallel series of fortuitous circumstances. But, Monsieur Bergson asks:

> What chance will there be that, through two completely different series of accidents being added together, two completely different evolutions would arrive at similar results? [...] The accumulation of accidental variations, such as is required to produce a complex structure, requires the convergence of, so to speak, an infinite number of infinitesimal causes. How could these entirely accidental causes reappear identically and in the same order at different points in space and time? [...] It is hardly out of the ordinary that two people out for a walk, leaving from different locations, and wandering through the countryside according to their individual whims, might eventually meet up. But it is entirely

implausible that, by wandering about in this way, they might trace out identical and perfectly superposable lines. Moreover, this implausibility will be even greater if the pathways travelled from beginning to end involve more complex detours. And it will become an impossibility if the zigzags of the two walkers are of an infinite complexity. Now, how will this complexity of zigzags compare to the complexity of an organ where thousands of different cells are arranged in a certain order and where each cell is itself a kind of organism?[5]

To reject mechanism is be oriented toward finality, and indeed all of the tendencies present in Monsieur Bergson's philosophy go in this direction. But we must expect that a thinker as personal as he does not adopt a doctrine without profoundly transforming it. And indeed, this is what takes place. The author of *L'évolution créatrice* wants to "go beyond" [*dépasser*] finalism.[6] Why? What he reproaches in this doctrine is above all a certain anthropomorphism—for having imagined the development of life, of which the mind is but a particular result, on the model of the works of the mind.[7] Life "transcends finality, if one understands finality to be the realization of an idea conceived, or at least conceivable, in advance."[8] And when it comes to life, Monsieur Bergson does not want us to speak of a "plan" or a "goal," or of means being adapted to the realization of them.

In effect, he says, life is not a combination of pre-existing elements, and it is wrong to liken life to the operations by which the mind coordinates elements into a "distinct multiplicity" in view of a certain goal.[9] Life is creative. It even creates the elements that it makes use of (you should understand by this that life transforms them in a radical way, by giving them precisely that special something that makes them apt to enter into its combinations). "[H]ere the causes, unique in their genre, are part of the effect, have come about at the same time as the effect, and are determined by it as much as they determine it."[10] We will never reach life by beginning from non-living elements and combining them in even the most ingenious ways. We will never construct the unity of a genuine individual by taking the isolated elements as given. As such, we must say that each moment of life is an absolute novelty, that life "goes beyond itself" at each instant. And this is why its effects are unpredictable. For we cannot know what does not yet exist in any way.[11]

In the second place, the order of the world does not resemble the order that would be established by an intellect such as our own. Radical finalism is too perfect to be applied to the world such as it is.

> The philosopher who began by positing, in principle, that each detail is linked to an overall plan will, the day he decides to examine the facts, go from disappointment to disappointment. [...] [F]ailure appears to be the rule, and success as both exceptional and always imperfect. [...] [T]he species and the individual think only of themselves, from which a potential conflict arises with the other forms of life. Harmony thus does not exist in fact.[12]

Finally, the very idea of looking for the reason behind the order of nature is itself deceptive. "Something that was not judged to be contingent would not require an explanation."[13] Now, it is precisely the case that the order of nature is not contingent. *There must be* some sort of order, and an absolute disorder is inconceivable.[14]

The conclusion of all of this is the affirmation of an ordering force, immanent to the world, that will be named "life," or "consciousness," or "supra-consciousness," and which, having been launched though matter, brings to it order and progress.[15] This force is not a thing: "There are no things; there are only actions."[16] Thus, it is an action, or better, a tendency to action, a "need for creation," an *élan* that continuously seeks to go beyond what it has attained. Moreover, it is essentially an "immaterial" and "extra-spatial" process, and even though it operates within matter, it is a reality of a "psychological order."[17] And these last terms give the reason why life can contain a plurality of virtual forms, which coexist within life, just as the multiple thoughts that exist within a single mind.[18] They also help us to understand why life's work can be fully assimilated to the actions of our freedom,[19] or to say it better, how these free actions are the conscious part of life. In fact, the creations of our inner freedom and those of nature emerge from the same current of life; it is the same effort that fashions organisms and that appears to itself in consciousness.[20]

Let me add that through an extremely subtle dialectical process, which I cannot fully analyze here,[21] Monsieur Bergson is ultimately able to derive the existence of matter from the *élan vital*. This is how we arrive at the "principle or source of all life as well as of all materiality."[22] So is this the Absolute, and are we here in the presence of the God of Bergsonian philosophy?

Yes, it seems so.

> If the same kind of action is being accomplished everywhere—an action that is sometimes unmaking itself and sometimes attempting to remake itself—then I simply express this likely similarity between worlds when I speak of a center from which these worlds would shoot forth, like the sparks shooting out from an immense firework [*bouquet*], so long as I do not present this center as a *thing*, but rather as a continuity of shooting forth. God, thus defined, has nothing of the ready-made; he is unceasing life, action, and freedom. Creation thus conceived is not a mystery; we experience it in ourselves the moment that we act freely. [...] The Absolute is then revealed to be very close to us and, to a certain extent, within us. The Absolute is psychological in essence, and not mathematical or logical. It lives with us. And like us—though in certain ways infinitely more concentrated and infinitely more condensed into itself—the Absolute endures.[23]

If my understanding of the sense of these passages is correct, then Monsieur Bergson seems to adopt the monist theory, with certain adjustments that make it somewhat more plastic and more supple. His is a dynamic and evolutionary monism. It is also a monism that is clearly finalist. Although he avoids qualifying his system in this way, Monsieur Bergson himself recognizes this to be its orientation. And what he ultimately rejects is simply the action of a superior intellect on the world, not a finality immanent to the things. Let us add that he denies himself the option of hypostasizing and localizing this finality in some distinct realities: vital principles, forms, entelechies, etc. For him, the finality is everywhere; it is diffused throughout the universality of becoming.

[...][24]

* * *

LETTER FROM BERGSON TO JOSEPH DE TONQUÉDEC (MAY 12, 1908)[1]

... I discuss God[2] as the *source* from which spring forth, one by one, through an effect of its freedom, the "currents" or "*élans*" that will each form a world. God would thus remain distinct from those worlds, and we cannot say of him that "most often this effort is cut short" or that he "remains subject to the materiality that it has had to take on."[3] Finally, the argument by which I establish the impossibility of nothingness is in no way directed against a transcendent cause of the world. To the contrary, I explained that it aims at the Spinozist conception of being.[4] It only ends up showing that *something* has always existed. True, when it comes to the nature of this "something," it offers nothing, no positive conclusion. But neither does this argument say that what has always existed is the world itself, and the rest of the book explicitly confirms the opposite.

[H. Bergson]

* * *

"IS BERGSON A MONIST?" (1912)[1]

By Joseph de Tonquédec

Readers of *Études* [*par des Pères de la Compagnie de Jésus*] have probably forgotten an article that I published four years ago on the topic of *L'évolution créatrice*.[2] That article having had the good fortune of receiving some comments from the author of the book, along with my loyalty to the article and my continued interest in the question, I am inspired to return to it here. It seems to me that the criticisms I formulated regarding the negative part of the work remain intact— the clarifications offered [by Bergson] do not at all touch the defense of finalism that I had presented. Rather, they bring to light Monsieur Bergson's positive conception of God. I will thus briefly explain the reasons that led me to interpret his conception as I did, examine if another interpretation is possible, and, finally, with the benevolent authorization of Monsieur Bergson himself, provide here the passages in which he offers his corrections.

The mention of God, of the Absolute, is rare in Monsieur Bergson's works, and it can be said that the difficult problems that come with this subject are directly addressed nowhere. By collecting together the hints, bringing together those passages, and looking in the direction that they seem to indicate, I believed that I had seen a monist hypothesis taking shape in his works.

In truth, Bergson places the mind, or consciousness, at the origin of everything. He continually speaks of creation, and he presents creation as an act of freedom. But we must decide upon the sense of these terms. As for creation, Monsieur Bergson attempts to bring it down from the heavens to the earth, from the transcendent to the immanent. He always defines it in relation to what creates itself, and not in relation to the one who creates. To read him, one would say that creation is an integral part of our world, without having any connection to a beyond. "Creation thus conceived is not a mystery; we experience it in ourselves the moment that we act freely,"[3] and we recognize creation outside of ourselves in the unexpected treasures and in the inexhaustible inventions of life. In fact, creation is not the passage from nothingness to existence, but rather the "idea of creation [...] merges with the idea of growth."[4] Creation is novelty. Everything that is new, that is not equivalent to some pre-existing elements, that in some way goes beyond what was previously given, is something created. Understood in this way, creation is the universe itself and the very stuff of all reality, since the concrete—that which alone exists—never repeats itself and continually changes.[5] "[R]eality is a perpetual growth, a creation that continues on endlessly. [...] The universe endures. [...] [D]urée signifies invention, the creation of forms, and the continuous development of the absolutely new. [...] [R]eality appears as an uninterrupted streaming forth of innovations where each one, having no more than sprung forth to make the present, has already receded into the past."[6]

If novelty is synonymous with creation, then creation is synonymous with freedom. Evolution is creative—and by that very fact, it is free. Taken in itself, and not at all in virtue of its relation to some transcendent cause, evolution merits this epithet.

> The *élan* of life [...] takes hold of this matter, which is necessity itself, and attempts to introduce into it the largest possible amount of indetermination and freedom. [...] the creation of a world is a free act, and life within the material world, participates in this freedom. The evolution of living species within this world represents what subsists of the initial direction of the original jet [that is, of freedom], and of an impulse that continues along in the inverse direction of materiality [that is, of determinism].[7]

In this way, consciousness is at the foundation of everything, by which I mean consciousness in *itself* and not necessarily consciousness for *itself*, since it "only shows itself to itself where creation is possible. Consciousness goes to sleep when life is condemned to automatism; it awakens the moment that the possibility of a choice is reborn."[8] The *élan vital*, that primitive jet is, in its very essence of the same nature as human consciousness. It is full of "psychical virtualities."[9] It is "consciousness, or supra-consciousness."[10] Here is the creative principle then. Consciousness is a *"need for creation* [...] consciousness, or better supra-consciousness, is at the origin of life. Consciousness or supra-consciousness is the firework whose extinguished debris falls into matter."[11] It is thus "the principle of all life as well as of all materiality."[12]

Such is cosmic reality. To grasp it is to reach the "absolute": the absolute, no longer in the conventional sense, relative to our needs or to our action, but to the real itself and as it is in itself. Speaking of concrete *durée*—in the sense of the time required for a lump of sugar to dissolve—Bergson writes: "It is no longer something thought; it is something lived. It is no longer a relation; it is something absolute."[13] And elsewhere he writes: "The Absolute is then revealed to be very close to us and, to a certain extent, within us. The Absolute is psychological in essence, and not mathematical or logical. It lives with us. And like us—though in certain ways infinitely more concentrated and infinitely more condensed into itself[14]—the Absolute endures."[15]

The doctrine presented up to here appears to be the precise account of experience, rigorously modeled upon the empirical givens. Even if something in it takes on the air of a transcendental speculation, perhaps due to the use of the words God and Absolute, the author is careful to alert us that he always holds himself to the same (empirical) region and that what he says is only justified there. It is experience, extended by hypotheses immediately grounded upon it, that allows us to see that "the same kind of action is being accomplished everywhere."[16] The metaphysical formulas and the metaphors mean nothing else: "I *simply express this likely similarity* between worlds when I speak of a center from which these worlds would shoot forth. [...] God, *thus defined*, has nothing of the ready-made."[17]

This last passage is of the utmost importance for the question we are considering. It is the most explicit passage that Monsieur Bergson has written on God, but we will not properly understand it except as the outcome or the crowning achievement of an entire system. I proposed an interpretation of that system that I believed to be precisely in continuity with the ideas whose summary you have just read. I thought that I was following it closely by translating it as follows. To speak of a center of springing forth is still not to leave experience behind, but rather to present in shorthand the image of a universal springing forth, continuous and everywhere similar to itself "simply express this likely similarity…" The word God, here, cannot mean anything else, since its meaning is circumscribed to mean just this: "God, thus defined…" And how has he been defined? By exactly the same characteristics as the *élan vital*: "Continuity of springing forth… unceasing life, action, and freedom, [He] has nothing of the ready-made." Nowhere do we perceive a creative act heterogeneous to what is created. The "immense reservoir of life" from which creative jets incessantly spring forth is not beyond the world and is not distinct from life itself. It is but life accumulated in view of future creations.[18]

> When considered in its initial impulse and prior to any scission, life was a tendency to accumulate something in a reservoir (as above all the green parts of plants do), with an eye toward an instantaneous and effective expenditure (such as the ones animals accomplish), that would have simply dissipated without it. [...] Imagine a container full of steam at a high pressure [*tension*] [...] jets must ceaselessly shoot forth from an immense reservoir of life, where each one, falling back, is a world.[19]

And so, this is what the "center from which these worlds would shoot forth" must consist in. The "continuity of springing forth" that constitutes it is but a collective noun to signify the jets that it continuously forms from its own substance. Such was my exegesis. It led Monsieur Bergson's thought toward a monist conclusion. But is another interpretation possible?

To begin with, it must be admitted that nowhere in Monsieur Bergson's works is the monist doctrine categorically affirmed. I acknowledged this by only presenting the monistic interpretation as probable.[20] Monsieur Bergson did not examine very deeply the question of God—he only made some brief allusions, in passing, that we probably should not push too far. Despite our curiosity, we should leave aside trying to find a theodicy in *L'évolution créatrice* when the author did not intend to put one there. The book is only a "cosmology": it treats of the world, of its inner genesis, of its constitutive elements, and does not directly explore its ultimate cause. Perhaps it is this concern to hold himself to the created reality of experience [*le créé expérimental*] that seems to suggest that the author admits no other reality.

It is not enough, however, that the book entirely leaves aside the supreme question. To lead me to definitively give up the monist interpretation, it would have to be shown that the little it does say, and above all what it suggests in terms of God, is not irreconcilable with a dualist doctrine. Thus, let me return to Monsieur Bergson's thought where it currently stands, where it is perhaps purely "cosmological," that is, below the level of the conclusions I have drawn, the ones that I have claimed to see taking shape, prior to approaching the question of God.

The characteristics of life immanent to the world have been recorded: a *durée* analogous to that of consciousness, an opposition to determinism and to the inertia of matter, a creative power higher than the faculty of assembling pre-existing materials. Let us work back up this stream. Everywhere along our journey we find nothing but continuity. Actual springing-forths prolong previous efforts, and they both emerge from the same "need for creation" put into motion by an "initial impulse." The center of life could not be heterogeneous to this; like it, the center will be a springing-forth too. But also, the closer we come to the source, the more we will see everything foreign to life being eliminated from it. We will grasp it more purely to the extent that the obstacles that block its pathway dissolve, i.e., the materials that is was obliged to carry along.

> If that life were pure consciousness, or better, supra-consciousness, it would be a pure creative activity. In fact, life is riveted to an organism that submits it to the general laws of inert matter. But everything takes place as if it does its best to break free of these laws. [...] The *élan of life* that I am discussing consists, in short, in a need for creation. This *élan* is not capable of absolute creation since it encounters matter, i.e., the inverse movement of its own movement.[21]

In our actual experience, we only witness a diminished or inhibited creation. Inventions of the human mind, free acts, the spontaneity of organisms:

> ...these are merely creations of form. But how could they be anything else? After all, we are not the vital current itself; we are merely this current insofar as it is already weighed down by matter, that is, by congealed parts of its substance that it carries along the length of its course. Whether in the composition of a work of genius or in a simple free decision, as much as we might tighten the spring of our activity to its highest degree and thereby create what no pure and simple assemblage of matter could have given (after all, what juxtaposition of already known curves could ever equal the pencil stroke of a great artist?), it is no less the case that here too there are elements that preexist and survive their organization.[22]

But matter, which is now seen to oppose the movement of life, nevertheless originally came out of that movement. Indeed, matter is less, not more, than life. Its characteristics—habit, mechanism, inertia—appear the moment that the activity relaxes or falls; as we have just seen, it is the "congealing" of the activity. Thus, if we were to find life in its pure state, if we could grasp it in its initial springing forth, we would see it to be completely free of material obstacles, and we would witness, no longer a half-creation operated upon pre-existing elements, but rather absolute creation that produces both form and matter. "If that life were pure consciousness, [...] it would be a pure creative activity."[23] "But if a simple halt in the action that generates the form were able to constitute its matter [...], then a creation of matter would be neither incomprehensible nor inadmissible."[24]

It is obvious where this philosophy is headed. At the origin of things, we must posit absolute creation, pure activity, and pure consciousness. Thought perceives the source, the center of springing forth, without any rupture or any interruption in the prolongation of the vital current. It imagines this center by purifying the very notion of life and activity and by freeing it from all that does not belong to it. "God, thus defined, has nothing of the ready-made; he is unceasing life, action, and freedom."[25]

Yes. Only, what is the nature of the union that connects this first reality to the ones of our universe? This is the point that remains obscure—and, nevertheless, it is the only one that could definitively resolve the fundamental question. At the center, is the springing forth itself of another nature than what we find in the jets? Would it be enough to liberate actual life, riveted as it is to matter, in order to thereby obtain the absolute activity? Is it merely through a degradation of energy, a relaxing of intensity, a falling back upon itself, that it *becomes* the world? Is it one and the same flow of action [*flux d'action*] that is God at the source and then the world afterwards? — Or rather, is there, in the transcendent regions, something radically distinct even from the purest current of created life? On these questions, Bergson's published texts do not provide the full clarification that we desire.

Nevertheless, by relying here on one of the fundamental principles of Bergsonian philosophy, which is so attentive to marking differences of quality and the irreducibility of essences, we reach a conclusion that is already quite considerable: there is no substantial identity between life and its source, nor could there be one. Let us not forget that Monsieur Bergson is the ardent adversary of Spencer's evolutionism, and of any system that composes the world out of homogeneous elements in various combinations, that is, out of some unique and accidentally modified substance. According to Monsieur Bergson, there is heterogeneity between the diverse actual creations of life. There would be an even more decisive heterogeneity between the cosmic life that experience shows to us and the pure life that we only grasp by way of an effort of thought.

And if Monsieur Bergson acknowledges, beyond even the pure current of life that creates itself, a source that is uncreated, a transcendent primordial impulsion, then this would really be infinitely different from all the rest.

Unfortunately, other aspects of Bergson's thought, and ones that are no less salient, make it impossible for me to settle on this interpretation. In fact, he is attached to continuity just as much as he is to heterogeneity. In the actual world, the varied forms of life do not block the unity of the *élan vital*. Creation, as we have read, is in its essence nothing other than growth. It consists in the fact that one reality—rather that producing a distinct reality by radiating, so to speak, existence out around itself—rather modifies itself and grows intrinsically.

> The idea of creation remains entirely obscure so long as we think [...] of a *thing* that creates and of *things* that would be created. [...] But

that action grows as it advances, that it creates as it progresses, this is something we can all see when observing ourselves acting. [...] But it is not inadmissible that a reality of an entirely different order—a reality that is to the atom what the poet's thought is to the letters of the alphabet—might grow through sudden additions.[26]

First of all, these passages protest against any attempt by the understanding to substitute constituted things or objects for the fluid becoming that is reality. But they go even further. For the opposition here is not merely that between the dynamic and the static, but rather between the different modes of genesis: one being the production of a reality entirely distinct from the producer; the other a simple growth within the producer. Monsieur Bergson absolutely rejects the first option. Thus, the supreme cause can only create by developing itself [en se développant]. True, we would no longer have a monism through identity and homogeneity[27]—since every new state is irreducible to the preceding ones—but rather a monism through growth and becoming.

Every philosophy of God must pass between two dangers: either isolating God from the world to the point of making him a foreigner there; or uniting him to the world to the point of misunderstanding the infinite difference that separates them. This is an antinomy: there must be a continuity between the current and its source; and there must be, a rupture that would be an abyss between the Absolute and the rest of the universe.[28] For my part, I find that the abyss is not deep enough in Monsieur Bergson's thought... but, on the other hand, the distinction that exists between God and the world is unique in its kind and thereby difficult to define. When it is boldly claimed that the distinction is tanquam inter rem et rem [as such, between two things], we forget that God is not res [a thing], in the ordinary sense, that the notion of being cannot be univocally applied to creations and to Him, and that there is, between them and Him, an extremely intimate and profound union, which is also unique in its kind and that could not exist inter rem et rem.

And after all, one cannot require that a few isolated words (found in a work treating of another topic) contain all of the precisions required of a completed theodicy. Nothing thus requires us to attribute an error to Monsieur Bergson that his work does not formally imply.

Let us conclude. (1) Bergson's thought advances in the direction in which it must encounter the primordial Cause, for it is surely by going back up the current of life that we find He of whom the Scriptures say: *apud te est fons vitae.*[29]

(2) According to the author of *L'évolution créatrice,* the source of life is certainly distinct from the current world and from everything that resembles it in the history of evolution. (3) But there are many types of distinctions, and Monsieur Bergson does not tell us which one that he adopts. We cannot divine, by simply reading him, whether he takes God to be the name for a reality that will *become* the world, or if the word designates something or someone more withdrawn into the beyond.

Behind the published texts, there is the author's personal thought. Monsieur Bergson did not recognize his own thinking in the interpretation that I offered, four years ago, of his paragraph on God—the interpretation that is summarized in the first paragraphs above. He was kind enough to send me a correction, which I thought of publishing, both out of loyalty to him and also as perhaps useful for readers of a work that is full of rays of light and shadows. I asked for his permission to do so, assuring him that I would happily add to them all of the complements and clarifications that he judged to be necessary to place around them. Along with the requested authorization, I received precious additional explanations that I believe must first be placed before the reader's eyes, since they clarify Monsieur Bergson's own general idea of philosophy and the value of the conclusions to which he has arrived.

[…][30]

The reader will decide if the second interpretation proposed in the current article coincides with the thought expressed in the two letters. It seems to me that my interpretation and these letters both lead to more or less identical conclusions. Beyond the cosmology presented in *L'évolution créatrice,* there is a place for a theodicy, and Monsieur Bergson seems to have already sketched out some of its main lines. I think that these ideas regarding God, as incomplete as they may be, or rather because of their very incompleteness, still lend themselves to equivocation. Whatever the case may be, according to Monsieur Bergson, they do not suffice (even if developed) to define God. They simply open a perspective in which the meditation on moral problems might finally bring out the great revelation.

* * *

LETTER FROM BERGSON TO JOSEPH DE TONQUÉDEC (FEBRUARY 20, 1912)[1]

… It goes without saying that you are fully at liberty to publish the passage to which you allude from the letter I sent you previously and in which I clarified the meaning of a couple of sentences from *L'évolution créatrice* relating to the nature of God. […] You would like to know if I have anything to add? As a philosopher, I do not see anything to add for the moment because the philosophical method, such as I understand it, is rigorously modeled upon experience (either interior or exterior) and does not allow the proclamation of a conclusion that goes beyond the empirical considerations it is grounded upon. That is why my works have (at least sometimes) been able to inspire some confidence in minds that had previously been left indifferent to philosophy—I have never allowed any place in my works for what is simply personal opinion, or for convictions incapable of *being made objective* through this particular method.

Now, the considerations presented in my *Essai sur les données immédiates de la conscience* were able to bring into the light the fact of freedom. And I hope that those found in *Matière et mémoire* directly touch upon the reality of the mind. The considerations found in *L'évolution créatrice* present creation as a fact. From all of that clearly emerges the idea of a creative and free God, generator of both matter and life, and whose creative effort extends, on the side of life, through the evolution of species and through the constitution of human personalities. But as a result, what emerges overall is the refutation of monism and of pantheism in general. But in order to further clarify these conclusions and to say more about them, it would be necessary to examine questions of an entirely different order, namely moral questions. I am not sure that I will ever publish something on this subject; I will not do so unless I arrive at some results that seem to me as demonstrable, or as "presentable," as those found in my other works. Everything that I would say in the meantime would be alongside, or even outside, of what I take philosophy to be, philosophy being for me something that is constituted according to a clearly defined method and that can, thanks to that method, claim an objectivity as large as that of the positive sciences, although this is an objectivity of an entirely different nature.

V

NOTABLE COMMENTARIES

CANGUILHEM AS READER OF *L'ÉVOLUTION CRÉATRICE*

French philosopher of science and historian Georges Canguilhem (1904–1995) was an influential figure in twentieth-century French thought, both for his own theoretical contributions and for the significant influence he had as a pedagogue and point of contact between French philosophers, such as Michel Foucault, Louis Althusser, Jacques Lacan, Pierre Bourdieu, Gilbert Simondon, Jacques Derrida, etc.[1] The following lecture notes from Canguilhem's course on Chapter III of *L'évolution créatrice* were written in 1941–1942 for a course he taught preparing students for the *Agrégation* in philosophy at the Université de Strasbourg (relocated at the time to Clermont-Ferrand because of World War II). As Giuseppe Bianco notes in his introduction to the French publication of these course notes, Canguilhem was also teaching two other courses that year: one on "Questions of Method in the Natural and in the Moral Sciences," and another on "The Causality of Time," which included long sections on time and causality in Bergson.[2] Thus, it is worth noting that Canguilhem's in-depth study of Bergson coincides quite closely with his work on his most famous book, *The Normal and*

DOI: 10.4324/9781315537818-12

the *Pathological*, published in 1943. This is even more remarkable given that Canguilhem had been influenced by (and even contributed to) the anti-Bergsonian atmosphere of the early 1930s.[3] As Bianco documents, Canguilhem offered significantly more positive commentaries on Bergson in the late 1930s, which culminated in his course. Bergson (and particularly *L'évolution créatrice*) would continue to be an important influence on Canguilhem work.[4]

* * *

COMMENTARY ON CHAPTER III OF *L'ÉVOLUTION CRÉATRICE* (1943)[5]

By Georges Canguilhem

Introduction[6]

It is a good when a project is accompanied by an awareness of the difficulties it faces. This is why I would like to say a few words about the difficulties that arise with any attempt to explain one of Bergson's texts, as well as about the help in this task that we can expect from Bergson's commentators and from Bergson himself.

Bergson was often asked (or often found himself obliged) to express his opinion regarding the value of some interpretation of his doctrine or regarding some element of his doctrine, a bit like Valéry, who was asked to give his opinion regarding the pertinence of some commentary on his poems. Bergson (again, like Valéry) always denied himself the power to recognize the precise embodiment of his own thought in the versions of it offered by his admirers or by his critics. But this refusal to recognize patronage did not have the same reasons for the philosopher as it did for the poet. For Valéry, the opinion of the artist in relation to his finished creation was valuable (or not valuable) in exactly the same way as the opinions of the spectators or the critics, since it is nothing but one more retrospective opinion among the others.[7] Bergson, to the contrary, always seemed to maintain that he himself—and probably he alone—held the key to the overall understanding [*intelligence totale*] of his work. In short, even the most intelligent interpreter does with Bergson's philosophy what this philosophy claims to do with regard to the real, namely an effort of sympathy. Now, since sympathy is not a transfer of love, but a respect of the originality and the alterity of its object, Bergsonian philosophy is condemned to be in part misunderstood by the very minds whose interest it most merits, in proportion to their own vigor and their own originality. Bergson writes in a letter:

> It is impossible for an original thinker to enter *fully* into another's viewpoints. You will thus not be surprised if I tell you that, in the general presentation that you have given of my work, there is not a single chapter to which I can entirely subscribe, not a single critique, therefore, that appears to me to bear exactly upon what I said or at least on what I thought (since we are never fully certain that what we think has really passed over into what we have said).[8]

Nevertheless, it is well known to what extent Bergson valued at least two of his commentators: Thibaudet and Jankélévitch.[9] Their works are thereby indispensable for any serious study of Bergson. Let me simply say at the outset that I own much to them, but for a project that is rather different from their own. They set themselves the goal of discovering the meaning of a doctrine and of estimating its value by situating it among other doctrines. My goal is to find the meaning of a fragmentary text and to situate it within the development of the doctrine itself.

To be sure, the first authorized interpreter, if not ultimately the only one, must be Bergson himself. And so our first obligation must be to clarify, as often as possible, one of Bergson's texts by another of his texts. Someone might object that this procedure is no more particularly appropriate here than anywhere else. Every doctrine tends to constitute something like a self-enclosed universe of concepts that is quasi-spherical and in which each point can be related to every other point. Certain philosophers, such as Spinoza in his *Ethics*, have even carried out this task on their own. But it must also be admitted that for many others, the discursive and hence necessarily successive exposition of their thought does not proceed without some variations, such that the question of the unity of their doctrine is often posed. After all, are we not led to ask of many philosophers (such as Comte, or Nietzsche) if they might not have several philosophies? For his part, Bergson always campaigned for a "philosophy that endures," that progressively and even unforeseeably creates itself, not only from one philosopher to the next, but even for the same philosopher. Thus, from the moment that philosophical discourse is nothing but an always inadequate approximation of the intuition that it masks at the very moment that it designates it, there is no *a priori* vice in the method of explaining one Bergsonian text by another text, even one that comes later, since the later one has some chance of leading us closer than the first to that source of springing-forth [*source jaillissante*], which is intuition. By discovering the source behind a curtain of words that is less dense [*touffu*] and less opaque than the first text was, we will not have altered the meaning of the text since to explain is less to substitute one discourse for another than to aim at (through the words and by their means and in converging directions) a unique and simple point of springing-forth [*jaillissement*].

In terms of at least five important passages in Chapter III of *L'évolution créatrice*, we have at our disposal important clarifications from Bergson himself:

(1) Concerning the relations between science and metaphysics, we will compare passages from *L'évolution créatrice* (212–18) with passages from *La pensée et le mouvant* (in particular: "De la position des problems," DPP, 42ff.);

(2) In terms of the question of the *absolute* value of science in a domain limited to inert matter, and inversely the *relative* value of a mode of knowledge illegitimately extended to what it was not designed for (*L'évolution créatrice*, 216ff.), we can refer back to a passage from the letter to Høffding available in *Mélanges* (particularly pages 1149–50);

(3) Concerning the geometry that is naturally inherent to the intellect (*L'évolution créatrice*, 230ff.), we can consult Bergson's 1908 response to Émile Borel: "À propos de l'Évolution de l'intelligence géométrique," in *Mélanges*, 753–58[10];

(4) Concerning the nature of the *élan vital*, we will find a very important clarifications: in a long passage in *Deux sources de la morale et de la religion* where the author notes the positive and negative ideas connected in this singular term (*DS*, 115–20); in a letter from Bergson to Floris Delattre, published by Delattre in his article "Samuel Butler et le Bergsonisme," in which the relationships between the concept and the image are clarified (*Mélanges*, 1522–28); and the report by Monsieur Jean de la Harpe of a conversation he had with Bergson in September 1936[11];

(5) And in terms of the final developments in Chapter III on individuality, the relations between consciousness and the body, and the metaphysical meaning of spiritual life, we will find passages capable of offering the best clarifications in *L'énergie spirituelle*, see particularly "La conscience et la vie" (1911) [CV], "L'âme et le corps" (1912) [AC], and "Le cerveau et la pensée" (1904).

[...]

The Genesis of Matter

Swiss philosopher Jean de la Harpe reports that Bergson made the following claim to him during a conversation in September 1936:

> I wrote each of my books by forgetting all of the others. I immerse myself in mediation upon a problem; I begin from *"la durée"* and I try to clarify the problem either through contrast or through similarity with it. Unfortunately, you see, my books are not always coherent with each other: "time" in *L'évolution créatrice* does not fully "adhere" to the "time" in *Essai sur les données immédiates de la conscience.*[12]

These words authorize the reader of *L'évolution créatrice* to formulate, in the same way, an impression that is impossible to easily reject: that "matter" in *L'évolution créatrice* does not "adhere" to the "matter" in *Matière et mémoire*. This fact was noted by Jankélévitch in Chapter IV of his book *Bergson*.[13] Or again, more recently, by H. von Balthasar:

> We could draw two diametrically opposed systems from Bergson. The first would have its center in the pure intensity of *durée*, freedom, and creation, with its circumference being spatial and intellectual disintegration. Creative Mind is, at the summit of the pyramid, the point that concentrates within itself the entire intensity of being; pure matter would serve as the base, but would remain always out of reach. Nevertheless, a second interpretation—which Bergson himself suggests to us through his image of an inverted cone[14]—would identify the maximum of intensity in pure perception, that field of action the least impregnated with memories, whereas the maximum of relaxation would be found in the spiritual memory, which would be the pure contemplation of time.[15]

Bergson himself warned us of this impossibility of going, without some effort of adjustment, from one notion borrowed from one of his works to the very same notion borrowed from a subsequent work. He gave the following reason:

> Tension, concentration, these are the words by which I characterized a method which required of the mind, for each new problem, a completely new effort. I should never have been able to extract from my book *Matter and Memory*, which preceded *Creative Evolution*, a true doctrine of evolution (it would have been one in only appearance); nor could I have extracted from my *Essay on the Immediate Data of Consciousness* [*Time and Free Will*] a theory of the relations of the soul and the body like the one I set forth in *Matter and Memory* (I should have had only a hypothetical construction).[16]

If it is thus impossible to reduce the matter of *L'évolution créatrice* to the matter of *Matière et mémoire*, the reason for this must be sought in the relation between matter and the evolution of life. In *Matière et mémoire*, Bergson *studies the relation between the matter of the individual organic body and individual consciousness, whereas in* L'évolution créatrice, *it is a question of the relations between cosmic matter and universal consciousness in its effort toward organic development* [organisation], something that Jankélévitch very clearly observed.

*

In *L'évolution créatrice*, just as in *Matière et mémoire*, Bergson defines matter in opposition to mind [esprit]. This opposition takes place here from two points of view: durée and extension. We have an experience of the mind by way of our personality, and so we must begin from personality.

I Personality

(A) *From the point of view of durée*,[17] our personality is simultaneously growth and tension, synthesis of memory and will, the bridge between the past and the future, in short, contraction, an effort to regain possession of itself and to coincide with itself. This effort is never completely successful, nor ever definitively completed; it admits of degrees and personality appears as a limit, a promise, and not a possession.

(B) From the point of view of extension,[18] our personality is a process of concentration parallel to this effort of contraction. We feel ourselves becoming "sharpened" more than merely pointed [punctiformes] (the dynamic and instrumental image of the sharp edge [pointe] wins out over the static and mathematical image of the point). To be a person is to be capable of piercing [poindre] into the future, like light into the night. A sharp edge [pointe] is an instrument for insertion or insinuation ("life proceeds by insinuation")[19] and of division or ruse ("It is as if freedom had to divide in order to rule"[20]; "It was an attempt to create out of matter…").[21] Two objects are suggested by the image of the sharp edge [la pointe]: the cone and the knife. The image of the cone is used in Matière et mémoire, but in order to illustrate the opposite theory to the one being presented here.[22] The image of the knife is used in L'évolution créatrice: "[T]he brain is the sharp tip by which consciousness penetrates into the compact fabric of events, but it is no more coextensive with consciousness than the sharp edge is coextensive with the knife."[23]

We will compare the analyses found on these pages[24] with Chapter III of Essai sur les données immédiates de la conscience, and in particular, the two aspects of the self [moi] with Bergson's lecture "La conscience et la vie" from May 29, 1911, where the relations between the tension of durée and the capacity for action [puissance d'agir] are discussed,[25] along with the substantial summary by J. Grivet of Bergson's course on La Personnalité (from 1910–1911) at the Collège de France, originally published in Études.[26] In these lectures, Bergson defines personality as a movement accumulating experience along its way and pushing this accumulated past into the future. If one looked to dynamics for an analogy, treating the personality as a living force, we could say that its alterations are the result either of its mass (memories [souvenirs]) or its speed (élan). "It will be exhausting to be a person; this privilege must be bought at the price of a continuous effort, to which, in certain psychological conditions, the subject will refuse to take part."[27] The rest of the time, personality disorders are most often caused by either a lifting of the inhibition of the intellectual functions or a corruption of logical thought. Inertia is the price of clear-sightedness. True, the intellect remains within its role by calculating possibilities, but they are infinite in number. At some point it must nevertheless move on, for there are other things to do: "The thought that one has other things to do is the great motivator for the action that is being done."[28] Personality is a possibility of continuous creation. And that is what life was seeking by tending towards the human form.[29]

II Materiality

(A) From the point of view of durée.[30]

The relaxation of the effort that is the personality is the abolition of memory and will, and ultimately this would be inertia or passivity and discontinuity or reiterated instantaneity. Matter is the inability to link the present to the past, to contract habits, or to use the present by pushing it into the future. Matter is nothing but falling—no past holds it up and it itself holds up to no future. There is an analogy here with Leibniz's conception in his *Theoria motus abstracti* and *Theoria motus concreti* ("...*corpus, mens momentanea, seu carens recordatione*" ["every body is a momentary mind, or one lacking in *recollection*"].[31] The analogy extends even into the way of presenting the experience of matter as an "interruption" of the experience of the mind: "We experience, in ourselves, a state where we remember nothing and where we have no distinct perception, as in periods of fainting, or when we are overcome by a profound, dreamless sleep."[32]

Matter is thus here first and foremost the product of oblivion [l'oubli]. In *Matière et mémoire*, Bergson cites Ravaisson's phrase on the materiality that begets oblivion in us.[33] But it seems that Bergson here reversed the relation between the two notions: it is oblivion that puts matter outside of us. In *Of Habit* [1838], Ravaisson conceives of matter as being outside of time: "[H]abit is not possible within the empire of immediacy and homogeneity that is the Inorganic realm. [...] Body exists without becoming anything; it is in some sense outside of time."[34] Bergson thinks that if the body becomes nothing, then it is nothing; and if it is outside of time, then it does not exist. This is why he says that "matter is a limit, the ideal term of an inclination."[35] This analysis surely resolves matter into a *durée* of vibrations that is fleeting, but it is not nothing [nulle].[36] Now, this idea of the fleeting limit is also there in Ravaisson, though in the later Ravaisson of the 1867 report on French philosophy (to which Bergson refers in a subsequent note on Plotinus).[37] Ravaisson writes:

> From the internal and central perspective of self-reflection, the soul does not exclusively see itself (as if in its very foundations) in the infinity from which it emanates; rather, it sees itself and recognizes itself as more or less different from itself, degree by degree, right up to these extreme limits where all unity seems to vanish into the

dispersion of matter and where all activity seems to disappear beneath the linking together of phenomena. From this perspective, and since we find in the soul everything that develops in nature, we can understand Aristotle's sentence that says the soul is the place of all forms. And since all objects appear to us as representing, through forms in space, the phases that the soul goes through in the succession of its states, we can understand the sentence in Leibniz where he claims that the body is a momentary mind.[38]

And yet in 1904, Bergson presented an homage to Ravaisson (recall that *Matière et mémoire* is from 1896 and *L'évolution créatrice* is from 1907), who he succeeded as a member of the *Académie des Sciences morales et politiques*.[39] In this homage, Bergson emphasizes, with much recognition, the profound influence that these pages from the 1867 *Rapport* on the nature of matter exerted upon the minds of his generation. The summary that Bergson gives of those pages[40] should be compared with the passage under consideration from *L'évolution créatrice*.[41] In particular, note how Bergson uses, and emphasizes, the term *distension*. It seems that this is the first appearance of this notion, which is destined to play such an important role in Chapter III of *L'évolution créatrice*.[42] In *Matière et mémoire*, it is only a question of *tension* and of *extension*,[43] but *L'évolution créatrice* inserts *relaxation* [détente] between tension and extension.

To summarize, from the point of view of *durée*, mind and matter do not appear as things but as movements, not as substances but as tendencies. The second of these movements comes from the first by means of interruption and since it is the inverse of it, we must posit that *interruption equals inversion*. The mind is not content to simply exist at the moment it stops acting; rather, it is canceled in the form of matter. The abdication of creation is not rest; it is the beginning of destruction. We can no longer say that "*sublata causa, tollitur effectus*" ["when the cause has been canceled, the effect is canceled"], but rather "*sublata causa, evertitur effectus*" ["when the effect has been canceled, the cause is canceled"].[44] To demonstrate that inversion is identical to interruption, Bergson will have to rely upon the following intermediaries: the spatial or geometrical order is a negative order, and as such, it is merely the suppression of a positive order.

What are the relationships, up to this point, between *L'évolution créatrice* and *Matière et mémoire* in terms of matter and mind? In *Matière et mémoire*, the life of the mind is also presented as a perpetual oscillation between two extreme limits[45]: the plane of action or of pure sensorimotor life, and the plane of dreams [rêve] or of pure imaginative life. Pure sensorimotor life is life in the pure present or pure impulsiveness; pure reverie is life in the past, for nothing but pleasure.[46] In fact, we always live between these two limits with more or less good sense and happiness, and these degrees correspond to degrees of "contraction" or "tension."[47] But in Matière et mémoire, personality is on the side of dilation, and impersonal banality is on the side of habit and the sensorimotor organs, that is, on the side of the contraction [resserrement].[48] These degrees of tension are the diverse mental dispositions.[49] In Matière et mémoire, it is not a question of inverse tendencies.

(B) Materiality from the point of view of extension.[50]

(1) Personality is an aggressive sharp edge [pointe], materiality is a withdrawal, a letting-go [un abandon]. Bergson here asks dreams [rather than action] to introduce us into matter. Personality is insertion into the future, dreams are withdrawal, "checking out," disinterestedness.[51] Here, in L'évolution créatrice, dreams are presented for the first time as a descent in the direction of space. We arrive at this idea in three stages: first, dreams are a scattering out [éparpillement], then a decomposition, and finally an exteriority. This idea remains established [in Bergson's thinking] in the years following 1907, since in 1908 he writes: "Cease desiring, disconnect yourself from life, become disinterested, and by this very act you go from the waking-self [moi de la veille] to the dream-self [moi des rêves], less tensed but more extended than the other one."[52] This idea does not appear in his 1901 article "Le rêve," in which the myself who dreams is merely a myself that relaxes [se détend]. And in 1896, in Matière et mémoire, it is the action of the living being that is responsible for the decomposition of exteriority, that is, it is perception and not dreaming that introduces us into matter.[53] By evolving toward activity, the state of the soul comes closer to extension.[54] Inversely, dream is the evocation of the past in the form of an image.[55] Thus, it is the function of uselessness and, for a being condemned to action and to using, dreaming requires an effort[56] (whereas in L'évolution créatrice, dreaming is a letting-go [laisser-aller]).[57] By connecting us to our past as past, dreams give us the experience of the unextended and of the spiritual, along with the experience of pure memory, which is impotent and useless.[58]

L'évolution créatrice presents materiality as a permanent threat that personality bears within itself, because personality is a *plus* and extension a *minus*, and because the passage from one to the other is simply a deficit. It seems to me that we must look in this idea for the explanation of Bergson's reversal of the relation between materiality and dreams. Some readers have noted the reversal without much amazement,[59] but its "curious and paradoxical" character did not escape the attention of Vladimir Jankélévitch.[60] Bergson's 1901 lecture "Le rêve" can, thanks to its intermediary position between *Matière et mémoire* and *L'évolution créatrice*, help us mediate the idea here. In *Matière et mémoire*, Bergson notes that space, the exteriority of parts, is not the support of movement but rather the deposit of movement, posterior and inferior to it.[61] We lodge movement within space, even though in principle space would be nothing but the exhaustion of our vision of movement. In the same way, dreaming, according to the 1901 lecture, is distinguished from perception, not through a plus which is in perception, but through a minus that is in the dream. "The dream is the entire mental life, minus the effort of concentration."[62] The dreamer is someone who is "fatigued," and it is his inability to muster an effort for the precision of adjustments that explains the "rapidity with which some dreams unroll themselves."[63] In the dream, images are above all visual, and the rapidity of the evocation of a multitude of successive events ultimately tends toward the simultaneity of a "panorama."[64] Everything happens as if, for a consciousness that is disinterested in the present and in action, everything tends toward becoming equally present. In short, it would be necessary to distinguish between the present that is the actual, i.e., the acting, the one that gives to *durée* a direction through the polarity of the past and the future, and presence, which is the presentation of all of the interior richness on a single plane of value, where the value is canceled out by equivalence. The dreamer's disinterestedness makes it such that everything becomes indifferent.[65] The attitude of indifference is what makes the indifference, that is, the equivalence or the homogeneity of objects. The spectator's position, being non-aggressive, makes the content of the myself [moi] into a spectacle. The relaxation [détente] makes the extension. Again, this kind of causality is very poorly expressed by the term "makes" [faire]. It is a question rather of letting extension be made by letting the personality be unmade.[66] This is not a question of efficient causality, but rather of "deficient causality" (to adopt Leibniz's term from the *Theodicy*, very aptly taken up again by Jankélévitch).[67]

This is one possible way of understanding the relation between space and dreams without having to violently break up the continuity of Bergson's thought. If reading the lecture from 1901 ["Le rêve"] suggests this hypothesis, it of course does not fully guarantee it. But between *Matière et mémoire* and *L'évolution créatrice*, Bergson had re-read Plotinus for his course offered in 1897–1898,[68] Ravaisson for the homage of 1904, had returned to the problem of "effort" in 1902,[69] and had reflected upon the problems posed by the conflict in physics between mechanism and energetics[70]—and his intellectual acquisition that belongs to this period is that of "relaxation" [détente], a polyvalent concept that is at once moral, psychological, and physical. Moreover, during this period Bergson elaborated the idea that consciousness abhors a void and that for it to renounce one mode of thought is to thereby pass over to the inverse mode of thought. In 1906, Bergson published in *Revue philosophique* his article on the idea of nothingness, which became part of Chapter IV of *L'évolution créatrice* in 1907.[71] But even before this, in 1904, the foundation of this thesis served as the principle argument in his refutation of the "psycho-physiological paralogism."[72] The doctrine of psycho-physiological parallelism is defined as an oscillation from realism to idealism and back, an oscillation so rapid that for vision it seems immobile, a fusion of the two systems.[73]

To summarize, *deficiency, relaxation* [détente], and *inversion* would be the notions that authorize bringing dreams and materiality together. What is common to the two is to have their origin in an undoing or failing [*une défaite*] of the mind, as well as their quality as a *substratum*. Dreams are the substratum of perception,[74] just as matter is the substratum of the personality,[75] but their quality of being *substratum* is simply identical to their quality as *substractum*—we do not go from dreaming to being awake nor from matter to mind through addition; we go from being awake to dreaming or from mind to matter by way of subtraction.[76]

(2) "In this way, our personality descends again in the direction of space. Moreover, it constantly mingles with space through sensation. But I will not insist here upon a point that I have developed elsewhere in depth."[77] Here Bergson refers back to *Matière et mémoire*.[78] It is important to distinguish clearly between *matter, the extended* [l'étendue], and *space*.[79] Matter is a possible direction of consciousness and, as such, it is extended more than it is spatial.

The extended is the content of perception, the direct product of the extension that follows upon relaxation [*détente*]. But extension, working within consciousness, always participates in the unity of consciousness—extension is undivided and this is why the extended is indivisible. On this subject, consider the following important passage from *Matière et mémoire*: "This extension, once attained, remains undivided and therefore is not out of harmony with the unity of the soul."[80] Here we encounter a certain kinship between Bergsonism and Spinozism.[81] But according to Spinoza, it is the understanding that unifies what the imagination had divided, whereas for Bergson it is the intellect that divides up the continuity of the extended.

Space, pure exteriority of homogeneous parts, or space, division into numbers, is the product of a conception that assigns to *extension* an ideal term. Space is a schema, and this schema is obtained by the extrapolation to the limit of the relaxation [*détente*] that engenders the extended by means of extension.

In short, we obtain, in the order of decreasing spiritual reality, *relaxation* [*détente*] (by which matter maintains a privative relation with mind), *ex-tension* (the material order, properly speaking, where the relation to the spiritual order is progressively forgotten), *the extended* (the past-participle signifying the complete oblivion [*l'oubli*] of past spiritual participation), *space* (exteriority of the parts to each other, and the exteriority of them all from the mind). Space is the completion of a sketch created by the intellect. We recognize here the violent character of action, the abusive character of logical thought. It is worth bringing together pages 157[–58] and 203–4 of *L'évolution créatrice* and 231–33 of *Matière et mémoire*.

The extended is thus a given of consciousness, a given by way of a letting-go [*abandon*]; space constructed by the intellect, and only suggested by consciousness. When we represent matter as divided, this is because the intellect has fully exploited a suggestion, and in this sense *spatial matter is a view of the mind*[82]; at root, matter is naturally closer to concrete unity.

Here we encounter again an idea that was already discussed in *Matière et mémoire*.[83] One might indeed bring Bergson's views here together with similar (and more recent) views that we find in Whitehead. Namely, an object is ingredient in its surroundings, it is ingredient throughout all of nature. For more on this point, see Jean Wahl's book *Vers le concret*.[84] In *La pensée et le mouvant*, Bergson himself refers to Whitehead, citing precisely

Wahl's book.[85] We have already seen that Berthelot compares Bergsonism and Stoicism on this particular point.[86] It will be noted that in *L'évolution créatrice*, Bergson is less definitive in his claim that the cutting up of matter into corpuscles, by passage from the extended to space, is a violence inflicted by the intellect upon matter: "[M]atter lends itself to this subdivision, and that by assuming matter to be divisible into parts that are external to each other, we construct a science that is sufficiently representative of the real."[87] In *Matière et mémoire*, Bergson uses images of fluids, such as whirlpools [*tourbillons*], or lines of force, rather than images of solids, such as atoms, in order to indicate the direction of the real.[88] One will find in *La pensée et le mouvant* a synthesis of these two theories that is, according to Bergson, in conformity with recent advances in quantum mechanics.[89] As such, we might say that matter, according to *L'évolution créatrice*, has more natural reality than it does according to *Matière et mémoire*. In *L'évolution créatrice*, matter is only a view or snapshot by the mind in terms of its spatial and geometrical form. According to its form in the extended, matter presents itself and lends itself to the habits of the intellect. In short, the extended is more real than space, but space is more intelligible than the extended.

(3) Here Bergson's analysis inevitably encounters Kant's theory of space.[90]

Against empiricism, Kant definitively established the *a priori* nature of the representation of space. But he did not interpret [the consequences of] this fact as aptly as he had established it.

Kant's doctrine is insufficient in that it takes space to be a ready-made form of the faculty of perception, whose necessary correlate is the thing-in-itself. Bergson reproaches Kant for a contradiction: why speak of things that you claim to have no knowledge of?[91] [...]

By equating matter and space, Kant is led to the antinomies. As soon as we consider matter to be a certain direction that is sketched out and space to be the completion of that direction, the antinomies disappear.

Kant did not understand that the intellect is inscribed within the mind, which surpasses it, and that spatial representation is only appropriate for the intellect alone. And he did not grasp that time is not at the same level as space, but higher or deeper than it. And indeed, these are the very conditions required for understanding the genesis of space. *Space is born in the mind from an interference and from a reciprocal accommodation.*

Kant invented transcendental idealism in order to escape from the problems of ontological realism (where the mind would be modeled upon the things) or from the expedients of skeptical empiricism (where there would be a natural harmony between the course of things and the procedures of the mind (cf. Hume)). In fact, another solution is possible that consists in opposing matter and the intellect, not as things and a form (that would be capable of deforming or informing), but rather as movements capable of modeling themselves on each other. The formation of the form would be correlated to the materialization of the matter. In his description of spatial form, Kant is focused on its use. Bergson thinks that we cannot understand the use if we do not understand the formation [of the form], since here as well it is the function that creates the organ. Earlier in the book,[92] Bergson qualified matter as an *organ*. We could say the same of space, which, moreover, Bergson had elsewhere qualified as *power*.[93] This is why, after the terms *form* and *schema* were used to define space,[94] it is the second of the two that is privileged by Bergson. By contrast, Kant, who defined space, in his "Inaugural Dissertation" from 1770, as "subjective and ideal; [space] issues from the nature of the mind in accordance with a stable law as a *scheme* [*schema*], so to speak, for co-ordinating everything which is sensed externally,"[95] ultimately privileged the term "form." Now, the *schema* is less a form than an indication or a direction of form.

[...]

MERLEAU-PONTY AS READER OF *L'ÉVOLUTION CRÉATRICE*

French philosopher Maurice Merleau-Ponty (1908–1961) is best known for having introduced Husserlian phenomenology into twentieth-century French thought. Like Canguilhem, Merleau-Ponty was part of a generation educated within the general anti-Bergsonism atmosphere of the 1930s, and this is reflected in the sometimes unfair criticisms of Bergson included in *Phenomenology of Perception* (1945).[1] Nevertheless, Merleau-Ponty's own reading of Bergson is better characterized as itself a "creative evolution."[2] By the time of Merleau-Ponty's Inaugural Address at the Collège de France in 1953, when he was appointed Chair of Philosophy (the position Bergson himself had previously held), his attitude toward Bergson had become more nuanced.[3] He suggests that "the best of Bergsonism" is to be found it what it promises beyond sub-jectivist readings of Bergson's notion of "intuition" as "coincidence."[4] For Merleau-Ponty, Bergson had caught sight of a means for "interro-gating mute being" and for reaching the creative or "expressive" evolu-tion of reality itself.[5] Thus, Merleau-Ponty identifies Bergson's thought with the foundational notions of his own emerging phenomenological ontology: "expression" and "reversibility."[6]

In 1959, Merleau-Ponty joined other notable figures to mark the centennial of Bergson's birth, and the title he gave to his presentation, "Bergson in the Making" [Bergson se faisant], again celebrates Bergson's ability to embrace creative becoming in all of its forms.[7] It is arguably in L'évolution créatrice that Bergson offers the most explicit elaboration of the distinction between the ready-made [tout fait] and the being made [se faisant].[8] Indeed, in the following selection from Merleau-Ponty's 1956–1957 course on "Nature" at the Collège de France, Bergson is presented as developing, from Schelling, a Romantic conception of Nature as both a "lost," undivided source prior to differentiation and an oriented movement of becoming that is nonetheless neither predetermined nor teleological.[9]

In the selection here, Merleau-Ponty offers a reading of "intuition" as going beyond merely subjective coincidence.[10] This leads Merleau-Ponty to work through some of the tensions in Bergson's earlier account of perception in Matière et mémoire so as to show intuition as a way back "to the things themselves" or to the "primordial ground" of Nature.[11] According to Merleau-Ponty, Bergson has the merit of studying percep-tion and experience not to find what neatly fits into the frameworks of intelligibility, but rather to catch sight of the creative movement of "what is at the limit of this experience," be it Nature or life.[12] Thus, L'évolution créatrice, not Matière et mémoire, provides us with Bergson's most important philosophical insight into reality and ourselves as in the making. The final discussion comparing Jean-Paul Sartre and Bergson is important for Merleau-Ponty's own position. He argues for the need to leave aside the extreme positions of positivism (Bergson) and nega-tivism (Sartre) and rather "find something at the jointure of Being and Nothingness."[13]

This lecture course demonstrates what Merleau-Ponty himself acknowledged in a 1959 interview, namely that his generation's attitude toward Bergson resulted in an important "missed opportunity"—had they read Bergson more closely, they might well have discovered much earlier many things that they later took to be their own inventions.[14]

* * *

"THE IDEAS OF BERGSON," FROM MERLEAU-PONTY'S COURSE ON NATURE AT THE COLLÈGE DE FRANCE (1956–1957)[15]

By Maurice Merleau-Ponty

Translated by Robert Vallier

Schelling and Bergson

At first blush, there is nothing in common between the Bergsonian and Schellingian theses. Schelling has the idea of a Nature irreducible to any philosophical principles like the Cartesian infinite, an obscure principle that, even in God, resists the light. He wants to retrieve the prereflexive, beyond idealism.

There is, on the other hand, a positivism in Bergson, as is shown by the critique of the negative ideas of the possible, of nothing, and of disorder. But the philosophical effort of intuition is not present in Bergson with the same character of tension as in Schelling, who passed by the school of reflexive philosophy, as the *System of Transcendental Idealism* (1800) shows.[16] In Bergson, the effort of intuition demands a tension that consists above all in postponing or rejecting habits of active life. These habits are practical obstacles that do not have the philosophical gravity of obstacles found in Schelling, for whom the subject is always tending to make itself object; it is necessary to go against Nature in order to attain intuition. Schelling's philosophy is anguished, whereas Bergson has the tendency to make problems disappear and to argue against the vertiginous anxiety of classical metaphysics. "Why is there something rather than nothing?"[17] This question that haunts Schelling has no meaning for Bergson. Finally, in Schelling, there had always been a tension between intuition and dialectic, between *positive philosophy* and *negative philosophy*. Bergson seems to install himself in intuition and to see in the dialectic only an empty play of concepts.

But Bergson's philosophy does not reduce itself to these themes.

Bergson's positivism is certain. There is in his work an effort to enter into contact with Being without encumbering himself with any negative idea. Nevertheless Bergson does not elude the idea of Nothingness. For that, Bergson would have to start from the Spinozist idea, according to which there is a force inherent to the truth by which Being is posited. But this Cartesian thinking, Bergson says, can bring about a passage only by the idea of Nothingness. Because the Cartesians want to overcome a threat of inexistence, they appeal to the existence of a logical idea. Bergson thus affirms the contingency of the world in order to be fully positive. But in doing so, the idea of Nothingness is less chased from than it is incorporated into the idea of Being.

Bergson presents philosophy as the end of anxiety and vertigo. But his tranquil affirmations are more a repression of vertigo than a true tranquility. Intuition is not always installation in Being. There is movement between the positive and negative. The intellect[18] is always necessary to pose questions, and instinct reduced to itself would not pose them: hence the work of one on the other. Within intuition, there is a tension between question and response. In *Creative Evolution*, Bergson insists on the positive value of the intellect by supporting the idea that consciousness owes its own mobility to language, and were there not language, the intellect would not have the possibility of displacing itself. Consciousness without language would be further from Being than consciousness endowed with language. On one hand, Bergson's philosophy is an addition of intuitions. But on the other hand, intuition reverts into its contrary. In this way Bergson posits at first that all perception is pure perception, since pure perception is but a limit never attained by perception. Following the movement of his own intuition, Bergson is led to reverse it. In this way, different from what appeared at first glance, his philosophy is not a philosophy of coincidence: to perceive is to enter into the thing, but to enter into the thing is to become Nature, but if we were Nature, we would "discern" nothing of the thing. The return of the subject on itself appears as an intuition of *discernment*.[19]

Bergson's philosophy is related to Schelling's because the whole of Bergson is in the idea of a unity as something that goes without saying and is primordial. Thus Bergson admits a unity of species at the origin, the unity of the vegetative and the animal; a native primordial unity, broken and expressed in what comes afterward and in relation to which the idea of Nothingness has no meaning. Being is what is primordially observed, that in relation to which all drawing back is impossible. Unity is given at the beginning more than in the development. In the same way, perception opens us to the things themselves and reveals a primordial order to us, which is a horizon that we can thus never elect as the domicile of our thinking, although it remains the obsession of the latter.

By this intuition of a primordial ground, philosophy is a *Naturphilosophie*. And it possesses all the characteristics of it:

- It posits first the problems in terms of time and not in terms of Being. Thus Bergson tells us that in terms of Being, it is impossible to think the relation of human being to the universe; in effect, we cannot understand that the universe is both transcendent and immanent to human being. On the contrary, the solution appears if we think these relations in terms of time. It is essential to my duration that it appear to me as mine and as the instrument of universal measure. When the sugar melts, in my waiting I grasp both my own duration and that of the physical phenomenon.
- It has next the idea of philosophy as empiricism, experience of the absolute: in one sense I am in it, and in another sense it is in me.
- It has finally the idea of a natural operation that is neither mechanism nor teleology, but which is analogous to that of a finite God.

Nature as the Aseity of the Thing

It is Nature as an ensemble of things that I perceive. In effect, Bergson starts from a universe of images; he decides not to confine himself to theses of realism or idealism and wants to come back to perception as a fundamental act that installs us in the things. He stands both against Berkeley's idealism, for which everything is a representation, and against a realism that admits that the thing has an aseity, but which posits that this is other than what appears. He wants to put an end to all the divisions between appearance and existence, to express the existence of the natural thing as something already there, which does not need to be perceived in order to be, and at the same time to affirm the natural unity of existence and appearance: our perception is in the very things; "it really is in [point] P, and not elsewhere, that the image of P is formed and perceived."[20] Presented with this affirmation, we said that Bergson was an animist, that he posited perceiving point P, that his universe of images was an ensemble of representations without a subject, and that he gave himself a diffuse consciousness floating in the things. Now in *Matter and Memory* he expressly rejects this interpretation, saying that it is a defect of psychologists to imagine that perception "could be in the perceived things only they had perception" and that in this way "sensation cannot be in the nerve unless the nerve feels."[21] Bergson excludes the idea that the nerve alone senses or that the things perceive; but that does not stop the things being perceived at a precise point. The contact between perception and the perceived is not a magical contact. To posit the universe of images does not mean to put souls in the things, or to take the things such as they are, then to slip souls into them. The perception of point P is at the point P when we are placed in the universe of the perceived world. The question, concerning perception, is to wonder how it appears to itself, or to think perception according to perception, and no longer according to a realist perspective. Bergson wants to make a phenomenology of it and presents it as it presents itself, independently of the concepts that metaphysics can otherwise give.

When I consider the relations of perception and things perceived by placing myself in the point of view of my perception, the things perceived do not appear to me different in kind from real things; they appear reduced only as to their possibility. "To transform its existence into representation, it would be enough to suppress what follows it, what precedes it, and also all that fills it, to retain only its exterior crust, its superficial skin."[22] The thing is not different from its representation, and its representation is not different from the thing; the unique difference between the two terms is that the thing is the representation swallowed up in effective existence, and that the representation is a thing that has lost its density. The passage from one to the other is made between homogeneous terms, by simple subtraction. The thing is total representation: "To perceive all the influences from all the points of all bodies would be to descend to the condition of a material object,"[23] or again: "In one sense we might say that the perception of any unconscious material point whatsoever, in its instantaneousness, is infinitely greater and more complete than ours, since this point gathers and transmits the influences of all the points of the material universe, whereas our consciousness only attains to certain parts and to certain aspects of those parts."[24] My perception appears to me as a decompression of total Being.

In truth, we here guess at, more than we grasp, the thought of Bergson. In order to understand it well, we have to discern the valid meaning of his thought, which is hidden behind less satisfying appearances. It seems to oscillate between a spiritualism, which would see analogues of souls in the things, and a materialism, which would make the awareness of extrinsic relations of matter come to the fore.

What guides Bergson when he posits the universe of images is the content of perceptive experience: the thing offers itself as preliminary, primordial, anterior to all perception, like a landscape that is already there before us and just as we will see it afterward Bergson, by positing the universe of images without a spectator, means that perception teaches us the things, and that in this sense, perception in its nascent state truly takes part in the things. But at the same time, I can posit this universe anterior to me only as I perceive it. Everything happens as if perception were preceding itself, as if the thing were a landscape, a spectacle that subtends consciousness. Every realist conception is constructed by borrowing from the perceived thing, from the universe of perception. Bergson thus posits consciously a paradox inherent to perception: Being is anterior to perception, and this primordial Being is conceivable only in relation to perception. But how does he think this reciprocal envelopment?

Bergson sees the decompression of total Being in the advent of my perception. There is this color, then this image of this color in perception. The relation between the color and the image must be conceived as the relation between *presence* and *representation*. On the one hand, there is the spectacle in itself; and on the other, there is the spectacle for me. We pass from one to the other by diminution, by obscuration, contrary to the philosophical tradition that wants knowledge to be light. On the one hand, there is the thing, full in all its parts, where everything counts equally; on the other, there is the tableau, where certain details are accentuated. On the one hand, there is an image that doesn't have a center; on the other, there is a putting-in-perspective, where elements appear closer and others further. The concern is still for the same image but passed over to inactuality in some of its parts, including the lacunae, the virtual regions.

By thus positing the relation of the world and perception as a relation between the full and the empty, the positive and the negative, Bergson can raise the paradox of his "universe of images." It is true that Being appears to me as primordial because perception appears to me as void of all initial presence and the perceiving subject as nothing as opposed to Being. Insofar as Nothingness is less than Being, the thing appears to me precisely as more real than perception. But on the other hand, there must be a positing of this thing; hence the legitimation of this hollow [creux], starting from which the thing is seen. We thus have to allow the priority or the simultaneity of the Nothingness that perceives.

If the exterior thing is thus posited as a sort of perception, it is not because Bergson places a soul in a material point, but rather because the thing is a more complete perception, since the representation is a less complete perception. Perception is already there, but, Bergson adds, it is neutralized in the same way, if the photo is already taken within things, it is not yet developed. Bergson thus avoids in fact the dangers contained in these formulae of a spiritualist style, but does he not risk falling into materialism? In effect, if the perception of things is a diminution of their being, a shadow of Being, then the things can only be perception—that is, lesser being. Does Bergson then divine the in-itself of the things in the being of consciousness?

In fact, by this double series of formulae, Bergson wants to signify that there is being anterior to all knowledge, which takes place at the same time as perception. We construct perceiving beings starting from the universe of "images"; but this world of "images" was already the world of a perceiving being. Idealism and realism only ever see half the things. Bergson wants to restore the entire circle, to describe a common medium of Being and perception—that is, this "universe of images" in itself, this perception in an impersonal third-person "On," without inherence to a touching individual, to one of its limits, to Being, and to the other of individual perception (partial being), and he wants to describe, within me, this ground of the real by which I "plunge" into things that have "deep roots."[25]

But does Bergson succeed in making this mystery clear by means of the intuition without movement, of vision as immobile? Perception is probably in some ways immobile intuition: when the world offers itself to us as presence and not as representation, when the thing is there, in its natural stupidity. But next to this intuition, Bergson posits that of perceived being, as the center of indetermination that introduces the possible into full Being. By this, nothingness comes into the world, and natural being loses the self-sufficiency that had at first been accorded to it. Can this second intuition simply be added to the first? Can one succeed the other in a relation of continuity? Pure perception, which coincides with the object, exists *de jure* rather than *de facto*: it would take place in instantaneity.[26] In our perceptions, there is a duration, memory, whence the budging or unsticking with respect to the real. But if our perceptions are absolutely distinct from pure perception, how can they keep pure perception in themselves and take root there? All that has been said of the relations of pure perception and our perceptions is reversed. My perception appeared at first as an impoverishment, "but there is, in this necessary poverty of our conscious perception, something that is positive, that foretells spirit: it is, in the etymological sense of the word, discernment."[27] What appeared as less, appears now as other and, in a sense, as more. The thing taken in itself would be at bottom altogether empty and inarticulate, were there not my exterior perception. Nothingness has a positive role. Without this virtual, the thing itself would be without content, without contour, indefinable, as an untaken or undeveloped photograph.

But then, not much of the initial intuition of the natural thing remains. How to conserve while overcoming it? We cannot reproach Bergson for this contradiction, but does he give it the place that it merits? One intuition has not just chased away another: there is at first a wholly positive perception (the world is there, the thing is there), then he says of it that:

> [W]e grasp, in perception, at one and the same time, a state of our consciousness and a *reality* independent of ourselves. This mixed character of our immediate perception, this appearance of a realized contradiction, is the principle theoretical reason that we have for believing in an external world which does not coincide absolutely with our perception.[28]

Here perception is contradiction realized, and Bergson is ready to make the contradiction thematic as the apparatus of our subjectivity, but he does not draw all the consequences from it:

> If there are actions that are really *free*, or at least partially indeterminate, they can only belong to beings able to fix, at long intervals, that becoming to which their own becoming clings, able to solidify it into distinct moments, and so to condense matter and, by assimilating it, to digest it into movements of reaction which will pass through the

meshes of natural necessity. The greater or lesser tension of their duration, which expresses, at bottom, their greater or lesser intensity of
life, thus determines both the degree of the concentrating power of
their perception and the measure of their liberty.[29]

It would thus be the concentration of perception that would permit
grasping another being, not by adhesion, but by contraction of its
rhythm. But then, is to perceive to espouse the rhythm of the brute
thing or to contract it?

Valid intuition according to Bergson is threatened with taking on a
positivist aspect, it places itself in danger. On the one hand, there is the
immediate intuition of the thing that is there, and in relation to which
all the rest is negation. There is in matter [something] like a presentiment of memory, because it is above the duration and does not need
it to be: it is always the same, at all times. Here is the revelation of
natural being. Except that, on the other hand, this intuition does not
suffice; we must double pure perception with Nothingness, and Bergson
is embarrassed by this relation of Being to Nothingness. The immediacy
of pure perception is no longer but a moment that awaits its own overcoming. There is not Being, on the one hand, and Nothing, on the other,
but rather there is a mixture. This natural immediacy must be considered
as a horizon that is signified to us by our perception, without our being
able to possess it. But does Bergson take account of this?

His merit is obvious. Bergson is one of those who seeks to find in the
experience of the human what is at the limit of this experience, be it
the natural thing or life. He wants "to seek experience at its source, or
rather above that decisive turn where, taking a bias in the direction of our
utility, it becomes properly *human* experience."[30] But this philosophical
and necessary effort is compromised by taking sides with positivism,
which makes of this prehuman a being with which we coexist. Does the
return of intuition to the immediate in coexistence translate properly
the profound thought of Bergson, which aims at being a return to the
primordial?

Nature as Life

Bergson wants to rediscover intuitively the natural operation of life,
in opposition to every type of human operation, or every type of

teleology. The *Essai sur les données immédiates de la conscience* is a book oriented entirely against both those who wanted to make of consciousness an ensemble of processes exterior to each other and those who wanted to reduce the unity of consciousness to a unity superimposed on these mechanisms. Bergson had had the intuition of subjectivity as duration. The whole reconstruction of our unity had been overcome by the cohesion of the duration that we are, and that is not in front of us as an object to construct. This intuition of duration guides Bergson in *Creative Evolution*, because life is also a natural operation that cannot be made up of assemblages, as mechanism makes it, and that sees there a multiplicity of processes; now finalism does give a unity to these assemblages by adding an end to them, but this end remains exterior to the living. It is only an idea. Yet life never gets caught there as we think it could: it is both above and below finalism.[31] Below, in the details, because it often creates monsters; above, because just as the idea results in works of art (the idea is no more the principal motor of these works than it is present in the author or discovered by the critic), so too does life not create by proposing the idea of an end. Certainly by looking against the grain, we can read a finality, discover an idea, but this idea is only the trace left by creative evolution; it is not creative evolution itself. This does not obey a vital principle: there is not a foreman who directs evolution, because there is not a worker, or rather, there is not a distinction between them.[32] In the natural operation, the end is immanent to the means.

Bergson thus retrieves the ideas of Kant. But the discussion of Kant's theses by Bergson rests on a misunderstanding: Bergson gets caught on the notion of internal finality, because, he says, the elements of natural being are not joined together, but are separated. Now Kant would gladly admit that he sees in the resemblance of father and son, and even in sexuality, a mark of internal finality. What defines internal finality for Kant is not that it is internal to an organism, but that there is an immanence between the end and the means. But Bergson adds a new element to these terms that we find already in the *Critique of Judgment*: the idea of a natural history. Bergson defines the living by history: the living organism is "through a unique series of acts that constitute a genuine history."[33] By describing the organism, Bergson leaves behind substantialist thinking, which saw in the end an immutable form, both at the origin and at the end of development. He defines the organism and life as types of temporality and thereby places them outside of every comparison with a physical system. The physical system is its past (Laplace). The organism, and the whole universe defined as natural system, is defined, on the contrary, by the fact that the present is not identical with the past. We can say of the physical system that it is re-created at each moment, that it is always new, or that it is uncreated and is identical to its past. On the other hand, the organism is never identical with its past, nor is it ever separated from it: it continues. Duration becomes the principle of the internal unity of it. *"Wherever something is living, there is a register, open somewhere, in which time is being inscribed."*[34] And this register is neither a consciousness interior to the organism, nor our consciousness, nor our notation of time. What Bergson thereby designates is an institution, a *Stiftung*, as Husserl would say, an inaugural act that embraces a becoming without being exterior to this becoming. This intuition of life as history brings out the value of several passages in *Creative Evolution*, in which Bergson posits that unity is at the origin and that it then tries to dissolve itself. For example, Bergson speaks of a reciprocal implication of tendencies at the beginning, but owing to an "unstable equilibrium of tendencies,"[35] this unity tends to undo itself in animals, vegetables, microbes. These three living forms are complementary functions that life contained in the state of reciprocal implication: life is like a bouquet that opens itself; the unity is at the beginning. In this way, because the vegetable is carried by the same *élan* that carries the animal, we find in the vegetable a sexed reproduction, which is a luxury on this scale and is explained only by the fact that animals were to come. Bergson wants to explain these strange consonances by a common *élan*. Complex mechanisms on divergent lines of evolution lead him to posit the unity as origin, to give credit to the idea of a living nature as impulsion, initial unity, but in no way excluding divergences in the result, unforeseeable at the start. Harmony brings up the rear, rather than making headway, is behind rather than ahead of it.

Bergson thus makes a scrupulous description of life as a blind and finite principle. It's only then that his intuition degenerates, when he wants to make of life an undivided principle pursuing a goal and accessible to a mystical intuition. At the start, the use of the expression *élan vital* served only to designate a thing that had begun, and which disposed at the start of a limited force but gradually lost its force because of duration. The *élan* of life is finite, it exhausts itself quickly: "But everything seems to indicate that this force is finite, and that it exhausts itself quite quickly when it manifests itself. This force finds it difficult to go very far in several directions at once. It has to choose."[36] Or again: "It is important to keep in mind that the force that evolves through the organic world is a limited force that always attempts to go beyond itself and that always remains inadequate to the work that it strives to produce."[37] This is in function of a contradiction constitutive of life: life is mobility and makes determinate forms appear out of itself, but this determination of living forms separates them from the *élan*. "Life in general is mobility itself, whereas the particular manifestations of life only receive this mobility reluctantly, and are always falling behind it."[38]

> In this way, the act through which life goes about the creation of a new form and the act by which this form takes shape are two different and often antagonistic movements. The first continues into the second, but it cannot be continued there without being distracted from its direction.[39]

Life is a distracted principle, capable of not pursuing what it had begun:

> From the lowest to the highest levels of the organic world, there is still but a single great effort; but most often this effort is cut short: sometimes paralyzed by the forces that oppose it; sometimes distracted from what it must do by what it is doing, absorbed by the form that it is in the process of becoming, and hypnotized by it as if gazing into a mirror.[40]

In this way, as in Schelling, Nature is never only a productive principle, but indissociably producer and product. It overcomes the product in the very act of creating it, but this overcoming most often is fictive, and creation of life is no more than the reproduction of one same being.

The success of creative life is ambiguous. Certainly there is a success, since there are living forms, but if we compare them to the movement from which they issue, nonsuccess is the rule.[41] Of the four directions of life indicated by Bergson, two are impasses, and in the other two, the effort is disproportionate in relation to the result.[42] As to human being, we cannot say that it is the goal and the end of evolution, nor is it "as pre-formed within the movement of evolution."[43] It is the result of a contingent struggle that this species sustained with others. If human being can be called a success, it is because there is in it the nonachieved and because it carries creation in it. But human being is not the point toward which evolution would have converged; it abandoned a part of the tendencies that life implied: "Everything takes place as if an indecisive and vague being, a being we could call, if you will, man or super-man, had attempted to realize himself and yet was only able to do so by abandoning a part of himself along the way."[44]

Living nature is thus a principle of finite unity that comes to terms with a contingency, which does not dominate it but is charged with realizing itself in this contingency and thus with undoing itself. By making itself, life undoes itself. It is thus not a principle of interiority that would yield to exteriority as it pleases. Between the producer and the product, there is a necessary discordance that we cannot regret, because it constitutes the very realization of life. Immanent finality, the finality of life, is thus a finality made heavy. If there is not a separation between the foreman and the worker, this does not mean only that the end dominates the means, but also that the means can denature the end, that their resistance, their inertia, gains the end. Bergson has the idea of a natural production whose contingency would not be a fault. If we compare evolution to a road, we must say that the accidents of the terrain are not impediments, but that "[a]t each moment, they provide it with what is indispensable: the ground itself upon which the road sits."[45] Living nature is a mixture, a mixed principle: its meaning "is to insert indetermination into matter,"[46] "to fabricate a mechanism that could triumph over mechanism,"[47] and this meaning is also a paradox. The negation that constitutes the matter within life has a positive value. Matter, as an obstacle to life, gives it not only the terrain on which it can be realized, but also the way in which it is realized. Language, which appears at first glance as an obstacle, is likewise a means for consciousness to realize itself:

> Now, this mobility of words, designed to allow them to go from one thing to another thing, also permitted them to be extended from things to ideas. Of course, language could not have given the faculty of reflection to an intellect that was entirely externalized and incapable of folding back upon itself. An intellect that reflects is an intellect that had a surplus of force to spend beyond its practically useful efforts. This is a consciousness that has already, virtually, won back control of itself.

But it is still necessary for this virtuality to become actual. Presumably, without language the intellect would have remained riveted to the material objects that it had some interest in considering. It would have lived in a state of somnambulism, outside of itself, and hypnotized by its work. Language greatly contributed to liberating it. [...] [The intellect] made use of the fact that the word itself is a thing in order to penetrate (being carried along by the word) into the interior of its own work.[48]

Bergson is very close to a philosopher who would define life not by rest, coincidence in itself, but by a labor of itself on itself, of which it could not complain, since it allows it to be realized. The analysis of concrete materials in Chapter II of *Creative Evolution* carries him down this path. But the metaphysical elaboration to which he subjects them in Chapter III turns him away from this path and leads him down another.

In Chapter III, life becomes pure creation, an undivided act that does not leave itself. And this conception is already visible in the passage where Bergson compares life to a road: "But if we consider the whole of the road and no longer each of its parts, then the accidents of the terrain appear as mere impediments or causes for delay since the road simply aimed at the village and would have preferred to be a straight line."[49] Life is going to become a principle in thought, separate from its operation. As Jankélévitch says in his first book on Bergson: "Life does not need the body; on the contrary, it would very much like to be alone and to go straight to its goal [...] [i]f life did not have to reckon with this yoke, what miracles would it not accomplish!"[50] Life becomes an eminent causality that contains all of evolution in its principle. There is a passage from life as an equivocal and dialectical principle to life as a univocal and intuitive principle, similar to the passage from our factual perception to pure perception: "[v]ision is a power that could, *in principle*, reach an infinity of things that are nevertheless inaccessible to our gaze."[51] The constitution of the apparatus of vision is, rather than a creation, a reduction of this power to see, caused by certain necessities of adaptation. The constitution of this apparatus is due to a work of canalization, to an act of piercing, and "the visual apparatus simply symbolizes the work of canalizing."[52] Just as it is not the mountain that makes a tunnel, neither does the apparatus of vision make vision. Again, as Jankélévitch says, "[T]he animal sees *despite* his eyes rather than by means of them."[53] Likewise the nervous system is compared by Bergson to a "vice"[54] that would impede consciousness from completely realizing itself and would allow it to pass only by deserting itself more or less completely.

In this way, the operation of life is ultimately attached to a principle of unity transcendent to the contingent manifestations of life, and thereby life is no longer thought but overcome. All the details of life lose their value; they must be considered only as a means of impeding constitutive reality. The constitutive act of life advances more or less far, following the resistance of matter met along the way or following the force of its *élan* (the explanation changes according to the texts, but this matters little: whatever negativity is in matter or in life changes nothing of the fact that life is conceived here as a transcendent reality). Order, if considered biologically or physico-mathematically, is no longer of interest and ceases to be a problem. The operation of life is only the reverse of reality. From now on, the *élan vital* is considered no longer as a principle, but as an attempt. The *élan vital* is going to be considered as a species of reservoir: "[L]ife is an immensity of virtuality, a mutual encroachment of thousands and thousands of tendencies."[55] And again: "This is how souls are continually being created that, nevertheless, in a certain sense, preexisted."[56]

If the *élan vital* becomes not an operation but a reservoir, then we have to blame it on Bergson's positivism, which transforms the varied unity of life into a principle adherent to itself. In Chapter III, Bergson no longer speaks of *élan vital* as a principle indissolubly [composed of] means and ends; he divides it into two terms: the physical and the psychical, attempting to derive the first from the second. "But imagine for a moment that [...] physical reality is simply an inverted psychical reality."[57] The creative action of life engenders matter by stopping it.[58] Life appears as what remains of the *élan* when it is relaxed: there was a creative gesture, and this gesture was undone, it became matter. Life is the effort of consciousness to find itself in matter.

The initial positions are reversed: at the beginning there was monism and the dialectical conception of the relations of matter and the living, according to which the *élan* is not analyzable into two elements, but is rather indissolubly active and passive. Now Bergson allows both a dualism and an emanatism, which is the negation of the dualism: matter is issued from the first element by the slackening of the latter; it is drawn from it by inversion. It was moreover inevitable that the analysis arrived at this conclusion. In effect consciousness is duration, it is not conceived without multiplicity. If we want unity, as positivism does, we must thus posit a unity beyond duration, a "supra-consciousness,"[59] a maximum of interiority symmetrical to the maximum of total exteriority, which is matter. The concept of Nature must burst and cede its place to God.

But Bergson hesitates to confuse God and evolution. Life is faced with matter, the slope of which it goes back up. Creative energy tries to regain itself. God is the same energy but drawn to its source. In this positivism, Bergson is thus going to see reborn the dualism he sought to escape. Bergson cannot realize it in a homogeneous principle. In a text on Ravaisson,[60] he allows that God creates non-Being, the void, at his expense. The dialectic is thus not suppressed but put back in God. Bergson hesitates, however, to make a theology of his philosophy: he sometimes sees in it the intuition of the total solidarity of the principle and its manifestation, the feeling of contact with an Absolute that "lives with us,"[61] absolute *naturans*, which lasts like us and of which we can wonder if it is a principle of the same kind as the Cartesian *naturans*. Jankélévitch anticipated this text on Ravaisson. Concerning the Bergsonian conception of life, he says: "[t]here are, present one to the other, one in the other, two inverse movement,"[62] one movement of decline and one of resurrection. To this extent, we see the dialectic surging forth again: "In order to affirm itself, life, in a singular derision, needs the matter that kills it," Jankélévitch says again.[63] We thus see the idea of Schelling, according to which the fall is an inseparable element from creation, appears again in Bergson.

We find again in this movement from the second to the third chapters of *Creative Evolution* Bergson's habit to pass to a defined positive reality, to perceive that in this reality there is a negation, hence the translation of this negation into positive terms (in this case, the physical and the psychic), and hence finally in order to conserve the incorporation of this new negation in the concepts of being and the positive, despite all the positive unity.

The Ontological Infrastructure of the Concept of Nature in Bergson: The Ideas of Being and Nothingness

Bergson polemicizes against negative concepts, but we will see that in his polemic he seems not to understand himself, because one part of his polemic cannot lead him to where he wants to go. In the whole of this polemic, Bergson wants to eliminate the idea of contingency, he wants to eliminate questions of the kind "Why is there something? Why this world and not another?" because these questions present themselves only when we give priority to fabrication over production, when we consider every species of being as fabricated either by X or by us. But they are no longer posed when we pass to the point of view of a natural production. We have to return quite simply to a being that exists.

The Idea of Disorder

According to Bergson, the idea of disorder is deprived of meaning: we believe only that there is disorder, because we find ourselves in the presence of a reality ordered otherwise than what we were expecting. But this notion can be only relative, never absolute. The idea of chaos is contradictory. So that chaos be chaos, I have to represent to myself a power within this chaos, which vigilantly never gives place to a legality-that is, reality keeps strictly to the law of being of chaos: hence the absence of absolute chaos.

Bergson supposes two orders, the physico-mathematical order and the vital order, and he posits that these two orders are not only contrary, but also contradictory. The physico-mathematical order consists in the constancy of certain laws: the same causes lead to the same effects. The vital order, on the other hand, consists in the fact that the same results are attained when the conditions are different. Life takes its permanence from a result to obtain, and nonlife is characterized by a permanence *a tergo*. Starting from this, Bergson reconstructs the feeling of disorder. The fragility of order I believe comes from the duality of orders. Disorder is only a way of speaking. The absence of order leads back to the presence of order, since the negation of one of the two orders is a way of designating the presence of the other order.

It is essential to this demonstration that the two orders be not only contrary but also contradictory. Are they indeed so contradictory? The presentation of the two orders does not square with the rest of the Bergsonian doctrine. Can we oppose order *a tergo* and order toward an end? As Lachelier argued,[64] the dogmatic idea of causality closes in on finality. Inversely, can Bergson define the order of life by this violence of ends that would unconditionally impose itself on the means in order to perpetuate their end, after having shown that life insinuates itself in mechanism, espouses the contours of it, before transforming them?

But let's agree that there are two orders, both positive and mutually exclusive. If these two orders form an absolute alterity, then they form a negativity with respect to Being. We have then not two things in presence, but one sole Being that has an absolute negativity in its flesh, which is sometimes one order, sometimes another. Radical positivism is in the end a radical negativism, since the two orders succeed one another without order, each being bound to the other by its own negation. The only way to eliminate disorder is to adopt Spinoza's position of a pure positivism, whereby we interpret his doctrine in the sense of an intrinsic necessity or an overfinalism (this is how Kant saw Spinoza).

Similarly, Bergson can pass by the idea of absence. There would be not absences, but alternating presences. For Bergson, everything is of the positive: the absence of a thing is for him only the presence of something else, there where we were waiting for that which is missing. We will respond that the observation of absence is not equivalent to the absence of the observation. But Bergson will then respond that the observation of absence is only in the mind. Absence is not in the things, since in the things we can find only the "il y a," the "there is." But in biology, absence has a signification: the death of an organism is not reduced to only the presence of a physical system. We have to allow the rivalry of the absent order and the present order—that is, a labor of one on the other. In the organic order, in a natural world where there are only individual beings, the absence of an order cannot be reduced to the presence of another order: it has an objective signification.

Finally, we have trouble considering that in Bergsonian philosophy, the physico-mathematical order is a positive thing. Does not Bergson himself affirm that the idea of a thing that undoes itself is as an idea no doubt one of the essential traits of materiality? It seems that positivism leads him to two demands: first, life is a positive reality from which matter is derived by simple stoppage; second, life and matter are two positive orders. On the one hand, Bergson posits the continual passage from one order to the other—that is, the continuity of two orders; on the other hand, he radically separates them, as two contradictory entities. But this last demand, taken literally, would be susceptible to making the Bergsonian concept of Nature burst, a concept according to which life is the retaking of the stopped creative movement, the stopping of which gives matter, which is a reality that is made in a world that undoes itself. As Jankélévitch underlined in the passage cited above, there are two contradictory principles that must be not only "in the presence of one another, but also one in the other."[65] So that these two movements are not canceled out, the movement must be bipolar. This contradiction, understood as interior to Nature, must be assumed. We must admit the idea of an operating negation in Nature, an idea without which this notion bursts.

The Idea of Nothingness

The true meaning of Bergsonian philosophy is not so much to eliminate the idea of Nothingness as to incorporate it into the idea of being.

Bergson shows at first that the idea of an absolute Nothingness is contradictory. There is not a void in the world, all thought of the void is the thought of a certain plenitude. If I annihilate the exterior world by thought, then this world takes refuge in the interior world of my consciousness. If my consciousness is at the point of annihilating itself, thought is at least filled by the idea of this void I cannot tear myself away from my thought. Each annihilation is irreducibly a positing of a certain term and an exclusion of certain other terms. The passage to the limit is impossible. We cannot totalize these negations. The idea of Nothingness comes from what we imagine ourselves to be able to contract in the instant of successive negations.

But Bergson not only does not allow the idea of Nothingness, he also does not recognize the value of negative thinking. For him, it is the same to say that something is not and to say that something is. All negation is only denegation, denial, *Verneinung* Freud would say. Negative thought appears only in the impersonal: it rests not on the things, but on a judgment we or others make, and it consists in forcing back an erroneous judgment. In its psychological profundity, negation is only an affirmation of an affirmation.

To say that something is not there is to say that something else is there. The absence of a thing leads back to the presence of another thing. There is an affirmation of thought in negative thought. I install myself in the irreal while trying to express the real starting from it. True thinking—philosophical or scientific—must be at the antipode of a likewise affected thought, because "the nonexistence of the nonexistent is not recorded."[66]

And Bergson, to trace a portrait of the positive mind, says:

> Imagine language abolished, society dissolved, and among humans all intellectual initiative and every faculty for splitting oneself or for judging oneself atrophied. The wetness of the ground would no less remain, capable of being inscribed automatically in sensation and of sending a vague representation to this dazed intellect. The intellect would thus still make an affirmation, though in implicit terms. And, as a result, the essence of affirmation is not to be found in distinct concepts, nor in words, nor in the desire to spread the truth around oneself, nor in the desire to improve oneself. But this passive intellect—who mechanically follows along a step behind the course of experience, neither getting out in front of the course of the real, nor lagging behind it—would not have even the slightest inkling of the act of negating. This intellect could not even receive the imprint of negation, since, again, although whatever exists can come to be recorded, the nonexistence of the non-existent is not recorded. [...] Such a mind would see facts following other facts, states following other states, and things following other things. At each moment, it would simply note the things that exist, the states that appear, and the events that take place. It would live in the actual [*l'actuel*] and, if it were capable of judging, it would never affirm anything other than the existence of the present [*le présent*].[67]

Such is the model that Bergson presents of the positive mind. But this idea of the in-itself, of the humid in-itself, if it is at the horizon of our thought, cannot constitute the mode of all our thoughts of Being. Such a positive being would not include history. With such a position, we wonder how Bergson can constitute his ontology of the past, the present, and the future. There is not a thought of the past without its negation. Bergson's analysis shows that thought is no less itself when it is empty than when it is full. As Jankélévitch underlines in his commentary: "The mind is never more totally present to itself than [...] when it denies [*nie*] and when it is wrong [*se trompe*]."[68]

The Idea of Being

The malaise is the same concerning the analysis of Being: "But we must become accustomed to thinking Being directly, without making a detour and without first directing ourselves to the phantom of nothingness that interposes itself between us and Being."[69] It thus seems that Bergson oriented himself to the Spinozist conception of a Being without fault. Yet Bergson himself criticizes this comparison. The Spinozist idea of Being as equality with itself seems to him to imply Nothingness as a contradictory idea in relation to Being, and which it would have to overcome. An existence that lasts does not seem to him strong enough to vanquish inexistence and to posit itself:

> This is the primary reason metaphysics tends to attribute a *logical* existence to true being, and not a psychological or physical existence. For such is the nature of a purely logical existence, namely, it seems to be self-sufficient and to establish itself simply as a result of the force that is immanent to the truth. [...] But the "logical essence" of the circle, that is, the possibility of tracing it according to a certain law—in short, its definition—is something that seems eternal to me.[70]

The idea of necessary Being would be the last straw of vertigo, the understanding of a radical contingency.

Bergson, not passing by the idea of an infinite producer since he does not pass by the idea of Nothingness, can affirm a natural productivity in the things, an Absolute in the phenomena:

> The Absolute is then revealed to be very close to us and, to a certain extent, within us. The Absolute is psychological in essence, and not mathematical or logical. It lives with us. And like us—though in certain ways infinitely more concentrated and infinitely more condensed into itself—the Absolute endures.[71]

But when we compare this text to what Bergson said earlier concerning the positive mind, always in the present, we perceive a contradiction. If Bergson blames Spinoza, he cannot be a positivist. To posit a being that endures is to posit a being that hesitates,[72] which cannot do everything at once, which places the negative in its being.

The Idea of the Possible

The ideas of Nothingness and disorder would never be taken seriously, being nothing, if did not conceive them as possible and if did not conceive Being as the victory over Nothingness, as a possible overcome.

Bergson shows that the idea of the possible, far from being the idea of a beginning of being, anterior to the actualization of this being, is an idea formed starting from the actual. Considering the past, it occurs to me both to project the present in the past and to believe that it [the present] was contained in a germinal state in the past, prepared by it; whatever the unfolding of things is, we can say and show that the past makes it possible. There is here a retrospective illusion, which is the difficulty of thinking the past independently of the present. Does this critique of the bad retrospective always liquidate the idea of the possible? If we take away the fictive possible, must we reduce Being to the actual?

The retrospective illusion is the illustration of a consciousness that does not coincide with Being, which is always late. But must we say that consciousness is vicious as soon as it does not coincide with Being? Is every valid knowledge a knowledge without distance?

If we must eliminate every idea of the possible, then we must see "discontinuous explosions" in the duration, in life, and in history.[73] But we do not see what the word "duration" or the word "life" could signify if there is no longer the envelopment of the past in the present. So that Bergson's descriptions remain valid, we must distinguish, as Jankélévitch does, next to a logical possible, which is nothing, an "organic possible," which is something, which is the seed.[74] The description of history is impossible if we suppress every envelopment. The Bergsonian instinct is described not as foresight, but as "prophecy," as Jankélévitch says.[75] It cannot foresee the future, but it is turned toward it. All these words concerning it imply its existence, but without the future being consciously present. Instinct is thinkable only as a function of what it will do, without it ever knowing what its future will be.

On the other hand, to say that there is not the possible amounts to saying that all is possible. For any given phase of creative evolution, not everything is possible, because the *élan vital* meets up with incompatibilities. If it chooses to realize something, it also chooses thereby not to realize other things. Matter means that at any given moment, not everything is possible. Finally, the objection that Bergson makes to the possible— namely, that it is only a retrospection made after the fact and not a being in the making—poses the question of knowing if the divergence between knowledge and the object is always a fault. Could there be a duration if there were not distance between us and Being? Bergson took account that there was not necessarily a fault in retrospection, and in the introduction to *The Creative Mind*, posterior to the work, he no longer speaks of the retrospective illusion, but of a "retrograde movement of the true": when we think something true, it is only retrospectively that the true appears to us as true.[76] The remodeling of the past by the present, this "metamorphosis" as André Malraux would say,[77] can designate an [arbitrariness] but can also indicate that contemporaries did not have a complete knowledge in their time. There are realities in the history of culture, about which we can say that they do not altogether exist in the present and that they need the future. In research like Galileo's, much more was implied than what Galileo found or even sensed. But is it arbitrary to say that Galileo opened a certain order of research? Briefly, it is indispensable to the Bergsonian conception of Nature to allow the possible as an ingredient of Being and to make of it something other than a psychological curiosity.

Note on Bergson and Sartre

There is a convergence between certain intuitions of Bergson and Sartre. The object taken in itself is all that it is. The "quarter of the moon" is only a human way of speaking, it is a complete figure. No principle takes it to the front. It is the same for the transfer of the possible of Being toward man Sartre had the idea that in the history of consciousness, there is not a preliminary lack: the human creates both the lack and his solution. Likewise, Bergson thinks in *Creative Evolution* that philosophers create problems and their solutions at the same time.

At first sight, it can seem paradoxical to compare two philosophies, where one is essentially a positivism and the other a negativism. Neither allows for a mixture of Being and Nothingness (see Sartre's critique of Hegel in Chapter 3 of *Being and Nothingness*). Of course, it is not the case that either of them make place just as well for the idea of Nothingness as for the idea of Being, but they do not allow their fusion. Thus in Sartre, Nothingness is the avidity of Being, and there is an equivocation in consciousness, but there is a vain effort of Nothingness to make Being. There is no place for a conception of Nature or for a conception of history in this philosophy. In Bergson, the official position of positivism also ruins the idea of Nature.

We can elaborate a valid concept of Nature only if we find something at the jointure of Being and Nothingness. Despite what Bergson says, there is a kinship between the concepts of Nature and radical contingency.

In order to elaborate this concept, we have to leave positivism or negativism, which maintains a separation between the objective and the subjective, and which thus makes impossible the subjective-objective that Nature will always be.

DELEUZE AS READER OF *L'ÉVOLUTION CRÉATRICE*

French philosopher Gilles Deleuze (1925–1995) presented six lectures on Chapter III of Bergson's L'évolution créatrice between March 14 and May 9, 1960 in a course at the École normale supérieure in Saint-Cloud, France. The following selection is an excerpt from Deleuze's lecture notes for the first two sessions. As Anne Sauvagnargues writes at the beginning of her introduction to the French publication of these course notes:

> When it comes to Deleuze, that "marvelous reader of Bergson," such a course was a genuine event. [...] [T]his course clarifies a stage in Deleuze's close reading of Bergson to which Deleuze will often return, and that continues to articulate the constitution of the successive stages in Deleuze's own system.[1]

Sauvagnargues describes how Deleuze's reading of Bergson was instrumental in his development of a philosophy of difference, especially insofar as difference becomes an operation of life, while the operations of matter and the intellect must be understood in terms of repetition. These are two main concepts for Deleuze's key text, Difference and Repetition.[2] Nevertheless, Sauvagnargues observes that even if the "interruption" [césure] between difference and repetition is originally Bergsonian:

> [T]he cleavage [clivage] between space and time, between matter and durée, is transcribed by Deleuze onto the methodological axis of difference and identity, and this deeply transforms it. Ultimately it is with Bergson that Deleuze takes aim at the philosophies of identity as his theoretical adversaries, and with Bergson that he will seek to articulate thought and becoming by developing the concept of difference.[3]

What Deleuze finds in L'évolution créatrice is the confrontation between thought and becoming, being worked out impersonally, on the cosmological terrain of the becoming of life and the genesis of matter, rather than within subjective consciousness. For Sauvagnargues, this implies the development of a term that is key for Deleuze's own metaphysics, namely différenciation.[4] As she writes:

> Deleuze understands *durée* as a virtual *différenciation* and as such enacts a certain torsion, visible though only barely explained in this course, of the spatio-temporal opposition in Bergson between Matter and *Durée*, an opposition Deleuze thus notably reshapes by substituting another duality for it, one between the virtual and the actual. The concept of *différenciation* that Deleuze presents as a biological concept here surely receives a specific impulsion from Bergson, and is at least theoretically implicit in the definition of *durée* that we find in *L'évolution créatrice*. But it is no less true that it is also a Deleuzian creation, and that it marks the passage from a philosophy of the *élan vital* (Bergson) to a philosophy of difference (Deleuze).[5]

* * *

LECTURE COURSE ON CHAPTER III OF BERGSON'S *CREATIVE EVOLUTION* (1960)[6]

By Gilles Deleuze

Translated by Bryn Loban

Ecole Normale Supérieure de Saint-Cloud

I. 14 March 1960

In the first part of this work, Bergson aims to present philosophy, and to show the necessity of conceiving of it as *genetic philosophy*. He thus comes to grips with something essential in philosophy. In effect:

a. philosophy has, prior to him, laid claim to be genetic;
b. cosmology—in ancient metaphysics—is portrayed as genesis;
c. Kantian inspired philosophy—representing modern metaphysics— is also portrayed as a genesis.

The third chapter of *Creative Evolution* is written counter to all these claims. In passing, it should be noted that for Bergson, to a certain extent,

Kantianism acts as a "reference point." To differing degrees, Kantianism claims to be a philosophy of genesis. To be precise, there is no genesis of the phenomenon, but in fact there is a genesis of the intelligibility of phenomena.

After Kant, with Maïmon and Fichte, the claim becomes explicit. In effect, they both say that it is necessary to pass from a transcendental philosophy to a genetic one.

But Bergson says that this genesis is badly enacted:

- either because it is a genesis of the intellect[7] derived from matter;
- or because it is a genesis of matter derived from the intellect.

In both cases, it is not a true genesis because, taking as a point of departure one of the terms, the other is immediately given, for there is a fundamental *reciprocal* relationship between the two.

In such a case, how are we to conceive a real genesis?

Bergson says that genesis must be double, in the sense that it must account for matter *and* the intellect at the same time, and consequently for their reciprocity. *How does Bergson present the problem in the first two paragraphs* of Chapter three?

He indicates first:

A. THE METHOD TO FOLLOW

1. What is to be "gained" from the *approach taken in the first chapter?* Bergson shows a *difference in kind* between the inorganic and the organic, between the inert and the living. In effect, one of the main functions of the method is to show differences in kind. How does Bergson understand this difference in kind? According to him, this does not arise from a special principle of life (many others have said this before him and, as such, were "anti-vitalists"), but from the fact that the living is a natural system—that is, one that has *duration*, while the inert is a system that is artificially—that is, approximately—closed. The former, on the contrary, is not closed, but *open.*

The theme of the first chapter thus highlights the fact that, in order to explain a difference in kind, it is not necessary to appeal to a special principle of the living. This does not resemble the second type of system, but "the whole of the universe." The living is a small "whole." Is this an idea inherited from Platonism? No, for Plato compares the Whole to the living, whereas the reverse is the case for Bergson.

No, since for Plato, it is a comparison that contains the idea of the Whole pre-existing the parts: totality implies interiority. For Bergson, it is the opposite: there is neither totality nor interiority within the Whole, for it would then be a closed system—i.e. inert—and consequently incomparable to the living, which is an open system.

The living is not a closed system (for Bergson, there is no finality, other than *external*); the living has a tendency to individualize itself, but without ever succeeding. It is this failure of individualization that characterizes the living.

Bergson thinks that in being guided by the comparison of the Whole to the living, one will find in the universe a principle of genesis that takes into account matter and its tendency to form closed systems. He has never linked life to interiority, to an *internal* finality. If there is finality, it can only be external, for the living system is never closed.

2. Same theme in the *second chapter*, difference in kind between instinct and the intellect. But there is another way of showing this. In fact:

a. the first chapter shows a difference in kind between the inorganic and the organized that consists of de-composing a composite [mixte];
b. in the second chapter, duration and the *élan vital* are of the same nature and therefore they cannot be de-composed. There is indeed a differentiation [*différenciation*], but it is due to their nature: it is in the nature of the *élan vital* and duration to differentiate themselves [*se différencier*].

II. 21 March 1960

How does Bergson pass from duration to the *élan vital*?

His philosophy is a philosophy of Life. He understood that this primary notion used by biologists is misunderstood by them. They speak of "differentiation" [*différentiation*]. Bergson posits that there is an evolution in the differentiation of species (later, he extends this to the embryo). This idea has not been understood because scientists do not understand duration, and it is duration that differentiates itself [*se différencie*].

In seeing that the philosophy of life needs to develop this concept, he realizes that all his philosophy must become a philosophy of life, and duration [durée] must become the élan vital.

The élan vital is duration [durée] that differentiates itself.

Scientists did not see that differentiation [différentiation] implies a virtual movement that actualizes itself—that is, a movement that creates at each instant two diverging lines [directives].

In the same way, in the domain of history, the "dialecticians" have substituted a simple opposition in place of a differentiation [différentiation]. They have "misconceived duration." In The Two Sources of Morality and Religion, Bergson refrains from constructing a philosophy of history, for the movement that runs through history is that of differentiation itself.

The living is essentially a being that has problems and resolves them at each instant.

Is it valid for history? Yes, but with one particularity. Humanity pursues a path as far as it can possibly go as long as it does not encounter an insurmountable problem (cf. diagram b [in Fig 1]). When humanity encounters such a problem, it effects a true qualitative leap and takes another path that leads further than the previous one, which it pursues as far as it can go, and so on.

We have here a differentiation [différentiation] of life as the élan vital. It is "consciousness" or "universal life," says Bergson. What does this mean? By right [en droit], Life and consciousness are one and the same. "By right" because the duration-consciousness identity is explained in this way: we are dealing with duration when the past is gathered into the

a) General diagram of differentiation

b) Diagram of differentiation in History

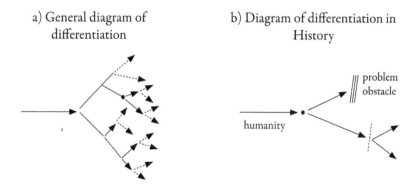

Fig 1 General Diagram of Differentiation [Différenciation]

present and the future is always new. It is this duration that is the condition of freedom and choice, conceived in opposition to "relaxation," where moments of time fall outside each other. In effect, in the latter case, there is no longer an organization of the three movements of time, but a pure repetition of the present. It is the very state of matter.

In *Matter and Memory*, Bergson recognized the value of certain of Freud's ideas on freedom. For Bergson, freedom resides in the new, not in the repetition of the past. Bergson, like Freud, has this same idea. Both affirm that memory is a *function of the future*, for repetition consists in the forgetting of the past. More past = more future, and thus freedom. Memory is always a contraction of the past in the present.

This is why duration is identified *by right* with consciousness.

In effect, duration falls back upon itself and becomes matter, as if *right* could not pass into *fact*. Consciousness cancels itself out in matter through a movement that is the inverse of the differentiation [*différentiation*] of duration. This rule is general, *except for one localized point* in which matter is opposed to itself: the human brain. In this case, the mechanisms of matter cancel each other out [Fig 2].

In reality duration only becomes self-consciousness in the human brain.

Thus, we can only hope to obtain a genesis of the intellect if we place ourselves in a universal consciousness. In effect, if we reunite the premises highlighted in the first and second chapters, we are led to identify the Whole with universal Consciousness.

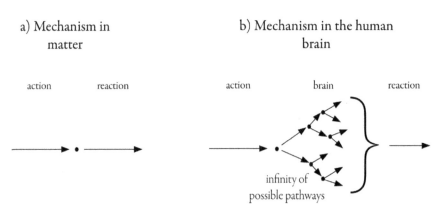

a) Mechanism in
matter

b) Mechanism in the human
brain

action   reaction

action   brain   reaction

infinity of
possible pathways

Fig 2 General diagram of mechanism

This is the point of departure for the third chapter. The problem raised is effectively this: Can we place ourselves in this Whole that is universal consciousness and *vice versa*? If this is the case, then the genesis of Matter and the Intellect can be achieved.[8]

The Third Chapter

Bergson affirms that philosophical genesis has not been understood before him. It has been done in two ways which, moreover, are not strictly identical, although both are worthless.

1. on the one hand, the genesis of matter is derived from the Intellect. In fact, says Bergson, we have posited the Intellect as a given;
2. on the other hand, the genesis of the Intellect from matter. In this case, both terms are given *at the same time*. Why "at the same time"? Because there is such an affinity between Matter and the Intellect, that Matter is split from itself in a way that only the Intellect sees fit to segment it.

In summary:

- materiality is the power [*puissance*] of being segmented, cut up;
- intellectuality is the power [*puissance*] to segment, to cut up.

If we attempt the second genesis (2), we assume "divisibility" [*découpabilité*] in order to finally discover the "act of cutting up" [*découpage*]!
    This is what Bergson, in his critique, pursues in three phases.

1 *Psychology*

When psychology attempts to be genetic, it bestows action on matter. Psychology starts with the explanation, in animals, of action-reaction cycles. At this point the Intellect can be engendered. No doubt the model gets complicated, but there is an infinity of possible paths. For Bergson, this is not a true genesis.

    In effect, if we stick to this model, there is complication, but there are only differences of degree between the forms. Everything is given from this initial model. By effecting the slightest action on matter, the Intellect has already been enabled with its power [*pouvoir*] to cut up [*découper*].

2 Materialist Cosmology

(Spencer in particular.) Bergson retains the following from his reading of Spencer: Spencer wants to construct a "philosophy of evolution." Of the first principles he says that "it is the purely physical interpretation of all the phenomena of the universe." Genesis is then: "evolution is an integration of matter accompanied by a dissipation of movement during which matter passes from an incoherent homogeneity to a coherent homogeneity." Or more simply: it is the passage from an undifferentiated state to a more differentiated one, from homogenesis to heterogenesis.

Bergson's response to this is: "Is this theory true?" In fact the Spencerian trend was killed off by another one. The latter trend arose in the context of a problem posed by the second principle of thermodynamics: the degradation of energy. This problem was posed through its extension to the cosmic scale, and many did not hesitate to approach it in this manner (Lalande, Meyerson...).

- Lalande takes as primary idea that the dispersion of energy comports an equalization of temperatures, therefore a homogenization, which goes counter to Spencer's philosophy of evolution.
- Meyerson affirms that Reason is the power to identify. In the second principle of thermodynamics, he sees a resistance to Reason. Reason has a tendency to equalize, but by an irreversible becoming, through a qualitative transformation. Thus there is no identification, hence the resistance to reason. Starting with the same facts, one ends up with the opposite of Lalande.

Bergson ponders whether matter has a tendency to pass from the homogeneous to the heterogeneous. In fact, in modern physics, matter presents systems that are more and more difficult to form, and to a certain extent, no longer signify anything.

If one grants Spencer the veracity of his thesis, no matter what, there is no genesis. He grants matter the power to be segmented in conformity with the manner by which the Intellect distinguishes systems in nature. In this way, he also has already taken the Intellect as a given.

3 Metaphysics

It is Kantianism that is especially targeted here. Bergson develops two arguments in two distinct paragraphs.

a. Kant wants to trace a *genesis of the understanding*. Granted, not explicitly in the Critique, but towards the end of his life he felt this genesis was necessary. Indeed, after him, Maïmon and Fichte wanted to fulfill this project. Kant aspired to find a principle from which the use of the categories, if not the understanding, can be comprehended. For example, the table of the twelve categories is a fact that cannot be deduced. Maïmon and Fichte wished to correct Kant on this point and to link the categories to the first principle by a genetic deduction.
    *   Bergson says that the whole of the Intellect is already present in this first principle.
b. Kantians are not satisfied with this claim: they also attempt to trace the genesis of matter itself, or at least the intelligibility of phenomena. This claim is explicit in Fichte and Maïmon, who wished to enact the genesis of the phenomena themselves.
    *   Bergson, in his critique, tells us: Kant posits space as an *a priori* form. And so space can be subjugated to understanding. Is this a genesis of space? In positing space, Kant posits Matter and the Intellect. The true problem of Kantianism is in fact the following: in what way are receptivity and spontaneity in harmony? Kant assumes that this problem is resolved.

In his *Time and Free Will* Bergson presents himself as the anti-Kant. For Kant, says Bergson, we see things in the forms that emanate from us. We see ourselves in the guise of forms emanating from things. The Intellect is more spatial than we think, in the sense that we spatialize matter itself. Matter has to be pushed further in its own direction than it would go of its own accord.

The process is as follows. Matter takes a step: exteriorization. It gives the Intellect an idea. Note that the Intellect could have already had this idea, but *virtually*. In dreaming, for example, I relax. Matter takes relaxation further. That is why Matter gives me an idea. With this idea I will be able to go further than matter itself. I form the idea of spaces, so

I spatialize matter. Matter in its pure state is relaxation. Space is matter in its ideal form. Matter is less spatial than one thinks; space is more "intelligent" than one thinks.

Space expresses the fundamental correlation of Matter and the Intellect (dream-matter-space: two steps in the same direction, expressing an essential affinity that has the form of space).

Kant is the first to have defined time on the basis of inner sense. Time does not presuppose movement but, on the contrary, it is movement that presupposes time (the opposite of the Greeks). Hence its reality and not its contingency [accidentalité]. Movement presupposes time in the sense that time is defined by inner sense. In this way it is conceived as being homogeneous.

Genesis is then also the correlation between Matter and the Intellect— namely space, because the latter refers to the fact that Matter takes one step further than the Intellect, and that the Intellect takes one more step than Matter. We have here a "progressive adaptation," and space is at each moment a form of this progressive adaptation.

How does Bergson proceed in order to operate his Genesis?

B THE BERGSONIAN SOLUTION

We have seen:

1. The *method consists* of reinserting ourselves in the Whole or universal consciousness. If we succeed, we have the feeling of being elevated to the genesis principle that would be different in kind from the engendered, and no longer different in *degree*.

2. *How does one reinsert oneself in the Whole?*

Philosophy has nothing to do with an assumption, a fulfilment of the human condition. Philosophy must "overcome" [dépasser] the human condition. Philosophy "will end up dilating humanity in us and making it such that humanity itself transcends itself."[9] This overcoming consists in fact of reinserting ourselves in the Whole or universal consciousness.

By what Means? —By taking a qualitative leap, by doing *violence* to the human condition in order to attain the principle of differences in kind.

All in all, Bergson's undertaking is modest: Philosophy is a collective enterprise. Humanity as a species needs to transcend the human condition. Why? —Because philosophy is a question of perception. But perception must not be constructed. The guiding rule of thinkers before Bergson is to have believed that philosophy aims at the concept, that it is individual. Individual, because there is a natural conception for which thought is insufficient, and the role of philosophy is to fill in the gaps of this natural conception. In short, it is necessary to extend perception through the concept, which is an individual task, as must be the construction of the concept.

A mistake, says Bergson, because this extension is understood as a correction and, therefore, conveys the idea of a limitation from the start. For Bergson, it is a matter of extending without correcting. Philosophy proceeds by extension without correction—that is, it extends the human present. The human condition is the maximum of duration concentrated in the present, but there is no co-exclusivity to being—that is, there is not only the present. If there were only the present, man would perceive an eternal present.

James also tried to define philosophy as perception, but with the same flaws as other philosophers. Bergson operates his genesis by saying that the genesis of the Intellect, Matter, and Space is but a movement of relaxation [*détente*] through which the Whole, contracted to the maximum, itself becomes relaxed [*se décontracte*].

And yet this direction is somewhat disquieting. In effect, if the essential theme has been until now, differences in kind, Bergson nevertheless affirms that Matter is produced by simple relaxation, "by simple interruption." Is Bergson not introducing here what he wants to deny—that is, differences of degree, of intensity, and the negative?

How can this be explained? How can this difficulty be resolved?

[...]

CRITICAL APPARATUS

EDITORIAL ENDNOTES TO CREATIVE EVOLUTION
By Arnaud François. Selected, translated, and edited by Donald A. Landes

Please note, editorial notes that refer to Bergson's footnotes are indented. Also, notes that end with an "*" have been modified or paraphrased for the purposes of this translation. Readers are thereby referred to the French critical edition for more information in those cases.

All page references to *L'évolution créatrice* refer to the French pagination, which is provided in the margins of this translation. French pagination in the critical French editions has also been privileged for citing Bergson's other works. Reference to the English translations is limited to where there are direct citations. For readers who do not have access to the scholarly editions in French, page correspondences for Bergson's major works are available on the translator's website.

INTRODUCTION

1 With this statement, Bergson confirms that *L'évolution créatrice* is an *essai* ("essay"), like his previous books: *Essai sur les données immédiates de la conscience* [1889]; *Matière et mémoire: Essai sur le rapport du corps à l'esprit* [1896]; and *Le rire: Essai sur la signification du comique* [1900]. [TN: In all of these cases, Bergson is playing with the double meaning of the word "essay" to mean either a "short piece of writing" or a "first attempt." Thus, although the word does not appear in the title of *L'évolution créatrice*, this book belongs to a series of descriptive "essays" or "attempts."]

2 For similar psychological descriptions of the intellect as feeling "at home" or "at ease," cf. below, vi, 154, 162, 164–66, 196, 199, 203, 214.

3 For a similar argument, cf. Henri Bergson, *Matière et mémoire* [1896], critical ed. Camille Riquier, series ed. Frédéric Worms (Paris: Presses Universitaires de France, 2008), 13–14 [*Matter and Memory*, trans. Nancy Margaret Paul and W. Scott Palmer (New York: Zone Books, 1991). Henceforth cited as *MM*, with pagination referring to the French edition]; Henri Bergson, "Le cerveau et la pensée: Une illusion philosophique [1904]," in *L'énergie spirituelle*, critical ed. Arnaud François, Camille Riquier, Stéphane Madelrieux, Ghislain Waterlot, Guillaume Sibertin-Blanc, and Élie During, series ed. Frédéric Worms, 191–210 (Paris: Presses Universitaires de France, 2009), 196–98 ["The Brain and Thought: A Philosophical Illusion," in *Mind-Energy: Lectures and Essays*, trans. H. Wildon Carr, 231–55 (New York: Henry Holt, 1920)]. The difficulty of this argument comes from the fact that the intellect is defined as a "part" of the "whole" of life, whereas we expect rather that the intellect, as a subjective faculty of knowledge, takes life as its "object." This is precisely Bergson's point: there is no subject-object relation between the intellect and life, but rather a part-whole relation. Bergson conceives of life in such a way that it is identified with consciousness (this is one of the goals of this book), whereas the intellect, for its part, is one of the two principal forms of consciousness, the other being instinct. (Another form of consciousness will be recognized in vegetative torpor.) In addition, the intellect and instinct are not faculties of knowledge, but rather, and above all else, faculties of action.

4 Even at this early stage in the book, Bergson is alluding not only to Immanuel Kant, but also (and above all) to Herbert Spencer (1820–1903). Indeed, Spencer names the absolute: "the Unknowable." See Herbert Spencer, *First Principles*, 2nd ed. (London: Williams and Norgate, 1867), 1–123. Mostly forgotten today, Spencer was a preeminent thinker who helped shape philosophy in the nineteenth century. He had the idea of evolution prior to Darwin, and defined it as "the passage from the homogeneous to the heterogeneous" (*ibid.*, 342). Despite various divergences between their theories, Darwin considered the evolution of living species to be a particular case of evolution in the more general sense envisioned by Spencer.

When Bergson was young, he was an adamant supporter of Spencer, whom he saw as offering a counterbalance to the dominant Kantianism of his day. Bergson's initial project was to devote himself to a career in the philosophy of science—an academic discipline not yet established in France at the time—in order to place Spencer's philosophy upon a solid foundation, in particular with regard to mechanics. Through his study of Spencerian mechanics, and specifically the conception of movement that it implies, Bergson became aware of the futility of traditional definitions of movement. These definitions reduced movement to a collection of positions, failing to account for its simplicity, i.e., for its *durée*. This new conception of movement, which animates his first book, led Bergson to develop an alternative to Spencer's evolutionism: a "true evolutionism" (cf. below, x, 367, 369). For Bergson's first book, see *Essai sur les* *données immédiates de la conscience* [1889], critical ed. Arnaud Bouaniche, series ed. Frédéric Worms (Paris: Presses Universitaires de France, 2007) [*Time and Free Will: An Essay on the Immediate Data of Consciousness*, trans. F. L. Pogson, 3rd ed. (Mineola, NY: Dover, 2001. Henceforth cited as *E*, with pages referring to the French edition]. Thus, *L'évolution créatrice* can be seen as the culmination of Bergson's early project, and this is why it is often considered to be the pinnacle of his corpus. This also helps to explain why Spencer is such an important figure in the present book. His influence lingers in the background of many of the analyses that follow.*

5   Cf. below, 199. This famous declaration, "*L'action ne saurait se mouvoir dans l'irréel*," constitutes, in many ways, a turning point in Bergsonism. Bergson's claim is that the intellect (which is an extension of action) does not merely produce fictions, but actually attains the absolute, though, of course, only from one "angle," namely in relation to matter. If the intellect can reach what matter contains of the absolute, this is because these two realities are constituted through a reciprocal adaptation. Bergson defends this position in Chapter III below, and it is the basis for his critique of Spencer throughout this book.

6   Cf. below, 199–200, as well as editorial note 47 to Chapter III.

7   For more on Bergson's characterization of life here, cf. below, 187. See also Henri Bergson, "La conscience et la vie [1911]," in *L'énergie spirituelle*, 8, 13 ["Life and Consciousness," in *Mind-Energy*. This chapter is henceforth cited as CV, with pagination referring to the French edition]. Note as well that the word "vision" belongs to the vocabulary that Bergson uses in relation to his notion of "intuition." For more, see "intuition, vocabulary of" in the Index.

8   This image of the intellect as a "core" that is surrounded by the "fringe" of intuition is a recurrent and absolutely decisive metaphor in *L'évolution créatrice*. See "core" in the Index.

9   Later editions of *L'évolution créatrice* correct a typo here.

10 This is the project Bergson attempts in Chapter III. Cf. below, 187.

11 For more on this idea of a collaborative philosophy, which is crucial for Bergson, cf. below, 192–93; CV, 4; Henri Bergson, *Les deux sources de la morale et de la religion* [1932], critical ed. Frédéric Keck and Ghislain Waterlot, series ed. Frédéric Worms (Paris: Presses Universitaires de France, 2013), 262–64 [*The Two Sources of Morality and Religion*, trans. R. Ashley Audra and Cloudesley Brereton, with W. Horsfall Carter (London: Macmillan, 1935). Henceforth cited as DS, with pagination referring to the French edition]; Henri Bergson, "Introduction (deuxième partie): De la position des problèmes [1922]," in *La pensée et le mouvant*, critical ed. Arnaud Bouaniche, Arnaud François, Frédéric Fruteau de Laclos, Stéphane Madelrieux, Claire Marin, and Ghislain Waterlot, series ed. Frédéric Worms, 25–98 (Paris: Presses Universitaires de France, 2013), 70 ["Introduction (Part II): Stating the Problems," in *The Creative Mind*, trans. Mabelle L. Andison, 33–106 (New York: Philosophical Library, 1946). This chapter is henceforth cited as DPP, with pagination referring to the French edition]; Henri Bergson, "La perception du changement [1911]," in *La pensée et le mouvant*, 144–45 ["The Perception of Change," in *The Creative*

Mind. This chapter is henceforth cited as PC, with pagination referring to the French edition].

12   Charles Dunan (1849–1931), who taught alongside Bergson at Collège Rollin, was often associated with Bergson primarily because of the three-part article mentioned in Bergson's note. If Bergson devotes the only footnote of his introduction to Dunan, this is surely because his work represents an important theory in Bergson's intellectual milieu that needed to be addressed. Indeed, there are some striking similarities between Dunan's approach and certain aspects of *L'évolution créatrice.* For instance, Dunan presents a double critique of mechanism and finalism, arguing that the living being cannot be accounted for via the *composition* of physico-chemical elements (mechanism) or via the positing of a *plan* that guides such a composition (finalism). Like Bergson, Dunan also discusses the life of the universe and the unity of all of its parts, the importance of psychological experience for understanding living beings, the shortcomings of the theories offered by Emil Du Bois-Reymond (1818–1896) and Pierre-Simon de Laplace (1749–1827), and the importance of understanding creativity in the human arts and in nature itself. These similarities are why Bergson recognizes the "profound insight" in Dunan's examination of life and in his insistence on adopting the point of view of time. Bergson's explicit reference to his own *Essai sur les données immédiates de la conscience* (E), however, is not merely a claim to have arrived at these ideas first. Rather, it would have reminded the contemporary reader of the unbridgeable gap between Dunan's and Bergson's respective theories. Dunan's manner of analyzing time fails to get beyond Bergson's criticism that time, conceived of as a synthesis of moments, is in fact space. In short, Dunan's theory that there is no pure succession is incompatible with Bergson's notion of *durée.* Ultimately an apologist for a Kantian philosophy, Dunan's theory of life falls prey to the same criticisms Bergson will offer of the Kantian conception of time. Dunan's analysis thus remains a Leibnizian attempt to establish a view from nowhere, rather than a new philosophy beginning from the concrete experience of lived duration.*

13   Bergson often uses the term *saisir* ("to grasp") in relation to his notion of intuition. For more, see "intuition, vocabulary of" in the Index.

CHAPTER I

1   The opening sentence of Chapter I thus makes it clear that one of Bergson's main goals in *L'évolution créatrice* is to provide nothing less than an analysis of the notion of existence itself. He addressed this topic in *Matière et mémoire* (cf. MM, 163–65) and will return to it again in Chapter IV (cf. the section "Existence and Nothingness," below, 277–98). Indeed, there the term "nothingness" is examined in "the sense in which we understand this notion to be opposed to that of existence" (*ibid.*, 277). For Bergson, "existence" signifies

both "life" and the "modal" category that Kant opposed to "possibility" and "necessity."*

2 Bergson is alluding to his first book, *Essai sur les données immédiates de la conscience (E)*.

3 Cf. PC, 170–73, where Bergson uses this same example to demonstrate that memories cannot reside within brain cells.

4 Throughout these pages, Bergson emphasizes this tripartite division of sensing, thinking, and desiring. In fact, Bergson often deploys this division when discussing the soul. Cf. Henri Bergson, "L'âme et le corps [1912]," in *L'énergie spirituelle*, 34 ["The Soul and the Body," in *Mind-Energy*. This chapter is henceforth cited as AC, with pagination referring to the French edition]; Henri Bergson (summarized by Jules Grivet), "Théorie de la personne d'après Henri Bergson [Cours au Collège de France, 1911]," in *Mélanges*, ed. André Robinet with Marie-Rose Mossé-Bastide, Martine Robinet, and Michel Gauthier (Paris: Presses Universitaires de France, 1972), 874; Henri Bergson, "On the Nature of the Soul [transcriptions and summaries of a conference given 1911]," in *Mélanges*, 952, 958–59; Henri Bergson, "Conférence de Madrid: La personnalité [transcription of a conference given in 1916]," in *Mélanges*, 1215; Henri Bergson (summarized by A. Birckel), "Conférence à Strasbourg: 'Sur l'âme humaine' [1919]," in *Mélanges*, 1317–18.

5 For this image of a thread, cf. DPP, 76; Henri Bergson, "Introduction à la métaphysique [1903]," in *La pensée et le mouvant*, 193, 207 ["Introduction to Metaphysics," in *The Creative Mind*. This chapter is henceforth cited as IM, with pagination referring to the French edition].

6 For Bergson's critique of the notion of a "substrate," linked as it is in what follows to the affirmation of the substantial nature of *durée*, cf. below, 300; DPP, 73–76; Henri Bergson, "L'intuition philosophique [1911]," in *La pensée et le mouvant*, 140–41 ["Philosophical Intuition," in *The Creative Mind*]; PC, 165–66, 173–74; and IM, 208–10.

7 This is perhaps the most definitive formulation of one of Bergson's most important, though rarely discussed, theses: that *durée* is not the product of a synthesis undertaken by a self that would be transcendent to it. Rather *durée is* this synthesis itself, an immanent synthesis that occurs between the moments that it itself constitutes as such. And since every act of synthesis must be the act of a consciousness, consciousness and *durée* must be identified. Consciousness *is durée*; it is not *in durée*. Cf. E, 89–90; Henri Bergson, *Durée et simultanéité* [1922], critical ed. Élie During, series ed. Frédéric Worms (Paris: Presses Universitaires de France, 2009), 41, 45–47 [*Duration and Simultaneity*, trans. Leon Jacobson, intro. Herbert Dingle (Indianapolis: Bobbs-Merrill, 1965)]. See also Frédéric Worms, "La conception bergsonienne du temps," *Philosophie 54, Henri Bergson* (June 1997): 76–79.

8 For Bergson, the term "symbol" designates the view [of moving reality] adopted by the intellect. Cf. IM, 177–82; as well as "symbol" in the Index. For the important image of "fabric" [*étoffe*], which allows Bergson to avoid the concepts of essence, nature, and, at least to a certain extent, substance, see the Index.

9   For Bergson's thesis regarding the substantiality of *durée*, see "substance, *durée* as" in the Index.

10   For this analysis, cf. *E*, 89; CV, 17; PC, 168; *Durée et simultanéité*, 46, 52. See also Worms, "La conception bergsonienne du temps," 79. [TN: For more on Bergson's embracing of the "virtual" despite his critique of the notion of "the possible," the pairings "possible/real" and "virtual/actual," and the differences between the verbs *réaliser* ("to realize") and *actualiser* ("to actualize"), see PR and Deleuze, *Bergsonism*, trans. Hugh Tomlinson and Barbara Habberjam (New York: Zone Books, 1991), 96–106.]

11   In Bergson's work, the term *progrès* ("progress" or "progression") almost never signifies "improving," a meaning that would be too teleologically tinged for him. Rather, it signifies a gradual development or an unfolding—a "progression." Cf. below, 105. [TN: Following Arnaud François's reasoning, I have translated *progrès* here and elsewhere as "progression" or "development" rather than "progress."]

12   For similar images related to the conservation of the past, cf. *MM*, 165; AC, 55; DPP, 80; PC, 170–73.

13   Cf. *MM*, 165–66; DPP, 80; PC, 152, 170–73.

14   Cf. Henri Bergson, "Le rêve [1901]," in *L'énergie spirituelle*, 95–96 ["Dreams," in *Mind-Energy*].

15   For the image of "contraband," cf. below, 49; Henri Bergson, "'Fantômes de vivants' et 'recherche psychique' [1913]," in *L'énergie spirituelle*, 79 ["'Phantasms of the Living' and 'Psychical Research,'" in *Mind-Energy*]. The same idea, but without the image, can be found below, 60, 120.

16   Cf. below, 6; IM, 183–84.

17   In 1907, Bergson was becoming increasingly interested in questions connected to the notion of personality. These questions would become a principal object of his reflection during the 1910s. See Bergson, "Théorie de la personne," in *Mélanges*; Bergson, "On the Nature of the Soul"; Henri Bergson, "Cours de Bergson à Columbia University: Sexuality and Liberty [1913]," in *Mélanges*, 981–89 ["Extracts of the Columbia University Lectures of Henri Bergson," *The Chronicle* XIII (1913): 214–20]; Henri Bergson, "Onze conférences sur 'La personnalité' aux Gifford Lectures d'Edinburgh/University of Edinburgh: Gifford Lectures, 1914: 'The Problem of Personality' [1914]," in *Mélanges*, 1051–71; Henri Bergson, "Conférence de Madrid sur l'âme humaine [transcription of a conference given in 1916]," in *Mélanges*, 1202–15; Bergson, "Conférence de Madrid: La personnalité," in *Mélanges*, 1215–35; Bergson, "Conférence à Strasbourg 'Sur l'âme humaine'," in *Mélanges*, 1316–19. In these texts, the question of personality is explored in the following ways. First, insofar as personality is *durée*, i.e., insofar as it is one and multiple (see IM, 197), though not in the sense in which there could be genuine "splits" in personality. Second, the problem becomes an occasion for meditating upon the relations between freedom and determinism. Third, it is the context for a reflection on the substantiality of the soul, understood as change (cf. PC). For additional occurrences of "personality" in *L'évolution créatrice*, see

the Index.* [TN: For more on personality in Bergson, see Donald A. Landes, "Personality and Character: From a Privileged Image of *Durée* to the Core of a New Metaphysics," in *The Bergsonian Mind*, ed. Mark Sinclair and Yaron Wolf (London: Routledge, 2021), 99–112.]

18 This image of a superhuman intellect, drawn from Pierre-Simon de Laplace's story of the all-knowing demon, recurs throughout the book (see the Index, "Laplace's Demon"). This image illustrates the question of whether the unforeseeable exists *in itself* or merely *for us*. Bergson, against Laplace and Emil Du Bois-Reymond (cf. below, 38–39), supports the first option. According to Bergson, and to adopt his formulation, we would be wrong to assume that "everything is given" in advance (cf. below, 38). [For more on Du Bois-Reymond, see editorial note 103 to Chapter I.]

19 Because of its continuous and creative nature, our personality can be compared to a "history." This is true for Bergson as early as his *Essai sur les données immédiates de la conscience* (E, 13, 125, 139–42). "History" is a constant theme in Bergson's work and particularly important in *L'évolution créatrice*, where life itself is compared to a history (see "history" in the Index). History is even the very first noun used in this book (see above, v).

20 Cf. Henri Bergson, "Le possible et le réel [1920/1930]," in *La pensée et le mouvant*, 110 ["The Possible and the Real," in *The Creative Mind*. This chapter is henceforth cited as PR, with pagination referring to the French edition].

21 Cf. above, 5; IM, 183–84.

22 For the image of a "creation of the self by the self," cf. CV, 23; AC, 31.

23 Cf. Bergson's letter: "Bergson à L. Brunschvicg [February 26, 1903]," in *Mélanges*, 586–87.

24 For a more detailed version of this analysis, cf. *E*, 124–30.

25 For other occurrences of the term "existence in general," though with slightly different connotations, see below, 279, 285.

26 Bergson intends the term "imponderable" here to refer to the classical concept of the "ether," a fluid substance that was believed to fill up the entire universe. For instance, Descartes explained the orbits of the planets as vortices in the ether; the Dutch mathematician, astronomer, and physicist Christiaan Huygens (1629–1695) and French physicist Augustin-Jean Fresnel (1788–1827) posited the concept of ether to explain the propagation of light; British physicist J. J. Thomson (1856–1940) extended the concept of ether to electrical and magnetic phenomena, and spoke of the ether as being comprised of molecular vortices; and Scottish physicist James Clerk Maxwell (1831–1879), who showed the unity of electrical and magnetic phenomena, also postulated an ether. English physicist Michael Faraday (1791–1867), whom Bergson often associated with Lord Kelvin (he discusses their concepts several times: see Index, "physics, Thomson and Faraday"), nevertheless called this notion into question (see editorial note 67 to Chapter III). It was Albert Einstein's theory of special relativity in 1905 that rendered the concept of the ether useless. It is worth noting that Bergson himself places quotation marks around "imponderable," even if, in 1907, physics had yet to fully integrate Einstein's critique of

the ether. Nevertheless, Bergson appears to maintain a certain pertinence for the notion for physics toward the end of this book (cf. below, 364).*

27   See "universal interaction" in the Index.

28   Cf. *E*, 86–89, 144–49.

29   Cf. below, 335–39. Bergson's two paragraphs here are very similar to the ones cited below. Bergson's practice of repetition, which always serves a philosophical purpose, is relatively common in *L'évolution créatrice*.

30   In other words, common sense, science, perception, and language all move in a similar direction: that of intellect and of action. Cf. "intellect, direction of" in the Index.

31   Cf. below, 338.

32   For this famous example, see "sugar" in the Index. Bergson claimed that the example first occurred to him during his course at the Collège de France in 1901–1902. As he stated the following year: "There is an example that I mentioned last year, a fact that is quite remarkable and, however you look at it, quite mysterious. It is the simple fact that if I drop a lump of sugar into a glass of water to make a glass of sweetened water, I must wait a certain amount of time for the sugar to dissolve. No matter what I do, I must wait. In other words, a certain process of maturation in exterior and material reality must take place. If there were no *durée* in things, no *durée* internal to things, and analogous to my own *durée*, how could I understand a fact of this kind? [...] It must be the case, then, that the things themselves endure, and endure in a manner analogous to the manner in which I myself endure. Such was the conclusion I reached on this point." Henri Bergson, *Histoire de l'idée de temps: Cours au Collège de France, 1902–1903*, ed. Camille Riquier, series ed. Frédéric Worms (Paris: Presses Universitaires de France, 2016), 254–55. [TN: The translation of Bergson's passage is my own.]*

33   To highlight the significance of this famous argument, which is repeated at the end of the book, consider the implicit discussion of matter in Bergson's first book, *Essai sur les données immédiates de la conscience*. Even if matter can be identified with space, defined as "exteriority without succession," and these together contrasted with *durée*, defined as "succession without exteriority," then it still remains impossible to explain why material objects seem involved in a becoming analogous to my own rather than as existing in a fixed world (cf. *E*, 81). In *Matière et mémoire*, Bergson thus distinguishes between matter and space, and even attributes a *durée* to matter itself (*MM*, 226–35). The example of the sweetened water confirms that material objects (and perhaps the universe as a whole) endure. As for the simultaneity of my *durée* and that of matter, cf. *Durée et simultanéité*, 41–67.*

34   For the later discussion of this tendency in matter, cf. below, 158, 190, 203–5.

35   For more on the notion of "continuous, creation," see the Index.

36   For other occurrences of this distinction, see Chapter III below, as well as "ascent/descent" in the Index.*

37   For the image of the mirror in this context, cf. *MM*, 15.*

38   For this important analysis, cf. below, 228, 301; *MM*, 220–24.

39 Bergson's argument here takes aim, in particular, at philosopher Félix Ravaisson (1813–1900). Cf. Félix Ravaisson, *Of Habit / De l'habitude*, trans. Clare Carlisle and Mark Sinclair (London: Continuum, 2008), 29.

40 For Bergson's internal reference, cf. below, 258–61. Although the question of individuality, already mentioned in the Introduction (see above, vi), is one of the most important and most difficult themes in *L'évolution créatrice*, it also remains relatively obscure and has not often been discussed by commentators. Nevertheless, it is a theme that allows Bergson to connect his philosophy of *durée* to the biological debates of his era, for example, over concepts such as the pertinence of cellular theory, the idea of an interindividual continuity implicit in the theory of heredity, and the very possibility of considering the organism to be a composition of physico-chemical elements. See "individual/individuality" in the Index.*

41 For Bergson's internal reference, see Chapter II below.

42 Here Bergson is largely influenced by the theories of German biologist August Weismann. Cf. below, 26–27, 43. For more on Weismann, see editorial note 68 to Chapter I.

43 In order to conform to the terminology universally used by the biologists of his era, Bergson usually capitalizes the names of animals and plants. [TN: In the text, I have maintained Bergson's capitalization wherever appropriate.] A hydra is a polyp measuring a few millimeters that can live in colonies comprising many individuals. The hydra was a commonly used example at the time. Notably, it is discussed in the work of zoologist Edmond Perrier (for more on Perrier, see Bergson's second note to page 260 and editorial note 241 to Chapter III). With the mention of sea urchins, Bergson is referring to the experiments on sea urchin embryos that made German biologist and philosopher Hans Driesch (1867–1941) famous. For more on Driesch, see Bergson's note to page 42 and editorial note 115 to Chapter I.

44 This is how Bergson defines determinism, namely in terms of time and causality.

45 This notion of "independence" is central in Weismann's theory. For more on Weismann, see below, 26–27, 79–83.

46 On the "facts of regeneration," cf. below, 42, 75–77.

47 This new and important thesis in Bergson's philosophy is nonetheless presented here in a formulation that is reminiscent of *Matière et mémoire* (cf. *MM*, 156).

48 Infusoria, or ciliated infusoria: these dated terms designate a sub-group of Protozoa that are characterized by the presence of cilia. Today we speak rather of the "ciliates." These organisms live in freshwater or saltwater, and some are parasites. The Stentor (discussed by Bergson below, 261) is a member of the ciliates.*

49 The journal *Archiv für Entwickelungsmechanik der Organismen* [*Archive for the Development Mechanics of Organisms*] was edited by the important German zoologist Wilhelm Roux (1850–1924), one of the founders of embryology. It is often called "*Roux's Archiv*," as Bergson himself does below (see Bergson's note on page 34). In the cited article by the zoologist

Gary N. Calkins (1869–1943), the goal was to refute Émile Maupas (1842–1916), Otto Bütschli (1848–1920), and Theodor W. Engelmann's (1843–1909) theory according to which conjugation brings about a "rejuvenation" of the cell, thwarting the effects of binary division. Claiming to follow Bütschli (cf. Bergson's note to page 33) and Charles Sedgwick Minot (cf. Bergson's note to page 17), Calkins argues that there exists a "potential" of "vitality" given at the beginning of life, and death comes once this potential is used up ("Studies on the Life-History of Protozoa," 166). The organism, however, alternates between phases of division and rejuvenation, and these rejuvenations can be caused by things other than conjugation. In the background of these discussions is Weismann's thesis according to which the Protozoa are, like germinal cells, in principle immortal. This last theory is shared by Calkins and Maupas, and Bergson implicitly subscribed to it as well. [TN: For more on Maupas, see editorial note 93 to Chapter I; for more on Minot, see editorial note 53 to Chapter I; and for more on Bütschli, see editorial note 87 to Chapter I.]*

50 For Bergson's internal reference, cf. above, 13.

51 For more on the image of an organism as a "society," see "society, organism as" in the Index.

52 Bergson is likely alluding to his discussion of the nature of the intellect and its difference from instinct (cf. below, 138–66). Indeed, the final section of Chapter II is titled: "Life and Consciousness: The Apparent Place of Man in Nature."*

53 Charles Sedgwick Minot (1852–1914) was an American scientist and a founding member of the American Society for Psychical Research. In this article, Minot focuses on a statistical and cytological explanation of aging, although he does call for a "general biology" that would study the "organism as a whole" and consider the organism in its milieu. Minot, "On Certain Phenomena of Growing Old," 286.*

54 "Interior milieu" is a concept established by physiologist Claude Bernard (1813–1878) that designates the milieu in higher animals in which cells are bathed. That is to say, essentially, in the blood and the lymphatic system. Cf. below, 124–25.

55 Félix Le Dantec (1869–1917) was one of the most important biologists of Bergson's day. He was a student of zoologist Élie Metchnikov (1845–1916) and above all of Louis Pasteur (1822–1895). Le Dantec had allegiances to both neo-Lamarckian theory and mechanism, and shared certain theoretical positions with Spencer, German embryologist Ernst Haeckel (1834–1919), and American paleontologist Edward Drinker Cope (1840–1897). For more on Haeckel, see editorial note 104 to Chapter I; for more on Cope, see editorial note 92 to Chapter I. Here Bergson is referring to Part II of Le Dantec's book, which is titled "Why We Become Old." The term "crusting over" is used throughout the section, and Le Dantec also uses the term "plastic substance," which Bergson himself adopts (cf. below, 34, 122–23, 256). Bergson mentions the "profound" nature of this theory because it grasps aging not as accidental to life, but as its very essence.

Le Dantec, "Pourquoi l'on devient vieux," in *L'individualité et l'erreur individualiste*, 84–104. For Le Dantec's critical response to *L'évolution créatrice*, see his essay included in this edition below.*

56   Metchnikov collaborated with Pasteur and is perhaps best known for his discovery of phagocytosis. The first article cited appeared in one of the most important journals in this domain, *L'année biologique: Comptes rendus annuels des travaux de biologie générale*, edited by zoologist Yves Delage. Like Calkins' article above, the conceptual framework of "Revue de quelques travaux sur la dégénérescence sénile" is based upon a Weismannian understanding according to which unicellular organisms are at least in principle immortal.*

57   For this argument in relation to continuity, which resonates with Bergson's criticism of Zeno of Elea, cf. below, 31, 92, 268, 299, 305, 307, 361.

58   For Bergson's internal reference, cf. below, Chapter IV.

59   For Bergson's internal reference, cf. below, Chapter III. As Bergson will argue, these intellectual tendencies are designed to allow us to think matter.

60   Note that in *L'évolution créatrice*, Bergson identifies that which is dead with the inorganic, an identity based on his philosophical position [concerning the relationship between matter and Consciousness]. Cf. "dead" in the Index.

61   From *Matière et mémoire* onward, Bergson defines matter as "a present that is continuously renewed." Cf. below, 202; MM, 154, 168, 236; CV, 5; AC, 32–33; "L'intuition philosophique," in *La pensée et le mouvant*, 142; IM, 184.

62   See René Descartes, *Principles of Philosophy*, in *The Philosophical Writings of Descartes*, vol. 1, trans. John Cottingham, Robert Stoothoff, and Dugald Murdoch (Cambridge: Cambridge University Press, 1985), Part I, § 21. Cf. below, 345; E, 156.

63   For the important metaphor of theme and variation, which Bergson uses in various senses, see "theme and variation" in the Index.

64   Paleontology will have to confirm what embryogenesis can only establish in a hypothetical manner. Comparative anatomy, the third science mentioned here, intervenes immediately afterward in the demonstration. Cf. CV, 18. Embryology plays a key role in *L'évolution créatrice*, and of these three sciences perhaps most naturally connects with considerations of temporality. Cf. below, 362.*

65   In the first edition of *L'évolution créatrice* (1907), Bergson writes "the recent experiments," rather than the "curious" ones. For more on Bergson's reading of the Dutch botanist and geneticist Hugo de Vries (1848–1935), see editorial note 174 to Chapter I.

66   Bergson has in mind here a form of "philosophical" probability obtained by drawing together a variety of "lines of facts." Cf. CV, 4; DS, 262–64.

67   Through the theory offered in *L'évolution créatrice*, Bergson renders conceivable this somewhat radical notion of "life in general" as something that transcends individuals. For more, see "life, in general" in the Index.*

68   German biologist August Weismann (1834–1914) was a key figure in debates on heredity (from the 1880s onward) and is an important reference in

L'évolution créatrice. His principal discovery was the "continuity of the germ-plasm," along with the related distinction between the *somatic cells*, the cells of the body of the individual, and the *germ cells*, the reproductive or sexual cells. Initially a follower of Haeckel, Weismann's discovery led him to reject the theory of the inheritance of acquired characters and to contribute to establishing neo-Darwinism. This marked a break from his previous adherence to the theory of acquired characters. It is difficult to define the exact nature of the "germ-plasm": Weismann situates it entirely in the reproductive cell, then limits its location to the kernel, and, more precisely, to the stained filaments that Haeckel's disciple Wilhelm Waldeyer (1836–1921) would, in 1888, name "chromosomes." Cf. August Weismann, *The Germ-Plasm: A Theory of Heredity*, trans W. Newton Parker and Harriet Rönnfeldt (New York: Charles Scribner's Sons, 1893). For more on Weismann, cf. below, 43, 79–83.*

69 Bergson here paraphrases the entirety of zoologist Louis Roule's (1861–1942) criticism of Weismann, even adopting some of his expressions. Roule's *L'embryologie générale* is presented as a manual for embryology.

70 For this model of an abrupt "spending" of energy, cf. below, 116–18, 121–25, 247, 252–53, 255–57.

71 For the various definitions and characterizations of life in this work, see "life, definition and characterization of" in the Index.

72 James Mark Baldwin (1861–1934) was trained as a philosopher but helped introduce experimental psychology to the United States and attempted to set it on an evolutionary foundation. Bergson knew the author personally and maintained a dialogue with him. Cf. Bergson, "Rapport sur *Le Darwinisme dans les sciences morales* de J. M. Baldwin [1913]," in *Mélanges*, 1020–23). Baldwin's idea of the irreversibility of evolution appears prominently in his book, particularly in the chapter "The Theory of Genetic Modes." Baldwin distinguishes a *genetic* science, which recognizes the irreversibility of certain phenomena, from an *agenetic* science. For Baldwin, the phenomena of life illustrate paradigmatically the necessity of associating the two perspectives, genetic and agenetic. Baldwin writes: "In the life processes there seems to be a real genetic series, an irreversible series. Each stage exhibits a new form of organization. After it has happened, it is quite competent to show, by the formulas of chemistry and physics, that the organization is possible and legitimate. Yet it is only by actual observation and description of the facts in the development of the organism, that the progress of the life principle can be made out. The former is quantitative and analytic science; the latter is genetic science." (Baldwin, *Development and Evolution*, 330). Drawing on Baldwin, Bergson's position is that, although the organism is a machine, it is one that remains unpredictable.*

73 Cf. *E*, 86–89, 144–49.

74 This idea of partial views taken by the intellect of phenomena that are in themselves "simple" is a key part of Bergson's overall argument in this book. Cf. for instance, below, 31, 90–98; IM, 177–82. See "views" in the Index. [TN: It

is also worth noting that in certain contexts the word *vues* ("views") has been translated as "views or snapshots" to emphasize the photographic metaphor for frozen moments or images of a movement that Bergson often invokes.]

75 Bergson is referring to the hypothesis of sudden mutations in the work of De Vries and other "mutationists." Cf. below, 65–70.

  76 Gabriel Séailles (1852–1923), Paul Janet's successor as Chair of History of Philosophy at the Sorbonne, focused primarily on aesthetics, ethics, and sociology. The book that Bergson discusses was dedicated to Ravaisson. Like Ravaisson, Séailles was particularly interested in the figure of Leonardo da Vinci. Séailles's main thesis is that between a genius and other men there exists a "difference in degree" and not a "difference in nature" (Séailles, *Essai sur le génie dans l'art*, viii). As a result, he argues that we can discover the "laws" of genius by drawing on the general laws of psychology (*ibid.*, vii–viii). For more on Séailles's interpretation of Leonardo da Vinci, see Gabriel Séailles, *Léonard de Vinci, l'artiste & le savant: Essai de biographie philosophique* (Paris: Perrin, 1892). Bergson, however, distances himself from Séailles by insisting that, although life must be thought of in terms of a "multiple unity" [*d'unité multiple*], this unity is not numerical; rather it is an indistinct multiplicity, that is to say, *durée*.*

77 For more on what we might call Bergson's "pragmatism," which has to do with science as well as the majority of our faculties, cf. below, 44, 85, 94, 192, 197, 249, 296, 298–300, 306, 329.

78 For the idea that our intellect has its instincts, cf. *E*, 100. Bergson uses the principle "like produces like," which is a classical formulation of determinism, to define the intellect. Cf. below, 45–46, 52, 201, 271.

79 Bergson uses precisely this phrase as a title for the section in Chapter IV that deals with ancient philosophy. For other occurrences, cf. "intellect, natural inclination of" in the Index.

80 As such, this role for philosophy (to go against the natural inclination of the intellect) also serves to carry us beyond the human condition (cf. below, 193, 368).*

81 Bergson's claim here is extremely important, although often forgotten. Cf. above, 12.

82 For Bergson's internal reference to his earlier discussion of reality as a whole, cf. above, 7–11.

83 For a similar discussion, cf. IM, 179–80, 191.

84 For more on this important comparison, cf. below, 214.

85 For the notion of a method of "qualitative integration," implicit in this discussion, cf. *MM*, 205; IM, 215.

86 For this interpretation of the history of mathematics, cf. below, 333–34; DPP, 29; IM, 214–15; and *DS*, 58–59.

  87 Otto Bütschli (1848–1920) was, alongside Roux (mentioned above), one of the founders of developmental mechanics. Bergson's note here and the following two refer to authors who collaborated in the final decades of

the nineteenth century. Their aim was to identify the possible areas where physico-chemical phenomena could imitate biological phenomena. They focused on unicellular organisms as a foundation for understanding more complex organisms. Bergson refers to the first part of Bütschli's book, which focuses on the structures of foam and protoplasm.*

88   Bergson appears to be drawing more specifically from Rhumbler, "Versuch einer mechanischen Erklärung der indirekten Zell und Kerntheilung," 548–54, where Bütschli's book is discussed.

89   Amoeba is a common term that designates several unicellular protozoan rhizopods that shoot out temporary extensions, or "pseudopods" (hence the term "rhizopod"), that help the organism to move or grasp food.

90   In *Studien über Protoplasmamechanik*, German botanist Gottfried Berthold (1854–1937) writes that the emitting of a pseudopod takes place when the attraction between the parts of the Amoeba and the foreign body are stronger that the reciprocal sum of attraction between the parts of the Amoeba itself, and stronger than the attraction between the parts of the Amoeba and the repressed surroundings, namely the water. The author merely notes the phenomenon.

In *Théorie nouvelle de la vie*, Le Dantec observes that when the Amoeba gets close to a solid body, the "molecular phenomena of attraction" causes the Amoeba to flatten (60). Consequently, the causes that habitually determine its movement in general, such as the movement of the water, can also provoke a deformation such as the appearance of the pseudopods. Le Dantec explicitly rejects any form of "teleological" explanation of the appearance of pseudopods (*ibid.*, 61).

91   For more on the opposition between plastic and energetic substances, cf. below, 122–23, 256.

92   American paleontologist Edward Drinker Cope (1840–1897) was a neo-Lamarckian and is an important reference in *L'évolution créatrice* (cf. below, 77–78). Cope was a critic of the Darwinian theory of natural selection and argued that the duration of different phases of embryonic development could account for variation, along with (as in Lamarck) a modification of the organs through their use. In the book cited here by Bergson, Cope critiques both Thomas Henry Huxley and Spencer (for more on Huxley, see editorial note 104 to Chapter I, below). Cope developed the distinction between *catagenetic* and *anagenetic* energies in his article "The Foundations of Theism," *The Monist* 3, no. 4 (1893): 630. There, for a more popular audience, is where Cope most explicitly associates these two energies with life and death. In *The Primary Factors*, Cope understands the "raising" action as one of "building machines" and the conversion of inorganic substances into complex compounds (478). As such, in this more scientific presentation of the distinction, he insists that the energies not be identified with simply organic or inorganic substances. Indeed, he had already specified that in living bodies we find both energies (although we only find *anagenetic* energies in living systems, which he equates with "consciousness" (Cope, "The

Foundations of Theism," 630). Cope's overall idea of two energies moving in opposite directions, characterized by the metaphors of raising and falling, indicates a deep resonance with Bergson's philosophy—and thus is in the same lineage of thought that passes from Plotinus through Ravaisson (cf. below, Bergson's footnote to page 211). The distinction also implies a certain conception of entropy as discussed by Bergson below (cf. below, 243–44).*

93 The presence of psychological activity even in the humblest manifestations of life is the philosophical thesis that Bergson will attempt to establish by taking up Cope's scientific considerations (see below, 77–78).

Émile Maupas (1842–1916) was an important zoologist in Bergson's time. Maupas does not use the term "psychological," suggesting that his inclusion in this group of thinkers is the result of Bergson's understanding of Maupas's anthropomorphic expressions. For instance, regarding *Ophryoglena magna*, Maupas writes: "We see it approach the objects that it encounters, palpating them with its lips [*lèvres buccales*] and engulfing them when it finds them suitable for nourishment" (471).

Paul Vignon (1865–1943) was an important biologist and philosopher of science who also deploys anthropomorphic descriptions that imply intelligence in the direction of trial and error. The book cited here was his thesis, undertaken in Delage's lab (for more on Yves Delage, see below, Bergson's note to page 60).

Herbert Spencer Jennings (1868–1947) was an important American zoologist. In *Contributions to the Study of the Behaviour of Lower Organisms*, he shows that the movements of lower organisms, in part, can be understood in terms of trial and error, and are not, as was believed until then, reducible to simple phenomena such as tropism (that is, the turning of the organism in response to external stimuli). Indeed, whereas tropism is a generalized and fixed reaction, the method of trial and error implies a progression (see Jennings, *Contributions to the Study of the Behavior of Lower Organisms*, 250–51). Reflecting upon which criteria the organism uses to recognize the error, Jennings invokes the analogy of pain, thus implying that lower organisms might in some way be capable of pleasure and pain (*ibid.*, 249–50). Jennings's argument is based upon a definition of intelligence as the ability to transform one's own behavior in response to success and error (*ibid.*, 237).*

94 Edmund Beecher Wilson (1856–1939) was an American cytologist and embryologist. He conducted his research from a purely evolutionist perspective and insisted on giving his research a philosophical dimension. *The Cell in Development and Inheritance* draws together Wilson's research on the relationship between cell theory and evolutionary theory, as well as on the respective roles of the nucleus and chromosomes in heredity. The phrase cited by Bergson occurs in the final chapter where Wilson compares Weismann's theory (which he names "preformation") to other theories of the inheritance of acquired characters (which he names "epigenetic"). See Wilson, *The Cell in Development and Inheritance*, 328.*

95   Albert Dastre (1844–1917) was a student of Claude Bernard, whom he eventually replaced as Chair of Physiology at the Sorbonne. *La Vie et la Mort* examines the principal schools in the philosophy of life (animism, vitalism, mechanism) and evaluates their relevance with regard to contemporary scientific data. Dastre argues that physico-chemical explanations of the living being cannot be complete without accounting for generation and evolution, which are the "most refractory and inaccessible to physico-chemical explanations." See Dastre, *La Vie et la Mort*, 43 [Dastre, *Life and Death*, 46].*

96   Bergson's use of the term "naturalist" perhaps follows Dastre, who in the passage previously cited uses it to refer to researchers studying the functions of generation, development, and evolution (in contrast to physico-chemical studies). Above, Bergson also uses the term to refer to those working in comparative anatomy and in paleontology (cf. above, 24–26). He again uses the term when talking about Cope (cf. above, 34).*

97   For this Kantian argument, cf. below, 284–86, 289–90; *DS*, 255; *DPP*, 50–51.

98   For Bergson's internal reference, see above, 11–23.

99   For Bergson's internal reference, see the preceding paragraph.

100   The idea that "that which does nothing is nothing" is an important premise for Bergson here (cf. below, 39, 198) and elsewhere (cf. *PR*, 102).

101   For the important locution "everything is given" [*tout est donné*], cf. below, 39, 46, 51, 241, 269, 339, 344, 348.

102   Cf. "Laplace's Demon" in the Index.

103   German physiologist Emil Du Bois-Reymond (1818–1896) was one of the founders of experimental physiology. Here Bergson provides a translation of a phrase that is found near the beginning of the German original of Du Bois-Reymond's talk (*Über die Grenzen des Naturerkennens*), and Du Bois-Reymond then goes on to discuss Laplace (18–20).*

104   Thomas Henry Huxley (1825–1895) was a morphologist, zoologist, and paleontologist, though he was equally interested in broader political and cultural topics. Uncharacteristically, Bergson does not provide a citation for this passage, which is drawn from Huxley's review of Haeckel's book. In fact, although Haeckel was a key figure in nineteenth-century biology and is mentioned several times in these editorial notes, Bergson himself does not cite Haeckel in *L'évolution créatrice*. Perhaps Bergson found his work outdated, given that it ultimately combined a theory of the inheritance of acquired characteristics with an associationist theory of the living being, resulting in a sort of mechanistic "monism." In addition, Haeckel had been accused of having falsified his results in the book (*Natürliche Schöpfungs-Geschichte*) reviewed by Huxley here, although some of his contributions survived these polemics. Finally, Haeckel advanced the theory of social Darwinism in Germany, and his "Monist League" has since often been viewed as a scientific birthplace of European fascism. The contrast between Bergson's silence regarding Haeckel and his relationship to Huxley could not be more striking. For example, Bergson

explicitly dedicated his lecture "La conscience et la vie" (CV) to Huxley. Huxley was a friend of Darwin, and ardently supported Darwin's theory. The title of Huxley's 1863 book, *Evidence as to Man's Place in Nature* (London: Williams and Norgate, 1863), resonates with the title Bergson gives to the final section of Chapter II below: "Life and Consciousness: The Apparent Place of Man in Nature."*

105 For this principle, cf. above, 37–38; below, 198; PR, 102.

106 For more on Bergson's terminology for "sensing" or "feeling," which should be compared with the terminology he uses around "intuition," see "feeling/sensing" in the Index.

107 This is perhaps a specific allusion to the Cartesian project of a *mathesis universalis* (cf. below, 344–48) or more generally to the series of philosophical "systems" rejected in Chapter IV below.*

108 For more on this Leibnizian doctrine, see the third (February 25, 1716), fourth (June 2, 1716), and fifth (mid-August, 1716) letters to Clarke, in Samuel Clarke and Gottfried Wilhelm Leibniz, *The Leibniz-Clarke Correspondence: Together with Extracts from Newton's* Principia *and* Opticks, ed. H. G. Alexander (Manchester: Manchester University Press, 1956). For more on Bergson's reading of Leibniz's understanding of time as "confused perception," see below, 351–52 and editorial note 207 to Chapter IV.*

109 For more on the "psychological nature" in question here, cf. below, 77–78.

110 Bergson clearly incorporates conflict into his understanding of life. See "conflict" in the Index.

111 The complexities of the notion of individuality are what led Bergson to adopt—admittedly in a recurrent and rather un-thematized manner—a model of the organism as a "society" or a "collectivity" [*collectivité*]. This model was relatively standard at the time and is adopted, at least to some extent, by most of the authors cited up to this point. In particular, Roux characterized the organism as a society and its parts as engaging in a Darwinian-style struggle for survival. See, Wilhelm Roux, *Der Kampf der Theile im Organismus: Ein Beitrag zur Vervollständigung der mechanischen Zweckmässigkeitslehre* (Leipzig: Wilhelm Engelmann, 1881). Roux's conception was influential on Nietzsche, but Bergson keeps his distance because he finds it overly mechanistic and associationist. For more on how Roux's conception was replaced by a form of dissociation, which is arguably closer to Bergson's own position, see below, 259–61 and editorial notes 242 and 243 to Chapter IV. See also "society, organism as" in the Index.

112 The distinction between germinal cells and somatic cells is drawn from Weismann's terminology. Cf. above, 26–27; below, 79–83.

113 Bergson's use of hierarchical vocabulary reveals his implicit adoption of the model of the organism as a "society" discussed in editorial note 111 to Chapter I.

114 Despite Bergson's claim that "life is, above all, a tendency to act upon brute matter" (below, 97), and the important role that the *élan vital* plays throughout this book, these lines demonstrate that he was not, at least according to the

classical connotation of the term, a "vitalist." According to a strict definition (in the tradition of physician and encyclopedist Paul-Joseph Barthez (1734–1806), and following Ravaisson's understanding of this tradition), "vitalism" denotes the doctrine according to which a "vital principle" or "vital force" exists as distinct from both the physico-chemical facts of the empirical world and from consciousness or thought. Bergson rejects what he takes to be a "superficial" vitalism that posits an external vital force capable of battling against physical forces. Cf. Henri Bergson, "La philosophie de Claude Bernard," in *La pensée et le mouvant,* 233; Félix Ravaisson, *La philosophie en France au XIXe siècle,* Recueil de rapports sur le progrès des lettres et des sciences en France (Paris: Imprimeries Impériale, 1867), 123–25. Given his affirmation that life is of a "psychological" nature, Bergson is perhaps technically closer to "animism" than he is to "vitalism," animism being the tradition that identifies life with the soul or with consciousness. But the novelty of Bergson's position prevents us from simply adopting any of these classical labels. For Bergson's discussion of vitalism, see below, 226–27.*

115 The expression "neo-vitalism" was used frequently at the time to designate the doctrines of Driesch and Reinke. German biologist and philosopher Hans Driesch (1867–1941) studied under Weismann and Haeckel. He became famous for his experiments in embryology on sea urchins in which the division of certain cells resulted in the development of complete (though smaller) larvae. Similar experiments had already been undertaken by Roux and had been done again since, notably by Wilson (cf. above, Bergson's note to page 36) and the mutationist Thomas Hunt Morgan (cf. below, Bergson's note to page 80). The results depend, above all, on the stage of division on which the artificial separation is carried out. Driesch named the forces driving such a rebuilding "entelechies," insisting upon the finalist (or teleological) sense of this Aristotelian term. Each organism, according to Driesch, follows a veritable *plan,* and that is probably what caused Bergson concern.

Botanist Johannes von Reinke (1849–1931) also promoted a philosophy of nature that is essentially theist. In reaction to Haeckel's "materialism," Reinke negates natural selection and makes evolution a simple "axiom." His "dominants"—a term already found in Leibniz, who discusses the soul as a dominant monad at the core of the aggregate of monads that make up the body—are the organizational principles of each living being and are at the service of an intellectually determined divine creative force. Here again the overly intellectualist aspect of this doctrine likely caused Bergson concern.*

116 This theory is important for Bergson because it asserts the continuity of life. According to the concept of *durée,* only a continuity—understood as an indistinct multiplicity—can be creative. But it should be noted that this is a radical interpretation of Weismann's theory of the continuity of germ-plasm. Cf. editorial note 68 to Chapter I.

117 This claim helps to justify why Bergson speaks of "life in general." This entire paragraph might be compared to what Bergson argued above on pages 26–27.

118 Cf. *MM*, 203–6; *DS*, 168–69, 173, 185; *DPP*, 34, 54–55; *PC*, 152. For more on the form of pragmatism that Bergson adopts in what follows, cf. editorial notes 77 and 78 to Chapter I.

119 For more on this analysis, cf. *E*, 153–58.

120 Cf. below, 211–17 and "intellect, natural inclination of" in the Index.

121 This description of habit bears some resemblance to Hume's, and Bergson elsewhere mentions Hume on this topic (cf., *E*, 152–53). In terms of Bergson's formulation here of the "principle of mechanism," cf. below, 57, 215.

122 This foreshadows Bergson's definition of humans as *Homo faber*. See below, 138–40.

123 Cf. above, 30 and "continuous, creation" in the Index.

124 For more on this belief, cf. *E*, 159–60; *DS*, 134–44.

125 For this formulation of the principle of mechanism, cf. above, 29; below, 52, 201, 271.

126 For other occurrences of the phrase "everything is given" [*tout est donné*], see editorial note 101 to Chapter I.

127 This characterization of the intellect will be contrasted with intuition. Cf. "intuition, vocabulary of" in the Index.

128 For this characterization of the intellect in terms of "retrospection" or "looking backward," and thus in terms of time, cf. below, 238, 267.

129 For this characterization of action, cf. *E*, 124–30. In terms of the role of retroactive explanation, cf. below, 48, 224; "Bergson à L. Brunschvicg [letter, February 26, 1903]," in *Mélanges*, 585–87.

130 Bergson is here alluding to an objection to his theory of free action in *Essai sur les données immédiates de la conscience* according to which the cost of his attempt to protect free action from mechanistic causality was too high, for it supposedly only left room for a purely capricious and lawless type of freedom. The objection was primarily formulated by philosopher Lucien Lévy-Bruhl (1857–1939) and by philosopher Gustave Belot (1859–1929). Bergson's impatience is explained by the fact that he had already responded to this objection (cf. *MM*, 207), and yet others continued to repeat it anew. See Lucien Lévy-Bruhl, Review of H. Bergson, *Essai sur les données immédiates de la conscience* (Alcan, 1889), *Revue philosophique de la France et de l'étranger* 29, no. 5 (1890): 519–38; Gustave Belot, "Une théorie nouvelle de la liberté," *Revue philosophique de la France et de l'étranger* 30 (1890): 361–92. For more, cf. "Bergson à L. Brunschvicg [letter, February 26, 1903]," in *Mélanges*, 585–87.

131 Cf. above, 47; below, 224; "Bergson à L. Brunschvicg [letter, February 26, 1903]," in *Mélanges*, 585–87.

132 The mention here of a conflict between systems already announces the main topic of Chapter IV below. One of Bergson's goals there will be to reveal the type of philosophy that is implied in the natural functioning of our intellect.

For Bergson's use of the metaphor likening concepts to garments as either "ready-made" [*vêtements de confection*] or "made to measure" [*taillés sur mesure*], cf. below, 362; IM, 188, 196–97, 213–14; Henri Bergson, "Introduction (première partie): Croissance de la vérité: Mouvement rétrograde du vrai [1922]," in *La pensée et le mouvant*, 1–2, 22–23 ["Introduction (Part I): Growth of Truth: Retrograde Movement of the True," in *The Creative Mind*. Henceforth cited as MRV, with pagination referring to the French edition].

133　This "critique" of knowledge as the establishing of the limits of its legitimate application is an allusion both to Spencer and to Kant.*

134　Bergson only capitalizes the word Life twice in this book: here and below, 339. Perhaps here he has in mind the distinction already mentioned between the lives of individuals and "life in general" (cf. above, 26). The second part of this sentence outlines the topic of Chapter II.

135　For more on this argument, cf. above, vi; below, 153.

136　As indicated, Bergson does not provide a citation for the quoted passage. It is from Spencer's formulation of the law of evolution. Spencer, *First Principles*, 330. Spencer later writes: "The more specific idea of Evolution now reached is—a change from an indefinite, incoherent homogeneity, to a definite coherent heterogeneity, accompanying the dissipation of motion and integration of matter." *Ibid.*, 380.

137　For more on the "good" vagueness in Bergson's philosophy, in contrast to a "bad" vagueness that results from the confusion between science and metaphysics, cf. below, 86. The vagueness comes from the fact that life is an *indistinct multiplicity*, and attempting to understand it involves having to resort to an image such as the *élan vital*.

138　It is worth noting that this is Bergson's definition of "adaptation," not to be confused with the various other definitions of this notion that he never tires of critiquing. Cf. below, 102–3; "adaptation" in the Index.

139　This is the first explicit occurrence of Bergson's important image of various tendencies diverging from a single source, which he will justify later. Cf. below, 89–98.

140　Thus, Bergson specifies that the *élan* is a *vis a tergo* (a "force acting from behind"). Cf. below, 104.

141　For other occurrences of the formulation "everything is given," cf. editorial note 101 to Chapter I.

142　On the exact sense of the word "progresses" here, cf. below, 105.

143　This is an important admission to keep in mind so as to grasp the exact tenor of the finalist and psychological formulations that Bergson will employ below. We will note these formulations where appropriate. Cf. editorial note 26 to Chapter II.

144　For more on the intellect as a faculty for linking the same to the same, cf. above, 29, 45–46; below, 201, 271.

145　Notwithstanding Bergson's critique of mechanism, Bergson has much to say about causality. Cf. "cause, causality" in the Index.*

146　For Bergson's internal reference, see above, 37.

147 Bergson goes *further* than finalism in that he assigns a psychological essence to life. Cf. below, 77–78.

148 TN: For Bergson's internal reference, see above, 51.

149 For more on this curious thought experiment, from which the following one appears to emerge as well, cf. below, 99, 257–58; *DS*, 273. It can be understood by the fact that the evolution of life is in essence identical to the evolution of individual character. As such, it becomes necessary to establish why the evolution of life, and not the evolution of the individual, necessitates a material division into separate individuals.

150 The hypothesis mentioned here should be understood as what Bergson intends to demonstrate in this paragraph (as indicated in its first line). The comparison with memories should be taken quite literally: if life is of a psychological nature, then it is, notably, a memory analogous to an individual's memory (cf. above, 19). For more on comparisons of this type, see "memory, as analogous to life" in the Index.

151 In this context, the term "mechanism" refers to both the Darwinian doctrine of imperceptible variations and the mutationist doctrine of sudden variations. For more, cf. below, 63–69.

152 Here Bergson proposes an objection that a supporter of zoologist Theodor Eimer (1843–1898) and of orthogenesis (cf. below, 69–77; editorial note 199 to Chapter I) might raise to the Darwinian and mutationist doctrines that were just discussed.

153 Given that Bergson does not introduce the concept of "adaptation" until this point in his argument, he seems to consider it more important in the theory of orthogenesis than it is in the theory of imperceptible variations or of sudden variations.

154 For Bergson's discussion of Eimer, cf. below, 69–77.

155 The two hypotheses in question here are: Darwin's or the mutationist accounts (insofar as they both rely upon the notion of adaptation), and Eimer's account.

156 For more on Bergson's interpretation of Darwinism, cf. editorial note 178 to Chapter I.

157 This idea is held not only by the neo-Darwinians, but also by the mutationists, at least on Bergson's interpretation of them. Cf. below, 63–69. But mutationism clearly tends to minimize the role of accidental variation. On this point, see below, Bergson's second note to page 63 and editorial note 172 to Chapter I.

158 For this principle of mechanism, cf. above, 44; below, 215.

159 For this comparison of the cell to an organism, as well as the conception of the organism that is thereby implied, cf. "society, organism as" in the Index.

160 The second hypothesis is Eimer's. Cf. below, 69–77.

161 For this characterization of adaptation in terms of the resolution of problems, cf. below, 71, 78.

162 Bergson is alluding here to the process of meiosis, which was discovered by Belgian embryologist and cytologist Édouard van Beneden (1846–1910) in 1883. In higher organisms, germ cells are the result of a division that reduces their number of chromosomes by half, transforming them into gametes. This

makes the union of two different genetic heritages possible, vastly increasing genetic variability while holding constant the number of chromosomes characteristic of a species. Mendel's laws show the relationship between these two parts and the concept of meiosis played an important role in the construction of genetic theories developed by mutationists. Cf. below, Bergson's notes to page 80 and page 82. For the notion of "half-germ cells," cf. below, 84.*

163 The proximity between the modes of reproduction in plants and animals was one of the great discoveries in biology at the end of the nineteenth century. Bergson here first cites botanist Paul Guérin's (1868–1947) comparison of fertilization in plants and animals: *Les connaissances actuelles sur la fécondation chez les Phanérogames*. The term *phanerogam* in the title refers to plants with visible reproductive organs, that is to say, seed plants [or *spermatophytes*].

Bergson's second citation refers to *L'hérédité et les grands problèmes de la biologie générale* by Yves Delage (1854–1920), who was an important French zoologist of the era and founder of the journal *Année biologique*. As a neo-Lamarckian of a mechanistic bent, Delage gave a new impetus to biology in France by taking into consideration the living being as a whole, rather than as a mere association of cells (cf. the editorial note 242 to Chapter III). This led him to reject the theories of Spencer, Haeckel, and Weismann (because of the fact that they must adopt the theory of association), the theory of pangenes (those elements supposedly constitutive of all living beings, according to Darwin); cf. the editorial notes 174, 219, 239 to Chapter I, and De Vries's theory (which, to a certain extent, adopted the Darwinian theory of pangenesis; cf. editorial note 174 to Chapter I). Delage instead supported Roux's theory of biomechanics, which, despite its mechanistic inspiration, gives more attention to the concept of dissociation than Weismann does. Unlike Bergson, Delage does not accept the identification between, on the one hand, associationism and mechanism, and on the other, dissociation and anti-mechanism. The cited pages provide a comparison between reproduction in seed plants and in animals. Guérin and Bergson both adopt, almost word for word, Delage's formulation of this comparison.*

164 In *Beiträge zur Lehre von der Fortpflanzung der Gewächse*, botanist Martin Möbius (1859–1946) identifies three advantages of sexual reproduction in plants: simple cross-fertilization facilitates the conservation of the species type; cross-fertilization doubles the growth of variability; and sexual reproduction in general allows for the elaboration of more complex forms (203–5). Möbius mentions *phanerogams* but concludes that "we nevertheless must not consider that sexuality plays a primordial role in the plant kingdom, as it does in the animal and human kingdoms" (206).

In the second article mentioned, "Sur les phénomènes de reproduction," Marcus Hartog (1851–1924), who was a specialist in developmental mechanics, only briefly mentions plant sexuality. Hartog suggests that cross-pollination is more of the order of a luxury or of acquired

necessities. For more on developmental mechanics, cf. above, Bergson's
note to page 33 and editorial note 87 to Chapter I.*

165 Paul Janet (1823–1899) was an important philosopher in the French
school of "eclecticism," established by philosopher Victor Cousin (1792–
1867). Bergson published a long review of one of Janet's major works (see
"Compte rendu des *Principes de métaphysique et de psychologie* de Paul
Janet [1897]," in *Mélanges*, 375–410), and he discusses *Les causes finales*
(*ibid.*, 399–400). Janet attempted to establish the "law of finality" through
two sorts of facts: facts related to the harmony between the organs and
their function, and facts related to the instincts. In terms of the first category of facts, Janet discusses the example of the eye. His description
draws on the work of German physiologist and comparative anatomist
Johannes Müller (1801–1858), who was Haeckel's teacher. Janet speaks of
the eye as a solution to a differential problem, and of nature as a chemist
or a physicist who solves that problem, as is apparent in the example here
of the "camera [obscura]" [*chambre noire*].*

166 The "accidental" nature of the fact that Bergson mentions here is according to
the neo-Darwinian and mutationist hypothesis under consideration.

167 Bergson himself refuses to put the problem in these terms. Cf. below, 64.

168 Bernard Balan has recently shown that the "analogous structure" between the
eye of the Scallop and that of humans was less complete than Bergson here
claims. For instance, the Scallop has a double-layered retina, and the inversion is not identical. See Bernard Balan, "L'œil de la coquille Saint Jacques—
Bergson et les faits scientifiques," *Raison présente*, no. 119 (1996): 87–106. It
is worth noting that Bergson's argument does not rely upon a perfect identity,
and the variations in the Scallop's eye structure do not negate its role in the
function of vision. For Bergson, what matters philosophically is that the same
"march toward vision" (see below, 97) leads, even on diverging evolutionary
lines, to the production of *analogous* organs, and Balan establishes in his article that this function is maintained (See Balan, "L'œil de la coquille Saint
Jacques—Bergson et les faits scientifiques," 89, 92). Nevertheless, perhaps
Bergson should have chosen the Squid instead, whose eye structure is closer
to that of the human. See Paul-Antoine Miquel, "De l'immanence de l'élan vital
à l'émergence de la vie," in *Annales bergsoniennes*, vol. 3, *Bergson et la science*,
ed. Frédéric Worms (Paris: Presses Universitaires de France, 2007), 221.*

169 To clarify, in the pages that follow, Bergson explores four main positions in
contemporaneous debates about the theory of evolution. First and second
are the neo-Darwinian theory of gradual accidental variation (below, 64–66)
and the mutationist theory of sudden mutations (below, 65–69). In the 1880s,
Weismann revised Darwin's theory so as to exclude the theory of the inheritance of acquired characteristics (which is likely why Bergson rarely refers
to this original aspect of Darwin's theory, cf. below, 170–71). The mutationist
approach emerged around 1900, thanks to the work of De Vries, Morgan,
and William Bateson (1861–1926). The next option mentioned in this short
paragraph is the third major position: orthogenesis, wherein variations are

the result or imprint of external physico-chemical conditions (below, 69–77). Eimer's theory of orthogenesis, which was more isolated and more difficult to distinguish from those of his contemporaries (drawing as it did from the Lamarckian theory of the inheritance of acquired characteristics), emerged in the 1890s. Finally, Bergson will turn to a theory not mentioned in this paragraph: neo-Lamarckism (below, 77–85). The Lamarckian principle of the inheritance of acquired characteristics is key to Haeckel's theory and to Spencer's ideas as well. It continued to have ardent support from Cope, Delage, and Le Dantec.*

170 Here Bergson uses Darwin's English term "sports," which in biology signifies "abnormal variations" from sudden mutations.

171 Bergson's citation is to the very beginning of Chapter II ("Variation Under Nature") of On the Origin of Species, where Darwin classified sudden variations as "monstrosities." Such variations can sometimes be seen in domestic animals. For example, in pigs that are "born with a sort of proboscis" (Darwin, On the Origin of Species, 34). But these variations would be incapable of preserving themselves in the state of nature: "their preservation would depend on unusually favourable circumstances" (ibid.). Against sudden variation, Darwin brandished an argument that is very close to the one that Bergson will use: "Almost every part of every organic being is so beautifully related to its complex conditions of life that it seems as improbable that any part should have been suddenly produced perfect, as that a complex machine should have been invented by man in a perfect state" (ibid., 33–34). For more, see below, 66. [TN: Note that Bergson was working with a French translation of the 6th edition of Darwin's On the Origin of Species, which is important given that the pig example just discussed did not appear in earlier editions.]*

172 William Bateson (1861–1926) was the founder of genetics, a term he coined. Bergson is citing Bateson's conclusion to Materials for the Study of Variation: Treated with Especial Regard to Discontinuity in the Origin of Species, which amounts to a diatribe against the unilateral explanatory power of natural selection.

The second work cited, "On Variations and Mutations," is by William Berryman Scott (1858–1947), an American paleontologist who criticized Bateson for "omitt[ing] any consideration of the paleontological facts" (358). He proposes in the article to compare Bateson's results and methods to the facts available in the study of the fossils of mammals. Scott argues against Bateson's foundational principle, "discontinuity" in variation, by introducing a distinction between "variation" (which is potentially sudden) and "mutation" (which occurs as a "steady advance along certain definite lines") (ibid., 372). As such, we find the beginnings of mutationist theory in the work of Bateson and Scott, built upon a foundation that is fairly hostile to Darwin's theory of natural selection (or at least to the "omnipotence" of that principle) (cf. ibid., 373). For Scott, variation is individual and spatialized, while mutation has to do with the species and is temporal (*ibid.*, 371–72). The causes of mutations are "deeper" (*ibid.*, 372) than those of variation, and Scott expresses himself in a manner that is extremely close to Bergson when it comes to describing the neo-Lamarckian doctrine (cf. below, 77–78). Ultimately, Scott recognizes that the current state of knowledge seems to invite the implicit positing of a "mystical directing force" ("On Variations and Mutations," 374) both in Bateson and in Weismann, but he hopes that science will soon clarify the nature of these forces. One can see here, as if in relief, the place left open for Bergson's notion of the *élan vital*.*

173 *Oenothera lamarckiana* (Evening Primrose) is a genus of herbaceous flowering plant that is found in the Americas and in Tasmania. Their flowers are large and red or yellowish, and they grow in clusters. They were a key example in the development of the theory of sudden variation. Nevertheless, De Vries was criticized for not being able to produce other examples. For more on this question and the significant contribution of De Vries's discovery for both *L'évolution créatrice* and biology in general, cf. below, Bergson's note to page 86 and editorial note 239 to Chapter I.

174 First, there are two adjustments to mention here regarding Bergson's note from the original version: the sentence itself originally contained the phrase "in a great number of different directions," and the final claim in Bergson's footnote does not appear in the first edition of *L'évolution créatrice*.

Hugo de Vries (1848–1935) was a Dutch botanist and geneticist. He is one of the most important figures to receive positive treatment in *L'évolution créatrice*. De Vries discovered "mutation" in 1886 when observing wild varieties of *Oenothera lamarckiana* (Evening Primrose) that differed in a stable way from the normal form. This being the case, De Vries (like Weismann) was able to reject the theory of the inheritance of acquired characteristics. Nevertheless, De Vries explained heredity via specific carriers that he called "pangenes" (not dissimilar to what we call "genes" today), which are not altogether incompatible with the inheritance of acquired characteristics, particularly given the influence of Darwin's idea of *pangenesis*. Mutations were understood to be the result of the appearance of new pangenes, and De Vries attempted to show this as hereditary, particularly through hybridization. He was later highly influenced by Mendelian-style statistical laws of heredity. The texts mentioned by Bergson are among the last great works by De Vries, and they focus on evolution taking place through sudden variations and independently of natural selection.*

175 Here the first edition of *L'évolution créatrice* includes the phrase "moreover, it is possible that both possess a certain part of the truth."

176 Bergson means the comparison between the Vertebrate's eye and the Mollusk's. See above, 62–63.*

177 For Bergson's discussion of the function in this sense, see above, 62.

178 Bergson is referring to Chapter VI of Darwin's *On the Origin of Species*, and likely the following sections: "Difficulties of the theory of descent with modification" and "Organs of extreme perfection and complication." As Darwin writes: "It has been objected that in order to modify the eye and still preserve it as a perfect instrument, many changes would have to be effected simultaneously, which, it is assumed, could not be done through natural selection; but as I have attempted to show in my work on the variation of domestic animals, it is not necessary to suppose that the modifications were all simultaneous, if they were extremely slight and gradual" (Darwin, *On the Origin of Species*, 145).*

179 It might be noted here that Bergson's understanding of natural selection is in harmony with the dominant interpretation of his day, or at least the one held by the authors he cites. In essence, natural selection is presented as a sort of filter, in a negative sense, operating solely via the elimination of poorly adapted organisms. Bergson's formulations are clear in this regard, such as when he speaks of variations that have some utility as providing natural selection a "grip" (above, 62, 64) or of an organism opening itself "to the play of natural selection" (just below, 66). But Darwin's work also suggests that natural selection preserves the *divergences*. Darwin introduces a "principle of divergence" or differentiation to explain how a single variety can become a genuinely different species, something he acknowledges cannot be the result of mere chance. For instance, he argues that: "The more diversified in habits and structure the descendants of our carnivorous animals become, the more places they will be enabled to occupy" (Darwin, *On the Origin of Species*, 88). By accentuating differentiation, Darwin seems to have at least suggested a principle that is not entirely incongruous with Bergson's position.*

180 It is interesting that Bergson here offers a definition of the concept of viability, given that he most often simply condemns the concept of adaptation. For in a certain sense, all organisms are viable, insofar as they succeed in living (cf. below, 130; CV, 18–19). It is that the viability in question here, in addition to referring to "conservation," also refers to "improvement."

181 The examples Bergson uses in this paragraph are discussed by Darwin on the same page where Darwin refers to the "mysterious laws of correlation" (*On the Origin of Species*, 9). Darwin also discusses other characteristics of inheritance on these pages, including the question of "acquired characters" (*ibid.*, 11). Darwin admits here that the laws of inheritance are "for the most part unknown" (*ibid.*, 10).

182 This definition of complementarity shares some similarities with the definition of "viability" offered by Bergson just above, 66.

183 In the article that Bergson cites, zoologist Alexander Brandt (1844–1932) studies the origin of hair and, against other theories, proposes that hair originates from the same substance as teeth. This kind of debate was quite common in post-Darwinian evolutionary theory.

184 From this point forward, Bergson resorts to a type of a Weismannian argument that allows him to circumscribe the area of the inheritance of acquired characteristics. Cf. below, 81–85.

185 For more on this important argument involving the "simplicity" of the function, cf. below, 88–98.

186 Here again Weismann's influence is clear. Cf. below, 81–85.

187 For the discussion of the two meanings of "adaptation," see above, 58–59.

188 Bergson will come back to this point about the link between form and function. Cf. below, 78, 86–87.

189 The distinction implied here between science and philosophy is not the one that Bergson himself proposes. For Bergson, philosophy and science do not differ in the manner of the principles and the consequences, but rather according to their methods and their areas of application. Cf. "science, and philosophy" in the Index.

190 Bergson will insist below on the "details" of the explanation. Cf. below, 362.

191 The expression "the genius of the species" [le génie de l'espèce] is from Schopenhauer. See Arthur Schopenhauer, "Supplements to the Fourth Book: On the Metaphysics of Sexual Love," in The World as Will and Representation, vol. 2, trans. E. F. J. Payne (New York: Dover, 1966), 549. [TN: I have preserved the English translation of Schopenhauer's le génie by "genius," though Bergson's formulation here perhaps also alludes to Descartes's le malin génie ("the malicious demon" or "the evil deceiver").]

192 Nevertheless, this is the claim that the neo-Darwinians and the mutationists would have needed, according to Bergson.

193 For the discussion of adaptation, cf. above, 55–61.

194 For more on this comparison, cf. above, 58–59; below, 78.*

195 This comment is important because it thematizes a difficult aspect of Bergson's theory of difference. For Bergson, there can be a gradual transition from one reality to another without that change thereby ceasing to represent a difference in nature. This is how a memory comes to "superpose" itself on perception, even though it remains radically distinct from it (MM, 248).

196 For more on the image of life "insinuating" itself within matter, cf. below, 93, 99–100; CV, 20–21.

197 For the expression "takes advantage of" [tirer parti de], cf. above, 58–59; below, 78.

198 This observation leads Bergson to take up again his notion of a "sensorimotor system," introduced in Matière et mémoire. Cf. below, 125–27.

199 For complete German and English references for the two principal texts by Eimer that Bergson cites, see Bergson's note to page 74.

Theodor Eimer (1843–1898) was a Swiss-German zoologist who popularized the term "orthogenesis," which he characterizes as "definitely directed evolution" (Eimer, On Orthogenesis and the Impotence of Natural Selection in Species-Formation, 20). Although he studied under Weismann, Eimer developed his theory in a long and bitter debate with Weismann, sardonically attacking his belief in "omnipotent natural selection" and labeling neo-Darwinian theory as "Weismannian Pseudo-Darwinism" (*ibid.*). Eimer argued that variation could not be explained by adaptation because "[e]verything is not adapted" (*ibid.*, 9) and thus that evolution moved in definite (and limited) directions according to laws and regardless of natural selection or use and inheritance. Since natural selection can only eliminate, Eimer accords it but a subordinate role to orthogenesis (*ibid.*, 21). Moreover, Eimer insists that the causes of orthogenesis are found in "the effects produced by outward circumstance and influences such as climate and nutrition upon the constitution of a given organism," and he quickly insists that this is not Lamarckism either (*ibid.*, 22). According to Eimer, the material constitution of the body shapes the limited number of possible directions, and the external circumstances shape the development. Given that the vast number of characters produced (systematically) in living beings are not "useful," natural selection and *accidental* variation as the engine of evolution must be rejected. In any case, the major difficulty of any doctrine of orthogenesis is to bring together external and internal influences—and in this sense, Bergson, goes right to the heart of the matter.*

200 Georg Dorfmeister was an entomologist in Freiberg. He discusses his experiments in Georg Dorfmeister, "Ueber den Einfluß der Temperatur bei der Erzeugung von Schmetterlingsvarietäten," *Mittheilungen des naturwissenschaftlichen Vereines für Steiermark* (1879): 3–8. Weismann repeated the experiments (cf. Weismann, *Ueber den Saison-Dimorphismus der Schmetterlinge*, vol. 1, *Studien zur Descendenz-Theorie* (Leipzig: Wilhelm Engelmann, 1875), which were later published in English with a preface by Charles Darwin himself (cf. August Weismann, *Studies in the Theory of Descent*, trans. and ed. Raphael Meldola, preface Charles Darwin (London: Sampson Low, Marston, Searle, & Rivington, 1882). For Eimer's discussion of this example, see Eimer, *Orthogenesis der Schmetterlinge*, 414–21 [see note 203 just below]. The variations have to do above all with the designs and colors of the wings. Eimer challenged Weismann's interpretations of these experiments.*

201 German zoologists Max Samter (1869–?) and Richard Heymons (1867–1943) extended research started by Russian zoologist Wladimir Schmankewitsch into the effects of the salinity of the water on the development of the family of small Crustaceans named *phyllopoda*, which they took as evidence in favor of neo-Lamarckism or orthogenesis. Bateson objects to their conclusions in the book cited by Bergson above (cf. above, Bergson's second note to page 63).

202 This specific analysis of causality is not found elsewhere in Bergson's corpus, though a number of texts prior to this book attest to his increasing interest in this topic. See Bergson, "Note sur les origines psychologiques de notre croyance à la loi de causalité [1900]," in *Mélanges*, 419–28; Bergson (summarized by J. C.), "Cours de M. Bergson sur l'idée de cause [*Cours au Collège de France*, 1901]," in *Mélanges*, 439–41.

203 TN: Bergson references the first two volumes of Eimer's three-volume collection of his works. The general title is: *Die Entstehung der Arten auf*

Grund von Vererben erworbener Eigenschaften nach den Gesetzen organischen Wachsens. To refer to volume I (*Ein Beitrag zur einheitlichen Auffassung der Lebewelt*), Bergson uses the shortened version of the general title, *Die Entstehung der Arten*. This first volume was translated into English as: Theodor Eimer, *Organic Evolution as the Result of the Inheritance of Acquired Characters According to the Laws of Organic Growth*, trans. J. T. Cunningham (London: MacMillan, 1890). To refer to volume II, Bergson uses a shortened version of that particular volume's title, *Orthogenesis der Schmetterlinge*. Although the entire second volume was not translated as a whole, the section Bergson cites from was translated into English as: Theodor Eimer, *On Orthogenesis and the Impotence of Natural Selection in Species-Formation*, trans. Thomas J. McCormack (Chicago: The Open Court, 1898).

Eimer often invokes the image of a "kaleidoscopic correlation" to insist upon the fact that correlation is not governed by function or use (Eimer, *On Orthogenesis*, 33). With his second citation, Bergson was surely referring to the following passage: "As soon as something or other in the original state, in the original arrangement of the parts of the organism, is changed, other parts also are set in motion, all arranges itself into a new whole, becomes—or forms—a new species," just "as in a kaleidoscope, as soon as on turning it one particle falls, the others also are disturbed and arrange themselves in a new figure—as it were re-crystallise" (Eimer, *Organic Evolution*, 49). It is worth comparing this to Bergson's own use of this image (cf. below, 306; *MM*, 20, 221; *DS*, 132, 297). [TN: Bergson mistakenly gives the publication information for *Die Entstehung der Arten* the second time it appears, rather than the first.]*

204 Bergson is thinking of the following passage: "Just as in inorganic nature from different mother lyes different crystals separate, as even simple mechanical shock can produce dimorphous crystallisation, so crystallise, if I may so express myself, in the course of ages, organic forms, to a certain degree different, out of the same original mass." (Eimer, *Organic Evolution*, 23).

205 Bergson truncates Eimer's argument. What Eimer wants to demonstrate here above all is the Lamarckian thesis according to which the use of the organ configures its structure. For Eimer, use or non-use is the principal cause of an organ's structure, and the structure thus formed is passed down through inheritance. There are therefore two causes rather than one alone that contribute toward the formation of organs, and Eimer discusses the example of the eye and other sense organs. Bergson does not analyse the rest of Eimer's argument, likely judging that it would either fall within his general attack on finalism here or the attack on the inheritance of acquired characteristics undertaken later.*

206 Wladimir Salensky (1847–1918) was a professor of zoology in St. Petersburg, Russia. Many of the scientists cited by Bergson in *L'évolution créatrice* also attended the congress where Salensky read this

paper. These include: Delage, Haeckel, Sedgwick, Henri Milne-Edwards (1800–1885), Perrier, Roule, Ernest Olivier (1844–1914), and Eugène-Louis Bouvier (1856–1944) (cf. the Index; for more on Bouvier, see editorial note 130 to Chapter II). As Bergson notes, Salensky's interest is in the different *origins* of organs for animals that are close to each other, rather than analogous organs appearing along distant lines of evolution. Salensky's view would have attracted Bergson's attention given that he attempted to understand the appearance of an organ in a diachronic, evolutionary sense, rather than in a synchronic or punctual one.

207 "Embryonic layers" are the superposed cellular lamina or layers that appear following the first stages of embryonic development in higher animals. The "theory of embryonic layers" was formulated by Karl Ernst von Baer (1792–1876), one of the founders of modern embryology.

208 German biologist Gustav Wolff (1865–1941) was an opponent of Roux and Bütschli's notion of developmental mechanics (cf. Bergson's note to page 33 and editorial note 87 to Chapter I). Wolff was also a vigorous critic of neo-Lamarckism and Darwinism (when interpreted as a "theory of chance"). In his article "Entwickelungsphysiologische Studien I: Die Regeneration der Urodelenlinse," in *Archiv für Entwicklungsmechanik der Organismen 1* (1895): 380–90, he discusses Urodeles, the order of amphibians to which notably the salamander and the newt belong. The question that animates his article is very close to Bergson's: "[I]s it possible that a developmental process (*Entwickelungsvorgang*) can be forced to occur in a manner that is totally different than normal, but nevertheless succeed in having the same results?" (*Ibid.*, 380).

209 Austrian embryologist Alfred Fischel (1868–1938) begins from Wolff's discovery, discussed in the previous editorial note. He does not, however, support the same general theoretical claims as Wolff.

210 Although this thesis is arguably somewhat secondary in Lamarck's own theory, it played an important role in evolutionary debates in the late 1800s, which discuss "Lamarck's law" or the "Lamarckian principle" as designating the idea of a hereditary transmission of characteristics acquired by the *individual*. The principle likely came to the fore as a result of Weismann's rejection of the inheritance of acquired characteristics at the beginning of the 1880s. [TN: For a helpful discussion of the era's neo-Lamarckian position regarding a more indirect transmission (that runs somewhat counter to Bergson's presentation here and is perhaps closer to what he will propose just below. Cf. below, 87–88), see also J. T. Cunningham's Translator's Preface to *Organic Evolution as the Result of the Inheritance of Acquired Characteristics*, by Theodor Eimer, xvi.]*

211 To grasp the implications, the content, and the significance of the analyses that Bergson proposes here, it is important to keep in mind that, at the end of the nineteenth century, the belief in the theory of the inheritance of acquired characteristics was neither outdated nor scientifically backward. Scholars, such as Delage or Le Dantec, continued to adhere to the theory, even after

1907. In fact, it was rather Weismann's *soma* versus *germen* distinction (cf. below, 79) and his idea of the continuity of the germ-plasm that were at first considered regressive insofar as they seemed to preclude evolutionary progress. This debate would have been very interesting to Bergson since it was a question of knowing whether the progressive accumulation of parts or of characteristics in fact allowed for a genuine conception of evolution, or, to the contrary, whether genuine evolution confirmed rather a continuity, philosophically defined as an indistinct multiplicity [TN: and hence not merely juxtaposition]. Bergson's goal was to demonstrate that the second option is correct, and for this he needed to attack the theory of the inheritance of acquired characteristics. Just as Bergson was introducing his philosophical interpretation, a paradigm shift was happening in the sciences, beginning from the critical question of individuality (cf. below, 259–61 and editorial notes 235–43 to Chapter III).*

212 Bergson associated "inherence in the germ" rather with the neo-Darwinian position. Cf. below, 86.

213 For more on the idea of problems "posed by the external circumstances," cf. above, 59, 71.

214 Note that Bergson's mention of *activité interne* ("inner activity") at the beginning of this sentence refers to *psychological* activity.

215 For more on the difficulty of making this shift from animals to plants, cf. below, 87–88, 112, 171.

216 For more on the notion of "functional changes," cf. above, 68; below, 86–87.

217 For more on Bergson's conception of the relationship between science and philosophy, cf. "science, and philosophy" in the Index. Even if the disciplines differ in terms of their methods and their domains of application, the domains are nevertheless connected, and the disciplines must in fact come together in experience.

218 Bergson criticizes Spencer for having thought of evolution as a progressive addition of parts (cf. below, 363–69). Now, Spencer could not have thought this if he did not consider that the parts could be preserved. He would have been forced to conceive of a genuine genesis. By contrast, we can grasp here the fundamental theoretical reason why Bergson rejects, to a large degree, the inheritance of acquired characteristics: because this notion leads, more and more, toward the "intellectualist" conception of evolution.

219 French biologist and geneticist Lucien Cuénot (1866–1951) was still at an early stage in his career in 1907, which perhaps explains why Bergson accords him little space in *L'évolution créatrice*. This article is generally a defense of Weismann's neo-Darwinism. Cuénot interprets the germ-plasm as identical with the nucleus of the germ cells, which is likely an influence on Bergson's own reading (cf. above, 43). It is also worth noting a resonance between Cuénot's position and Bergson's discussion of a chemical explanation for alcoholism and more generally the Weismannian attempt to account for inheritance of acquired characteristics (cf. Cuénot, "La nouvelle théorie transformiste," 79; below, 82–85).

Bergson's second citation is to Thomas Hunt Morgan (1866–1945), who, in *Evolution and Adaptation*, provides a very similar reversal in the mole example as described by Bergson. Morgan writes: "Thus, from the point of view that is here taken, an animal does not become degenerate because it becomes parasitic, but the environment being given, some forms have found their way there; in fact, we may almost say, have been forced there, for these degenerate forms can only exist under such conditions" (357). This helps to establish "mutation theory," and indeed Morgan, who was a disciple of Wilson (cf. above, Bergson's note to page 36), is in effect a mutationist.*

220 Charles-Édouard Brown-Séquard (1817–1894) was a specialist in the physiology of the nervous system who worked both in France and in the United States. He succeeded Claude Bernard in the Chair in Experimental Medicine at the Collège de France and was a cofounder of the journal in which the article that Bergson cites appears (*Archives de physiologie normale et pathologique*).

221 Lesions of the "restiform body" refer here to a swelling in the cerebral trunk.

222 The first work mentioned here, *Aufsätze über Vererbung und verwandte biologische Fragen*, is a republication of a previous work from 1886: August Weismann, *Die Bedeutung der sexuellen Fortpflanzung für die Selektions-Theorie* (Jena: Gustav Fischer, 1886). In this text, Weismann wants to show that the inheritance of illnesses that are artificially produced is not proof in favor of the general principle of the inheritance of acquired characteristics.

In the second text cited by Bergson, *Vorträge über Deszendenztheorie*, Weismann responds to two objections raised in Brown-Séquard, "Hérédité d'une affection due à une cause accidentelle," 686–88.*

223 In "Hérédité d'une affection due à une cause accidentelle," Brown-Séquard argues that, by examining the evidence and the logical limitations of Weismann's position, Darwin's theory survives Weismann's criticisms. It is worth noting that the lively debates over the inheritance of acquired characteristics primarily involved the *orthodox* Darwinians, who in fact allowed for a certain form of this transmission, a topic on which Bergson remains notably discreet.*

224 Jules Voisin (1844–1920) and Albert Peron's (1870–19?) experiments consisted in injecting human urine from epileptic patients into guinea pigs and rabbits. The results were that these animals nearly always died following a violent congestion of the lungs. The authors hypothesized a correlation between certain toxic substances present in the urine and the beginnings of epilepsy, particularly through the accumulation of toxins in the blood. The second text cited by Bergson is Voisin's report on the repetition of the same general experiment, claiming thereby to confirm the earlier conclusions.*

225 In terms of the first article, the first edition of *L'évolution créatrice* also cites page 202. Albert Charrin (1856–1907) had occupied the Chair of General and Comparative Pathology at the Collège de France, where he worked with Gabriel Delamare (1875–19?) and Gustave Moussu (1864–1945). In the article, the authors are less interested in the transmission of traits than in the homology between the damaged organs across generations. As Bergson explains, for them, transmission was to be explained through chemicals particular to certain organs that acted upon successive generations, rather than through biological transmission.

For more on Bergson's second citation (Morgan, *Evolution and Adaptation*, 257), see Bergson's note to page 80 and editorial note 219 to Chapter I. Morgan discusses these experiments and connects them to Brown-Séquard's experiments.

Bergson's third citation refers to *L'hérédité et les grands problèmes de la biologie générale* by Yves Delage. Cf. Bergson's note to page 60. Delage also focuses on particular substances produced by specific organs and is thus surely thinking of the experiments of Voisin and Peron (discussed in the previous editorial note).*

226 Charrin and Delamare's article "Hérédité cellulaire" is a case study of the effects of eclampsia and diabetes in a mother on the liver and red blood cells of her fetus. The authors claim that the two generations present the same modifications. In conformity with their overall theory, they prefer the explanation according to which a purely chemical effect had acted on the two subjects.*

227 In "L'hérédité en pathologie," Charrin discusses experiments that involved injecting rabbits with a pyocyanic bacterial secretion and studying their descendants, who often had reproductive troubles. The author held this to show the transmission of acquired characteristics. This article is very important for Bergson's understanding of the argument for transmission of acquired characteristics through the effect of chemicals on the germ.*

228 The argument can thus be summarized as follows: there are cases where the soma can act on the germen, since here it is a question of accounting for Brown-Séquard's experiments. But this takes place in a global way (such that there is very little chance that the homologous part of the soma of the offspring would be damaged) and through a *chemical* process (that is, not a biological one, properly speaking), which has nothing to do with heredity. Bergson seems to adopt this position until the end of this section. By refusing in principle the inheritance of acquired characteristics, Bergson is able to remove from life every principle of direction that would not be an *internal* one in the sense of a psychological one.*

229 For example, the change would be "negative" in the case of a "tendency toward the loss of vision" that we find in the mole. See above, 80.

230 In these lines, Bergson is somewhat borrowing this vocabulary from the article just cited above: Charrin, "L'hérédité en pathologie." He is, however, giving this vocabulary a somewhat larger meaning.

231 For more on this argument, cf. above, 59 and editorial note 162 to Chapter I.

232 For more on substances capable of modifying the germen, see above, 81–83. The second aspect of this claim relates to the negative changes that can take place in the germen, discussed above, 83.

233 In the first edition, Bergson moderates this claim with a "perhaps."

234 Alfred Giard (1846–1908) was an authoritative figure in evolutionary biology in France. Giard wrote the preface to Le Dantec's book *L'individualité et l'erreur individualiste* and criticized Weismann's distinction between modifications in the soma and those in the germen.*

235 The relationship between science and philosophy here is one between a simple movement and the snapshots that we might take of that movement. This recalls a distinction that Bergson developed earlier in "Introduction to Metaphysics" (IM, 177–82) and that he will take up again below.

236 For more on the pragmatic aims of science, cf. editorial note 77 to Chapter I. Here the notion of "vagueness" (cf. above, 50; editorial note 137 to Chapter I) is not considered a defect in comparison to scientific precision, but rather a function of a difference in method and object for metaphysics.*

237 The "three" main forms of evolutionary theory that Bergson has in mind here are: (1) theories of accidental variation (neo-Darwinian and mutationist theories, taken together) (see above, 63–69); (2) Eimer's theory of orthogenesis (see above, 69–77); and (3) neo-Lamarckism (see above, 77–85).

238 The first edition contains the phrase "according to De Vries" rather than "as De Vries holds." In general, the changes to later versions of the text are less favorable to De Vries's doctrine, which can be seen by comparing some of the differences. See the editorial notes 65 and 174 to Chapter I.

239 In "La notion d'espèce et la théorie de la mutation," botanist Louis Blaringhem (1878–1958) recalls De Vries's discovery of mutation in the *Oenothera lamarckiana*. The "analogous facts" mentioned by Bergson refer to Blaringhem's discussion of mutations in the *Solanum lycopersicum*, the *Cocos nucifera*, or the *Solanum commersonii* (Blaringhem, "La notion d'espèce et la théorie de la mutation," 111).

The second text cited by Bergson, *Species and Varieties*, by De Vries himself, draws on the example of the *Solanum Lypopersicum* in the section titled "Systematic Atavism." Bergson discusses a similar example in his course from December 26, 1902, suggesting there that he had actually introduced it the previous year. See Bergson, *L'idée de temps: Cours au Collège de France, 1901–1902*, ed. Gabriel Meyer-Bisch, series editor Frédéric Worms (Paris: Presses Universitaires de France, 2019), 84.*

240 For Bergson's internal reference, see below, 120. For more on the argument regarding form and function in variation, see above, 68, 78.

241 Bergson's internal reference is to his discussion below, 121–27.

242 What Bergson here calls *détermination* ("determination") is the effect of the thrust, which is itself free, and which leads to the appearance of analogous organs along diverging lines of evolution.

243 For this argument, see above, 78–79; below, 112, 171.

244 For Bergson's internal reference, see above, 26–27.

245 Here Bergson begins presenting the metaphysical doctrine of life that seems to him implicit in the scientific discussion just summarized. This doctrine is primarily based on the opposition between that which is simple and the various views that one can take of that which is simple (cf. IM, 177–82), represented here by the simple function of vision and the highly complex structure of the eye. The paradigmatic example of this distinction, to which Bergson returns below, is the one between a movement itself and the static positions that it leaves behind as it proceeds. By "simple," Bergson does not mean that the function of the eye is easy to accomplish or reproduce, but rather that its function is quantitatively indivisible. For more on the "simple functioning" of the eye, see above, 68. For Bergson, life, insofar as it is *simple*, is a vague or indistinct multiplicity, that is, it is neither fully a static one nor merely a collection of many, so long as we understand the terms "one" and "many" to be numerical designations.

246 The first option alludes to Eimer's theory (cf. above, 69–77), while the second refers to Darwinism and Bergson's interpretation of mutationism (cf. above, 63–69).

247 This principle of dissociation and splitting is fundamental throughout *L'évolution créatrice*. If life in fact proceeded via association and addition, then it would function exactly like our intellect, and thus there would be elements already given in advance (such is, according to Bergson, Spencer's position). If this were the case, then life would rise up through various degrees. By contrast, dissociation is the movement of creation, and there is a difference of nature between two diverging tendencies.

248 Here the expression *degré de réalité* ("degree of reality") merely signifies that the simple aspect is more real than the composite one. Bergson does not intend to hypostatize these "degrees of reality" in nature itself.*

249 For more on this analysis, cf. IM, 177–82.

250 This is Bergson's adaptation of Ravaisson's critique of a method for teaching drawing that had been proposed by the Swiss educational theorist Pestalozzi (1746–1827). Ravaisson was a great admirer of Leonardo da Vinci, cf. Henri Bergson, "La vie et l'œuvre de Ravaisson [1904]," in *La pensée et le mouvant*, 264–66, 277–80 ["The Life and Work of Ravaisson," in *The Creative Mind*. This chapter is henceforth cited as VOR, with pagination referring to the French edition]. Cf., Henri Bergson, *Histoire des théories de la mémoire: Cours au Collège de France, 1903–1904*, ed. Arnaud François (Paris: Presses Universitaires de France, 2018), 23–25. Bergson's mention here of an "indivisible intuition" corresponds to what Ravaisson called the "generating axis" of the figure. See VOR, 264.

251 For more on this important affirmation, which has as much to do with reality as it does with movement, cf. below, 156.

252 For more on "mobility" as the "essence" of movement, cf. below, 302; E, 86; MM, 213, 219; DPP, 26.

253 For other occurrences of this argument regarding the absence of parts in a simple movement, cf. the editorial note 57 to Chapter I. [TN: Bergson's claim that the simple movement is also *less* than a series of positions should be compared to his later arguments regarding order and disorder or being and nothingness. The simple movement is *less* because the series of positions requires the movement *plus* the intellectual work that divides it into positions. Cf., for example, below, 232, 286.]

254 Bergson's use of the term "single idea" here is perhaps an allusion to Claude Bernard's concepts of a "directing" or "creative" idea (cf. "La philosophie de Claude Bernard," in *La pensée et le mouvant*, 232–35; VOR, 274). Bergson interprets Bernard through Ravaisson, and attributes to this "directing idea" the sense of a simple movement by which the organism is engendered. Here, in particular, he is discussing the "idea" that is the visual function or the "march toward vision."

255 The distinction between "fabrication" on the one hand, and "organic development" [*organisation*] or "creation" on the other, is an essential part of this book. Cf. "fabrication" in the Index. [TN: See also the Translator's Introduction. Given the interchangeability in Bergson's French between "organic" and "organized," it should be recalled that I have translated *organisation* as "organic development." Indeed, "organization" in English would risk suggestion more of an arrangement of pre-existing parts or components, and hence *fabrication*, the very opposite of Bergson's point here.]

256 Throughout this book, Bergson puts into question the Platonic and Plotinian opposition between the One and the many, particularly in relation to his conception of "simplicity." Cf. "unity/multiplicity" in the Index. [TN: It is worth noting that in French, Bergson writes *multiple* for "many," which would have had a certain resonance for his readers to his own reflections on "multiplicities," a link that should be kept in mind since fabrication and the "many" in question here are particularly multiplicities of "juxtaposition." Cf. *MM*.]

257 Cf. above, 71; below, 99–100; CV, 20–21.

258 For more on what we might call here Bergson's pragmatism, cf. editorial note 77 to Chapter I.

259 It is worth noting, however, that Bergson himself, in certain precise argumentative contexts, adopts this identification between the body and a machine. Cf. above, 12, 30.

260 This argument is difficult to understand without taking into consideration the metaphysical arguments that Bergson will explore beginning in Chapter III. For Bergson, the material organism or one of its organs is simply the negation of a positive reality, namely the deeper movement of the *élan vital* itself. In other words, the organism must be understood as the *reverse side* of something else, of a deeper movement. The organism or organ represents the collection of obstacles that the *élan vital*, at a particular point, had to avoid or overcome. As Bergson will argue below, materiality itself is generated by the inversion of a deeper movement that he calls Consciousness (of which the *élan vital* is a continuation, cf. Chapter III). The organism represents

how this immaterial movement falls into matter at a certain point, without thereby exhausting the original movement that continues along its course. With this in mind, we can understand why Bergson here rejects the idea that the organism is the result of a work of fabrication aiming in advance at a goal, and hence a positive reality built up out of ready-made parts toward some pre-conceived end. Rather, the organism is a negation of a movement in the sense of the "reverse side" of the movement, its imprint. It is how the movement falls into (or better, falls as) matter, matter also being the obstacle that this movement encounters. Vladimir Jankélévitch (1903–1985) captures all of this in a striking phrase: "In this sense one could say, without vain paradoxology, that the animal sees *despite* its eyes rather than by means of them" (Vladimir Jankélévitch, *Henri Bergson*, ed. Alexandre Lefebvre and Nils F. Schott, trans. Nils F. Schott (Durham and London: Duke University Press, 2015), 138). It is thus worth returning to these pages after having read Chapter III. For more on the vocabulary around "negation" and the "negative," cf. below, 95–96. More generally, the opposition between the "positive" and the "negative" runs throughout this book and can assume a variety of meanings, although they are all quite tightly interconnected. Cf. "negation" in the Index.*

261 Here Bergson is referring to *MM*, 27–45.

262 In *Matière et mémoire*, Bergson writes: "[This type of perception] would suit a phantom, not a living, and, therefore, acting, being," *MM*, 43 [*Matter and Memory*, 44].

263 For the notion of a "symbol" taken in this double sense, as at once the "view" taken of an activity and the anatomical structure as it relates to the function, cf. below, 300; *MM*, 40, 57, 249; IM, 177–82.

264 Where there is "resistance," there is matter. Bergson will argue in Chapter III that matter is simultaneously the "inversion" of the movement that continues in the *élan vital* and the "obstacle" that the *élan* encounters. Cf. "resistance" in the Index.

265 Here again Bergson uses the image of a blueprint or plan to characterize finalism. Cf. "plan" in the Index.

266 Similarly, the brain canalizes and thereby limits the life of the mind. As such, the brain sketches out the principal articulations of the mind. Cf. below, 181–82, 253, and editorial note 257 to Chapter II.*

267 TN: Again, Bergson is foreshadowing some of his later arguments. This particular sentence, for instance, alludes to his argument regarding "interruption" below. Cf. 201–11, 218, 221, 238, etc.

268 Bergson here (and just below) is discussing the "march toward vision," which he uses to characterize the vital thrust as a whole.

269 Bergson concentrates his argument on the neo-Darwinians, leaving aside the respective doctrines of Eimer and the mutationists.

270 The Serpula is a marine worm from the *Annelida phylum* (which includes ragworms, earthworms, and leeches), of the Polycheata class that has two eyespots. The *Alciopidae* is a family of *Polycheata annelida*, marine worms characterized by the large size and lens structure of their eyes.*

271 The expression "the march toward vision" is absolutely decisive. It contains the idea of indivisibility: the march is indivisible, which is to say that it consists of a simple movement from which our intellect abstracts positions. And it also contains the idea that the function alone, vision, pertains to the vital thrust from its origin, although in a mode other than that of a goal. A certain simple movement that we call the exercise of vision, has, in a variety of circumstances, extracted all that it can from the matter available to it. The variety of structures that we find in eyes are so many systems of views on this simple movement or march toward vision. Finally, it is along such a movement, such as the "march toward vision," that we can conceive of "obstacles." That is, the notion of an "obstacle avoided" that characterizes the organ itself gets its sense from this image as well. Cf., above, 94–95.

CHAPTER II

1 For more on the thought experiment of life following a single pathway, cf. above, 53; below, 257–58; *DS*, 273.

2 For the notion of "insinuation," cf. above, 71, 93; CV, 20–21. Bergson uses the railroad track image, though in a slightly different way, in *MM*, 250.

3 For this vertiginous thought, cf. below, 130–31.

4 According to this quite lovely passage, evolution (which is at the same time creation) is always a form of differentiation and effort. In general, these key paragraphs present the double conditions necessary for differentiation. As Bergson tells us, differentiation has to do both with [the resistances of] matter and with life as a bundle of simple tendencies that, while somewhat distinct, are nevertheless interpenetrating.

5 It is worth noting, this whole chapter is devoted to identifying the tendencies of life and evaluating their respective importance.

6 Cf. *E*, 139–42; IM, 178–79. Bergson also applies this same analysis to tragedy in his book *Le rire*, 127–29.

7 Cf. CV, 25–27. The famous opposition that Bergson draws in *Les deux sources de la morale et de la religion* between the "closed" society (here: the "fixed") and the "open" society is already, in a fairly explicit way, at work here. Cf., for instance, *DS*, 283ff.

8 For more on Bergson's notion of vagueness, see above, 50, 86.

9 Or, to be more precise, this claim should be attributed to Eimer. See above, 69–77.

10 Bergson is here referring to microscopic marine unicellular organisms that develop shells around themselves. Their fossil record extends upward of 500 million years. Nevertheless, their immutability is today in part contested. The Silurian Period is a middle period in the Paleozoic Era.*

11 The Lingula is a genre of mollusks that have a bivalent shell and that are furnished with a pedicle valve (such as brachiopods). Fossil records indicate

that they have been around since the Cambrian Period, the first part of the Paleozoic Era.

12 In his review of *On the Origin of Species*, F. Marin regularly deploys Bergsonian formulations regarding the continuity of life and the way that our intellect, obsessed with action, divides life up into parts (Marin, "Sur l'origine des espèces," 577). He mentions the philosopher Édouard Le Roy (1870–1954), a disciple of Bergson, critiques Le Dantec's mechanistic theory, and even claims that "determinism gives too much importance to the milieu" (*ibid.*, 580). For more on Le Roy, see editorial note 111 to Chapter III. According to Marin, biological evolution takes place, according to two factors: an "essentially active factor" that exerts an effort and as such is "spiritual" and "intelligent"; and a "second factor that is passive, inert, a principle of resistance" (*ibid.*, 585–86). There is, however, an important break with Bergson in this characterization since Bergson would never admit that the finality at work in life could be considered "intelligent," [TN: since the intellect is rather associated with materiality (see below)].*

13 For a similar argument, see above, 56–57.

14 Bergson's reasoning here echoes his claims about free acts that, *once accomplished*, create "as much intelligibility as one would like" (see above, 48; cf. "Bergson à L. Brunschvicg [letter, February 26, 1903]," in *Mélanges*, 585–87). In the following years, Bergson will develop this idea into the notion that an event retrospectively creates its own possibility (see MRV, 13–19; PR, 99–116) and toward a certain sympathy for William James's thesis by which truth is an invention (see "Sur le pragmatisme de William James: Vérité et réalité," in *La pensée et le mouvant*, 244–52 ["On the Pragmatism of William James: Truth and Reality," in *The Creative Mind*].

15 One might compare this important claim with the related passage above, 51.

16 For Bergson's internal reference, cf. below, 128, 255. See also "hypnotism" in the Index.

17 Later editions have corrected a small typo here.

18 In the following pages, Bergson focuses on defining the criterion that allows us to make out the "main directions" of evolution.

19 For more regarding this difficult and important argument, cf. above, vi, 49–50; below, 153.

20 Bergson will present his results concerning this central question at the end of both Chapter II and Chapter III.

21 For more on this claim, cf. above, 13.

22 For more on the methodology implied in these lines, cf. above, 102.

23 The Drosera is an insectivore plant with tentacles, found in the bogs of Europe. The *Dionaea muscipula*, or the "Venus flytrap," is an herbaceous carnivorous perennial plant in the Droseraceae family. *Pinguicula* is the generic name for the grass known commonly as butterworts, which is a small plant found in marshes. Its sticky leaves hold onto the insects that land on them.

24 The term "saprophyte" designates heterotrophic organisms (organisms incapable of synthesizing their own cellular compounds that are thus dependent, for their dietary and energy needs, upon other animals or plants) that feed upon decomposing organic substances.

25 Gaston de Saporta (1823–1895) was the founder of evolutionary paleobotany in France. He was a follower of Darwin, with whom he corresponded. He contributed to the introduction of the embryological point of view in botany.

Antoine-Fortuné Marion (1846–1900) was Professor of Zoology at the University of Marseille. His book *L'évolution du règne végétal: Les cryptogames* attempted to retrace the entire evolution of the plant kingdom. Bergson copies (without citing) several exact phrases from this source.*

26 Bergson often resorts to psychological or finalist formulations (here, the image of a species "giving up") when speaking about an organism or species. This linguistic gesture alludes to his theory regarding the essential identity of consciousness and life. Cf. below, 110, 113–14, 124, 126, 131–32. For Bergson's initial justification, see above, 51.

27 For this example of the amoeba, cf. below, 121. For more on animals with a sensorimotor system, cf. below, 125–27.

28 "Sensitive plant" is a common name for the species *Mimosa pudica*.

29 Frédéric Houssay (1860–1920) was a zoologist. He taught at the *École normale supérieure* until 1897 and later became the dean of the Faculty of Sciences in 1912. An epistemologist of biology and a pioneer of what would later become morphogenesis, Houssay stood out for his interest in the temporal character of development. He was a partisan of Spencer's and criticized both Weismann and Darwin for their understanding of the "random" character of variation. He later drew closer to a Lamarckian-inspired reductionism.*

30 Bergson here employs finalist and psychological metaphors. Cf. editorial note 26 to Chapter II.

31 It should be noted that the term "kingdom" [*règne*] in Bergson's text refers more to a tendency than to a domain or a region.

32 Thus, after his discussion of the difference with regard to nutrition (cf. above, 107–9) and the difference in terms of mobility (cf. above, 109–11), Bergson here turns to a third difference.

33 For Bergson's understanding of "choice" in this context, cf. *MM*, 24–27; *CV*, 8–11.

34 Cf. *MM*, 24, 28; *CV*, 7–9.

35 This "absurdity" can be found in Schopenhauer's *The World as Will and as Representation*, 65. For the same objection, cf. *CV*, 7.

36 This is the first occurrence in the book of this distinction between the reflexes and voluntary action, which will play a very important role. Cf. "will, vs. reflex" in the Index.

37 For other occurrences of the image of "crossroads" [*carrefours*], see the Index.

38 This foreshadows Bergson's critique of Spencer, developed below, 364–65.

39 *Rhizocephala* is an order of Cirripedia crustaceans that are parasites on the stomachs of crabs. Deprived of limbs and digestive organs, they draw their food directly from the body of the host through a peduncle or root-like set of tubes.

40 Here we find the basis for one of the arguments that Bergson offers against neo-Lamarckism. Cf. above, 78–79, 87–88; below, 171.

41 TN: Cope does not explicitly discuss the animal "going to sleep," rather, it seems that Bergson is citing a general discussion that touches on many of the themes at issue in these paragraphs. See Cope, *The Primary Factors of Organic Evolution*, 74–77. For more on Cope, cf. Bergson's note to pages 35, his discussion on pages 77–78, and editorial note 92 to Chapter I.

42 Once again Bergson deliberately uses finalist and psychological formulations. The same is true for the three following paragraphs. Cf. editorial note 26 to Chapter II. "Zoospores" are reproductive plant cells with flagellum.

43 Here Bergson establishes the distinction between "awakened consciousness" (including both instinct and intellect) and "dormant consciousness" (or "torpor"). Note that these three kinds of consciousness figure in the title of this chapter. To the extent that they refer to the directions of consciousness, they also refer to the directions of life or "evolution." Bergson will later distinguish between instinct and the intellect (cf. below, 136–80).

44 The theme of "fatigue" that results from the effort of maintaining a certain existence is an increasingly important theme in Bergson's works (particularly after 1910). Cf. Bergson, "Le rêve," in *L'énergie spirituelle*, 100–104; "Le souvenir du présent et la fausse reconnaissance [1908]," in *L'énergie spirituelle*, 128, 144–52 ["Memory of the Present and False Recognition," in *Mind-Energy*]. See also Bergson, "Théorie de la personne," in *Mélanges*, 849; "Cours de Bergson à Columbia University," in *Mélanges*, 987; "Onze conférences sur "la personnalité" aux Gifford lectures d'Edinburgh," in *Mélanges*, 1082, 1085; "Conférence de Madrid sur la personnalité," in *Mélanges*, 1225.*

45 For more on this claim regarding convenience, cf. below, 131.

46 For more on this distinction, cf. above, 107–9; below, 131; CV, 14.

47 The connection between sensation and volition takes place because sensation itself is a function of the "indetermination of will" (see MM, 39). For more on the characterization of the nervous system as a "sensorimotor" system, cf. below, 125–27. In terms of the detailed explanation of Bergson's allusion to a *"sui generis* chemical activity," cf. below, 254 and editorial note 216 to Chapter III.

48 Maria von Linden (1869–1936) was a German bacteriologist and zoologist as well as the first female Professor in Germany and the editor of Eimer's posthumous texts. The first discovery of the type discussed here was attributed to Engelmann in 1883.

49 For Bergson's internal reference, see above, 97–98. For the precise characterization of life mentioned here, cf. below, 127, 252; CV, 12–17.

50 In relation to this argument, cf. AC, 34–36; Bergson (course notes by Paul Fontana), "Cours au Collège de France: Théories de la volonté [1907]," in Mélanges, 686; "Conférence de Madrid sur l'âme humaine," in Mélanges, 1207.

51 Here the term "work" should be understood in the technical sense it has in mechanics, namely the quantity of energy received by a system that is being moved under the effect of a force. Work is a scalar quantity.

52 For this sense of desire, cf. below, 122; MM, 25–26, 124.

53 For the precise meaning of these remarks and their contemporary relevance, cf. editorial note 216 to Chapter III.

54 For Bergson's various uses of the metaphors of "triggering," "explosives," "spark," and others of the same type, cf. below, 121–22; CV, 14; AC, 35–36; "Cours de Bergson à Columbia University" in Mélanges, 986–87; "Conférence de Madrid sur l'âme humaine," in Mélanges, 1207. Also, see the Index.

55 Cf. above, 100; and on Fungi, cf. above, 108. The Euglena is an order of photosynthetic single-celled flagellated protists that manifest both plant and animal characteristics.

56 In other words, there is an often-overlooked division within the plant world such that the kingdom of "torpor" is not completely uniform. Similarly, the animal kingdom splits between instinct and intellect.*

57 Here Bergson marks out the limits of the image of the "division of labor," an image that he (following the biologists of his day) regularly uses to describe organisms. If the image has its limits, it seems to be because the notion of work itself is charged with anthropomorphic and finalist implications (cf. above, 60, 89–90, 92, 106). For more on his critique of the image of the division of labor, see below, 255.

58 For more on this Leibnizian-inspired formulation, cf. below, 258; E, 124; "Fantômes de vivants," in L'énergie spirituelle, 77–78; IM, 190.

59 For Bergson's internal reference, see above, 107.

60 Here is where Bergson provides the criterion by which we can recognize that we are dealing with an original tendency. The whole paragraph can be understood as an elaboration of this principle.

61 In the same way, the function of the brain is to inhibit—that is, to push back into the "unconscious"—all of those aspects of the individual life (memories, perceptions, etc.) that would be useless or harmful in the present situation. Such is the mechanism Bergson calls "attention to life." See editorial note 217 to Chapter II. We are thus here dealing with one of the most profound structures of Bergsonian thought, directly linked back to the structure of durée.

62 For the mention of "luxury," see above, 60.

63 The allusion here is surely to De Vries, and the "recent experiments" are likely the ones discussed above, 86–87. See also editorial note 239 to Chapter I.

64 For Bergson's internal reference, cf. above, 115–17.

65 For more on pseudopods, see above, 109–10.

66 Cf. *MM*, 27–45.

67 For this definition of "sensorimotor system," cf. below, 125–27. For the mention of desired movements, see above, 116; *MM*, 25–26, 124. For more on explosives, see above, 115–17, as well as editorial note 54 to Chapter II.

68 This presentation again seems to imply a conception of the organism as a collectivity or society (see "society, organism as" in the Index). Thus, the comparison that immediately follows between the State and the organism is hardly accidental.

69 With regards to "potential energy," see below, 242–43.

70 For the distinction between "plastic substances" and "energetic substances," see above, 34 and below, 256. See also editorial note 216 to Chapter III.

71 Bergson is here thinking of Claude Bernard's discovery of the glycogenic function of the liver, which was the point of departure for the research examined in the following discussion. Cf. also the works mentioned in Bergson's note to page 124.

72 Later editions change the phrasing here slightly.

73 With the notion of exchanging favors, Bergson again offers a finalist and psychological formulation. Cf. editorial note 26 to Chapter II.

74 An important article by Brown-Séquard, which Bergson cites above, also appears in this journal (see above, Bergson's third note to page 81). Dufourt regularly focused his research on questions pertaining to the liver's production of glycogen and the independence of glycogen levels in the blood in relation to food intake and muscle use.*

75 Thus, it is "quite literally" that the organism is said to be a collectivity. Cf. "society, organism as" in the Index.

76 In the first reference, Bergson is thinking of the discussion on page 325 of physician Marie de Manacéïne's (1841–1903) "Quelques observations expérimentales sur l'influence de l'insomnie absolue," where the author examines the conservation of the brain despite significant changes in the other parts of the body during starvation. The opposite is observed in animals that have died from insomnia. In the second reference, Podwyssozki explains this microscopic study of the organs and the effects of starvation on the cells.

77 Georges Cuvier (1769–1832) was a French zoologist and statesman. In this famous article, "Sur un nouveau rapprochement à établir entre les classes qui composent le Règne animal," which is important in the history of comparative anatomy, he attempts to solve the taxonomy problem of determining kingdoms by focusing on the nervous systems of animals. His focus on nervous systems had important repercussions in philosophy, for instance in the work of Maine de Biran (1766–1832). For his discussion below, Bergson adopts a slightly different (though not unrelated) division to the one proposed by Cuvier between vertebrates, mollusks, articulated organisms (that is, Arthropods, or "Articulata" [*Articulés*]: Crustaceans, Insects, etc.), and zoophytes (Infusoria, etc.). Cf. below, 131, 135.

Bergson's second reference is to Michael Foster (1836–1907), a British physiologist and biologist. In the cited article, Foster writes: "The master tissues and organs of the body are the nervous and muscular systems, the latter being, however, merely the instrument to give effect and expression to the motions of the former. All the rest of the body serves simply either in the way of mechanical aids and protection to the several parts of the muscular and nervous systems, or as a complicated machinery to supply these systems with food and oxygen, *i.e.*, with blood, and to keep them cleansed from waste matters throughout all their varied changes..." (Foster, "Physiology," 17). Foster calls for further study of the nervous system, a science that was still "in its infancy" (*ibid.*). This article is an important starting point for Bergson's arguments here.

It is also noteworthy that Bergson takes the time in this passage to develop his definition of the sensorimotor system, a concept first proposed in *Matière et mémoire* and used regularly by him afterward. Bergson is likely attempting to respond to criticisms of his position in light of the contemporaneous debate over "muscular sense" and the difference between sensory nerves and motor nerves, perhaps not sufficiently emphasized in his initial work. This debate has important consequences for our awareness of our body, as discussed already by Bergson (*E*, 15–20; cf. *MM*, 218). See also Henri Bergson, "Discussion à la Société française de philosophie [1904]," in *Mélanges*, 647–48.

Bergson's mention of "interior milieu" should be understood to imply constancy and the phenomenon of autoregulation. The constancy of the interior milieu was one of the great discoveries of Claude Bernard, whose work was influential on Bergson here (cf. above, 17).*

78   This passage again offers a finalist and psychological formulation. Cf. editorial note 26 to Chapter II.

79   Bergson is referring here to the first chapter of *Matière et mémoire*.

80   For more on this double necessity, cf. below, 252; *MM*, 24–27.

81   Anastomosis is the intercommunication between two channels, canals, or branches. The term was originally used to describe cross communication among arteries and veins. Bergson occasionally uses the term in a figurative manner, cf. *MM*, 130; *DS*, 6, 109, and 133. For the difficult question regarding the choice of actions, cf. *MM*, 24–27; *CV*, 8–9.*

82   "Monera" is the name given by Haeckel to the most basic form of life, a simple mass of protoplasmic jelly, lacking any nucleus, which in this way contrasts with the Amoeba. The term was not particularly current, and Haeckel had forged it from a fully speculative perspective—the Monera being conceived as an element of all living matter, and the organism being conceived as an association of parts, all from within a general "monism."

83   For the internal reference, Bergson is most likely referring to above, 97–98. For more on the explicit formulation regarding the insertion of indetermination, cf. above, 116; below, 252.

84 For more on this image of the nervous system in relation to pathways and questions, cf. below, 262–63; MM, 43, 45. The characterization here of a nervous system as a "reservoir of indetermination" is at once original, difficult, and decisive. For Bergson's use of "indetermination," cf. MM, 28–29, 35–36, 39–41; CV, 12–13, 15–17.*

85 For more on Bergson's important claim regarding the finitude of the *élan*, cf. "finitude, of the *élan*" in the Index. For the most explicit discussion of this claim, see below, 254.

86 This is an allusion to Leibniz's *Theodicy*. Cf. below, 351.

87 Regarding this double limitation between opposition and distraction, cf. above, 99–101. For the image of hypnotization, see "hypnotism" in the Index. For the image of the mirror, which Bergson inherits from Plotinus, cf. "Le rêve," in *L'énergie spirituelle*, 96–97. As for the egoism of each individual species, cf. below, 255.

88 Later editions correct a small typo here.

89 For the essential opposition between "things" and "progressions," as well as some distinctions of the same type, cf. below, 249, 309; E, 82, 97–98, 136, 147, 149, 165; MM, 135, 139, 142, 180, 211.

90 For more on this theme of "self-consciousness," which is charged with a double sense—at once liberation and distraction in relation to a principle—cf. "consciousness, self-" in the Index.

91 For more on this important Spinozist principle, cf. CV, 18–19. For more on the relation between "viability" and adaptation, cf. above, 51, 55–59, 66, 70–72, 102–4.

92 Cf. above, 100.

93 The first edition contained a typo here, corrected in later editions.

94 Albert Gaudry (1827–1908) was responsible for establishing evolutionary paleontology in France. Although Darwin references him and held him in high esteem, Gaudry notably criticized the mechanism in Darwin's theory and argued for the common origin of all species. *Essai de paléontologie philosophique* is the synthesis of Gaudry's work on the relation between paleontology and evolutionary theory, and it is worth noting that he characterizes his work as "philosophical" paleontology. He claims that the animate world as a whole evolves like an individual, according to a plan. For Gaudry, the function of paleontology was to replace philosophy by studying this plan. He posits an "unchanging organizer" or creator who would thus be distinct from its creation.*

95 Cf. above, 114, and editorial note 26 to Chapter II.

96 Cf. above, 107–8, 114–15; CV, 14.

97 The Echinoderm is a phylum of organisms that possess a hard and spiny skin. Echinoderms include, for example, sea urchins and starfish.

98 Bergson usually reserves the term *torpeur* ("torpor") for the plant world.

99 For more on this mention of "human societies," cf. CV, 25–27. For the theme of "risk" in Bergson, cf. the Index.

100 For Bergson's internal reference, see above, 102–4.

101 Nathaniel Southgate Shaler (1841–1906) was a geologist at Harvard University, but his book *The Individual: A Study of Life and Death* went far beyond the limits of geology by incorporating biology, examining the place of the human individual within nature, and exploring their relationship to death. Bergson's interest in this book could be explained by many of its themes, including the close link that Shaler establishes between the development of the intellect and the configuration of the skeleton.*

102 For more on Bergson's understanding of the hand, cf. DPP, 92–93.

103 Bergson, having already identified the two principal lines of evolution, still has to identify their "high points." To do so, he must begin by identifying a criterion. Cf. above 106, 120 and editorial notes 18 and 60 to Chapter II.

104 For Bergson's earlier discussion of "precision," see above, 13, 107, 113.

105 Bergson returns to this claim below, 265–66; cf. CV, 25.

106 French physiologist and biologist René Quinton (1866–1925) contested the evolutionary theories of Lamarck and Darwin, and he developed a theory based on the "law of the original thermal constancy," dating the appearance of a species by its internal temperature. The idea was that, in response to the cooling down of the earth, life would have produced organisms increasingly capable of generating (through "effort") their own internal warmth. He was thus critical of using the intellect as a classificatory tool, and this reminds us that Bergson's insistence that human intellect is what sets humans apart was not universally accepted at the time *L'évolution créatrice* was published.*

107 Bergson formulates this conclusion in the two sentences that conclude the paragraph. The "different paths" he mentions leading up to that conclusion are the ones outlined in the two previous paragraphs: extent of a species' dominance and timing of its appearance.

108 From this point forward, the distinction between instinct and intellect will be the focus of this chapter.

109 For the expression "philosophy of nature," which captures Bergson's project in this book, see VOR, 266. Bergson's goal is to set up an opposition with Spencer's philosophy of nature. For Bergson's important claim regarding the nature of a tendency, which comes directly out of his notion of *durée*, cf. above, 90, 118.

110 Similarly, in his earlier work, Bergson argues that empirically lived perception is always mixed with memory (*MM*, 67–68). And again, in his later work, moral obligation such as we actually encounter it is the result of both the closed and the open (*DS*, 98). This distinction between the "pure" and the "mixed" also describes the nature of *durée*, and the "mixed" serves to define Bergson's concept of the "concrete," a term used just below, 137.

111 Cf. "fringe" in the Index.

112 For more on rigid definitions, see above, 13, 107, 113, 134.

113 For more on this definition of the organism, see above, 97.

114 This is a key claim throughout Bergson's work: the intellect is, first and foremost, a faculty of *action*, not of speculation.

115 In 1863, French archaeologist Jacques Boucher de Crèvecoeur de Perthes (1788–1868) discovered a human jawbone next to carved flintstones near Abbeville, France. In 1864, he published his book *De la mâchoire humaine de Moulin-Quignon: Nouvelles découvertes en 1863 et 1864*. This book was criticized by his contemporaries, and it turned out that the jawbone was a fake. Nevertheless, by considerably pushing back the date of the appearance of man, Boucher de Perthes helped to establish the science of prehistory [Boucher de Perthes, *De la machoire humaine de Moulin-Quignon: Nouvelle découvertes en 1863 et 1864* (Paris: Jung-Teuttel, [1864])].*

116 Note that the first edition reads *infléchissement* ("shifting" or "reorienting") rather than *fléchissement* ("bending").

117 Reflections on the functioning of the steam engine gave birth, in particular, to thermodynamics, an important science for Bergson. See below, 244 and editorial note 182 to Chapter III; cf. also, below, 185–86.

118 Bergson develops the relationship between creation and feelings more fully in *Les deux sources de la morale et de la religion*, see DS, 36–39.

119 Paul Lacombe (1834–1919) was a French historian. In *De l'histoire considérée comme science*, Lacombe focused on how history might be treated as a science and how historical causality might be determined, placing the emphasis on technological inventions rather than religion, metaphysical systems, or governmental institutions. Lacombe also suggests that a historical age can be named after its technical inventions (*ibid.*, 173). The influence of this book on Bergson can be detected below, 187, 242; cf. DS, 324–25.*

120 George John Romanes (1848–1894) was a physiologist and naturalist. Darwin encouraged him to apply the theory of natural selection to the process of mental evolution. The distinction between "primary" and "secondary" instincts is found in Romanes, *Mental Evolution in Animals, with a Posthumous Essay on Instinct by Charles Darwin*, 180–99. For Romanes, primary instincts are those born through natural selection, while secondary instincts are those that develop through the intervention of the intellect. Unlike Bergson, Romanes posits a difference of degree rather than of nature between animal intellect and human intellect.*

121 Bergson often uses the notion of an "ideal limit," and indeed develops a novel usage of it. See the Index.

122 One might compare Bergson's formulation here to his review of Paul Janet's *Principes de métaphysique et de psychologie* ("Compte rendu des *Principes de métaphysique et de psychologie* de Paul Janet [1897]," in *Mélanges*, 375–410). See also Janet, *Les causes finales*, 137. Given his familiarity with Janet's work, it is perhaps noteworthy that Bergson does not refer to him in these pages. The reason might be attributed to how critical he was of Janet's "finalism."*

123 *Pseudoneuroptera* and *archiptera* are insects whose metamorphosis is incomplete. Scientists used to include termites, dragonflies, and mayflies in this group. For more on polymorphism, see below, 158–59; DS, 296.

124 For this important characterization of instinct and the intellect as modes of activity, see above, 137; see below, 144.

125 For more on the infinite in relation to the intellect, see "infinite/indefinite" in the Index.

126 For more on this theory of the organ as a tool, see below, 162.

127 For more on the creation of a "new need," see below, 146. The same analysis is developed prodigiously in *Les deux sources de la morale et de la religion* (*DS*, 317–28).

128 For the idea of an "unlimited force," see below, 150.

129 See "finitude of the *élan*" in the Index.

130 Eugène-Louis Bouvier (1856–1944) was a prominent entomologist. Bergson's sentence here offers a general summary of the article "La nidification des Abeilles à l'air libre." For Bouvier, the inventions in question here are essentially "architectural" in nature. Bouvier compares the work of the Bees with the work of an artisan, invokes the idea of a "hive-mind" (*esprit de la ruche*) and suggests that there is an implicit exchange of ideas between the workers (*ibid.*, 1020).*

131 Here Bergson is alluding to the myth of Prometheus and Epimetheus. See *DS*, 63, 302–3.

132 In terms of "psychical activity," see above, 137, 141.

133 For Bergson's internal allusion, see above, 133.

134 This is because life *resolves* problems, whereas intuition (which nonetheless prolongs life) *poses* problems. Deleuze emphasizes this in his commentary on Bergson (see Deleuze, *Bergsonism*). For more on life as the resolution of problems, see above, 59, 71, 78.

135 For Bergson's internal reference, see below, 171.

136 Although the distinction between two kinds of unconsciousness is surely in harmony with Bergson's philosophy, this is the only place he develops it explicitly.

137 Bergson writes in *Matière et mémoire*: "Extended matter, regarded as a whole, is like a consciousness where everything balances and compensates and neutralizes everything else" (*MM*, 246–47 [*Matter and Memory*, 219]). The direction, however, of this relationship seems to be inverted in *L'évolution créatrice*.

138 Here Bergson is taking up and modifying a discussion found in one of Spinoza's Letters. See Baruch Spinoza, "Letter 58," in *Ethics, Treatise on The Emendation of the Intellect, and Selected Letters*, trans. Samuel Shirley, ed. Seymour Feldman (Indianapolis: Hackett Publishing Company, 1992), 285–86.

139 For the allusion to habitual actions, see CV, 10–11; VOR, 266–67. For the example of the sleepwalker, see *Le rire*, 141; *DS*, 20.

140 For this definition of consciousness, see *MM*, 24–27; CV, 10–11; VOR, 266–67.

141 See below, 264. With the "zone of indetermination" and "possible" or "virtual" action, Bergson here returns to some major concepts from the first chapter of *Matière et mémoire*, see *MM*, 16–60.

142 Cf. CV, 10–11; VOR, 266–67.

143 See above, 142; DS, 317–28.

144 For more on this opposition, see "enacted vs. thought" in the Index.

145 A botfly is an Insect that lives as a parasite of horses. It is related to the fly and the horsefly.

146 For more on this example, see Bergson's discussion of the Sphex, below, 173–75.

147 The Sitaris is a Beetle (*Coleopteran*) from Southern Europe. For the "story," which Bergson outlines in here, see Fabre, *Nouveaux souvenirs entomologiques*, 262–303. French entomologist Jean-Henri Fabre (1823–1915) devotes two chapters to the Sitaris. In fact, many of Bergson's examples of insects are taken from Fabre's work (see the notes and editorial notes below, 173–74). Darwin himself was interested in the Sitaris, an interest he too learned from Fabre.

148 Hymenopteran insect of the Apidae family.

149 For this formulation, which serves as a definition here, and for the problem that corresponds to it, see "internalized/externalized" in the Index.

150 Here Bergson is thinking of Kant. But by deliberately enlarging the Kantian sense of the words "matter" and "form," he is also alluding to Aristotle. See below, 316, 326, 357–58.

151 For more on the "force immanent to life," see above, 142. For Bergson's internal reference, see above, 127, 142.

152 Revealing another underlying debate with Kant, Bergson here alludes to the Kantian distinction between categorical and hypothetical judgments from Kant's famous section "On the logical function of the understanding in judgments." See Immanuel Kant, *Critique of Pure Reason*, trans. and ed. Paul Guyer and Allen W. Wood (Cambridge: Cambridge University Press, 1998), A70/B95.

153 On Bergson's use here of the term "indefinite," which is importantly a characteristic of the intellect, and perhaps also of intuition, see "infinite/indefinite" in the Index.

154 As "internal" and "full," instinct is in close kinship with intuition, but intuition is disinterested. Cf. IM, 177–82.

155 But perhaps humans alone are capable of valuing what is "useless" (*MM*, 87). Cf. see "infinite/indefinite" in the Index.

156 Bergson here lays out the precise problem to which the notion of intuition responds.

157 For Bergson's internal reference, see above, 146–49.

158 This is the critique that Bergson will offer against Fichte and Spencer (see below, 190–92). For more on this critique, see DS, 168–69, 173, 185; DPP, 34, 54–55; PC, 152.

159 This is indeed, according to Bergson, the function of the intellect: the intellect analyzes and spatializes.

160 Although Kant is not named here, this is indeed one of Bergson's most developed and thoroughly argued presentations of his critique of a certain Kantian "relativism" (see below, 198, 231, 361).

161 As Bergson announced earlier: "Action cannot move in the unreal," above, vii.

162 It is in this sense that the intellect can, itself, be called a "part" of life (cf. above, vi, 49–50, 106).

163 For more on this connection between matter and intelligibility, see below, 189, 365–67.

164 To establish the link between matter and intellect, it will be necessary to critique the idea of disorder. See below, 209–38.

165 Indeed, Bergson occasionally appeals to common sense. See below, 277; MM, 1–3, 41, 46, 61, 76, 154, 158, 202. [TN: note that the first pages cited here (MM, 1–3) were only added to Matière et mémoire as the "Avant-propos de la septième édition" in 1910, after the publication of L'évolution créatrice. See the "Introduction" to Matter and Memory, 9–11.]

166 For the image of "feeling at home," see above, v, and editorial note 2 to the Introduction.

167 Bergson is surely alluding here to Descartes and his phrase "partes extra partes."

168 Cf. MM, 220–26.

169 Cf. below, 302.

170 For an important discussion of the connection between reality and movement, see above, 92. For the claim regarding immobility and the relative, see DPP, 77; PC, 159–60.

171 See "violence" and "intuition, conversion to" in the Index.

172 Bergson's internal reference is to his discussion of disorder (below, 209–38), of nothingness (below, 273–98), and of the "cinematographic mechanism of thought" (below, 298–307), as well as to the pages devoted to the history of philosophical systems (below, 308–69).

173 This is the general formula of what Bergson calls a "pseudo-problem." See the Index.

174 For Bergson's internal reference, see below, 298–315.

175 For more on Bergson's use of this image, drawn from Plato's Phaedrus, see DS, 109.

176 Here, merely "in passing," Bergson defines the very essence of the intellect, which is space!

177 With the question as to whether space is perceived or conceived, we again find Bergson working through an entire underlying debate with Kant. In Essai sur les données immédiates de la conscience, he suggested that space is a "form of our sensibility" (E, 177, cf. also, 70). Nevertheless, Bergson also suggests there that space is the object of a "conception" (E., 71), and reveals his own hesitation when he characterizes the act of grasping space as "the intuition, or rather the conception, of an empty homogeneous medium" (E., 70). Bergson proposes to distinguish the "perception of extension" and the "conception of space" (E, 72). Whereas Matière et mémoire leans in favor of an intermediate solution, since that book considers space to be the "mental diagram [schème] of infinite divisibility" (MM, 232 [Matter and Memory, 206]), L'évolution créatrice here seems to come back toward the formulations proposed in the Essai.*

178 This character of being able to be manipulated "as we please" forms, for Bergson, the very nature of space. See below, 307, 309, 331–32, 337; *E*, 61–63; *MM*, 231–32.

179 In relation to our possible action upon things, cf. *MM*, 235–36, 260. For Bergson's internal reference, see below, 203–5.

180 For this allusion to animal spatiality, cf. *E*, 70–73.

181 This enumeration began above, 138.

182 But this is not merely an "addition": life in society is certainly found at the endpoints of the two main lines of evolution, which attests to the fact that it is indeed the goal of life, although in a non-teleological sense. As such, it is a sort of "success." Cf. *CV*, 25–27.

183 See below, 328–29. Cf. *E*, 82–89; *Le rire*, 117–21; *IM*, 177–82. Bergson develops a general theory of the sign in *Histoire de l'idée de temps*, 28–57.

184 Questions regarding insect language are debated, for instance, in the articles mentioned in Bergson's note to page 177.

185 For Bergson's discussion of polymorphism, see above, 141; cf. *DS*, 296.

186 For this analysis, which is an important one for Bergson, see below, 265; *DS*, 168.

187 For the idea that animals generalize, see *MM*, 177–80.

188 Bergson's famous model of an inverted cone as an image for memory is implicit in this sentence. See *MM*, 179.

189 This use of ready-made concepts stands in opposition to Bergson's call for individually tailored concepts that would be "cut to the exact measure of the object" (MRV, 23 ["Introduction (Part I): Growth of Truth: Retrograde Movement in Truth," in *The Creative Mind*, 17]). See also IM, 197.*

190 For Bergson's internal reference, see above, 152.

191 Bergson proposes an alternative type of clarity that is worth comparing to the Cartesian variety. See DPP, 31–32. Bergson is here completing his underlying debate with Descartes, which begins above when he discusses what the intellect can conceive of "clearly." Cf. above, 155–57.*

192 This sums up Bergson's critique of the Cartesian *cogito*: the *cogito* is not grasped by the intellect as a temporal continuity.*

193 For Bergson's internal reference, see below, 212–17.

194 For Bergson's internal reference, again see below, 212–17. These paragraphs reveal Bergson's somewhat tentative position in the contemporary debate regarding the foundations of mathematics, which in part circled around the question of the primacy of either mathematics or logic. Édouard Le Roy, Bergson's disciple, actively participated in this discussion, alongside contributions by Russell, Whitehead, Poincaré, and others. These paragraphs invited attack, notably from mathematician Émile Borel (1871–1956). See Borel's essay included in the **Correspondence, Reception, and Commentaries** section below. For more on Le Roy, see editorial note 111 to Chapter III.*

195 For more on the notion of "feeling at home," see editorial note 2 to the Introduction.

196 Bergson clarifies below that, for him, "good sense" [*le bon sens*] refers to "the continuous experience of the real" (214; see also *MM*, 170; *Le rire*, 140–41, 149; "Le rêve," in *L'énergie spirituelle*, 103; *DS*, 109–10, 241–42; "Le bon sens et les études classiques [1895]," in *Mélanges*, 359–72).

197 Cf. above, 142.

198 For Bergson's use of the term "fascination" here, cf. "Le rêve," in *L'énergie spirituelle*, 96–97; cf. also "hypnotism" in the Index.

199 For this image of "twisting back upon itself" [*se tordre sur elle-même*], see below, 238–40, 251. This sentence also contains the first of only two occurrences of the term "creative evolution" in this book itself. For the other, see below, 224–25. For other occurrences of the phrase in Bergson's work, see CV, 27; *DS*, 99, 115, 264; MRV, 13; DPP, 95.

200 Theodor Schwann (1810–1882) completed much of this research in the years 1830–1840, followed by Rudolf Virchow (1821–1902). The ambiguity in Bergson's praise here likely comes from the fact that, given his commitment to *durée*, it was impossible for him to simply account for the organism as composed of cells. And yet, this paradigm was considered incontestable by the majority of those he references from the positive sciences. By the 1890s, thanks to the work of Weismann and Delage, the idea of the organism as a whole and the cells as the result of a process of dissociation was increasingly viable. This is one of the points upon which we find a remarkable convergence between Bergson's philosophical interrogations here and debates within contemporaneous scientific research.*

201 Bergson's reference to below, 281, is surely erroneous. He is almost certainly thinking of Chapter III, 261. The discussion of the point begins on 259. The scientists he alludes to here are most likely the same ones mentioned in Bergson's fourth note to page 260.

202 For more on this new "discontinuity," see above, 31–33.

203 This definition of evolution is also a definition of life. See the "life, definition and charaterization of" in the Index.

204 See below, Chapter IV. This is one of Bergson's main goals in the pages devoted to the so-called "history of systems" (below, 308–69).

205 For related images, cf. below, 307, 312; IM, 206.

206 For more on this important point, see "reconstruction, critique of" in the Index.

207 For the idea of circularity between cause and effect, cf. Bergson "L'effort intellectuel," in *L'énergie spirituelle*, 190 ["Intellectual Effort," in *Mind-Energy*]. This idea was central for Bergson in his 1906–1907 course at the Collège de France on "volition/will." For more on this course, see: "Cours au Collège de France: Théories de la volonté [1907]," in *Mélanges*, 693n1, 711, 715, 721–22. On the notion of "sympathy" and its relation to intuition and instinct, see "sympathy" in the Index. For more on the objects that Bergson suggests are accessible to intuition, see DPP, 27–29.

208 For more on this highly significant remark, see "genius" [*génie*] and "springing forth" [*jaillissement*] in the Index.

209 In contrast to the clumsiness, cf. the "feeling at home" or "being at ease," above, v, and editorial note 2 to the Introduction.

210 For these questions, which we often forget are extremely important ones to Bergson, cf. *DS*, 99–102; *DPP*, 93–95; Bergson, "La spécialité [1882]," in *Mélanges*, 257–64; "Le bon sens et les études classiques [1895]," in *Mélanges*, 359–72; "De l'intelligence [Discours au Lycée Voltaire, 1902]," in *Mélanges*, 554–60; "Bergson à A. Hébrard [letter, January 19, 1915]," in *Mélanges*, 1136–39; "Les études gréco-latines et la réforme de l'enseignement secondaire [1922]," in *Mélanges*, 1366–79.

211 See above, 140–41, and the editorial notes to those pages.

212 For more on the distinction, offered by Romanes, between primary and secondary instincts, see above, 140 and editorial note 120 to Chapter II.

213 For more on the notion of "vitality," which Bergson criticizes (along with that of "vitalism"), see the Index.

214 For this example, see below, 172 and editorial note 229 to Chapter II.

215 For other occurrences of the metaphor of "theme and variation," which Bergson regularly uses in relation to instinct, see the Index.

216 For more on this analogy between life and consciousness as well as Bergson's understanding of memory, see "memory, as analogous to life" in the Index.

217 Bergson is referring here to the mechanism of "attention to life." See *MM*, 192–98; *AC*, 47–49, 57; "Fantômes de vivants," in *L'énergie spirituelle*, 75–79; "Le rêve," in *L'énergie spirituelle*, 94–100; "Le souvenir du présent et la fausse reconnaissance," in *L'énergie spirituelle*, 144–46.

218 In the publication of *L'évolution créatrice* in the French collected volume of Bergson's works (*Œuvres*), editor André Robinet notes that Bergson's allusion here is to Plotinus. Cf. Plotinus, *The Enneads*, IV, 4, §32. Robinet, following Émile Bréhier's 1927 translation of Plotinus into French, connects Plotinus's use of the notion of "sympathy" to describe a living whole with the Stoics' use of the same term. See Bergson, *Œuvres*, ed. André Robinet, intro. Henri Gouhier, 6th ed. (Paris: Presses Universitaires de France, 2001); Plotin, *Quatrième Ennéade*, trans. Émile Bréhier (Paris: Les Belles Lettres, 1927). Bergson addresses some of these issues in *L'évolution du problème de la liberté: Cours au Collège de France, 1904–1905*, ed. Arnaud François (Paris: Presses Universitaires de France, 2017), 129–61, and particularly 149.*

219 For the distinction between "primary" and "secondary" instincts, see above, 140 and editorial note 120 to Chapter II. [TN: Bergson earlier refers to a plurality of primary instincts, but here refers to a single one admitting of several modalities.]

220 The term "analysis" possesses here, and in many similar occurrences (see the Index), the sense that Bergson gives to it in "Introduction to Metaphysics": an operation by which the mind, insofar as it is intellect, reduces a thing down to its elements rather than penetrating into it, i.e., rather than joining its simple movement via an effort of intuition. See *IM*, 177–82.

221 These formulations allude to what is known as "Molyneux's problem." The problem was first posed to John Locke in a letter from an Irish scientist and

politician named William Molyneux (1656–1698). Molyneux asked Locke whether a person who is blind from birth, and who is used to distinguishing between a cube and a sphere by touch, would be able to distinguish between the two by vision alone after having regained their sight. (See John Locke, *An Essay Concerning Human Understanding*, ed. Kenneth P. Winkler (Indianapolis: Hackett Publishing Company, 1996), Book II, paragraphs 8–9.) This question invited reflections from many thinkers in the empiricist tradition, including Berkeley, Diderot, and Condillac, and came down to Bergson through its continuing influence on nineteenth-century English empirical psychology.*

222 For Bergson's critique of relying on images drawn from the sense of touch, see *MM*, 223–24.

223 Cf. above, 63–65.

224 Cf. above, 66.

225 Here Bergson feigns agreement on precisely the claims he had refuted above when discussing Darwinism. Cf. above, 65.

226 For this characterization of the plant world, cf. above, 78–79, 87–88, 112.

227 Bergson formulated exactly the same question with regard to the coordination among the cells of a living body. Cf. above, 167.

228 In this paragraph, Bergson applies to instinct the conclusions he drew above regarding the various forms of evolutionary theory. Cf. above, 86–88.

229 Apidae is a family of Hymenopteran insects. The group includes honeybees, bumblebees, *meliponines* (stingless bees), *anthophorini* (digger bees), etc. *Meliponines* belong to the tribe *Meliponini*, and *Bombini* is the tribe of apid bees to which the bumblebees belong.

In the passage cited by Bergson, German biologist Buttel-Reepen (1860–1933) concludes that there is no strict relation of kinship between the bumblebees and the *Meliponines*, but that their formation of social groups marks an important degree between the bumblebees and the bees ("Die phylogenetische Entstehung des Bienenstaates, sowie Mitteilungen zur Biologie der solitären und sozialen Apiden," 108).*

230 The meaning of the recurring comparison (see above, 23, 167–68) to musical themes and variations becomes clear here: Bergson is challenging the model of a difference of degree ("the addition of elements") and supporting the model of a difference of nature, in which it is nevertheless possible to perceive degrees of intensity, that is, the different "keys" into which the theme is transposed (cf. *MM*, 7, 189). As such, "variation," alongside divergence and inversion, is one of the several ways that Bergson describes a difference of nature in *L'évolution créatrice*.

231 Bergson here begins a series of examples drawn from Fabre's *Souvenirs entomologiques*, published to great renown in ten "series" between 1879 and 1907. Fabre was generally against evolutionism and tended to attribute to Insects a sort of intellect higher than mere instinct. Bergson might have been criticized for according so much weight to these relatively non-scientific studies. Note that the Scolia, referred to in this sentence, is a large Hymenopteran insect found in warm weather countries.

232 The golden digger wasp is a relatively large Hymenopteran insect of the *Sphecidae* family. [TN: Bergson uses Fabre's name for this species, *Sphex à ailes jaunes*, which is further specified by Fabre as *Sphex flavipennis*.]

233 The *Ammophila hirsuta* [properly named *Podalonia hirsuta*] is a Hymenopteran Insect of the Sphecidae family (the thread-waisted wasps). This example is one of the most well-known ones used in *L'évolution créatrice*, to the point that (and in part because) philosopher Raymond Ruyer (1902–1987) dedicated an entire article to it. See Raymond Ruyer, "Bergson et le Sphex ammophile," *Revue de métaphysique et de morale* 64, no. 2 (1959): 163–79. [TN: An English translation of this article is included in the **Correspondence, Reception, and Commentaries** section below.]

234 This "theme" plays exactly the same role as the "march toward vision" at the end of the first chapter (cf. above, 97): a function implicit in the *élan vital*, of which the organic form is but a collection of "views."

    235 In the chapter that Bergson cites, George William Peckham and Elisabeth G. Peckham contest somewhat the specific observations of Fabre regarding the conditions of the larva's ability to feed on the caterpillar (see *Wasps: Social and Solitary*, 28–30). [TN: Note as well that in this sentence, Bergson seems to erroneously merge the two names from his two examples into "Sphex Ammophila," as noted by Tano S. Posteraro, translator of Raymond Ruyer's article "Bergson et le Sphex ammophile," which repeats this formulation in its title. The Sphex and the Ammophila are two separate names for genera in the family of thread-waisted wasps and not an individual species. Cf. Raymond Ruyer, "Bergson and the Ammophila Sphex," trans. Tano S. Posteraro, in "Instinct, Consciousness, Life: Ruyer contra Bergson," by Tano S. Posteraro and Jon Roffe, *Angelaki* 24, no. 5 (2019): 134–47, which can be found in the **Correspondence, Reception, and Commentaries** section below. The original French appears as: "Bergson et le Sphex ammophile," *Revue de métaphysique et de morale* 64, no. 2 (1959): 163–79.]

236 Cf. above, 79–85.

237 For more on Bergson's use of the expression "from the outside" [*du dehors*] for designating intellectual knowledge, see IM, 177–82.

238 For this important characteristic of instinct and intuition, see "sympathy" in the Index.

239 For more on Bergson's account of the relationship between philosophy and science, see the Index.

240 With the term "compound reflex," Bergson is referring to Spencer's understanding of instinct: "Using the word, not as the vulgar do to designate all other kinds of intelligence than the human, but restricting it to its proper signification, Instinct may be defined as—compound reflex action. Strictly speaking, no line of demarcation can be drawn between it and simple reflex action, out of which it arises by successive complications." Herbert Spencer, *The Principles of Psychology* (London: Longman, Brown, Green, and Longmans, 1855), 432. This passage in Spencer follows a chapter devoted to the reflex, which Bergson cites later in *L'évolution créatrice* (see below, 364).*

241 For more on the relation between "intellect" and what is "intelligible," see below, 191, where Bergson again takes aim at Spencer.

242 On the mention of "unreflective sympathies and antipathies," cf. DPP, 28.

243 Bethe's work, *Dürfen wir den Ameisen und Bienen psychische Qualitäten zuschreiben?*, is a response to Forel's article "Un aperçu de psychologie comparée." Albrecht Bethe (1872–1954) was a German physiologist who worked on the nervous system of invertebrates, noting its "plasticity." Auguste-Henri Forel (1848–1931) was trained as a neuroanatomist, psychiatrist, and naturalist who worked on the structure of the brain. Influenced by Haeckel's version of monism, Forel's article here posits a certain reductionist panpsychist theory proposing the existence of small brains and souls in all of the elements of the living world, and indeed in the inert world too.*

244 For more on "nullified" consciousness [*conscience annulée*], see above, 144–45. [TN: Here Bergson does not write *annulée*, but rather *tombée* ("fallen" or "diminished").]

245 Bergson uses precisely the formulations he previously used in "Introduction à la métaphysique." See IM, 177–82.

246 Here Bergson begins to thematize his notion of intuition, which emerges as a response to the problem set out above, 152. For more on "self-consciousness" and "reflection," cf. DPP, 95 and the Index.

247 For this argument relating to aesthetic experience, see *Le rire*, 115–25; PC, 150–53.

248 Bergson again bases his analysis upon his interpretation of Ravaisson (see above, 90–91; VOR, 264–66; *Histoire des théories de la mémoire*, 23–26.

249 Cf. *Le rire*, 115–25.

250 Bergson here returns to the stakes that he laid out as early as the first pages of this book. Cf. above, vi.

251 Bergson will turn his attention more fully to the theory of knowledge at the start of Chapter III. The difficulties he likely has in mind are those related to the concepts of disorder (below, 209–38) and nothingness (below, 273–98). In relation to "phantoms" or "pseudo-problems," see the Index.

252 For more on the idea of a "metaphysical problem," see below, 187–93.

253 For more on this metaphor of consciousness coursing through matter, see CV, 19–21. For more on the idea of a "common origin," see "principle or source" in the Index. Bergson returns to this argument in *Les deux sources de la morale et de la religion*.

254 Indeed, the next chapter is titled: "On the Meaning of Life, the Order of Nature, and the Form of the Intellect."

255 For Bergson's internal reference, cf. above, 97–98, 126–27, 145.

256 For more on this argument, cf. below, 262–63; MM, 78–80.

257 Bergson devoted an entire article to the question as to whether a strict parallelism between the brain and thought is intelligible: "Le cerveau et la pensée," in *L'énergie spirituelle*.]

258 For more on this difference, cf. *MM*, 87–88.

259 "Equivalence" is Bergson's technical term for positing in a critical manner the idea of a parallelism between the brain and ideas, as well as of the idea of consciousness as an epiphenomenon. See below, 354; *MM*, 41, 264; *AC*, 36, 38–39, 49, 51, 55, 59; "Fantômes de vivants," in *L'énergie spirituelle*, 71; "Le cerveau et la pensée," in *L'énergie spirituelle*, 191, 195, 200–201, 204, 206–7, 209–10; *DS*, 335.

260 For Bergson's interjected definition of life, cf. *CV*, 12–13 and "life, definition and characterization of" in the Index.*

261 For Bergson's internal reference, see 263–65. In a general way, the two sections in question are extremely similar, and the second one responds to the first one by offering it a foundation for its claims. See also *CV*, 19–20.

262 For more on technical invention or "fabrication," see above, 138–40.

263 For Bergson's internal reference, see above, 140–41, 154.*

264 For more on "releasing" [*déclenchement*] as a form of causality, see above, 73–75.

265 British engineer Thomas Newcomen (1664–1729) invented the first steam engine in 1712. Newcomen was attempting to find a more efficient way of pumping out the water that flooded tin mines. His machine took the form of a pendulum, with one side attached to a pump at the bottom of the mine and the other to a moving piston in a cylinder. By injecting low pressure steam into the cylinder, the resulting vacuum caused the piston to descend in the cylinder and, inversely, the parts linked to the pump to rise. But it was necessary to manually operate both the injection of steam and the spraying of water. Bergson's anecdote about the young boy is well known: his name was Humphrey Potter, and he performed the improvement described by Bergson in 1713. Then, between 1769 and 1784, James Watt (1736–1819) perfected Newcomen's machine by adding a centrifugal governor, and this new version became one of the driving forces behind the Industrial Revolution.

266 The expression "sudden leap" [*un saut brusque*], which here designates a difference of nature, was already employed when Bergson was reconstructing the arguments of the mutationists (above, 66). It is also found in a related passage in "La conscience et la vie" (cf., *CV*, 20). Above all, Bergson uses the idea of a leap in important ways in *Les deux sources de la morale et de la religion* (cf. *DS*, 73, 120, 208, 229).

267 For more on this important claim—which can only be justified by what follows and by attenuating the teleological character of this presentation of it—see below, 265–67; *DS*, 223–24, 273.

268 Cf. "principle or source" in the Index. For more on Bergson's use of the term "source" or "principle," see Bergson's letters to theologian Joseph de Tonquédec (1868–1962) included in the **Correspondence, Reception, and Commentaries** section below.

CHAPTER III

1   For Bergson's internal reference, see above, 7–11.

2   For Bergson's internal reference, see above, 136–80.

3   For Bergson's internal reference, see above, 180–86. This use of "coextensive" is important for Bergson: cf. above viii; CV, 8, 13.

4   Note that "intellectuality" and "materiality" should be distinguished, to a certain extent, from the related distinction between the intellect and matter. Cf. the Index.

5   Cf. above, 186.

6   Here Bergson announces the order of the topics that he will address over the next three paragraphs: psychology, cosmology, and metaphysics.

7   For this important argument against Spencer, cf. above, 154; below, 365–67.

8   Cf. above, 8–9; MM, 220–26; DS, 274–75.

9   Here Bergson offers a tentative version of a claim he makes elsewhere regarding each body's connection to the whole universe. See below, 204, 209; MM, 19–20, 31–44, 224. Bergson's claim drew criticism, documented by philosopher René Berthelot (1872–1960) in his book Un romantisme utilitaire: Étude sur le mouvement pragmatiste, vol. 2, Le pragmatisme chez Bergson (Paris: Félix Alcan, 1913). It might be noted that in L'évolution créatrice, Bergson somewhat limits the influence of a terrestrial body to the solar system since he insists that it constitutes a "relatively closed system" within the universe (cf. below, 215, 242).*

10   See "physics, Thomson and Faraday" in the Index.

11   Cf. MM, Chapter I.

12   See below, 366; cf. MM, 176–78; Le rire, 116.

13   For more on the notion of tendency, see above, 10, 158; below, 203–5.

14   For this engagement with the history of metaphysics, cf. DPP, 48–50.

15   Bergson gave a course on Fichte at the École normale supérieure in 1898, when a selection of Fichte's works was on the reading list for the national exam (the Agrégation) that year. See Octave Hamelin and Henri Bergson, Fichte: Deux cours inédits, ed. Fernand Turlot and Philippe Soulez (Strasbourg: Presses Universitaires de Strasbourg, 1989).

16   Bergson here identifies his targets for Chapters III and IV as those doctrines that presuppose a system of truth existing in itself and encompassing all of reality. Such a system of truth will be characterized by the notions of "unicity," "eternity," and "immutability." Plato's philosophy is a paradigmatic case of such a doctrine (see below, 320) since he elevated kinds into the Ideas and organized the Ideas into a system. But Kant's philosophy is equally paradigmatic since his notion of a law reintroduced the very same general conception of science as Plato (see below, 358–60). Bergson's dualism between life and matter is a weapon against this conception, and it has not been recognized often enough that this dualism is precisely Bergson's critical contribution in Chapter IV below (cf. Martial Gueroult, "Bergson en face des philosophes," Les études bergsoniennes, vol. 5, Bergson et l'histoire de la philosophie (1960): 9–35). See also the Index, "systems, critique of." For more on Bergson's critique of

the classical conception of systematic truth, see IM, 221–23 and, for example, *Histoire des théories de la mémoire*, 256.*

17 Here Bergson provides his summary of Spencer's position (differences of complexity) and of Fichte's position (differences of intensity). Cf. below, 361–62.

18 In other words, to follow Spencer in taking for granted the intelligible structure of the world as it is already actualized, one takes for granted the intellect itself. Recall that for Bergson, in general, the intellect is a part of life itself (above, iv). The critique outlined here will subsequently be directed toward Kant in a more developed way (see below, 207, 358–59).

19 For more on the pragmatic character of intellect, see editorial note 77 to Chapter I.

20 This rather enigmatic image of a "beneficent fluid" can only be understood if we consider it alongside other passages where the intellect is compared to a core surrounded by a "fringe" of intuition, a core that would have been formed by way of condensation and solidification precisely within the "fluid" of life itself. Bergson himself returns to this comparison in what follows.

21 Bergson's important definition of philosophy here should not be separated from his conception of the exigency to transcend the human condition, an idea mentioned in what immediately follows. Cf. above, 30–31.

22 See "surpassing the human condition" in the Index. For more on Bergson's apparent allusion here to the figure of the super-man, cf. below, 266–67; DS, 223–24, 273.

23 In what follows, Bergson implicitly distinguishes his own project (the genesis of the intellect) from Kant's critical project (which takes the intellect for granted). Cf. CV, 2–4.

24 For the same analysis, but applied to the example of the waltz, see Bergson, "L'effort intellectuel," in *L'énergie spirituelle*, 178–81.

25 See "risk" in the Index. Bergson returns in detail to this analysis in his 1910–1911 course at the Collège de France. See: "Théorie de la personne d'après Henri Bergson," in *Mélanges*, 863–64.

26 Cf. above, 136–86.

27 For more on this argument, cf. above, ix, 46–47, 49–50, 136–37, 178, 193.

28 See "leap" in the Index.

29 Cf. CV, 2.

30 Bergson is alluding to Kant here.

31 For the term "inconsistent," see AC, 37–41; Bergson, "Le cerveau et la pensée," in *L'énergie spirituelle*, 192–93.

32 For a longer version of this line of attack, see Bergson, "L'intuition philosophique," in *La pensée et le mouvant*, 134–38.

33 Here Bergson radicalizes one of his key claims. Cf. for instance, IM, 217–18.

34 For its part, matter is "weighed down with geometry." See below, 221.

35 For more on the relationship between intuition and vision, see "intuition, vocabulary of" in the Index. For the form of pragmatism implicit in this paragraph, see editorial note 77 to Chapter I.

36   Cf. above, 191.

37   For Bergson's interpretation of Kant's "skepticism," see DPP, 69; IM, 213.

38   For the relationship between being and doing, at issue in the first part of this sentence, cf. above, 37–39; PR, 102.

39   Here Bergson primarily describes the possible oscillation between Platonic metaphysics (of pure Form) and Kant's Critical Philosophy (the "form of nature" having become the "form of thought," and the "unity of nature" having become the "unity of the sciences"), thereby breaking down the border between ancient and modern philosophy. To be precise, Bergson's idea here is that the two methods are fundamentally of the same nature since Kantianism is merely the movement from what is "necessary" to what is "sufficient" in order to ground science. That is, their underlying theory of truth as a system is identical. See below, 356.

40   It is not by accident that Bergson later returns to this same image when discussing Spinoza and Leibniz. Cf. below, 346.

41   For more on Kant's "relativism," see below, 231, 361. This criticism of Kant is apparent as early as Bergson's first works, see E, 174–80.

42   This claim regarding the "unknowable" aims as much at Kant as it does at Spencer. For more on what Spencer takes to be unknowable, see above, vii.

43   Cf. above, 79 and "science, and philosophy" in the Index.

44   This is, according to Bergson, Kant's conception of science. Cf. below, 358–60.

45   For Bergson's image of the intellect feeling "at home," see editorial note 2 to the Introduction.

46   For Bergson's internal reference, see above, vii.

47   Bergson often formulates this key claim in similar terms: cf. above, vii, 49; below, 208, 231. Cf. also, DPP, 33, 36, 43, 69, 71, 84; PR, 104.

48   Here Bergson introduces his own conception of metaphysics, such as he defines it in his 1903 article "Introduction to Metaphysics." Cf. IM, 181–82.

49   Bergson is attached to this description of knowledge, and in particular of scientific knowledge. Cf. below, 208; DPP, 33, 42–44, 70–76.

50   This is an implicit citation of Saint Paul, from The Acts of the Apostles 17:28. Bergson also discusses this in PC, 176. Here Bergson modifies Saint Paul's formula by reversing the first and the third terms, placing the emphasis on life rather than on being.

51   Cf. below, 207, 357; MM, 203–6; MRV, 21–22.

52   See "intuition, vocabulary of" and "surpassing the human condition" in the Index.

53   Throughout these paragraphs, Bergson's vocabulary circles around the notion of "tension" (i.e., tension, extension, détente ["relaxation"], etc.). This usage should be understood in relation to the fourth chapter of Matière et mémoire, where these terms take on a cosmological and metaphysical meaning. Cf. MM, 226–35.

54   Cf. E, 124–30.

55   For more on the linking of the same to the same, see above, 29, 45–46, 52; below, 271.

56 The first edition of *L'évolution créatrice* reads "and even remains incommensurable."

57 For more on this double movement, cf. Bergson, "L'effort intellectuel," in *L'énergie spirituelle*, 153.

58 Note that Bergson often uses the phrase "endlessly begins anew" as a definition of matter. Cf. editorial note 61 to Chapter I.

59 For Bergson's internal reference, see below, 209–38. To understand this argument, it is important to keep in mind that only two processes exist, defined below as two "orders," and that disorder, properly speaking, does not exist. As a result, the *interruption* of the spiritual or vital process immediately entails the appearance of the material process, that is, of its own *inversion*, the material process being nothing other than the inversion of the vital process.

60 Cf. *MM*, 170–73; "Le rêve," in *L'énergie spirituelle*, 94–96, 102–4; "L'effort intellectuel," in *L'énergie spirituelle*, 160.

61 See *MM*, 235–45.

62 For this psychological characterization of feeling "at ease," cf. editorial note 2 to the Introduction.

63 Bergson often attributes the function of spatial representation to the "imagination" rather than directly to the "intellect" or the "understanding." See "imagination" in the Index.

64 For this way of characterizing space, cf. *MM*, 235–36, 260.

65 Bergson is likely alluding to: *MM*, 235–45.

66 See editorial note 9 to Chapter III.

67 Michael Faraday (1791–1867) was an important English physicist and chemist who discovered electromagnetic induction. Bergson often mentions Faraday alongside Lord Kelvin (Sir William Thomson, 1824–1907), with whom Faraday collaborated. As early as *Matière et mémoire*, Bergson takes inspiration from their conceptions of matter (see *MM*, 224–26). But in their work, Bergson finds less a viable position to adopt and more a certain undoing of contradictions in atomism, along with the negative demonstration of the need to adopt a philosophical perspective on matter. It should not, however, be forgotten that Bergson is quite critical of Thomson and Faraday. In "A Speculation Touching Electric Conduction and the Nature of Matter," Faraday argues that the phenomenon of electric conduction requires that we abandon the idea of atoms as impenetrable and separated in space. (For Faraday's redefinition of atoms, see Faraday, *ibid.*, 140–43). Although Bergson is extremely close to Faraday's position in terms of seeing matter as a continuity of energy, he cannot accept his conclusion, since Faraday here tends toward effacing the distinction between matter and space.*

68 For the idea that matter lends itself to this division, cf. above, 10, 157–58, 190, 204.

69 See Kant, *Critique of Pure Reason*, beginning on A19/B33. [TN: "The Transcendental Aesthetic" is the section in the *Critique of Pure Reason* where Kant provides his arguments regarding space and time as forms of pure

intuition.] For more on Bergson's reading of Kant, see below, 357–61. Cf. also
E, 68–70, 174–80; MM, 208, 237, 244n2; MRV, 21; PC, 154–56.*

70   Bergson frequently associates this use of "atmosphere" with Kant. See
the Index.

71   Cf. Kant, "Chapter III: On the ground of the distinction of all objects in general into *phenomena* and *noumena*," in *Critique of Pure Reason*, A254–56/
B310–12.

72   For Bergson's critique of this crucial element in Kant's philosophy, see below,
221, 233; DPP, 68–69; PR, 108–9.

73   Cf. Immanuel Kant, "To Marcus Herz, February 21, 1772," in *Correspondence*,
trans. and ed. Arnulf Zweig (Cambridge: Cambridge University Press, 1999),
132–38; and "To Marcus Herz, May 26, 1789," in *ibid*., 311–16. In terms of
Bergson's reading of Kant's rejection of "preestablished harmony," cf. MM, 23.

74   Bergson, by contrast, recognizes degrees of spatiality. See MM, 201–3; 235–45.

75   TN: Kant's presentation of the "Antinomy of Pure Reason" begins on A405/
B432 of *Critique of Pure Reason*.

76   Cf. Kant, "To Marcus Herz, February 21, 1772," in *Correspondence*, 132–38; and
"To Marcus Herz, May 26, 1789," in *Correspondance*, 311–16.

77   For the first error, cf. above, 191; below, 358–59. For the second, cf. below, 360;
E, 174–80; MM, 208, 237; PC, 155–57.

78   For the question of relativism, cf. above, 200; below, 357; MM, 203–6;
MRV, 21–22.

79   Cf. MM, Chapter I.

80   Bergson is here taking a position on the debates over "conventionalism,"
which in some ways were tied to the dispute over the foundations of mathematics (see above, 162 and editorial note 194 to Chapter II). The term designates
doctrines put forward by thinkers such as Duhem (see below, 243), Poincaré,
and Bergson's own disciple Édouard Le Roy (who studied the role human
conventions play in the formulation of scientific laws, see below, 219), though
these authors did not explicitly use the term "conventionalism." Cf. below,
219–21. In other words, here Bergson is continuing his attack on the reduction of all knowledge to a relativism (i.e., a relativism to human conventions,
versus a different knowledge that touches upon the absolute).*

81   For more on the positive sciences in relation to reality, cf. editorial notes 47
and 49 to Chapter III.

82   Cf. above, 200; DPP, 33, 42–44, 70–76.

83   Cf. "L'effort intellectuel," in *L'énergie spirituelle*, 153, 160, 189–90. The
model envisioned in *L'évolution créatrice* (and already developed in "L'effort
intellectuel") is not fully congruent with the one adopted in *Matière et mémoire*
(MM, 170–73). Here consciousness is no longer pictured as a cone but rather
as a pyramid, and the most considerable effort should no longer be attributed
to the adopting of an intermediary position between the base and the summit,
but rather to the adopting of a position at the summit. In "Le rêve," a talk
written after *Matière et mémoire* and before "L'effort intellectuel," Bergson still
uses the cone image, but the point with the highest tension has already been

displaced from the middle to the summit. Cf. "Le rêve," in *L'énergie spirituelle*, 94–97, 102–4.

84   For more on this "presence" of the whole to its parts, cf. above, 189, 204; *MM*, 19–20, 31–44, 224.

85   What Bergson means by this classic image is difficult to grasp in this context. It should be recalled that consciousness is already matter, and that matter is already consciousness. This is the double argument found in *Matière et mémoire*, 226–45. Consciousness tends toward matter; matter tends toward consciousness. But the zone where they meet up remains to be explored. [TN: I would suggest that an alternative interpretation of this image is also possible. Perhaps here Bergson intends for us to understand the "two ends of the chain" to be matter (which we grasp as such) and space (which we grasp as an idea of that endpoint toward which matter tends without ever reaching it). We thereby have both the current realization (matter) and an idea of where it is headed (space), without necessarily knowing the steps in between.]

86   The "incline" [*pente*] in question here is the same one that the intellect descends. Cf. "intellect, natural inclination of" in the Index.

87   For the expression "true reality," cf. below, 296.

88   For this important and recurrent example, see "poem" in the Index.

89   It is difficult to grasp the exact relevance of this passage (as well as a previous one where Bergson discusses concentration and relaxation, above, 201–2) unless we keep in mind the implicit opposition between the *simple* (here, the poet's inspiration) and the various "views" or snapshots of the simple that he places under that inspiration (here, the words and lines that make up the poem). Cf. above, 90–98. By analogy, this opposition holds for spirituality (consciousness) and materiality (along with intellectuality).*

90   This theory is the subject of the lengthy and difficult demonstration that follows.

91   For these allusions to Ravaisson, cf. VOR, where Bergson discusses Ravaisson's *Of Habit / De l'habitude* (VOR, 266–67) and alludes to Ravaisson's famous report on French philosophy, *La philosophie en France au XIXe siècle* (VOR, 274–75). As Bergson notes in his essay, "twenty generations of students have learned these pages by heart" (VOR, 275).

In terms of Bergson's critique of elevating mathematical essences into absolute realities, see Chapter IV. Although Plotinus and Bergson both speak of a simple reality being diffracted, the simple reality for Plotinus is the One, which is a principle or source outside of time and movement. For Bergson's analogy between *durée* and extension, cf. PC, 153–54; IM, 217. For Bergson's internal reference to the next chapter, see below, 308–28. Also, cf. Bergson's discussion of Plotinus and the λόγος, in *L'évolution du problème de la liberté*, 177–207.

92   Here Bergson is here thinking of where Spinoza defines the circle by its mode of being engendered, such that the definition would include the figure's proximate cause. Cf. Spinoza, *Treatise on the Emendation of the Intellect*, §95–96. Cf. also below, 276–77, 334.

93 The formulation here is clearly Cartesian. See, for instance, Descartes' comments in his "Dedicatory Letter to the Sorbonne," which opens his *Meditations on First Philosophy*. René Descartes, *Meditations on First Philosophy*, in *The Philosophical Writings of Descartes*, vol. 2, trans. John Cottingham, Robert Stoothoff, and Dugald Murdoch (Cambridge: Cambridge University Press, 1984), AT VII, 6.

94 Cf. *E*, 56–63.

95 Henry Charlton Bastian (1837–1915) was a physiologist and neurologist who Bergson also cited in *Matière et mémoire* (*MM*, 109n4, 125, 125n3, 137n5). Bastian discusses an issue that had long interested Bergson (cf. *E*, 70–73), namely how members of non-urban nomadic groups would be alerted to the fact that they were lost by a specific feeling of inner "upheaval," a sense that they were off track (Bastian, *Le cerveau, organe de la pensée chez l'homme et chez les animaux*, 167n1). In addition, Bergson was interested in how some animals also demonstrate a "sense of direction" (see *E*, 72).*

96 This formulation contains Bergson's primary theory on the foundations of mathematics (see above, 162). He clarifies it at the end of the current paragraph.

97 This sentence begins with an important definition of "spirituality," which might be compared with Bergson's various definitions of the mind or spirit [*esprit*] and uses of "psychological," "mental," and "moral." See the Index for these terms.

98 Cf. above, vi, 165–66.

99 For this notion of "good sense," see editorial note 196 to Chapter II.

100 Cf. *MM*, 11–28, 176–79.

101 Bergson wrote a lengthy note on the psychological origins of our belief in the law of causality (Bergson, "Note sur les origines psychologiques de notre croyance à la loi de causalité [1900]," in *Mélanges*, 419–28) and he gave a course on causality in 1900–1901 at the Collège de France ("Cours de M. Bergson sur l'idée de cause," in *Mélanges*, 439–41). Cf. also above, 44, 57; *E*, 149–64.

102 Cf. above, 189; below, 242.

103 Bergson refers us here to a passage according to which mathematical necessity, which is moreover reducible to logical necessity, is one of the two origins of our notion of causality, the other being psychological dynamism.

104 This problem occupies Bergson at length in his *Essai sur les données immédiates de la conscience*. See, for instance, *E*, 2, 24, 47, 53.

105 See "physics, Thomson and Faraday" in the Index. Again, Bergson will reject the identification of matter with space.

106 An interruption can be partial, meaning that there can be *several* interruptions (see below, 240). This leads to the idea of a plurality of worlds (see below, 242).

107 See "Le rêve," in *L'énergie spirituelle*, 96–97.

108 The first part of this complex passage provides an important formulation of the positive part of Bergson's position here, namely the creative movement of

the whole of reality, which should not be confused with his ongoing critique of any classically teleological position according to which this movement is the unfolding of a predetermined plan. See "teleology and plan, realization of" in the Index. For Bergson's allusion to free actions, see *E*, 124–30.*

109 Here again Bergson is responding to "conventionalism." See above, 208.

110 Scientific progress consists in coming closer and closer to the spatiality immanent in matter; this is the source of Bergson's favorable (and even enthusiastic) embracing of Einstein's theory. See Bergson, *Durée et simultanéité*, 30–33, 180–81.

111 A selection of studies by Édouard Le Roy that Bergson is likely alluding to here are as follows: "Science et philosophie," *Revue de métaphysique et de morale* 7, no. 4 (July–August 1899): 375–425; no. 5 (September–October 1899): 503–62; no. 6 (November–December 1899): 708–31; 8, no. 1 (January–February 1900): 37–72; "Un positivisme nouveau," *Revue de métaphysique et de morale* 9, no. 2 (March–April 1901): 138–53; "Sur quelques objections adressées à la nouvelle philosophie," *Revue de métaphysique et de morale* 9, no. 3 (May–June 1901): 292–327; no. 4 (July–August 1901): 407–32. See also Édouard Le Roy and Georges Vincent, "Sur la méthode mathématique," *Revue de métaphysique et de morale* 2, no. 5 (September–October 1894): 505–30; no. 6 (November–December 1894): 676–708; and Édouard Le Roy and Georges Vincent, "Sur l'idée de nombre," *Revue de métaphysique et de morale* 4, no. 5 (November 1896): 738–55.

Édouard Le Roy (1870–1954), who had trained as a mathematician, can be considered the first and perhaps only true "disciple" of Bergson. He succeeded him at the Collège de France (in 1921) and at the Académie Française (in 1945). Alluding to these articles and describing them as "profound" is hardly an insignificant gesture, coming from Bergson. They were in fact important contributions to debates over "conventionalism" and responses to philosopher and mathematician Louis Couturat (1868–1914), who accused Le Roy and Bergson of "nominalism." See Louis Couturat, "Contre le nominalisme de M. Le Roy, *Revue de métaphysique et de morale* 7, no. 1 (January–February, 1900): 87–93. This designation of "nominalism" would stick to them in this debate. It might also be noted that Le Roy was, within Bergson's entourage, someone who encouraged Bergson's turn to questions of mysticism (see Bergson's discussion of "psychological parallelism" and "positive metaphysics" in: "Le parallélisme psycho-physique et le métaphysique positive [1901]," in *Mélanges*, 492–95). Despite Le Roy being an ardent Catholic, his works were nevertheless placed on the Index of Prohibited Books (*Index Liborum Prohibitorum*), along with those by Bergson.*

112 Cf. *E*, 2, 24, 47, 53.

113 As we saw above, for Bergson, the well-founded part of conventionalism resides in its idea of a necessarily successive order of posing scientific questions. See above, 208.

114 Cf. below, 231, 368. Bergson's images here are perhaps a bit misleading. They should be understood to suggest that with matter, although there is no particular set of laws at its foundation, there is a certain general lawfulness. To grasp this, one must remember that for Bergson lawfulness (or, more generally, "geometricity" or spatiality), is a tendency, not a state. By contrast, to consider that a "definite system of mathematical laws" would be at the "base of nature" would be to confirm a Platonic and Kantian conception of the truth that Bergson has been attacking since the beginning of this chapter. Let me add that the ballasted or weighed down [lestée] nature of matter here leads one to imagine a weight that indeed tends to "fall," which foreshadows images that Bergson will invoke below (cf. 246).*

115 One of Bergson's main targets here is Kant, for whom the *a priori* forms of intuition and the categories of understanding come to give order to a "sensible manifold." Cf. above, 206; DPP, 68–69; PR, 108–9.

116 Again, the idealist perspective in question here is primarily Kant's. Cf. above, 206; below, 233.

117 For Bergson's internal reference, see below, 272–98.

118 For the characterization of illusion as a false problem, which replaces Kant's characterization of the transcendental illusion, cf. above, 156; below, 273, 296; PR, 104–9.

119 The first time that Bergson uses this argumentative strategy (of criticizing the hypostasizing of what is actually an oscillation) is in 1904, in his article "Le cerveau et la pensée," in L'énergie spirituelle, 191–210. Cf. Bergson's critique of nothingness, below, 277–98.*

120 By adding the analogous relationship between form and matter, Bergson makes it clear that his criticism of Aristotle's opposition between form and matter (along with the Platonic theory of Forms) are also at stake in this discussion. Cf. "Aristotle, matter" in the Index.*

121 For Bergson's internal reference, see above, 201–4.

122 We might wonder how an order has a "limit." Bergson is implying that an order itself is a movement, not a thing. Geometrical order is the movement of materiality, and the limit toward which this movement tends is space. Cf. above, 203.

123 Cf. above, 47–48; Bergson's letter to Brunschvicg: "Bergson à L. Brunschvicg [February 26, 1903]," in Mélanges, 585–87.

124 For the characterization of life as a "creative evolution," see above, 163.

125 On the framework of finality as both "too broad" and "too narrow," cf. above, 52.

126 For the example of Beethoven, cf. DS, 268.

127 Cf. below, 329; DPP, 57–64. The analysis that Bergson is setting up here is closely tied to Chapter IV below.

128 For Bergson's internal reference, see above, 59–88.

129 With this important declaration, which retrospectively throws light on the place of the first two chapters in his argument, Bergson shows that he rejects Newton's proof of the parallelogram of forces as a way of understanding biological phenomena.

130 Bergson's critical allusion to a "vital principle" here again shows why he rejects vitalism (cf. above, 42–44, as well as editorial notes 114–17 to Chapter I; "La philosophie de Claude Bernard," in *La pensée et le mouvant*, 232–33).*

131 For the distinction between organic destruction and organic creation, see above, 19–21, 34–36. For the notion of an indivisible act, see above, 90–98.

132 Bergson notably criticizes this Kantian idea (see above, 191; below, 358–60). More generally, this idea presides over the construction of all the philosophical systems he examines in Chapter IV.

133 For this characterization of the "problem of knowledge," cf. PR, 105–9. For more on Bergson's attack on all "theories of knowledge" that originate in Kantian philosophy, see DPP, 68–69.

134 Cf. below, 329.

135 On the possibility of "natural general ideas," cf. DPP, 57–64.

136 For the image of cutting from the whole, see above, 11–23; below, 301, 306; MM, 220–24. For Bergson's own reference to *Matière et mémoire*, see in particular MM, 173–81, 220–26.

137 Cf. below, 330; IM, 217.

138 Bergson is citing the following distinctions: *Physics* IV, 215a2 is where Aristotle distinguishes between "compulsory movement" and movement that is "according to nature"; *Physics* V, 230b12 is where Aristotle distinguishes between "upward and downward" movement (fire naturally moving upward, and only moving downward when forced to do so); *Physics* VIII, 255a2 is where Aristotle distinguishes between the "light" and the "heavy" (light things moving naturally upward, heavy things moving naturally downward). In terms of the passages cited from *On the Heavens*: in Book IV, Chapters 1–5, Aristotle focuses on the heavy and the light; Book II, 296b27 is where Aristotle discusses how earth moves toward the center (while fire moves toward the extremities), and Book IV, 308a34 is where Aristotle distinguishes between an absolute heavy and light versus a relative heavy and light (something being absolutely heavy if it naturally carried toward the center, and absolutely light if it is naturally carried toward the extremities, with relative heaviness and lightness being found in those bodies that move faster or slower than others in these conditions).

139 Kepler and Galileo are important for Bergson's argument. Cf., below, 334; "Fantômes de vivants," in *L'énergie spirituelle*, 70–71; IM, 217; "La philosophie de Claude Bernard," in *La pensée et le mouvant*, 229.

140 Here again Bergson is alluding to "conventionalism," cf. above, 208, 219–21.

141 This conception will be later connected to Kant and Fichte. See below, 356.

142 The "certain cases" of independently existing terms alluded to here are surely those of the biological genera. See DPP, 57–64.

143 In other words, reality is divided into domains, and science and philosophy are each appropriate for one of those domains. See "science, and philosophy" in the Index.

144 See editorial notes 47–51 to Chapter III.

145 This characterizes the metaphysics that Bergson draws out of Weismann's scientific doctrines. Cf. above, 26–27, 43, 79–83.

146 Cf. above, 221.

147 For Bergson's uses of the term "mystery," which often simultaneously connotes a pseudo-problem and an anxiety, see the Index.

148 The first allusion is to Aristotle (see "Aristotle, matter" in the Index) and the second to Kant (see above, 206, 221).

149 For this representation of the universe, see DS, 144–220.

150 The idea of chance is one of the five main pseudo-ideas that Bergson critiques, the others being disorder, nothingness, possibility, and the all-powerful. For more on chance, see DS, 154–57. In fact, in Les deux sources de la morale et de la religion, Bergson notes that he developed his critique of chance as early as 1898. DS, 154n1.*

151 For the roulette wheel example, cf. DS, 146–47. For the wind example, cf. DS, 154–55. This second, canonical example of chance comes from Aristotle's Physics, 197a36–198a13), and it appears again in the appendix to Part I of Spinoza's Ethics. For the allusion to an "evil genius," cf. above, 69.

152 Bergson's Latin term here comes from a well-known verse in Juvenal's Satires that often serves as an adage for designating arbitrary decisions: "Hoc volo, sic jubeo, sit pro ratione voluntas" ["I want, I command it: let my will suffice as reason"]. See Juvenal, The Sixteen Satires, trans. Peter Green (London: Penguin Books, 1998), no. 6, I, verse 223.

153 For Bergson's internal reference, see above, 233.

154 According to Bergson's analogy, this absence of order would correspond precisely to Plato's "non-Being" or to Aristotle's "matter," such as these are defined below in Chapter IV.

155 Cf. above, 201–9.

156 For the role of the term "Consciousness," cf. above, 187, where we find the word capitalized and where Bergson already uses the phrase "for lack of a better word." The conclusion of Chapter II also explores the pertinence of the term as applied to life. Cf. above, 180–86.

157 For Bergson's earlier discussion of retrospective vision, see above, 47. For more on distinctness in thought, cf. above, 161.

158 See "intuition, conversion to" in the Index.

159 For more on free action, cf. E, 124–30; The letter to Brunschvicg: "Bergson à L. Brunschvicg [February 26, 1903]," in Mélanges, 585–87.

160 For more on the method of "installing ourselves within," see "intuition, vocabulary of" in the Index.

161 A "pure will" [pur vouloir], which would be nothing other than durée, is the most suitable name for what Bergson describes here as "the principle or source of all life as well as of all materiality."

162 Above, Bergson suggested that art fulfills this function (cf. above, 178). Later, it will be the mystic who goes the "furthest" in this direction: see DS, 255–66.

163 Bergson's claims in this paragraph thus offer another reason for devoting an entire section to the "History of Systems" in Chapter IV. For more on how his claim here is applied to Spinoza and Leibniz, cf. below, 346.*

164 For the description of this experience, cf. *DS*, 43–44, 268–70; Bergson, "Intuition philosophique," in *La pensée et le mouvant*, 119–20; IM, 225–27.

165 As Spinoza writes: "*Sentimus experimurque nos aeternos esse*" ("we feel and experience that we are eternal"). Spinoza, *Ethics*, Part V, Proposition 23, Scholium.

166 On the "miracle" carried out by the will, cf. AC, 31; "De l'intelligence [Discours au Lycée Voltaire, 1902]," in *Mélanges*, 560.

167 It is difficult to fix the precise definition of the word "form" here, especially given that Bergson himself often defines life as a creation of forms (cf. above, 11, 45, 87). In *L'évolution créatrice*, form in fact has several meanings: a form of life (a species); the shape or configuration of an object; Platonic forms; form as opposed to matter in Aristotle; the forms of intuition in Kant. This plurality of meanings is exacerbated by a similar plurality with regard to the term "matter," a term that is again complicated through the complex relationship set up by Bergson's idea of a simultaneous genesis of the intellect and of matter. In the current context, form seems to indicate the figure of the work of art, insofar as the artist arranges the matter he finds at hand without himself or herself having the power to create that matter. For its part, life too creates forms (species, organic structures, etc.). The pure will, however, is perhaps capable of creating matter itself, as is suggested by the mutual adaptation between our intellect and the solids it finds around itself. For more on form, see the Index.*

168 To understand this somewhat enigmatic argument, it is important to keep in mind first that the "action that generates the form" is the simple movement of which "matter" is but a set of views, and second, that the "halting" of this simple movement is indeed the "interruption" discussed above (cf. above, 202), which is equivalent to an inversion. For other moments where Bergson discusses the artist as he does in this paragraph, cf. above, 90–91; VOR, 277–79.

169 For more on the idea of a momentary interruption, cf. above, 218; below, 242.

170 To understand this analogy, which again relies upon the distinction between the simple and our multiple views of something that is simple, it is important to keep in mind that the tracing of the letters is always the other side of the simple poetic intention. No letters arise by themselves, they are the other side of a simple movement, and as such are a static view of that deeper movement of poetic creation. Concerning the analogy of the poem, and how Bergson is inspired by Plotinus here, see "poem" in the Index.*

171 See "physics, Thomson and Faraday" in the Index.

172 For Bergson's argument against the *ex nihilo* creation of the universe, cf. below, 249.

173 For Bergson's internal reference, cf. below, 298.

174 This critique is not only of Plato and Aristotle, but also of Fichte's brand of Kantianism, cf. below, 321, 356. For more on Bergson's general critique of positions that assume "everything is given once and for all," see editorial note 101 to Chapter I.

175 In this important section, Bergson returns to a claim he first defended in *Matière et mémoire*: that all the points in the universe influence each other. He somewhat moderates his position here. Cf. editorial note 9 to Chapter III.

176 Based on observations made between 1783 and 1805, astronomer William Herschel (1738–1822) demonstrated that the sun moves toward a point in the constellation Hercules. In fact, the idea of a displacement of the sun already existed in the eighteenth century, notably in the work of Swiss German mathematician and astronomer Johann Heinrich Lambert (1728–1777).

177 For the distinction between naturally and artificially closed systems, a distinction Bergson refines here, cf. above, 7–11.

178 Indeed, this law does not tell us *what* is preserved (kinetic energy or potential energy); nevertheless, *something* is preserved.

179 Pierre Duhem (1861–1916) was an important physicist in France at the beginning of the twentieth century. He was a leading theoretician in the tradition of "conventionalism" (although he rejected the term), and some of Bergson's arguments address this manner of thinking (cf. above, 208, 219, and Bergson's first note to page 219). Alluding to Duhem helps Bergson to establish that any measurement of a *total* energy that would be conserved must be artificial. If it makes sense, within the limits traced out by Duhem, to measure a quality in relation to itself, there would still be no way to give a global measurement of qualities in relation to each other. As Bergson insists, there are different types of energies and they differ according to their quality.*

180 Bergson is again thinking of Duhem's book, *L'évolution de la mécanique*. Duhem explains the distinction between kinetic energy and potential energy, arguing that the total energy of an isolated system is invariable. And yet, argues Duhem, no system is fully isolated or closed. For Duhem, this shows the conventional nature of physical principles. His conventionalism, mixed with pragmatism and supplemented by the thesis that no system in nature is fully isolated, comes close to Bergson's position. Bergson breaks with Duhem, however, insofar as Duhem places the second law of thermodynamics on the same level as the principle of conservation of energy.*

181 In fact, the idea of giving a philosophical meaning to the second law of thermodynamics, and of integrating it into a doctrine that opposes, within the universe, a movement of "ascent" [*montée*] with a movement of "descent" [*décente*], was not entirely new at the time. It can be seen in the work of philosopher and psychologist Joseph Delbœuf (1831–1896) and philosopher André Lalande (1867–1963). See Joseph Delbœuf, *La matière brute et la matière vivante* (Paris: Félix Alcan, 1887); André Lalande, *La dissolution opposée à l'évolution dans les sciences physiques et morales* (Paris: Félix Alcan, 1899). Berthelot discusses this connection in *Le pragmatisme chez Bergson*, volume 2

of *Un romantisme utilitaire* (138–39). The ascent and descent model, which has its origins in neo-Platonism and later in Ravaisson, is also embraced by philosopher Elme-Marie Caro (1826–1887). See Elme-Marie Caro, *Le matérialisme et la science* (Paris: Hachette, 1867). We also encountered it above in Cope (see editorial note 92 to Chapter I) and Marin (see editorial note 12 to Chapter II). As Berthelot notes, these doctrines tend to merely invert Spencer's doctrine, leaving the problem itself intact, whereas Bergson attempts to reconfigure the problem itself (particularly against Lalande, cf. below, Bergson's note to page 247).*

182 Nicolas Léonard Sadi Carnot (1796–1832) was a French physicist and mathematician who contributed to the founding of thermodynamics. He worked on how steam engines transform heat into "motor power," that is, "work" (for the steam engine example, see above, 185–86 and editorial note 265 to Chapter II). His discovery informs part of the second principle of thermodynamics, sometimes known as "Carnot's Principle," "Clausius's Principle," or the "Carnot-Clausius Principle." Bergson discusses Clausius just below.*

183 Rudolf Julius Emanuel Clausius (1822–1888) was a German mathematical physicist and an important contributor to the founding of thermodynamics. In 1850, he reformulated Carnot's Principle by affirming that heat could not pass from a colder body to a warmer body without some external intervention. This provided the foundations for the measurement of entropy, and so the second law of thermodynamics shows phenomena all marching in a direction that clearly interests Bergson in these passages.*

184 The second principle of thermodynamics is "metaphysical" because it indicates merely a *direction* in which all phenomena tend to move. This makes it possible to apply it to phenomena of entirely different types. In some of its formulations, this also allows us to posit the irreversibility of *time* (see below, 245 and editorial note 187 to Chapter III). For Bergson, metaphysics is always concerned with time.*

185 Here it is worth recalling Bergson's definition of matter as the "present that is continuously renewed" (see editorial note 61 to Chapter I). Or when, in *Matière et mémoire*, Bergson writes: "Matter thus resolves itself into numberless vibrations, all linked together in uninterrupted continuity, all bound up with each other, and traveling in every direction like shivers through an immense body," *MM*, 234 [*Matter and Memory*, 208].

186 For Bergson's internal reference, see above, 202. Note that Bergson himself does not necessarily appeal to this reciprocal influence. See editorial note 9 to Chapter III.

187 Ludwig Boltzmann (1844–1906) was an Austrian physicist and a philosopher of science. Along with Maxwell and Clausius, Boltzmann was one of the principal founders of the kinetic theory of gases. He is known for his probabilistic interpretation of the second principle of thermodynamics. Moreover, he understood the movement from an organized to an unorganized state not as a law, but as an extreme probability, and the reverse passage not as impossible, but as extremely improbable (indeed,

an improbability beyond all imagination). For Boltzmann, the irreversibility of time is not a necessity, but a result of our restricted view of probabilistic processes. The theoretical distance between Bergson and Boltzmann is significant, especially on the question of time.*

188 See "physics, Thomson and Faraday" in the Index.

189 For Bergson's internal reference, see above, 201–4.

190 Cf. above, 209–38.

191 In terms of the first part of this claim, recall that *durée* names what is "in-the-making" or "making itself" or "being made" [*se faisant*], in opposition to the "ready-made" [*le tout fait*] (see above, 238). According to this passage, that which is "being made" is also opposed to that which "unmakes itself." In terms of Bergson's characterization of matter, see below, 248–49, 251, 272, 342. Here Bergson is quite close to the cosmologies of Plotinus and Ravaisson (alluded to in Bergson's own note on page 211), the difference being that the principle or source that "unmakes itself" [*se défait*] on Bergson's account is not an unchanging One, but rather has something to do with movement itself. This difference makes the doctrines incompatible.

192 For the notion of a supra-consciousness, see below, 261; *DS*, 264. On the difficulties of the idea of a pure creative activity, see *DS*, 223–24, 273.

193 See above, 115–27; below, 252–54; *MM*, 33–34; *CV*, 14–15.

194 For Bergson's internal reference, see above, 238–40.

195 The scission Bergson is alluding to here is the one between plants and animals.

196 André Lalande (1867–1963) was one of the founders of *Revue de métaphysique et de morale* and the Société française de philosophie. The book that Bergson cites, *La dissolution opposée à l'évolution dans les sciences physiques et morales*, was Lalande's doctoral thesis. Lalande adopts Spencer's definition of evolution as "the passage from the homogeneous to the heterogeneous," and then demonstrates, step by step, that the world actually follows the opposite pathway, toward dissolution, which is the passage from the heterogeneous to the homogeneous, or rather, that the world is the result of the opposition between these two movements and that ultimately "dissolution" always triumphs over evolution. This anti-Spencerian philosophy attracted Bergson's attention, and it should be noted that Lalande too bases his position on the second law of thermodynamics. Nevertheless, unlike Bergson, Lalande's orientation is Kantian. Moreover, rather than accepting Spencer's definition of evolution, Bergson forges a *new* concept of evolution that itself integrates the second law of thermodynamics. Bergson's position is in fact a reformulation of the problem.*

197 For this important conclusion, see editorial note 191 to Chapter III.

198 Cf. above, 241–42.

199 For Bergson's internal reference, see below, 298–315.

200 Bergson is thinking of a certain version of pragmatism here. See editorial note 77 to Chapter I.

201 This is a fundamental claim for Bergson's argument. For more on this type of opposition, see editorial note 89 to Chapter II.

202 This definition of God, offered in passing, opened Bergson's position up to many attacks and misunderstandings. The first and best-known analysis came from Joseph de Tonquédec. Tonquédec found that Bergson was too ambiguous as to whether God is or is not immanent to creation. Bergson responded in two important letters (included below in the **Correspondence, Reception, and Commentaries** section below), and he comes back to the question in his final book, *Les deux sources de la morale et de la religion* (see, *DS*, 272).

203 Bergson is criticizing Aristotle here (cf. below, 316). He is repeating a formulation he used in *Matière et mémoire* (cf. *MM*, 18).

204 For the image of an undivided flux, see above, 187.

205 The only passage in which Bergson comes close to providing an explicit definition of the organism, an important moment in the text. See also CV, 13.

206 This important definition of the mind implies a conversion. Cf. above, 238.

207 Cf. above, 248.

208 For Bergson's internal reference, cf. above, 180–86.

209 Indeed, this captures what will be Bergson's main concern through to the end of the chapter. In fact, together Chapters II and III lead up to this point.

210 For more on the phrase "a need for creation" [*une exigence de création*], see below, 262.

211 For this definition of matter in terms of necessity, see below, 264. For the characterization of life, see above, 116, 127; CV, 12–17.

212 For Bergson's internal reference, cf. above, 125–26.

213 For the first model involving signals, cf. *MM*, 26; CV, 8–9. In terms of the will itself setting up the mechanism, Bergson has in mind his notion of habit, cf. CV, 10–11; VOR, 266–67.

214 For the development of this image in Bergson's work, cf. *DS*, 245.

215 For more on this position, see above, 115–27, 246–47; CV, 14–15.

216 The *"sui generis* chemical reaction" discussed by Bergson here is what we now call photosynthesis. The process consists in the transformation of light energy into chemical energy. It makes possible the conversion of inorganic carbon into an organic form of carbon that living beings can use, and it takes place thanks to "chloroplasts," which are elements of the plant cell that contain chlorophyll, a pigment sensitive to light. The missing "key" to the process, to echo Bergson's terminology here, is the synthesis of adenosine triphosphate (ATP) out of inorganic phosphate and adenosine diphosphate (ADP). ATP has the function of "storing up" energy temporarily, to again recall Bergson's terminology, and of then setting it free through a sort of degradation, which is the inverse process of the synthesis just mentioned, into ADP and inorganic phosphate. This degradation takes place in metabolic processes such as muscle contraction. These muscle contractions mobilize what Bergson calls, in what follows below, the "energetic substances" in opposition to the "plastic substances."

217 For this myth of an unlimited *élan*, cf. above, 246.

218 This principle is, according to Jankélévitch, the most important one in Bergsonism. See Jankélévitch, *Henri Bergson*, 137–50. See also "finitude of the *élan*" in the Index.

219 Cf. above, 107–27. Here again Bergson calls into question the model of a "division of labor."

220 Cf. above, 116–21.

221 For this double limitation [of tendencies in the *élan* and of the resistance of matter], cf. above, 99–101.

222 See above, 127–36.

223 For more on this "egoism," cf. above, 127–30.

224 The insistence upon the inherently conflictual dimension of life leads Bergson, *a contrario*, to discover here the style of a Leibnizian theodicy.

225 For this type of conjecture, cf. *DS*, 223–24, 273.

226 For this distinction between energetic substances and plastic substances, cf. above, 34, 122–23.

227 For more on Carnot's law, see above, 243–45, and editorial notes 182–83 to Chapter III.

228 In other words, nothing material could come to limit (that is, to canalize) the vital tendencies. Bergson is here thinking (by analogy) of the brain as an organ of what he calls "attention to life." See editorial note 217 to Chapter II.

229 Cf. above, 53, 99; *DS*, 273.

230 The reasons justifying the image of life as an *élan* will be explained again in *DS*, 115–20. For some less perfect images (according to Bergson), see above, 246, 248. [TN: For the impossibility of any one image providing an adequate understanding of *durée*, cf. 182–85.]

231 That life is "of the psychological order" is one of Bergson's central claims in this book. The description he offers here of the psychological as a "confused plurality of interpenetrating terms" is indeed one of the main discoveries of Chapter II of *Essai sur les données immédiates de la conscience*, as well as the primary meaning of the notion of an "indistinct multiplicity" that he discusses here.

232 Bergson first addresses this double challenge of unity and multiplicity as early as page vi.

233 For this Leibnizian consideration, see editorial note 58 to Chapter II.

234 For more on Bergson's interjection about language, cf. *E*, 96–104. For his discussion of unity and multiplicity in "Introduction à la métaphysique," see in particular IM, 189.

235 An individuation from two series of causes, as above, 99–101.

236 Cf. above, 210–11, 240–41; below, 319–20.

237 For more on the question of individuality at stake in this paragraph, see "individual/individuality" in the Index.

238 For more on Bergson's use here of "society," cf. CV, 26–27. The second part of this claim sheds light on Bergson's adherence to the idea of a "society of nations." Cf. CV, 26–27; *DS*, 305–8; Bergson, "Conférence de Madrid: La personnalité," in *Mélanges*, 1233–34. For more on Bergson's later

role in international politics and his contributions to the precursors to the United Nations, see Philippe Soulez and Frédéric Worms, *Bergson: Biographie* (Paris: Presses Universitaires de France, 2002), 153–204.

239 Bergson here cites a French review by Wladimir Podwyssozki of *Sur la structure des colonies bactériennes*, a book by Stanislaw Serkovski originally in Russian. The review was published in Delage's journal, *Année biologique*, which Bergson refers to regularly in *L'évolution créatrice* (see, for instance, Bergson's note to page 125 above). According to Podwyssozki, Serkovski claims that bacterial colonies form an "essential core" or "organ," making the colony a "composite organism." See Podwyssozki, "Compte-rendu de *Sur la structure des colonies bactériennes* par S. Serkovski," 317.

240 Protophytes are unicellular plants and the Protozoa are unicellular animals.

241 Edmond Perrier (1844–1921) was an important French zoologist and one of Bergson's friends (see Conférence de Madrid: La personnalité," in *Mélanges*, 1224). Critical of Darwinism over its explanation of variations (see above, 63–65), Perrier proposed the "colonial theory," which presupposes a version of polymorphism (see above, 141, 158–59) and the model of a "division of labor" (see above, 118 and the Index). Several of Bergson's examples are drawn from Perrier's book, including his discussion of the Foraminifera (above, 104) and the hydra (above, 13–14). In the preface to the second edition of *Les colonies animales et la formation des organismes*, Perrier focuses on protoplasm as the originary substance of life and responds to Delage's criticisms and the political implications some had attributed to his work.*

242 For other citations of Delage, see Bergson's first note to page 60 and second note to page 82 above. Bergson's citation here of *L'hérédité et les grands problèmes de la biologie générale* is to a section titled "L'individu." Delage breaks with the theory that was dominant at the end of the 1890s according to which there is an "essential difference" between unicellular and pluricellular organisms. Bergson also cites Delage's "La conception polyzoïque des êtres," which is an edited version of an important lecture given by Delage at the Faculté des sciences de Paris. Devoted primarily to case studies, including of the hydra, Delage argues that the facts of polyzoism are true, but "exceptional" ("La conception polyzoïque des Êtres," 644). Delage argues the evidence in relation to ontogenesis shows that Haeckel's theory must, at least in part, be rejected. He claims that his own theory alone can account for experimental results (see editorial notes 111–15 to Chapter I). Delage characterizes the opposing doctrine as "mystical" since it amounts to attributing little souls to the ultimate constitutive parts of matter (see Bergson's note to page 177 and editorial note 243 to Chapter II). This reversal of epithets indicates a change in paradigm: up until this point in history, it was Haeckel's disciples who treated as "mystics" those who claimed to explain the constitution of the living being other than by the addition of mechanical parts. Bergson in

turn will find himself being characterized as a "mystic," this time because of his theory by which evolution is the dissociation of a whole that is first given as an indistinct multiplicity.*

243 Paul Busquet (1866–1930) participated in debates about "cell theory," arguing that the unity of the cell is at first physiological and *only later* morphological (Busquet, *Les êtres vivants*, 57). Given his overall argument and the various researchers cited, Busquet's book was likely a key resource for Bergson. Busquet participated in a network of researchers working on these questions, which included his teacher Joseph Kunstler, as well as Yves Delage, zoologist Adam Sedgwick, and Alphonse Labbé. See the works by Delage, Sedgwick, and Labbé referenced in Busquet's *Les êtres vivants* as well as Chapters III and IV of his book, which discuss Perrier and polyzoism (see also the two editorial notes just above). In Busquet, Bergson found support for his own insistence on dissociation over association, nevertheless Bergson gives this thesis a metaphysical weight. Insofar as he finds it necessary to return in this chapter to these references to contemporary biology, this is likely because he understood the problem of individuation to be the most *metaphysical* problem in biology.

244 See editorial note 2 to the Introduction.

245 A Stentor is a horn-shaped freshwater, heterotrophic ciliate protozoan measuring about a millimeter.

246 Bergson is alluding to his own discussion above, 258–59. Cf. also "unity" and "multiplicity/multiple" in the Index.

247 For more on Bergson's use of *réflexion* ("reflection," "thought," "meditation"), see above, 158–66 and "consciousness, self-" in the Index.

248 For Bergson's analysis of consciousness, cf. above, 180–87, 238. For the use of supra-consciousness, see above, 246.

249 For this image, cf. above, 251.

250 Cf. above, 252.

251 Bergson is referring to *Matière et mémoire*, and perhaps to "Le cerveau et la pensée," in *L'énergie spirituelle*, 191–210.

252 Cf. *MM*, 14, 47, 153, 256–57, 261.

253 For more on Bergson's use here of a "sum of contingency," see above, 252; cf. above, 116; "Bergson à L. Brunschvicg [February 26, 1903]," in *Mélanges*, 585.

254 For the role of the muscles in this analysis, cf. above, 122–26.

255 Cf. above, 180–82; *MM*, 78–80.

256 For the important notion of a "nascent action" in Bergson's thought, cf. below, 270; *MM*, 6, 19, 26, 29, 66, 89, 168, 262; *AC*, 42, 44, 46; "Fantômes de vivants," in *L'énergie spirituelle*, 74–75; "Le cerveau et la pensée," in *ibid.*, 200.

257 For Bergson's reference to this image elsewhere in his work, cf. *MM*, 82, 169, 193, 196, 198, 274; *DS*, 280.

258 For more on this remark and the entire argument that follows, see the final section of Chapter II (above, 180–86) and *CV*, 19–20.

259 Here again we find the repeated phrase "any ... whatsoever" [*n'importe quelle*], which Bergson often employs to characterize the *indefinite* or *indeterminate* form of humans (see "infinite/indefinite" in the Index). The theme is also present in the following paragraph.

260 For the use of the term "real action," cf. above, 145; *MM*, 16–60.

261 For the use of "necessity" as the definition of matter, cf. above, 252. For the image of a net, see *MM*, 235–36, 280.

262 For the image of "dividing and conquering," cf. *CV*, 20.

263 For this way of thinking about society, cf. above, 159; *DS*, 168.

264 For the theme of "success" (which should be distinguished from that of a "goal" set in advance), see above, 134–35; *CV*, 19.

265 In terms of Bergson's internal reference, his claim here summarizes one of the main objectives of Chapter I as a whole.

266 Cf. "struggle" in the Index.

267 For this image of "accidents," cf. above, 103–4. For the idea that we might have turned out differently, cf. above, 256–57; *DS*, 223–24, 273.

268 This is the only occurrence of the term "super-man" [*sur-homme*] in Bergson's published corpus (other than his use of the adjective "superhuman" when discussing Laplace's demon as a "superhuman intellect," see the Index). The idea, if not the term itself, can be found again *Deux sources de la morale et de la religion* (see *DS*, 223–24, 273). It seems as if Bergson, between the present book and *Deux sources de la morale et de la religion*, attempted to strip away any overly Nietzschean elements from his vocabulary and indeed from his thinking.

269 For the imagery of an "inversion" of direction, cf. above, 47, 238.

270 Cf. *DS*, 224.

271 For the notion of "winning back" or "reconquering," cf. above, 160.

272 For more on the idea of the "habits" in relation to matter, cf. *VOR*, 266–67.

273 For the allusion to "the whole of nature," it is worth recalling that the title of the last section of Chapter II was: "Life and Consciousness. The Apparent Place of Man in Nature." Here, in Chapter III, these final two sections ("Humanity" and "The Life of the Body and the Life of the Mind") appear to offer Bergson's own response to the question of "man's place" in nature. For more on the allusion to "our destiny" [*notre destinée*], cf. *CV*, 1–3, 27–28; *AC*, 57–60; *DS*, 279–82.

274 For more on this definition of the mind, cf. above, 251.

275 For this argument, cf. above, 239–40.

276 Recall that the subtitle of *Matière et mémoire* is "Essai sur la relation du corps à l'esprit." [TN: "Essay on the Relation of the Body to the Mind." Note that this subtitle does not appear in the English translation of that book.]

277 Bergson is here thinking of a tradition that runs from Descartes to philosopher Jules Lachelier (1832–1918).

278 See the references provided in editorial note 101 to Chapter I.

279 For this formulation of the problem of the relation between consciousness and the brain in terms of "interdependence" [*solidarité*], cf. *MM*, 4–5; *AC*, 36–37; "Le cerveau et la pensée," in *L'énergie spirituelle*, 209–10.

280 It is perhaps worth emphasizing that Bergson discusses the "survival" [*survivance*] rather than the "immortality" of the soul. Cf. CV, 1–3, 27–28; AC, 57–60; DS, 279–82. For the mention of the bodies of the two parents, cf. above, 43.

281 The "philosophy of intuition" is how Bergson would have characterized his own philosophical method at this time.

282 Cf. above, vi, 258–59; IM, 189; as well as "unity" and "multiplicity/multiple" in the Index.

283 Cf. above, 99–101.

284 Bergson here hints at the consequences of his philosophy for the theological question regarding the nature of the soul.

285 For a similar image, cf. above, 103–4.

286 For the idea of the "beginnings of a performance," see the references provided for "nascent action" in editorial note 256 to Chapter III.

287 Cf. above, 29, 45–46, 52, 201.

288 Cf. DPP, 64–70; PR, 104–9.

289 For more on Bergson's idea of "living well," which is implicit here, cf. PR, 116–17; "L'intuition philosophique," in *La pensée et le mouvant*, 141–42; PC 175–76.

290 In terms of this feeling of "isolation," cf. DS, 274–75.

291 The final words of this concluding flourish—which is admittedly both enthusiastic and enigmatic—are often poorly interpreted. It is important to keep in mind that the focus of this entire chapter is the relationship between life and matter. Can life subsist without matter, or is it rather constituted within the resistance provided by matter? Passages from this chapter seem to provide support for either interpretation. For instance, Bergson refers to life as a "pure creative activity" but also (and on the very same page) as "an effort for going back up the incline that matter descends" (above, 246). It would thus perhaps be most faithful to the text to stick to the most explicit definition of life that Bergson gives: life is "consciousness launched through matter" (above, 183; cf. CV 13). Nevertheless, Bergson is clearly interested in the various forms that life may take, including in bodies that are more docile and malleable than the human body (cf. above, 256–57; DS 223–24, 273). He also occasionally considers the possibility of life as breaking free of corporeality altogether, suggesting something of an eschatological perspective (cf. CV, 1–3, 27–28; AC, 57–60; DS, 279–82).

CHAPTER IV

1 During the years mentioned in this footnote, Bergson held the Chair in Greek and Latin Philosophy at the Collège de France. Summaries of his courses from those years are available in *Mélanges*. [TN: Relatively complete transcripts of some of the courses have been recently published in French as well: Bergson, *L'idée de temps, 1901–1902*; *Histoire de l'idée de temps, 1902–1903*; *Histoire des théories de la mémoire, 1903–1904*;

L'évolution du problème de la liberté, 1904–1905. These documents provide important insights into Bergson's images and arguments, which are often presented here in a significantly condensed form.]*

2   For Bergson's internal reference, cf. above, 156, 163–64, 179, 222, 241–42.

3   For more on the terms used here, see editorial note 191 to Chapter III.

4   For nearly identical expressions, see *E*, 100; *Le rire*, 115; cf. also AC, 57.

5   It is worth recalling here that the title of this chapter includes the phrase: "Real Becoming and False Evolutionism."

6   Here Bergson is referring to the problem of disorder (cf. above, 179, 224, 227, 232). In his article "Le possible et le réel," he suggests that the pseudo-problem of disorder engendered "theories of knowledge" and the pseudo-problem of nothingness engendered "theories of being." See PR, 105.

7   For Bergson's internal reference, cf. above, 222. Note that a small typo that occurs here in the first French edition was corrected in later editions.

8   The differences between Bergson's earlier article "L'idée de néant" and the slightly revised version presented here are indeed minimal, consisting in small clarifications or in necessary adjustments to transform an article into a chapter within a larger book. André Robinet identifies these changes in Bergson, *Œuvres*, 1513–15. What I take to be the more substantial or philosophically interesting changes are mentioned in the following editorial endnotes.*

9   In the previously published version of this material, Bergson begins by noting that there are latent ideas behind the explicit arguments offered by philosophers. Since they are latent, no one thinks to examine the important role that they nevertheless play. The idea of nothingness is one such idea (see "L'idée de néant," 449). For more on Bergson's notion of "anxiety" in this context, see *DS*, 267; DPP, 65; PR, 105, 107. For more on "vertigo," see below, 307; *DS*, 238, 276; DPP, 65; PR, 106.*

10 This entire sentence did not appear in the previously published version of this material. For the images of "ascent" and "descent," see above, 246, 269, as well as the Index. For more on the idea of a Principle [or Source], cf. above, 238–39, 249. In the final part of this sentence, Bergson is clearly referring to Leibniz's *De rerum originatione radicali*, which he discussed in a course in 1898. Bergson, "Cours de Bergson sur la *De rerum originatione radicali* de Leibniz [1898]," ed. Matthias Vollet, in *Annales bergsoniennes*, vol. 3, *Bergson et la science*, ed. Frédéric Worms (Paris: Presses Universitaires de France, 2007), 25–52.*

11 Here Bergson comes close to the ancient thesis of an eternal existence (see above, 251–52). In his allusion to a receptacle, he is thinking of Plato's notion of the *chora* (cf. Plato, *Timaeus*, 52b) and more generally of Aristotle's notion of matter (see below, 316, 326).*

12 For this distinction, see below, 298. Bergson's goal in this section is, to the contrary, to show that "true being" can be psychological in nature. Note that this paragraph begins differently in the previously published version of this material (cf. "L'idée de néant," 450).*

13   Bergson is clearly thinking of Fichte, particularly Part I of Fichte's *The Science of Knowledge*. Cf. J. G. Fichte, "Fundamental Principles of the Entire Science of Knowledge," in *The Science of Knowledge*, ed. and trans. Peter Heath and John Lachs (Cambridge: Cambridge University Press, 1982), I, 93ff. He is also likely thinking of Schelling's developments from this Fichtean starting point.

14   It is Spinoza who conceives of the definition of the circle in this way. Cf. Spinoza, *Treatise on the Emendation of the Intellect*, § 95–96. Cf. also above, 212–13; below, 334. Note, in the previously published version of this material, Bergson discusses conceivability, not possibility (cf. "L'idée de néant," 450). On the fact that the possible is necessarily eternal, cf. MRV, 14; PR, 111.*

15   Although Bergson only names Spinoza and Leibniz here, this whole section on the idea of nothingness is full of allusions to other philosophers and thereby maintains an ongoing dialogue with many philosophical systems.*

16   It should be kept in mind that Bergson's critique of the idea of nothingness is precisely and exclusively insofar as he considers it to be a pseudo-idea.

17   Bergson uses the word "eminently" here in a classical philosophical sense to mean "to the highest point or degree." Thus, he is clearly alluding to something like a God, or at least to a "Principle." Cf. above, 249; IM, 210–11.

18   For more on Bergson's appeal to common sense, see editorial note 165 to Chapter II. Note that there are a few differences in these sentences between the previously published version of this material and this current version. For instance, Bergson writes "immediate observation" rather than "intuition" in "L'idée de néant" (cf. 450–51).*

19   There is a notably Cartesian tone to this passage, which continues on the following pages. As Descartes famously writes at the beginning of his Third Meditation: "I will now shut my eyes, stop my ears, and withdraw all my senses. I will eliminate from my thoughts all images of bodily things" (Descartes, "Third Meditation," in *The Philosophical Writings of Descartes*, vol. 2, AT 34). Bergson also makes use of Descartes' example of the chiliagon, the thousand-sided polygon discussed in the Sixth Meditation (cf. Descartes, "Sixth Meditation," in *ibid.*, AT 72; below, 280). Or again, Descartes seems present in Bergson's discussion of dreaming (below, 282) and in his invocation of pure understanding (below, 284). But Bergson aims to go beyond Descartes so as to demonstrate the futility of the idea of nothingness and to show that thought cannot exist (as Descartes believed) prior to time.*

20   For this exact definition of the present, see MM, 152–54.

21   For the distinction between external and internal perception, cf. MM, 52–59.

22   Descartes, "Sixth Meditation," in *The Philosophical Writings of Descartes*, vol. 2, AT 72.

23   It is worth recalling here that "regret" and "desire" play an important role in the section devoted to the idea of nothingness. See the Index.

24   In relation to the lovely phrase "consciousness lagging behind itself" [*la conscience retardant sur elle-même*], cf. below, 292.

25 Bergson is likely here referring to a famous phrase from Descartes' correspondence with Marin Mersenne (1588–1648), where Descartes dismisses readers who are unable to "withdraw their minds from their senses" (in Latin: *abducere mentem a sensibus*). Cf. René Descartes, "To Mersenne, 27 February 1637," in *The Philosophical Writings of Descartes*, vol. 3, trans. John Cottingham, Robert Stoothoff, Dugald Murdoch, and Anthony Kenny (Cambridge: Cambridge University Press, 1991), AT I, 350. Cf. also René Descartes, *Discourse on Method*, in *The Philosophical Writings of Descartes*, vol. 1, AT VI, 37.*

26 Bergson rarely considers space in a way that is comparable with time. Cf. *E*, 141.

27 Here Bergson is alluding to the section titled "On the impossibility of an ontological proof of God's existence" in Kant, *Critique of Pure Reason*, A592–602/B620–32. See editorial note 97 to Chapter I.

28 This is the first time in *L'évolution créatrice* that Bergson uses the word "possible" as a noun. His ongoing dialogue with Kant's critique of the ontological argument (cited in the previous endnote) necessarily mobilizes a modal reflection upon possibility, existence, and necessity. Cf. below, 289–90, 292, 296; MRV, 1–23; PR, 99–116. [TN: Following Kant's translators, I have opted for "the merely possible" or something "merely possible" to translate Bergson's various uses of "*le simple possible*." See Kant, *Critique of Pure Reason*, A599/B627.]

    29 Christoph Sigwart (1830–1904) was professor of philosophy at the University of Tübingen, and the book that Bergson cites was one of the most important books on logic in Germany at the end of the nineteenth century. Sigwart, who was influenced by Kant, offered an analysis of negation (via the negative proposition and not via the concept of negation itself) that appears to have been significant for Bergson's argument. Sigwart claims that negation is not symmetrical with affirmation: "The object of a negation must be either a completed or an attempted judgment, and for this reason we cannot regard the negative judgment as a species equally primitive with the positive judgement" (Sigwart, *Logic*, vol. 1, *The Judgment, Concept, and Inference*, 119).*

30 The previously published version of this material names here the "intellect" rather than the "mind" (cf. "L'idée de néant," 458). In *L'évolution créatrice*, Bergson clarifies the sense of the word "mind" [*esprit*], and thus simultaneously the difference between "mind" and "intellect." Cf. above, 213, 251.

31 See above, 284–86 and editorial note 97 to Chapter I.

32 For this characterization of the possible, cf. above, 285, 292, 294, 296; MRV, 1–23; PR, 99–116.

33 The phrase "or philosophical" does not appear in the previously published version of this material (cf. "L'idée de néant," 460). For more on the image of a "mind" that is disinterested in others, which will occupy Bergson in the following pages, see above, 281; below, 317

34 In terms of Bergson's theory of language, cf. *E*, 96–104.

35 For the image of "lagging behind," cf. above, 282.

36   For the highly suggestive phrase "turn our back on reality," cf. above, 294. In the previously published version of this material, Bergson wrote "ascend the current of reality" rather than "turn our back on reality" (cf. "L'idée de néant," 462).

37   In this sentence in the previously published version of this material, Bergson writes "intellect" [*intelligence*] rather than "mind" [*esprit*] (cf. "L'idée de néant," 462). The same is true for the next two occurrences of the term. Cf. above, 288.

38   Thus, following what we might call the *genealogical* moment of Bergson's analysis of negation—namely, his attempt to *determine*, by working backward, the origin of negation in the basic emotions of regret and desire—we find its *genetic* moment. We *follow* the right way around, so to speak, the step-by-step process by which the mind generates the act of negation.

39   Of course, we should recall that for Bergson "to *imagine* is not to *remember*" (*MM*, 150), and a perception differs in nature from the memory of this perception. But the image's mode of being, in the sense in which the imagination is a reproductive faculty, is the same as the memory's mode of being: they are both virtual. There is a gradation in intensity between the image as perception and as pure memory, and this allows Bergson to employ the seemingly oxymoronic expression "memory-image" (cf. *MM*, 147).

40   Cf. above, 293.

41   In the previously published version of this material, Bergson uses the term "merely possible" [*simple possible*] in this sentence (cf. "L'idée de néant," 463). For more on the expression "merely possible," cf. above, 285, 289–90, and editorial note 28 to Chapter IV.

42   Bergson will also characterize "the possible" as a "phantom" (cf. PR, 111). For more on the use of the term "phantom" in *L'évolution créatrice*, see the Index.

43   For a similar expression, cf. above, 209.

44   In the previously published version of this material, Bergson italicizes the phrase "judgment of a judgment" ("L'idée de néant," 464).

45   For Bergson's internal reference, cf. above, 156, 222, 273. In the previously published version of this material, Bergson does not include the phrase "I said above," but rather refers to "previous works" (Bergson, "L'idée de néant," 464). Bergson is probably thinking of pages 203–6 in *Matière et mémoire*, where he refers to his *Essai sur les données immédiates de la conscience*. In addition, in the previously published version, the phrase "venture beyond their proper limits" reads, rather: "the forms of human action *go astray*, in some sense, in the domain of speculation" ("L'idée de néant," 464).*

46   The influence of Bergson's interpretation of pragmatism can be seen here. Cf. in particular above, 192–93, and editorial note 77 to Chapter I.

47   For Bergson's internal reference, see above, 273–74. This internal reference of course does not appear in the previously published version of this material.

48   The previously published version of this material concludes with the following lines: "Once this illusion has dissipated, the problems that sprang up around it evaporate. / The child whose dreams carry him up to the moon finds himself

suspended so high above the earth, so surrounded by emptiness, that he is stricken with vertigo and feels himself falling. A few concepts drawn from cosmography, or even a simple reflection on the nature of the high and the low, would convince him that his vertigo is illegitimate and would thereby cure him of it by way of demonstrative reason, if I can speak this way. Cases of illegitimate vertigo also appear in philosophy, and it might be useful to point them out, at least when this is merely to set aside the illusory difficulties and the artificial causes of the agitation [*le trouble*] prior to facing the genuine problems head on." "L'idée de néant," 465–66. The vivid conclusion, which has a certain Kantian resonance (cf. Kant, *Critique of Pure Reason*, A235–60/B294–315), helps to show the kinship between Bergson's notion of pseudo-problems and Kant's notion of a transcendental illusion.

49   This can be understood as the formal conclusion of the argument that began above, 276.

50   This comment reveals that Bergson, up to this point, has been generally attacking Laplace's idea that "everything is given." One of the main justifications for challenging Laplace's idea is precisely the critique of the idea of nothingness found in the previous paragraphs. See editorial note 101 to Chapter I.

51   Cf. above, 276.

52   For this formulation, cf. IM, 210–11.

53   For this expression, cf. below, 359. Here Bergson is referring to intuition, which is also implicit in his subsequent use of the image of "placing ourselves within" [*s'installer*].

54   This phrase anticipates the title of Bergson's last published book, the collection of essays titled *La pensée et le mouvant* (1934). [TN: This connection is lost, however, in the title of the English translation of that collection: *Creative Energy*. A more literal translation of the title would be *Thought and Moving Reality*, where "moving reality" should be understood as encompassing whatever is marked by becoming, and hence *durée*. Given Bergson's section title to this current section ("Becoming and Form), another (less direct) translation for the title of his later collection could be: *Thought and Becoming*.]

55   Here again we can sense Bergson's reading of pragmatism. See editorial note 77 to Chapter I.

56   For this movement from idea to act, see *E*, 158–59.

57   Here Bergson is implying that movement is not built out of the positions of the moving object. See editorial note 57 to Chapter I.*

58   Cf. above, 187, 250.

59   Cf. *MM*, Chapter I. For more on Bergson's pragmatism, see editorial note 77 to Chapter I.

60   For this use of "symbolize," cf. above, 95; *MM*, 40, 57, 249; IM, 177–82.

61   Cf. *MM*, 226–35; CV, 15–17.

62   Again, Bergson is thinking here of Thomson and Faraday. See "physics, Thomson and Faraday" in the Index.

63   For more on this critique of the substrate, see editorial note 6 to Chapter I.

64   Cf. *MM*, 232–33, 249–50, 279–80; *CV*, 15–17.

65   For additional images of this type, cf. above, 264–65; *MM*, 264, 279.

66   Cf. *MM*, 170; *CV*, 15.

67   For more on this important doctrine regarding the "living body," cf. above, 11–23, 228; *MM*, 220–24.

68   Here the term "image" seems to have approximately the same sense as it does in the first chapter of *Matière et mémoire* since it signifies a piece of matter.

69   For more on this distinction between the "surface" and the "depths" of matter, see *MM*, 226–35.

70   For Bergson's internal reference, see above, 155–56. For more on the notion of mobility, see above, 92; *E*, 86; *MM*, 213, 219; *DPP*, 26.

71   This tripartite analysis of qualitative, evolutionary, and extensive "movements," "becomings," or "changes" is systematic in the current chapter. Cf. below, 313. Bergson worked through this distinction at length in his 1902–1903 course at the Collège de France (Bergson, *Histoire de l'idée de temps*). There he began from the Aristotelian classification of different movements: according to quality, according to quantity, or according to place, to which it is necessary to add change according to generation and corruption.

72   Here Bergson uses the slightly awkward plural form "becomings" [*devenirs*]. He will insist on this point in the following pages. If, by contrast, we sufficiently examine becoming, we will notice that we must reject the idea of an essence of becoming. For the important phrases "becoming in general," "movement in general," and "change in general," see below, 305–7, 326.

73   TN: The cinematograph was a relatively portable early motion picture device that could serve both as a camera and as a projector. Bergson finds in this example the perfect illustration of the mechanism of the intellect insofar as it captures movement by taking a series of views or snapshots and then artificially reconstitutes it by projecting them in succession. Such a mechanism is unable, according to Bergson, to put us into contact with concrete *durée*, a contact reserved primarily for "intuition." Cf. Bergson's response to Le Dantec in the **Correspondence, Reception, and Commentaries** section below and "intellect" in the Index.

74   See the citations provided in editorial note 57 to Chapter I.

75   For this precise use of the opposition between inside and outside, cf. *IM*, 177–82.

76   For the expression "the reality that passes by," cf. above, 294.

77   Cf. above, 126–27.

78   For the image of a kaleidoscope, cf. *MM*, 20, 221; *DS*, 132, 297.

79   These lines indicate Bergson's understanding of the pragmatic function of the intellect. See editorial note 77 to Chapter I.

80   For Bergson's use of "rhythm," cf. *MM*, 226–35.

81   For the term "pulsation," cf. above, 300.

82   The actions of "placing oneself back within" [*se replacer*]," "placing oneself within [*s'installer*]," and "grasping" [*saisir*] in these two sentences evoke the vocabulary Bergson uses for his notion of intuition. See the Index.

83   See editorial note 55 to Chapter I.

84   For Bergson, this ability to manipulate the quantitative aspects of space "as much as you like" is an important part of the definition of space. See editorial note 178 to Chapter II.

85   For images related to smoke, see above, 164; below, 312; IM, 206.

86   Even though there is no break in the original text here, the part of the chapter devoted to "The History of [Philosophical] Systems" (as announced in the chapter title) begins here with the consideration of the Presocratics.

87   For Bergson's other discussions of the important example of Zeno of Elea, cf. E, 84–86; MM, 213–15; DS, 32, 51–52, 72, 207–8; "Mouvement rétrograde du vrai," in La pensée et le mouvant, 8–9; PC, 146–47, 156–57, 160–61; "Théorie de la personne," in Mélanges, 864–65; "Conférence de Madrid: La personnalité," in Mélanges, 1221–223. Zeno's "intention" was, of course, to negate movement itself. Bergson makes this explicit in the months following the publication of L'évolution créatrice. See Bergson, "À propos de 'L'évolution de l'intelligence géométrique' [Réponse à un article d'É. Borel, 1908]," in Mélanges, 758. [TN: Included below in the **Correspondence, Reception, and Commentaries** section.].

88   For this example, cf. MM, 214; DS, 32, 207–8.

89   Recall that Bergson uses a similar vocabulary in the example of moving one's arm from A to B. See "arm" in the Index.

90   For this example, see IM, 184.

91   Cf. editorial note 57 to Chapter I.

92   For other distinctions of this type, cf. editorial note 89 to Chapter I.

93   For instance, Bergson discusses Zeno's paradoxes at MM, 213–15.

94   It is important to note that durée does not form an absolutely indivisible whole. Such a whole would be indiscernible from the neo-Platonic idea of the One. Rather, durée is articulated, and the [intuitive] philosopher, to continue the metaphor from Plato's Phaedrus, will be the one who knows how to cut it up at the joints [articulations]. Cf. above, 157; DS, 109; DPP, 51–53. See also "articulations of the real" in the Index.

95   Cf. E, 84–86; MM, 214; PC, 160–61; and "Conférence de Madrid: La personnalité," in Mélanges, 1222–223.

96   François Évellin (1835–1910) was a philosopher and a friend of Bergson's. Bergson asked Évellin to read Essai sur les données immédiates de la conscience before submitting it to his dissertation committee. Since the seventeenth century, the attempt to refute Zeno's paradoxes via infinitesimal calculus was a classical approach and one that was supported in Bergson's era by mathematician and philosopher Paul Tannery (1843–1904). The 1880s saw a renewed interest in refuting Zeno's paradoxes, and Bergson in fact already mentions Évellin's book in his Essai sur les données immédiates de la conscience (E, 85n1). Here Bergson is mostly interested in Évellin's discussion of how geometrical progression gives an "infinitesimal" solution to Zeno's paradox (cf. Évellin, Infini et quantité, 73–74). In order to root out presuppositions in Zeno's formulation,

Évellin's strategy is to separate the question of *when* Achilles overtakes the tortoise from that of *how* he goes about it (cf. Évellin, *Infini et quantité*, 71). Infinitesimal calculus can only ask the first question. Bergson similarly shifts his focus to *how* Achilles goes about overtaking the tortoise.

Bergson's second reference is to a response by Évellin to a letter by Tannery that was itself written in response to a critical review of *Infini et quantité* published by Victor Brochard (see Paul Tannery, "Letter to Ribot, April 10, 1881," *Revue philosophique de la France et de l'étranger* 11, no. 5 (May 1881): 561–63; Victor Brochard, Review of *Infini et quantité* by François Évellin, *Revue philosophique de la France et de l'étranger* 11, no. 4 (April 1881): 421–31). In his letter, Évellin comes back to the relations between multiplicity and space, and again takes up his argument regarding Zeno's paradox. For his part, Tannery judged the infinitesimal solution to be satisfactory, and, for the rest, considered this paradox to be nothing but a "tired sophism" about which there was nothing at all left to say in philosophy! (Tannery, "Letter to Ribot," 562).

For Bergson's allusion to his own previous work, see *E*, 82–89.*

97 For similar formulations, see above, 164, 307; IM, 206.

98 For more on this claim, see "intellect, natural inclination of" in the Index.

99 Cf. Bergson, *Le rire*, 37.

100 Here Bergson recalls the tripartite account of "becomings" [*devenirs*] that he introduced above, 303.

101 Cf. above, 303.

102 For more on Bergson's suggested alternative translation of εἶδος ["Idea"], see below, 316. In these passages, it becomes clear that a "view" or "snapshot," as opposed to the simple reality itself (cf. 90–98), is to a certain extent the εἶδος.

103 For Bergson's internal reference, see above, 302–3.

104 For Bergson's internal reference, see above, 313. In this sentence, Bergson again assumes the tripartite account of *becomings*. See above, 303, 313.

105 On this "fact," see above, 9; below, 338; *E*, 166; PR, 116.

106 For Plato's "non-being," see Plato, *The Sophist*, trans. Nicholas P. White, in *Complete Works*, ed. John. M. Cooper, with D. S. Hutchinson, 235–93 (Indianapolis: Hackett Publishing Company, 1997), 237a–259d. Having questioned the idea of nothingness, Bergson was led to consider Plato's idea of non-being. For Bergson's criticism of Plato, cf. below, 323, 326.

Bergson vigorously rejects Aristotle's notion of "matter," contrary to what is sometimes believed, because this notion signifies, notably, *indetermination*. See the Index. Bergson identifies matter with "becoming in general," and thus as a tributary of the same illusions faced by its opposite, namely form. He interprets matter as the disorder to which order (form) would have to be added, or again the non-being that would have to be overcome by a logical being.

107 The allusion to "truth" in this passage is not incidental; it has to do with how Plato and Aristotle, according to Bergson, imagine truth as the immediately given logical structure of reality. See "history, of systems" in the Index.

108 For the meaning of Bergson's use of the term "view" here, cf. above, 314.

109 For this striking image, cf. below, 324–25, 343; "L'effort intellectuel," in *L'énergie spirituelle*, 161; IM, 180. [TN: Note that Bergson does not intend us to understand "spare change" in the sense of that which can in fact equal the piece of gold (in the sense that four quarters or one hundred pennies equal a dollar). Rather, in his image here, the value of the piece of gold can never be equaled by the juxtaposition of parts, in the way the Platonic Idea is never exhausted by the particulars that participate in it, or in the way the movement can never be adequately reconstructed out of the immobile positions that it passes through. Moreover, beginning from the piece of gold, we can go to all of the possible subdivisions into "spare change," but the opposite procedure is impossible.]

110 In terms of this Platonic definition of time, cf. IM, 217. As Bergson mentions in his first footnote to this chapter (page 272), Bergson also discusses this notion in his courses at the *Collège de France*, notably in *Histoire de l'idée de temps, 1902–1903*.

111 Bergson repeatedly returns to this thought experiment in the section above devoted to the idea of nothingness. Cf. above, 281, 290–94.

112 For more on this notion of "distension," which Bergson attributes to Plotinus, cf. Bergson's note to page 211 and below, 320. Given his association of this notion with the philosophy of Form, we can now see why Bergson refused to incorporate it into his own account.

113 For this image of the oscillations of a pendulum, cf. above, 316.

114 It is worth recalling that the term "deficit" also appears above in Bergson's characterization of matter [as an "interruption"]. See above, 248.

115 For this notion of a self-positing [and hence necessary] Idea, cf. above, 276–77.

116 This is Bergson's first explicit attempt to synthetize his understanding of Platonism, which he identifies as a certain conception of truth that he has been attacking since Chapter III and that he will continue to criticize throughout this chapter. Cf., in particular, below, 328, 353.

117 Aristotle, *Metaphysics*, trans. W. D. Ross, in *The Complete Works of Aristotle*, ed. Jonathan Barnes, 1552–728, the revised Oxford translation, vol. 2, Bolligen Series LXXI.2 (Princeton: Princeton University Press, 1995), 1074b34. Cf. below, 323; DS, 257–59; DPP, 48; *Histoire des théories de la mémoire*, 253–67. [TN: The cited English translation of Aristotle renders this phrase "thinking on thinking," and the idea is often alluded to by the phrase "thought thinking itself."]

118 Cf. Bergson, *Histoire des théories de la mémoire*, 265.

119 Bergson's allusion to the Alexandrian philosophers is in fact to Plotinus. This passage shows how Bergson intentionally runs together his interpretations of Plato, Aristotle, and Plotinus, belonging as he does to a tradition of Plotinian interpretation that sees Plotinus as offering important insight into Aristotelianism and Platonism. For instance, in a course from 1903–1904, Bergson suggested that Plotinus could serve as a magnifying glass in the study of Plato and Aristotle (*Histoire des théories de la mémoire*, 261). This description also aligns with Bergson's conception of Greek philosophy more generally as having a certain unity insofar as it represents the "natural metaphysics of the human intellect" (above, 314–15) and provides a clear expression of the "cinematographic mechanism of thought" (below, 325–28).*

120 On the idea of causality by "impulsion," cf. above, 73–75.

121 Bergson elaborates on this argument, as well as the following comparisons with the Alexandrian philosophers [that is, Plotinus], in his January 30 and February 6, 1903, course lectures at the Collège de France (*Histoire de l'idée de temps*, 135–66). It is also worth noting that Bergson's Plotinian interpretation of Aristotle began with his first thesis on "the idea of place in Aristotle" (see "Idée de lieu chez Aristote [*Quid Aristoteles de loco senserit*, 1889]," in *Mélanges*, particularly 48–50. [TN: This appears in English as: "Aristotle's Concept of Place," trans. J. K. Ryan, *Studies in Philosophy and the History of Philosophy* 5 (1970): 64–66. Cf. also, Bergson, *Histoire des théories de la mémoire*, 261–65.])*

122 "Procession" and "conversion" are two central notions in neo-Platonism, particularly for Plotinus. Bergson explains their significance just below.

123 In other words, Bergson's critique of systematism here is a continuation of his criticism of the idea of nothingness, and vice versa.

124 In the first edition of *L'évolution créatrice*, this reads: "of the entire interval."

125 Cf. Bergson, *Histoire des théories de la mémoire*, 264.

126 For more on generation and corruption, see above, 319.

127 Cf. Bergson, "L'idée de lieu chez Aristote," in *Mélanges*, 49–50 ["Aristotle's Concept of Place," 65–66]; *Histoire des théories de la mémoire*, 261–67.

128 For this theory regarding ancient philosophy, see above, 314–15.

129 For a nearly identical phrase, though in a wholly different context, cf. MM, 37.

130 Cf. above, 318–19.

131 This is the second time Bergson attempts to provide the concept of truth that he takes as natural for our intellect (cf. above 320; below, 353).

132 According to the conception of truth that Bergson is attacking here, knowledge or science is contained within being itself. It is found at the foundations of reality and not in the simple subjective act by which we relate to that reality. This point can be difficult to grasp. According to this theory, sensible reality is but a degradation of this knowledge and of this knowledge alone. There is thus an identity (already present above, cf. above, 316) between truth and reality, which is also personified in the identity of the subject and the object in Aristotle's God (see above, 323). Bergson himself (and particularly in *L'évolution créatrice*) defends a similar identity: individual, distinct consciousness is a part of life and it is not something that holds itself outside of life so as to scrutinize it. But then, the sense of this identity, as required by the very notion of *durée*, is entirely different—here we find the reason for Bergson's lengthy attack against the systematic conception of truth that is notably implicit in Greek philosophy.

133 TN: In his Table of Contents, Bergson lists this chapter as beginning on page 323. The discussion of modern science and the comparison with ancient science, however, clearly begin here.

134 Although there is no paragraph break here, and although Bergson returns to the discussion of ancient philosophy just below, this is where the sections on modern philosophy [and science] begin.

135 For more on the role of "signs," see editorial note 183 to Chapter II.

136 For more on Bergson's interpretation of pragmatism, see editorial note 77 to Chapter I.

137 Cf. *MM*, 226–35.

138 Cf. above, 299.

139 This characterization of knowledge comes from Auguste Comte (1798–1857), as developed in the second lesson of his *Cours de philosophie positive*. August Comte, *Premiers cours de philosophie positive: Préliminaires généraux et philosophie mathématique* (Paris: Presses Universitaires de France, 2007).

140 For Bergson's internal reference, see above, 227–32.

141 In terms of Bergson's inclusion of ἀκμή [acme, "highest point"] here, consider the following passage from Aristotle where it is translated as "maximum": "If the movement is uneven, clearly there will be acceleration, maximum speed, and retardation, since these appear in all irregular motions. The maximum may occur either at the starting-point or at the goal or between the two; and we expect natural motion to reach its maximum at the goal, unnatural motion at the starting-point, and missiles midway between the two." Aristotle, *On the Heavens*, 288a17–22. For Bergson's arguments here, cf. above, 229.

142 Again, it is worth noting that Bergson's first thesis *Quid Aristoteles de loco senserit* dealt with these questions. See "L'idée de lieu chez Aristote," in *Mélanges*, 1–56 ["Aristotle's Concept of Place"]. [TN: Cf. Bergson's discussion and citations of Aristotle above, 229.]

143 For more on this aspect of Galileo's position, see above, 229; IM, 217.

144 For more on these "natural articulations," cf. above, 309–11 and "articulations of the real" in the Index.

145 For more, see note 176 to Chapter II.

146 Cf. *E*, 86–89, 144–49.

147 Nevertheless, according to Bergson, the Greeks were the ones who invented "precision." See Bergson, "Fantômes de vivants," in *L'énergie spirituelle*, 82–84.

148 For more examples of this type, see *MM*, 211, 234.

149 Cf. above, 227–32.

150 This theory regarding ancient and modern sciences is found in Duhem's *L'évolution de la mécanique* (1903), which Bergson discusses in Chapter III. One finds the thematic opposition between quality and quantity throughout Duhem's work. Cf. above 243 and editorial notes 179–80 to Chapter III.

151 Again, this rather Bergsonian affirmation (cf. above, 225–32) can be connected to Duhem's *L'évolution de la mécanique*. Cf. above 243 and editorial notes 179–80 to Chapter III. The claim also connects in important ways to Comte. Cf. also, Bergson, *Histoire des théories de la mémoire*, 209–11.*

152 For Bergson's internal reference, cf. above 331, where, prior to the discussions of the difference between quality versus quantity and concepts versus laws, Bergson notes a fundamental difference in the conception of time.*

153 For more on this theory, cf. above, 228–31, 330–31; below, 369; AC, 39; "Fantômes de vivants," in *L'énergie spirituelle*, 80; IM, 217.

154 Here Bergson is paraphrasing Kepler's second and third laws. The first law (which Bergson alluded to a few lines earlier) is that the orbits of the planets around the sun describe elliptical pathways. The second law, the "law of area," holds that the radius between the planet and the Sun sweeps across equal areas (through its displacement; this is the notion of a "radius vector") during equal intervals of time. This explains why planets move faster when they are closer to the Sun. The third law indicates a mathematical relationship between the major axis of an ellipsis and the time it takes to cross it. In both the second and third laws, gravitational phenomena are being described, and indeed Kepler's laws would prove important for the development of Newton's theory of universal gravitation.*

155 Bergson is probably thinking here of Newton's theory of fluxions (cf. DPP, 29; IM, 214). More generally, Bergson often compares modern and ancient geometries in terms of this inclusion of time and laws of generation (cf. above, 31–33; DS, 58–59). In philosophy, it is Spinoza who defines a shape by the law of its generation (Spinoza, *Treatise on the Emendation of the Intellect*, §95–96). These passages reveal how important geometrical considerations were for Bergson's philosophy (cf. above, 212–13, 276–77).

156 The first edition of *L'évolution créatrice* reads "a single moment" [*un moment unique*] rather than "a single movement" [*le mouvement unique*].

157 On this image of kinship, cf. above, 230; "Fantômes et vivants," in *L'énergie spirituelle*, 70–71; "La philosophie de Claude Bernard," in *La pensée et le mouvant*, 229.

158 For this famous image of heaven and earth, cf. IM, 217.

159 For more on Laplace's formulation of the problem, see "Laplace's Demon" in the Index.

160 See "intellect, direction of" in the Index.

161 For other key formulations regarding "mobility," see editorial note 252 to Chapter I.

162 For Bergson's reference to his previous work, cf. E, 86–89, 144–49.

163 See above, 7–11; cf. also, 11–44.*

164 For the voluntary nature of this operation, see editorial note 178 to Chapter II.

165 For Bergson's internal reference, see above, 8–9.

166 For a critical use of this expression, see above, 9.

167 For this thought experiment in relation to time, cf. E, 86–89, 144–49.

168 For Bergson's use of "sign" here, see above, 328 and editorial note 183 to Chapter II.

169 For this argument, cf. E, 145–46.

170 The allusion here to time as a "force" is important, specifically in terms of understanding what Bergson means by "force" elsewhere. Cf. E, 93, 158–62.

171 Cf. editorial note 101 to Chapter I.

172 For another example of the word "Life" being capitalized, see above, 49. In terms of the image of a pathway in relation to life and matter, see above, 246. For Bergson's use of "interdependent" [solidaires], see below, 342, 368.

173 Bergson's claim here is grounded on the idea that durée changes nature through self-differentiation. Cf. E, 62–63; MM, 231–32.

174 Cf. Bergson, "L'effort intellectuel," L'énergie spirituelle, 189–90.

175 For more on "resemblance," cf. E, 129. For a claim very similar to this striking phrase ("that unforeseeable nothing that is everything" [cet imprévisible rien qui est le tout]), cf. PR, 99.

176 For more on this "ideal space," cf. E, 56–63; MM, 167–73, 186–92.

177 For more on this key theoretical claim, cf. PR, 101–2.

178 This summary serves as a conclusion for the argument that began above, 335.

179 Here Bergson identifies the first of three missed opportunities in which philosophy might have broken with the cinematographic mechanism of the intellect and with ancient thought. The second will be in Kant's philosophy (see below, 357–58) and the third in Spencer's (see below, 362–63). In other words, from this point forward, Bergson's so-called "history of systems" will be the staking out of missed opportunities.

180 Bergson uses the phrase "to be transported" or "to transport oneself" [se transporter] as part of the vocabulary around "intuition." See "intuition, vocabulary of" in the Index.

181 Bergson's use of the expression "kind of knowledge" here is clearly in the style of Spinoza. See Spinoza, Ethics, Book II, proposition 40, scholium 2. He repeats this expression several times in the current chapter.

182 For more on ce qui se défait ("what is being unmade") and ce qui se fait ("what is being made"), see editorial note 191 to Chapter III.

183 Here the term "metaphysics" should be understood in the fully Bergsonian sense of the term, namely having to do with intuition and durée.

184 See above, 316–18, 321, 323–25.

185 See, for example, Aristotle, On the Soul, 431b24–432a8 and 429b10–22. See also "Aristotle, matter" in the Index.

186 For more on τέλος ("end") and ἀκμή ("highest point" or "maximum"), see above, 330.

187 These pages provide an important insight into Bergson's understanding of the relationship between philosophy and science (for more, see the Index). For the ancients, there could be no radical distinction between science and philosophy since there was but a single system of knowledge immanent to the things themselves. Bergson, for his part, wants to distinguish between domains, regions, and objects—with science being concerned with inert matter, and philosophy (once it becomes "metaphysical") reserving for itself everything that, by essence, endures. For Bergson, such a division of disciplines became possible with the emergence of modern science.

188 See editorial note 101 to Chapter I.

189 For more on this Cartesian "oscillation," see AC, 40.

190 For a similar discussion, cf. MM, 216–17.

191 Cf. AC, 40. Bergson develops this theme in *L'évolution du problème de la liberté*, 191–241.

192 In a 1905 course lecture at the Collège de France, Bergson examined a section in which Descartes distinguishes between the movement of a thing and its direction. According to Descartes, the direction "can be altered, while the motion remains constant" (Descartes, *Principles of Philosophy*, Part II, §41). Nevertheless, Bergson does not go so far as to turn this distinction into the foundation of a Cartesian theory of the insertion of freedom into the world. See *L'évolution du problème de la liberté*, 225–41.]

193 On the Cartesian notion of "continuous creation," cf. above, 22; *E*, 156. See also Descartes, *Principles of Philosophy*, Part I, §21.

194 Here Bergson is likely thinking of Descartes' *Principles of Philosophy*, Part II, §36–63, where Descartes presents the laws of movement. For Bergson's interpretation of Descartes' position, cf. MM, 216–17.

195 According to Bergson, we cannot conceive of the will without referring to the notion of time. See Bergson, "Bergson à A. O. Lovejoy [letter, May 10, 1911]," in *Écrits philosophiques* (Paris: Presses Universitaires de France, 2011), 396–97.

196 Here Bergson summarizes an argument that he previously developed in his 1902–1903 course at the Collège de France. See *Histoire de l'idée de temps*, 18–39.

197 Here Bergson returns to observations first offered in his 1904–1905 course at the Collège de France: "Leibniz is a pure intellectualist. No one had ever been more of an intellectualist than he was, nor even equal to him. He is more of an intellectualist than even Spinoza, since Spinoza was only an intellectualist through an effort of will and there are passages in the *Ethics*, above all in the last part, in which Spinoza's intellectualism seems somehow ready to break apart under the influence of some interior thrust or pressure. But in Leibniz, nothing like that can be found." *L'évolution du problème de la liberté*, 259–60.

198 For more on Bergson's analysis of the presupposition that "everything is given," see editorial note 101 to Chapter I.

199 In terms of Aristotle's God, cf. above, 321.

200 For more on Bergson's distinction between genres and laws, cf. above, 227–32; 329–30.

201 This is in opposition to Greek philosophy, which aimed to be "comprehensive" of all of reality. Cf. above, 314.

202 To be precise, Aristotle defines the soul as "an actuality of the first kind of a natural body having life potentially in it" (*On the Soul*, 412a27). For more on this important comparison between the ancient and modern conceptions of body and soul, cf. Bergson, *Histoire des théories de la mémoire*, 265–67.

203 For more on this model of "translation," which Bergson uses to characterize the parallelism doctrine in Spinoza, Leibniz, and others, see Bergson, *Histoire des théories de la mémoire*, 313–35.

204 See Spinoza, *Ethics*, Part II, prop. 1 and prop. 2.

205 See *ibid.*, prop. 7. Cf. Bergson, *Histoire des théories de la mémoire*, 316f.

206 This description could be compared to when Bergson discusses the implicit conception of causality present in ancient philosophy, cf. above, 323.

207 In terms of "confused perceptions," Bergson is likely alluding to his own interpretation of the famous Leibniz-Clarke letters. Cf. Bergson, *Histoire de l'idée du temps*, 314–21. For more on Bergson's interpretation of space and time in Leibniz, see just below.

208 This is a highly compressed version of an argument Bergson developed more patiently across several lectures at the Collège de France. See *L'évolution du problème de la liberté*, 259–75. As Bergson notes, he is thinking of the invention of the stereoscope here. If we place two images of the same object before our eyes, the two images having been taken from different positions corresponding to the spacing between our eyes, then the object takes on the beginnings of depth. Bergson imagines a stereoscope that would allow us to bring together an infinite number of views. But his intention here is critical. The term "view" possesses (as everywhere else in Bergson) the sense of a symbol with which the intellect hopes in vain to reconstitute the things. This being the case, it is hardly a reality that Leibniz embraces when he imagines his God. As Bergson argues elsewhere: "Though all the photographs of a city taken from all possible points of view indefinitely complete one another, they will never equal in value that dimensional object, the city along whose streets one walks" (IM, 179–80 ["Introduction to Metaphysics," in *The Creative Mind*, 134]).

   In terms of the notion of substance, Bergson offers the following definition in his course: "We associate all the monads with points of view. God is simultaneously at each point of view, or rather he is not at any one point of view; he must be defined as follows: God is the monad that has no point of view." *Histoire du problème de la liberté*, 280.*

209 For this notion of laws being "immanent," cf. Bergson, *Histoire des théories de la mémoire*, 298f.

210 Cf. above, 40. Again, Bergson's interpretation here is of the Leibniz-Clarke correspondence. In a course at the Collège de France, Bergson considers the definitions of space and time in this exchange between Leibniz and philosopher Samuel Clarke (1675–1729) to have been "purely polemical" (Bergson, *Histoire de l'idée de temps*, 316). He considers Leibniz's true position on the question of time to be articulated in article XIII of his *Discourse on Metaphysics* and in his *Correspondence with Arnauld*, that is, as related to his theory of substance. See Bergson, *Histoire de l'idée de temps*, 316–20.*

211 A passage from Bergson's 1904–1905 course at the Collège de France can help clarify this point, and indeed connects it to the just mentioned question of eternity. Bergson says: "Spinoza's aim is the same as Aristotle's since if we open Spinoza's *Treatise on the Emendation of the Intellect*, we find his understanding of the goal of philosophy mentioned on the very first pages: *Our happiness depends solely on the quality of the objects to which we are bound by love, for that which is not loved can never cause us strife, nor sorrow if it is lost, nor envy if it is possessed by another—in a word, no emotional agitation of the soul, all of which, however, occur in the case of the love of perishable things. But love toward a*

thing eternal and infinite feeds the mind with joy alone, unmixed with any sadness. As such, the aim of philosophy is to tell us what we should love. And what we should love is that which nourishes our soul with a joy that is without sadness, without mixture, an infinity, which is to say the love of an eternal and infinite thing, as Spinoza puts it. // And this is exactly what Aristotle said in Book X of the *Nicomachean Ethics*: the goal of philosophy is precisely that, "so far as we can, make ourselves immortal" (Aristotle, *Nicomachean Ethics*, trans. W. D. Ross and rev. J. O. Urmson, in *The Complete Works of Aristotle*, ed. Jonathan Barnes, the revised Oxford translation, vol. 2, Bolligen Series LXXI.2 (Princeton: Princeton University Press, 1995), Book X, 1177b33). The word ἀθανατίζειν is untranslatable. It means: "to make oneself of immortality; that we must make ourselves out of immortality." Bergson, *L'évolution du problème de la liberté*, 244–45. For more on Spinoza's notion of adequacy, see Spinoza, *Ethics*, Book II, definition IV. [TN: Note that in the passage in italics above, Bergson is only loosely quoting Spinoza. I have nevertheless followed relatively closely the English version of the paragraph in question (cf. Spinoza, *Treatise on the Emendation of the Intellect*, 234–35). I have also followed Bergson's translation of ἀθανατίζειν from where he offers his translation just above.]

212 Here Bergson is referring to a course titled "La psychologie de Plotin" [Plotinus's Psychology] that he offered at the same time as a course on Plotinus's *Fourth Ennead* (see the listing of the "Cours du Collège de France, année 1897–1898," in *Mélanges*, 413). He was particularly interested in a double analogy between Plotinus with Leibniz on the one hand, and Aristotle with Spinoza on the other. These themes reappear in his courses between 1901 and 1904. In terms of the first analogy, Bergson is particularly interested in how each "intelligible" contains all of the other "intelligibles" for Plotinus, and how each monad is representative of all other monads for Leibniz.*

213 TN: I have followed the logic of Bergson's prose here, but, based on the content of the description just above regarding laws (and hence modern metaphysics) as being imminent to reality and the ancient notion of Ideas as being above sensible reality, it seems he inadvertently reverses his attribution of modern and ancient metaphysics.

214 For more on the relation between truth and reality, cf. above, 316.

215 Here we arrive at the third major passage on the conception of truth that Bergson attacks throughout these pages. Cf. above, 320, 328.

216 The opposite of this "so-called empiricism," of course, would be a "true" or "genuine" empiricism (see IM, 196–97), which Bergson is proposing (one that would account for "integral experience," *ibid.*, 227), or again the "radical empiricism" of James (cf. Bergson, "Sur le pragmatisme de William James," in *Mélanges*, 241).

217 Cf. above, 11–44, 177, 224.

218 For more on the resolution into elements, see above, 347–48. For the idea of an "integral translation," see above, 348–50.

219 For the idea of an equivalence between the brain state and the psychological state, see above, 181–82 and editorial notes 257–59 to Chapter II.

220 Bergson insists upon the fact that contemporaneous parallelism and epiphenomenalism are deeply dependent upon seventeenth-century metaphysics. He discusses this theme, in nearly identical terms, in "L'âme et le corps" (AC, 38–41) and in "Le cerveau et la pensée" (in L'énergie spirituelle, 192–93). And this theme serves as a leitmotif for the final three lectures in his course Histoire des théories de la mémoire.

221 For analogies of this type, cf. above, 26; MM, 5–6; AC, 36–37; "Le cerveau et la pensée," in L'énergie spirituelle, 209–10.

222 Bergson's allusion to a previous work is to Matière et mémoire.

223 Haeckel brandished the term "monism" like a flag, going so far as to form a "Monist League," which was as much political as it was scientific, and to which several of the scholars named by Bergson in the first two chapters belonged (see also editorial note 104 to Chapter I). Supporters of this apparently reductionist thinking included philosophers Hermann Lotze (1817–1881), Gustav Theodor Fechner (1801–1887), and Wilhelm Wundt (1832–1920), who were to varying degrees influenced by Leibniz. Bergson's account here of the monadological turn of "monism" primarily targets Lotze (cf. E, 69; MM, 51). Fechner can be seen as a target of Bergson's earlier work (see E, 45–54), while Wundt's name also appears a number of times in Bergson's work (E, 16, 32, 69; MM, 109n3, 133n4, 143n1). Bergson had already associated these authors with Leibniz as early as 1894: "Leçons d'histoire de la philosophie moderne et contemporaine," in Cours, vol. 3, Leçons d'histoire de la philosophie moderne: Théories de l'âme, ed. Henri Hude (Paris: Presses Universitaires de France, 1995), 126).*

224 Bergson is thinking of philosophically oriented physicians or materialists such as Julien Offray de La Mettrie (1709–1751), Claude Adrien Helvétius (1685–1755), Charles Bonnet (1720–1793), and Pierre-Jean-Georges Cabanis (1757–1808) (see AC, 38–41; "Le cerveau et la pensée," in L'énergie spirituelle, 192–93). As is typical, Bergson condenses, into a single phrase, long and very precise developments that he had presented to his audience at the Collège de France. Bergson discusses, for instance, Condillac, Bonnet, La Mettrie, Cabanis, Taine, Hartley, Bain, and Wolff in Histoire des théories de la mémoire, 321–37. Bergson is committed to bringing out the debt, acknowledged or not, of each of these thinkers with regard to Cartesian philosophy, Spinozism, or Leibnizian metaphysics.*

225 For this eminent "being," see above, 320–22.

226 [TN: In other words, Bergson here suggests that Aristotle was most interested in the act of thinking thought and less in thought as having been taken as an object for thought.] Bergson follows the same analysis (using the very same terms) during his course from 1903–1904. See Histoire des théories de la mémoire, 266–67.

227 Cf. above, 352.

228 Cf. above, 347–48; 352–53.

229 This is a tricky point. Not only does Kant refuse to ground the *materiality* of things in the manner of the ancient philosophers (or even in the manner of Leibniz and Spinoza), namely in a transcendent principle, but he even refuses to ground their *intelligibility* (to which Leibniz and Spinoza limited themselves) through such a principle, preferring (or at least intending) to make do with human intellect alone.

230 With this somewhat elliptical and suggestive remark, Bergson is likely alluding to Fichte's account of the "self" in the *Doctrine of Science*. For more on this aspect of Fichte's doctrine, and on Bergson's interpretation of him, see Jean-Christophe Goddard, "Introduction à Fichte," in *La destination de l'homme*, by Johann Gottlieb Fichte, trans. Jean-Christophe Goddard (Paris: Flammarion, 1995), 9–13; and Jean-Christophe Goddard, "Bergson: Une lecture néoplatonicienne de Fichte," *Les études philosophiques* 4, no. 59, Bergson et l'idéalisme allemand (October–December 2001): 465–77.*

231 This is Kant's general project in the section "The Transcendental Aesthetic," in his *Critique of Pure Reason*.

232 Cf. above, 344–46.

233 For these "two forms of reality," cf. above, 348–50.

234 This, of course, characterizes a major part of Bergson's own project. See above, 200, 207; MM, 203–6; MRV, 21–22.

235 This is the second of three missed opportunities where philosophy could have learned all there was to learn from the upheaval in science at the arrival of modernity, and that might have thereby led to the development of an intuitive philosophy, i.e., a "metaphysics" in the Bergsonian sense of the term. Cf. above, 341–46; below, 362–63.

236 This question regarding the genesis of the intellect is reminiscent of earlier passages devoted to Fichte and Spencer, above, 188–91.

237 Bergson is here paraphrasing Kant, *Critique of Pure Reason*, B19–21.

238 As early as Chapter III above, Bergson proposes the distinction between different domains of reality as an alternative to Kant's relativism (cf. above, 199). These paragraphs on Kant resonate with Bergson's argument from above, 191–201.*

239 Cf. above, 191, 207.

240 See Kant, *Critique of Pure Reason*, A19–22/B33–36.

241 On the "intellectual intuition" that Kant rejects, cf. PC, 154–57.

242 Cf. above, 298 and "intuition, vocabulary of" in the Index.

243 On this notion of "extension," cf. above, 347–50; 353–54.

244 For what is shared between sensible intuition and the intellect (or in Kant's terms, the understanding), see "intellect, direction of" in the Index.

245 This is established in Chapter III. Cf., in particular, above, 188, 207.

246 For this critique of Kant's conception of time, cf. above, 207; E, 174–80; MM, 208, 237; PC, 155–57.

247 For more on the geometry immanent to space, cf. above, 209–21. For this general doctrine, cf. above, 203–5.

248 Bergson adopts this conception of experience (cf. Bergson, "Fantômes et vivants," in *L'énergie spirituelle*, 67–70), which implies that "an existence can only be given in an experience" (DPP, 50–51; cf. above, 36–37, 289–90; DS, 255). But he refuses the Kantian correlate, namely the distinction between, on the one hand, the categories and time (space presents different problems) as *a priori* conditions of representation, and, on the other hand, existence as pure position.

249 For Bergson's argument against a non-temporal intuition, cf. DPP, 25–27; PC, 153–54; IM, 194–96.

250 For more on the critique of "dogmatic metaphysics," see above, 276–77. For more on the relationship between consciousness and life, see Bergson's 1911 essay, "La conscience et la vie" (CV), 1–28.

251 For this Bergsonian inversion, which is analogous to the one between movement and positions, see editorial note 57 to Chapter I.

252 This is likely an allusion to the pantheism controversy, which began in Germany in 1785 with the publication of *Concerning the Doctrine of Spinoza in Letters to Herr Moses Mendelssohn* by Friedrich Jacobi (1743–1819). Jacobi attacked modern rationalism, primarily in the figures of Gotthold Ephraim Lessing and Moses Mendelssohn, accusing them of leading inevitably to Spinozism (interpreted as radically deterministic). Kant, but also later Fichte, Schelling, and Hegel, were all forced, each in their own way, to take a position with regard to Spinozism. See Jacobi, *Concerning the Doctrine of Spinoza in Letters to Herr Moses Mendelssohn* [1785], in *The Main Philosophical Writings and the Novel* Allwill: *Friedrich Heinrich Jacobi*, trans., George di Giovanni, 173–253 (Montreal: McGill-Queens University Press, 1994).

253 For Bergson, this notion of a unified science has its roots in Plato. Cf. above, 320, 328, 353.

254 The notion of an "Idea" here is an allusion to Hegel [rather than Plato], while that of the "Will" is an allusion to Schopenhauer. Cf. DPP, 49.

255 This paragraph bears on what is traditionally called German Idealism—notably, Fichte, Schelling, and Hegel. Bergson also associates Spinoza and Schopenhauer with at least the spirit of that movement (DPP, 49). It is possible to read Bergson's "Introduction à la métaphysique" (1903) as in part a critique of German Idealism (cf. IM, 177n1, 194–95).*

256 This claim alludes to Bergson's use of the metaphor of "made-to-measure" garments. See editorial note 132 to Chapter I. [TN: Also implicit here is Bergson's criticism of Kant's relativism versus the kind of absolute knowledge involved in intuition.]

257 For the use of the word "detail" in this sentence, cf. above, 69.

258 It is worth recalling that Bergson was a member of the *Académie des sciences morales et politiques*, where he had succeeded Ravaisson. In terms of Bergson's understanding of the place of embryology in the biological sciences, cf. above, 23–24, 36.

259 This sentence characterizes Bergson's own goal in Chapter III (see above, 187–93). As such, we can understand here in what sense Bergson wanted to succeed where Spencer failed—namely by developing a "true evolutionism"

(above, x; below 367, 369)—as well as in what sense Chapter III is crucial to Bergson's project in *L'évolution créatrice*.

260 This is the third (of three) opportunities where, according to Bergson, philosophy might have discovered the need to break with the cinematographic mechanism of the intellect and with ancient thought so as to develop an intuitive philosophy. The other two are discussed above, 341–46, 357–58.*

261 For this puzzle example, cf. above, 339–40.

262 Bergson is perhaps thinking of this definition from Spencer: "Evolution is always an integration of Matter and dissipation of Motion" (Spencer, *First Principles*, 542). This passage comes from Spencer's conclusion, but the terminology used here in fact occurs several times in the book. Cf. also editorial note 136 to Chapter I.

263 Bergson is alluding to Thomson and Faraday here. See "physics, Thomson and Faraday" in the Index.

264 Cf. *MM*, Chapter I.

265 Cf. above, 111–12. See also Spencer, *Principles of Psychology*, particularly Part IV, Chapters 4, 5, and 9.*

266 Similar images relating to a "fluid" reality also appear in *Matière et mémoire*, although Bergson's intended meaning there is somewhat different. Cf. *MM*, 67, 135.

267 For this argument about determinate mechanisms, cf. above, 109, 111–12, 126–27; *MM*, 24–27; *CV*, 9–10.

268 For the distinction between the brain and the spinal cord, cf. *MM*, 17–19; *CV*, 8–9.

269 Cf. above, 154, 189. By seeking a justification for Spencer's philosophy in terms of the sequential questions of matter, mind, and the "correspondence" between the two, Bergson summarizes the different points of his own positive position as presented in Chapter III. There it was a question of establishing the simultaneous genesis of the mind (being determined as intellect) and of matter, whereas Spencer is rendered incapable of doing this for the reasons explained in these paragraphs as well as for other reasons explained earlier (cf. above, 187–93).*

270 See Spencer, *Principles of Psychology*, Part IV, Chapter 1 ("The Nature of Intelligence").

271 Rather than this correspondence between thought and reality, Bergson argues that there was a reciprocal adaptation between the two. His arguments here also target Kant, for whom there is "no kinship" between our intellect and the sensible manifold or "matter of knowledge" that is presented to it (cf. above, 358).

272 Cf. above, 190; *MM*, 176–78; *Le rire*, 116.

273 An alternative form of this argument is already directed at Spencer in Chapter III, specifically in the section where Bergson clarifies his own guiding principle. See above, 188–89.

274 In these lines, Bergson formulates his interpretation of the equivalency between Spencer's and Kant's positions.

275 For the expression "true evolutionism," which captures Bergson's overall project, see above, x, below, 369.

276 Bergson usually introduces Thomson and Faraday in these terms, but in the current context he is likely referring to the birth of quantum mechanics. This would be the first reference to this new science in Bergson's publications, a development that was to be received positively in his later published works (cf. DPP, 61n1, 77–78; PR, 100). His interest in these theories, however, can be found as early as his 1904–1905 course at the Collège de France. See *L'évolution du problème de la liberté*, 164. He was interested in how this theory might allow physics to recognize indetermination in the things themselves. He would, however, later admit in 1937 that it was likely that a new determinism had come to obscure the indetermination that had been discovered. See Bergson, "Message au congrès Descartes [1937]," in *Mélanges*, 1578.*

277 Cf. *MM*, 220–26. This formulation recalls almost precisely a phrase from Faraday's article already cited in Bergson's note, above, 204. See Faraday, "A Speculation Touching Electric Conduction and the Nature of Matter," 1414.

278 For the respective formulations that Bergson mentions here, cf. above 187, 250, 299.

279 For Bergson's internal reference regarding geometry, cf. above, 221, 231. For the image of ascending and descending movements, cf. above 339, 342.

280 Here Bergson is drawing on the vocabulary of intuition, which again appears in his use of "inserting oneself" [s'insérer dans] just below. See "intuition, vocabulary of" in the Index.

281 For Bergson's thoughts on the "proper function of philosophy," which includes taking us beyond the limits of the human condition, cf. above, 29–30, 193.

CORRESPONDENCE, RECEPTION, AND COMMENTARIES

Introduction

1 Henri Bergson, *L'évolution créatrice* [1907], ed. Arnaud François, series ed. Frédéric Worms (Paris: Presses Universitaires de France, 2013).

2 *Ibid.*, 373–88.

Correspondence

James–Bergson Correspondence

1 TN: William James, *Pragmatism: A New Name for Some Old Ways of Thinking; Popular Lectures on Philosophy* (New York: Longmans, Green, and Co., 1907).

2 TN: Henri Bergson, "Introduction: Vérité et réalité," in *Le pragmatisme*, trans. E. Le Brun (Paris: Ernest Flammarion, 1911), 1–16. This preface was later printed as "Sur le pragmatisme de William James: Vérité et réalité," in *La pensée et le mouvant*, 239–51.

3　TN: This letter is from *The Letters of William James*, ed. Henry James (Boston: The Atlantic Monthly Press, 1920), 290–94.

4　TN: Bergson, Letter to James, June 27, 1907, in "William James et M. Henri Bergson: Letters (1902–1910)," *Revue des deux mondes* (October 1933): 808–9. This also appears in *Mélanges*, 726–27.

5　*Pragmatism*, 257. [TN: Bergson quotes this sentence in English, and it is in italics in the original.]

6　TN: Bergson quotes and italicizes this phrase in English, and it is found unitalicized on page 226 of *Pragmatism*.

Letter to H. Wildon Carr

1　H. Wildon Carr, "Bergson's Theory of Knowledge," *Proceedings of the Aristotelian Society, 1908–1909*, New Series, 9 (1908–1909): 41–60.

2　TN: For Arnaud François's full introduction, see *L'évolution créatrice*, 633–34.

3　*Ibid.*, 45–46.

4　*Ibid.*, 47–48.

5　*Ibid.*, 49.

6　*Ibid.*, 53–57.

7　*Ibid.*, 57–59.

8　*Ibid.*, 43.

9　TN: This extract of a letter is published at the end of Carr's article, "Bergson's Theory of Knowledge," 59–60.

10　*Ibid.*, 47.

Letter to F. Znaniecki

1　TN: Henri Bergson, *Ewolucja twórcza*, trans. Florian Znaniecki (Warsaw: Gebethner and Wolf, 1913).

2　TN: Bergson perhaps also had in mind here a negative review of Mitchell's English translation from 1911. See J. Solomon, Review of *Creative Evolution*, *Mind* 20, no. 79 (July 1911): 432–33.

3　TN: Henri Bergson, "Bergson à F. Znaniecki," in *Mélanges*, 960.

Critical Reception in Biology

Bergson's Biology (Le Dantec)

1　François's note: Félix Le Dantec, "La biologie de M. Bergson," *Revue du mois* 4 (August 10, 1907): 230–41.

2　À Paris, chez Félix Alcan.

3　TN: Le Dantec is referring to Bergson's "Introduction à la métaphysique." See also IM.

4　In relation to what? Monsieur Bergson does not tell us.

5　Same comment.

6   TN: IM, 177–78; "Introduction to Metaphysics," in *The Creative Mind*, 133–34. Le Dantec adds italics to "despite apparent differences" and "effort of imagination."

7   TN: This is a loose quotation from IM, 181.

8   TN: This sarcastic quotation mark is in the original.

9   See above, v.

10 Félix Le Dantec, *Les influences ancestrales* (Paris: Ernest Flammarion, 1904), 134. See also Félix Le Dantec, *Les lois naturelles* (Paris: Félix Alcan, 1904).

11 Félix Le Dantec, *Le conflit: Entretiens philosophiques* (Paris: Armand Colin, 1901).

12 Félix Le Dantec, "L'ordre des sciences," *Revue philosophique de France et de l'étranger* 64 (July 1907): 1–21; (August 1907): 248–71.

13 This is a precise summary of the theory I developed in *Les influences ancestrales* and in *Traité de biologie* (Paris: Félix Alcan, 1903).

14 Nevertheless, this is what happens in photography.

15 Here we see the true difference between Bergson and mechanism.

16 See, nevertheless, the error that we can make by placing ourselves (in thought) into an inert moving object in order to grasp movement in itself.

17 Witness the law of habit, the most fundamental law of biology and the most familiar one to men. [TN: This passage is from *L'évolution créatrice*, vi.]

18 TN: See Bergson, *L'évolution créatrice*, xi.

19 *Ibid.*, 91–92.

20 I attempted to demonstrate the artificial aspect of these methods of analysis in my *Éléments de philosophie biologique* (Paris: Félix Alcan, 1907).

21 *L'évolution créatrice*, ix.

22 *Ibid.*, ix-x.

23 Félix Le Dantec, *La lutte universelle* (Paris: Ernest Flammarion, 1906).

24 *L'évolution créatrice*, 252.

25 *Ibid.*, 89–90.

26 *Ibid.*, 88.

27 But if it does not require such a representation, then why not simply use objective language when one narrates evolution? Why introduce these finalist and subjective considerations into it?

28 Given that the individuals along these lines do not die along the way; this is something that we can only know *après coup*.

29 *Ibid.*, 97.

30 Gaston Rageot, "*L'Évolution créatrice* d'après H. Bergson [Review]," *Revue philosophique de la France et de l'étranger* 64, no. 7 (1907): 73–85.

Letter to the Editor (Bergson)

1   Henri Bergson, Letter to the Editor (Émile Borel), August 20, 1907, *Revue du mois* 4 (September 10, 1907): 351–54. Reprinted in *Mélanges*, 731–35.

2   TN: IM, 178; "Introduction to Metaphysics," in *The Creative Mind*, 133. Note that Bergson here only loosely quotes his own text. I have reflected Bergson's changes in this translation.

3 TN: Again, this is more of a paraphrase of Le Dantec than a direct quote. See above, 333.

4 TN: Henry More, Letter to René Descartes, March 5, 1649. Bergson also cites this passage in *Matière et mémoire*, noting that More's comment is somewhat in jest (MM, 217). Bergson provides the following citation: H. Morus, *Scripta philosophica* (1679), vol. II, 248. I have used the English translation from *Matter and Memory*, 194.

5 TN: Bergson is here paraphrasing Le Dantec's argument.

6 TN: See above, 334.

7 TN: See *MM*.

Bergson and the Ammophila Sphex (Ruyer)

1 Posteraro's note: Bertrand Russell, *The Analysis of Mind* (New York: Macmillan, 1921), 56.

2 TN: Tano S. Posteraro and Jon Roffe, "Instinct, Consciousness, Life: Ruyer contra Bergson," *Angelaki* 24, no. 5 (October 2019): 124–47. For additional introductory information, see in particular 128–34.

3 Posteraro's note: This article originally appeared in French in 1959, as "Bergson et le Sphex ammophile," *Revue de métaphysique et de morale* 64, no. 2 (1959): 163–79. The title is somewhat confusing. Ammophila is the name of a genus in the thread-waisted wasp family; Sphex is the name of another.

4 *L'évolution créatrice*, 173–74. [TN: The cited passage is where Bergson summarizes Jean-Henri Fabre, *Nouveaux souvenirs entomologiques*, 14ff.]

5 G. W. Peckham and E. G. Peckham, *Wasps, Social and Solitary*, 28.

6 *L'évolution créatrice*, 169–70.

7 Posteraro's note: Ruyer seems to run together the following two passages: (1) "the instinct is everywhere complete, but it is more or less simplified and above all simplified *in diverse ways*"; (2) "Rather, it is much more likely that we find ourselves before a certain *musical theme* that would first be itself transposed, as a whole, into a certain number of keys, and then upon which, also as a whole, diverse variations would have been executed" (*Ibid.*, 172–73).

8 Cf. Niko Tinbergen, *L'étude de l'instinct* (Paris: Payout, 1953), 280 [*Social Behaviour in Animals, with Special Reference to Vertebrates* (London: Methuen, 1962)].

9 *L'évolution créatrice*, 174–75.

10 Bergson, *MM*, 41 [TN: *Matter and Memory*, 43. Ruyer has modified Bergson's words in the cited passage]. Note that by rejecting Bergson's theory we do not return to the theory of "projection." Our thesis is that the visual sensation is in our head, and that it stays there. It has no more to be projected than to be straightened since it is the whole of our visual consciousness.

11 *L'évolution créatrice*, 168.

12 *Ibid.*

13 *Ibid.*

14   TN: Ruyer is here freely combining phrases and paraphrasing from *L'évolution créatrice*, 173–74.

15   *Ibid.*, 175. [TN: Ruyer is here freely adjusting Bergson's text to fit his example.]

16   *Ibid.* [TN: Again, Ruyer is here freely adjusting Bergson's text to fit his example.]

17   *Ibid.*

18   Cf. especially Tinbergen, *L'étude de l'instinct* [*Social Behaviour in Animals*].

19   Posteraro's note: Ruyer here paraphrases Bergson from *L'évolution créatrice*, 172–73. See also Peckham and Peckham, *Wasps, Social and Solitary*, 28.

20   See *L'évolution créatrice*, 144–45.

21   Posteraro's note: Ruyer is referring to *ibid.*, 266.

22   *Ibid.*, 300.

Critical Reception in Mathematics

Geometrical Intellect (Borel)

1   Émile Borel, "La logique et l'intuition en mathématiques," *Revue de métaphysique et de morale* 15, no. 3 (May 1907): 273–83.

2   *Ibid.*, 282.

3   TN: Émile Borel, "L'évolution de l'intelligence géométrique," *Revue de métaphysique et de morale* 15, no. 6 (1907): 747–54.

4   Borel, "La logique et l'intuition en mathématiques," 273–83.

5   Paris, Alcan, 1907. [TN: This is an eliptical reference to *L'évolution créatrice*.]

6   *L'évolution créatrice*, 333–34.

7   *Ibid.*, 212.

8   Nevertheless, it is worth citing Clairaut's *Éléments de géométrie* [1741] as a (premature) attempt in this very direction. Among our contemporaries, the first that should be cited is Monsieur Méray who was an initiator. But, for several different reasons that are not important to mention here, his new geometry was hardly used for teaching; its influence was felt more among mathematicians. [TN: Alexis-Claude Clairaut, *Éléments de géométrie* [1741], new ed., ed. Honoré Regodt (Paris: Jules Delalain, 1853); Charles Méray, *Nouveaux éléments de géométrie* (Paris: F. Savy, 1874).]

9   TN: Here Borel shifts to considering more specialized mathematical questions that he addressed in his previous article "La logique et l'intuition en mathématiques." These pages are not important for understanding his exchange with Bergson in relation to *L'évolution créatrice*. Bergson does, however, praise the position taken by mathematicians Gösta Mittag-Leffler (1846–1927) and Karl Weierstrass (1815–1897) in their letters included by Borel, and, in particular, Weierstrass' suggestion that the "genuine mathematician is a poet" (cited by Borel, "L'évolution de l'intelligence géometrique," 752). As the rest of the exchange below makes clear, Borel fundamentally refuses to accept Bergson's distinction between intellect and intuition.

In Response to Borel (Bergson)

1 TN: Henri Bergson, "À propos de 'L'évolution de l'intelligence géométrique'," *Revue de métaphysique et de morale* 16, no. 1 (1908): 28–33.

2 1901–1902 and 1902–1903.

3 *Revue de métaphysique*, January 1903. See in particular pages 25–36 [IM, 211–27]. Cf. le *Bulletin de la Société française de philosophie* (June 1901): 44–45. ["Séance du 2 mai 1901: La parallélisme psycho-physique et la métaphysique positive," *Bulletin de la Société française de philosophie* 1 (June 1901): 33–71. Reprinted in *Mélanges*, 463–502.]

4 TN: Bergson is quoting this phrase from the part of Borel's article not included in this section. For Borel's citation of Weierstrass, see: Borel, "L'évolution de l'intelligence géometrique," 752. In other words, Bergson is happy to embrace a notion of an "imaginative vision" and invention in mathematical thought (i.e., the intellect), but will not concede that this is the same as what he means by "intuition," which, as he shows above, moves in the opposite direction as intellect. Cf. *L'évolution créatrice*, 240–41.

5 *Ibid.*, 212.

6 TN: Here Bergson loosely quotes his own sentence. I reflect his changes in the translation.

7 TN: This is a loose quotation of the claim Borel makes just above.

8 TN: Émile Borel, *Leçons sur les series divergentes* (Paris: Gauthier-Villars, 1901). Borel invokes this concept of a field of integration, for instance, on pages 22 and 81.]

9 And I am not even citing more recent discussions, but a special mention is owed to Dunan, Évellin, and Tannery, who were able to fully bring out the importance of Zeno's arguments.

10 TN: This quote is from Borel's article above.

Response to Bergson (Borel)

1 Émile Borel, Letter to the Editor (Xavier Léon), *Revue de la métaphysique et de morale* 16, no. 2 (March 1908): 244–45.

Critical Reception in Theology

Introduction (Bergson-Tonquédec)

1 TN: See Bergson's second response below, 400.

2 TN: Bergson's two responses were originally published (out of order) by Tonquédec at the end of the final article below. I have extracted Bergson's letters so as to recreate here the chronology of the dialogue.

3   TN: This preface (from 1912), alongside the two articles from 1908 and 1912, were all republished in Joseph de Tonquédec, *Sur la philosophie bergsonienne* (Paris: Beauchesne, 1936).

4   March 5, 1908, and February 20, 1912.

5   TN: Tonquédec is here quoting an unpublished letter from Bergson.

6   TN: That is, Bergson may not be a "monist" on some particular interpretation of the term, but he could still be a "monist" in other senses.

7   Édouard Le Roy, *Une philosophie nouvelle: Henri Bergson* (Paris: Félix Alcan, 1912).

The Order of the World (Tonquédec)

1   This article focuses on: Henri Bergson, *L'évolution créatrice* (Paris: Alcan, 1907). Unless otherwise indicated, all page references are to that work. [TN: This article originally appeared as: Joseph de Tonquédec, "Comment interpréter l'ordre du monde?" *Études par des Pères de la Compagnie de Jésus* 114 (March 1908): 577–97. As always, citations to *L'évolution créatrice* have been updated to reflect the current French pagination.]

2   Recall that for Bergson, the image is a reflective procedure for philosophical exposition. An image does not claim to define or to express the inexpressible, but it is suggestive of it. Cf. IM, 185.

3   *L'évolution créatrice*, 56.

4   *Ibid.*

5   *Ibid.*, 54, 56, 57.

6   *Ibid.*, 50, 199–200.

7   *Ibid.*, vi.

8   *Ibid.*, 225.

9   *Ibid.*, Bergson's note to page ix.

10   *Ibid.*, 165.

11   *Ibid.*, 103–6.

12   *Ibid.*, 105, 130, 51.

13   *Ibid.*, 233.

14   *Ibid.*, [233–35].

15   [*Ibid.*, 246, 261.]

16   *Ibid.*, 249.

17   *Ibid.*, 245–46, 258.

18   *Ibid.*, 270.

19   *Ibid.*, 255, 264, [270].

20   *Ibid.*, ix–x, 270–71.

21   TN: Tonquédec's note, added to the reprinted version of this article: I have attempted this analysis in a later study. See Joseph de Tonquédec, "La clef des 'Deux Sources'," *Études par des Pères de la Compagnie de Jésus*, 213 (December 1932): 516–43; 667–83.

22   *Ibid.*, 239

23 *Ibid.*, 249, 298. See also what is said in on page 35 of *Introduction à la métaphysique* concerning the various kinds of *durée*: "In advancing in the other direction [i.e., in the opposite direction of materiality], we go toward a duration which stretches, tightens, and becomes more and more intensified: at the limit would be eternity. [...] It would be a living and consequently still moving eternity where our own duration would find itself like the vibrations in light, and which would be the concretion of all duration as materiality is its dispersion." [IM, 210; "Introduction to Metaphysics," in *The Creative Mind*, 158].

24 As such, Monsieur Bergson, who has renewed so many questions, does not offer a novel response to the supreme question [...]. [TN: The author here goes into the details of a debate with Le Roy that are not necessary for the purpose of understanding the exchange between Bergson and Tonquédec.]

Bergson's First Response to Tonquédec

1 TN: This extract of a letter was included by Tonquédec himself at the end of the article below, but I have separated it to help illustrate the chronology of this exchange. It was later published in *Mélanges*, 766–77.

2 *L'évolution créatrice*, 248–51.

3 *Ibid.*, 128.

4 *Ibid.*, 276–77, 298.

Is Bergson a Monist? (Tonquédec)

1 TN: Joseph de Tonquédec, "M. Bergson est-il moniste?," *Études par des Pères de la Compagnie de Jésus* 130, no. 1 (February 1912): 506–16.

2 Tonquédec, "Comment interpréter l'ordre du monde?," 577–97.

3 *L'évolution créatrice*, 249.

4 *Ibid.*, 241–42.

5 In order to determine the integral causes of current reality, we must, according to Monsieur Bergson, combine the movement of creation with an inverse movement, which nevertheless also comes out of creation and which results in the constitution of matter.

6 *L'évolution créatrice*, 240, 11, 47.

7 *Ibid.*, 252, 248.

8 *Ibid.*, 262.

9 *Ibid.*, viii.

10 *Ibid.*, 261.

11 *Ibid.*

12 *Ibid.*, 239.

13 *Ibid.*, 10.

14 Because our materiality scatters us out into space.

15 *Ibid.*, 298.

16 *Ibid.*, 249.

17   *Ibid.* My italics.

18   TN: These fragments of Bergson's prose are drawn from *L'évolution créatrice*, 247–49 and 298.

19   *Ibid.*, 247–48.

20   See "Comment interpréter l'ordre du monde?" above, 390.

21   *Ibid.*, 246, 252.

22   *Ibid.*, 240.

23   *Ibid.*, 246.

24   *Ibid.*, 240.

25   *Ibid.*, 249.

26   *Ibid.*, 249, 250, 241.

27   Spinoza's pantheism, Spencer's evolutionary theory, Taine's monism, etc.

28   The antinomy is resolved when we consider that the unique *source of being* is necessarily interior to everything and infinitely different from everything. Interior to everything since everywhere that there is even a drop of existence, it is the source that had poured forth; infinitely different from everything since it alone is the *source*.

29   TN: "For with You is the fountain of life," Psalm 36.

30   TN: Here is where Tonquédec includes first Bergson's second letter (presented below), followed by the first letter (included above). I have again separated out Bergson's letter for clarity and to help recreate the chronology of this exchange.

Bergson's Second Response to Tonquédec

1   TN: This extract of a letter was included by Tonquédec himself in the article just above, but I have separated it to help illustrate the chronology of this exchange. It was republished in *Mélanges*, 963–64.

Notable Commentaries

Canguilhem

1   TN: For more on Canguilhem's influential role, see Paola Marrati and Todd Meyers, "Foreword: Life, as Such," in *Knowledge of Life*, by Georges Canguilhem, trans. Stefanos Geroulanos and Daniela Ginsburg (New York: Fordham University Press, 2008), vii–xii.

2   TN: Giuseppe Bianco, "Présentation du 'Commentaire' de Georges Canguilhem," in *Annales bergsoniennes*, vol. III, *Bergson et la science*, ed. Frédéric Worms (Paris: Presses Universitaires de France, 2007), 99–111.

3   TN: *Ibid.*, 102.

4   TN: Marrati and Meyers, "Foreword," ix.

5   TN: This is a translation of a selection from Georges Canguilhem, "Commentaire au troisième chapitre de *L'évolution créatrice*," *Bulletin de la Faculté des lettres*

de Strasbourg 21, nos. 5–6 (March–April 1943): 126–43; no. 7 (May 1943): 199–214. The majority of this selection is taken from the second part of this work. Additional editorial comments are from Camille Limoges's annotated version of this course, which is found in: Georges Canguilhem, *Résistance, philosophie, biologique et histoire des sciences, 1940–1965, Œuvres complètes*, IV, ed. Camille Limoges (Paris: Vrin, 2015), 111–70.

6 Chapter III of *L'évolution créatrice*, Chapter III: "On the Meaning of Life" was on the program for the *Agrégation* in philosophy for 1943, as it had also been for 1942. Having believed it necessary last year to attempt to guide the students in their preparation for reading this difficult text, I have decided to publish here (with the same pedagogical goals in mind) the essential parts of my lessons. [...]

7 TN: See the Preface by poet Paul Valéry (1871–1945) in the edition of *Charmes* that includes a commentary by Alain. Paul Valéry, *Charmes*, commentary Alain (Paris: Gallimard, 1929), 11–18.

8 TN: Letter dated March 15, 1915 to Danish philosopher Harald Høffding (1843–1931), who was the author of *La philosophie de Bergson: Exposé et critique*, trans. Jacques de Coussanges (Paris: Félix Alcan, 1916). Høffding published this from Bergson in his book. This letter also appears in Henri Bergson, *Écrits philosophiques*, ed. Frédéric Worms (Paris: Presses Universitaires de France, 2011), 441–45. For this passage, see page 442. See also Limoges's notes in Canguilhem's *Résistance, philosophie, biologique et histoire des sciences*, 113.

9 TN: Albert Thibaudet (1874–1936) was a philosopher and the author of *Trente ans de vie française*, vol. III, *Le bergsonisme*, 2 vols. (Paris: Gallimard, 1923).

10 TN: This article is included on pages 376–81.

11 TN: Floris Delattre (1880–1950), Bergson's nephew, was a translator of William James and taught languages and literature at the Faculté de Lettres de Paris. His article is: "Samuel Butler et le bergsonisme, avec deux lettres inédits d'Henri Bergson," *Revue anglo-américaine* 13, no. 5 (1936): 385–405. Jean de la Harpe (1882–1947), Professor in the Faculté des Lettres de Neuchâtel, published: "Souvenirs personnels d'un entretien avec Bergson," in *Henri Bergson: Essais et témoignages inédits*, ed. Albert Béguin and Pierre Thévenaz (Neuchâtel: La Baconnière, 1941), 362. See also Camille Limoges's footnotes in Canguilhem, *Résistance, philosophie, biologique et histoire des sciences*, 116.

12 *Henri Bergson: Essais et témoignages*, 360.

13 See Jankélévitch, *Henri Bergson* [Fr.], 183, 237, 240, 252, 282–84.

14 Cf. *MM*, 165.

15 This passage is from an article by Swiss philosopher and theologian Hans Urs von Balthasar (1905–1988): "La philosophie de la vie chez Bergson et chez les allemands modernes," in *Henri Bergson: Essais et témoignages*, 276.

16 DPP, 97 [*The Creative Mind*, 71].

17 *L'évolution créatrice*, 201.

18 *Ibid.*, 202.

19 *Ibid.*, 71.

20 CV, 20.

21   *L'évolution créatrice*, 264.

22   *MM*, 181.

23   *L'évolution créatrice*, 163.

24   *Ibid.*, 201–2.

25   CV, 16–17.

26   TN: Jules Grivet (summary), "La théorie de la personne d'après Henri Bergson," *Études par des Pères de la Compagnie de Jésus* 129, no. 16 (November 1911): 449–85. Grivet's summary also appears in *Mélanges*, 847–75.

27   TN: Bergson, "La théorie de la personne," in *Mélanges*, 849.

28   TN: *Ibid.*, 864.

29   *L'évolution créatrice*, 264.

30   *Ibid.*, 201–2.

31   TN: English translation from O. Bradley Bassler, "Motion and Mind in the Balance: The Transformation of Leibniz's Early Philosophy," *Studia Leibnitiana* 34, no. 2 (2002): 221–31, here 224.

32   See Leibniz's *Monadology*, §20. [Gottfried Wilhelm Leibniz, *Discourse on Metaphysics* and *The Monadology*, trans. George R. Montgomery (Amherst, NY: Prometheus Books, 1992), 71.]

33   *MM*, 198. [*Matter and Memory*, 177. TN: In *Matière et mémoire*, Bergson provides a reference to the third edition of Ravaisson's report: Félix Ravaisson, *La philosophie en France au XIXe siècle* [1867], 3rd ed. (Paris: Librairie Hachette, 1889), 176.]

34   Ravaisson, *Of Habit / De l'habitude*, 29; 33–35.

35   *L'évolution créatrice*, 202. [TN: Note that Canguilhem is here paraphrasing rather than quoting Bergson.]

36   *Ibid.*

37   *Ibid.*, Bergson's note to page 211.

38   Ravaisson, *Philosophie en France au XIXe siècle*, 309–10.

39   VOR, 253.

40   VOR, 275–76.

41   *L'évolution créatrice*, 203ff.

42   TN: Bergson uses *détente* on 203. Ravaisson's term "distension" appears in Bergson's note to 211, 318, and 320. Canguilhem is somewhat drawing an identity between these terms.

43   *MM*, 202, 203, 278.

44   TN: Canguilhem does not provide a specific reference for these Latin phrases, which often appear in philosophical discussions of causality. For instance, the two phrases (and the English translations included here) appear in Kant's "*Metaphysik L$_2$* [1790–1791]" lectures, see: Immanuel Kant, *Lectures on Metaphysics*, trans. and ed. Karl Ameriks and Steve Naragon (Cambridge: Cambridge University Press, 1997), 28: 573.

45   *MM*, 187, 193.

46   *MM*, 170.

47   *MM*, 187, 188.

48   *MM*, 189.

49   *MM*, 189.

50   *L'évolution créatrice*, 203–4.

51   Cf. Bergson's lecture "Le rêve," in *L'énergie spirituelle*, 95.

52   Henri Bergson, "Le souvenir du présent et la fausse reconnaissance [1908]," in *L'énergie spirituelle*, 128.

53   *MM*, 222–25.

54   *MM*, 247–48.

55   *MM*, 90–91.

56   *MM*, 90–91.

57   *L'évolution créatrice*, 203.

58   *MM*, 152, 267.

59   For example, Émile Rideau, *Les rapports de la matière et de l'esprit dans le bergsonisme* (Paris: Félix Alcan, 1932), Chapter I, and particularly page 4.

60   TN: Canguilhem appears to be referring to a passage in Jankélévitch's book *Henri Bergson* [Eng.], where Jankélévitch writes: "In a curious reversal, Bergson even ends up treating dreams, with their jumble of useless recollections, as the irruption of veritable materiality in the mind." (141)

61   *MM*, 244.

62   "Le rêve," in *L'énergie spirituelle*, 104 [*Mind-Energy*, 127]. [TN: Note that I have added quotation marks in these lines to help indicate the concepts that Canguilhem is drawing out of Bergson's lecture.]

63   *Ibid.*, 103–4; 106 [*Mind-Energy*, 126–26, 128].

64   *Ibid.*, 106 [*Mind-Energy*, 129–30].

65   *Ibid.*, 103 [*Mind-Energy*, 126].

66   TN: *...de laisser se faire l'étendue en laissant se défaire la personnalité.*

67   François's note: Leibniz speaks of those who "call the cause of evil *deficient.*" Gottfried Wilhelm Leibniz, *Theodicy: Essays on the Goodness of God, the Freedom of Man, and the Origin of Evil*, ed. Austin Farrer, trans. E. M. Huggard (London: Routledge and Kegan Paul, 1951), Part I, §20, 136. See Jankélévitch, *Henri Bergson* [Eng.], 148.

68   François's note: This course was in fact given during the 1898–1899 academic year. See further, Henri Bergson, "Cours sur Plotin," in *Cours*, vol. IV, *Cours sur la philosophie grecque*, ed. Henri Hude (Paris: Presses Universitaires de France, 2000), 17–78.

69   See "L'effort intellectuel," in *L'énergie spirituelle*.

70   TN: The theory that considers energy to be the fundamental principle of all things.

71   TN: See Bergson, "L'idée de néant" and the section "The Idea of Nothingness" in *L'évolution créatrice* above.

72   See Bergson, "Le paralogisme psycho-physiologique," 895–908. [TN: This was republished as "Le cerveau et la pensée: Une illusion philosophique," in *L'énergie spirituelle*.]

73   *Ibid.*, 906 ["Le cerveau et la pensée," 206].

74 Bergson, "Le souvenir du présent et la fausse reconnaissance," in *L'énergie spirituelle*, 128.

75 *L'évolution créatrice*, 203.

76 *Ibid.*, 210–11.

77 *Ibid.*, 203.

78 *MM*, 230ff., and in particular to 244–45.

79 TN: Note here that Canguilhem rigorously distinguishes between *l'étendue* ("the extended" or "extended reality") and *l'extension* ("extension" or the activity of "extending"). I am also reflecting Canguilhem's differentiation here between *distension* and *détente* ("relaxation" or "distension"), which he connects just above in the discussion of Ravaisson.

80 *MM*, 248 [*Matter and Memory*, 220].

81 See Spinoza, "Letter 12," in *Ethics, Treatise on the Emendation of the Intellect, and Selected Letters*, 267–71.

82 *L'évolution créatrice*, 157[–58].

83 See *L'évolution créatrice*, 190; *MM*, 19ff.

84 Jean Wahl, *Vers le concret: Études d'histoire de la philosophie contemporaine* (Paris: Vrin, 1932), 131–32.

85 TN: The reference appears in Bergson's second introduction to *La pensée et le mouvant*, which was written in 1922. See DPP, 78n2.

86 TN: Canguilhem is alluding to an earlier passage from his course notes that is not included in the selection published here. He is referring to René Berthelot, *Un romantisme utilitaire*, 39–49. See Limoges's footnote in Canguilhem, *Résistance, philosophie, biologique et histoire des sciences*, 118.

87 *L'évolution créatrice*, 204.

88 *MM*, 226.

89 DPP, 76–77.

90 *L'évolution créatrice*, 205–7.

91 *Ibid.*, 205–6.

92 *Ibid.*, 162.

93 *Ibid.*, 157.

94 *Ibid.*, 203.

95 Immanuel Kant, "Inaugural Dissertation [1770]," in *Theoretical Philosophy, 1755–1770*, trans. and ed. David Walford, with Ralf Meerbote (Cambridge: Cambridge University Press, 1992), 397 [2:403].

Merleau-Ponty

1 TN: Maurice Merleau-Ponty, *Phenomenology of Perception*, trans. Donald A. Landes (London: Routledge, 2011), 57–60, 63–64, 81, 186, 444, 514n23, 543n60.

2 TN: See Donald A. Landes, "Merleau-Ponty: A Bergsonian in the Making," in *The Oxford Handbook of Modern French Philosophy*, ed. Dan Whistler and Mark Sinclair (Oxford: Oxford University Press, forthcoming).

3   TN: Maurice Merleau-Ponty, *In Praise of Philosophy* [1953], trans. John Wild and James M. Edie (Evanston: Northwestern University Press, 1963).

4   *Ibid.*, 29–30.

5   *Ibid.*, 19.

6   For more on the importance of expression in Merleau-Ponty (including in his reading of Bergson), see Donald A. Landes, *Merleau-Ponty and the Paradoxes of Expression* (New York: Bloomsbury, 2013).

7   TN: Maurice Merleau-Ponty, "Bergson in the Making," in *Signs*, trans. Richard C. McCleary (Evanston: Northwestern University Press, 1964), 182–91.

8   TN: See, for instance, *L'évolution créatrice*, 238, 248–49, 272, 342.

9   TN: See Robert Vallier, Translator's Introduction to *Nature: Course Notes from the Collège de France*, by Maurice Merleau-Ponty (Evanston: Northwestern University Press, 2003), xviii.

10 TN: Merleau-Ponty's focus on Bergson's term "discernment" is perhaps a precursor to his own identification of Bergson with "interrogation" in his posthumously published manuscript: Maurice Merleau-Ponty, *The Visible and the Invisible*, trans. Alphonso Lingis (Evanston: Northwestern University Press, 1968), Chapter 3.

11 TN: Below, 419.

12 TN: Below, 424.

13 TN: Below, 438.

14 TN: Maurice Merleau-Ponty, "La philosophie de l'existence [1959]," in *Parcours deux: 1951–1961* (Paris: Verdier, 2000), 253. See also Donald A. Landes, "Merleau-Ponty: A Bergsonian in the Making."

15 TN: Maurice Merleau-Ponty, *La Nature: Notes Cours du Collège de France*, ed. Dominique Séglard (Paris: Seuil, 1995); *Nature: Course Notes from the Collège de France*, trans. Robert Vallier (Evanston: Northwestern University Press, 2003).

16 TN: F. W. J. Schelling, *System of Transcendental Idealism* (1800), trans. Peter Heath, intro. Michael Vater (Charlottesville: University Press of Virginia, 1978).

17 Karl Jaspers, *Schelling: Größe und Verhängnis* (Munich: Piper, 1955), Chapter 3, Part 1, 122 ff. [Vallier's note, from earlier in the text: Merleau-Ponty here cites Karl Jaspers, *Schelling*. [...] At the time of its publication in 1955 (one year before [Merleau-Ponty's] course), Jaspers's book was one of the only widely available commentaries on Schelling, and has not yet been translated into English. Merleau-Ponty used it extensively in the preparation of the materials on Schelling.]

18 TN: Note that I have updated the translation of *intelligence* by "intellect" to reflect my translation of *L'évolution créatrice*.

19 Bergson, *Matter and Memory*, 38 [*Matière et mémoire* (MM), 35].

20 *Ibid.*, 43 [MM, 41].

21 *Ibid.*, 60 (translation adjusted) [MM, 61].

22 *Ibid.*, 36 [MM, 33].

23 *Ibid.*, 49 [MM, 48].

24 *Ibid.*, 38 (translation adjusted) [MM, 35].

25 *Ibid.*, 69 [MM, 71].

26 *Ibid.*, 37–38 [MM, 35].

27 *Ibid.*, 38 [MM, 35].

28 *Ibid.*, 203–4 [MM, 229].

29 *Ibid.*, 210 [MM, 236].

30 *Ibid.*, 184 (translation adjusted) [MM, 205].

31 Bergson, *L'évolution créatrice*, 225.

32 *Ibid.*, 226–27.

33 *Ibid.*, 36.

34 *Ibid.*, 16. [TN: Bergson's italics in the original were not included by Merleau-Ponty.]

35 TN: *Ibid.*, 99.

36 *Ibid.*, 142.

37 *Ibid.*, 127.

38 *Ibid.*, 128–29.

39 *Ibid.*, 130.

40 *Ibid.*, 128.

41 "From this new perspective, failure appears to be the rule, and success as both exceptional and always imperfect." *Ibid.*, 130.

42 *Ibid.*, 127–30.

43 *Ibid.*, 266.

44 *Ibid.*, 266–67. [TN: Bergson's italics in the original were not included by Merleau-Ponty.]

45 *Ibid.*, 103.

46 *Ibid.*, 127.

47 TN: *Ibid.*, 264.

48 *Ibid.*, 159–60.

49 *Ibid.*, 103–4.

50 Jankélévitch, *Henri Bergson* [Fr.], 237–38 [*Henri Bergson* [Eng.], 139–40].

51 *L'évolution créatrice*, 94.

52 *Ibid.*, 94–95.

53 Jankélévitch, *Henri Bergson* [Fr.], 235 [*Henri Bergson* [Eng.], 138].

54 *L'évolution créatrice*, 180.

55 *Ibid.* 259.

56 *Ibid.*, 270.

57 *Ibid.*, 203.

58 *Ibid.*, 240.

59 *Ibid.*, 261.

60 Bergson, VOR, 286–91 ["The Life and Work of Ravaisson," in *The Creative Mind*, 212–216]. The passage concerns the report on French philosophy.

61 *L'évolution créatrice*, 298.

62 Jankélévitch, *Henri Bergson* [Fr.], 245 [*Henri Bergson* [Eng.], 144].

63 *Ibid.*, 246 [*ibid.*, 145].

64 Cf. Jules Lachelier, *Fondement et l'induction* [1871], ed. Thierry Leterre (Paris: Agora, 1993).

65   TN: Here Merleau-Ponty is paraphrasing the citation given above: Jankélévitch, *Henri Bergson* [Fr.], 245; *Henri Bergson* [Eng.], 144.

66   *L'évolution créatrice*, 292.

67   *Ibid.*, 91–92; 93–94.

68   Jankélévitch, *Henri Bergson* [Fr.], 269 [*Henri Bergson* [Eng.], 175].

69   *L'évolution créatrice*, 298.

70   *Ibid.*, 276–77.

71   *Ibid.*, 298.

72   See Bergson, PR ["The Possible and the Real," in *The Creative Mind*, 73–86].

73   Jankélévitch, *Henri Bergson* [Fr.], 188 [*Henri Bergson* [Eng.], 122–23].

74   *Ibid.*, 297 [*ibid.*, 180–81].

75   *Ibid.*, 219 [*ibid.*, 129–30].

76   TN: *See* MRV.

77   André Malraux, *La metamorphose des dieux* (Paris: Gallimard, 1957).

Deleuze

1   TN: Anne Sauvagnargues, "Deleuze avec Bergson: Le cours de 1960 sur *L'évolution créatrice*," in *Annales bergsonniennes*, vol. 2, *Bergson, Deleuze, la phénoménologie*, ed. Frédéric Worms (Paris: Presses Universitaires de France, 2004), 151. That "marvelous reader of Bergson" is a phrase from: Alain Badiou, *Deleuze: The Clamor of Being*, trans. Louise Burchill (Minneapolis: University of Minnesota Press, 2000), 39.

2   TN: Gilles Deleuze, *Différence et répétition* (Paris: Presses Universitaires de France, 1968).

3   TN: Sauvagnargues, "Deleuze avec Bergson," 153.

4   TN: Deleuze's developing distinction between *différenciation* and *différentiation* was lost in the English translation of this course, and is difficult to detect in the English publication of his 1966 book *Bergsonism* (see Deleuze, *Bergsonism*, 96–109). It is, however, important to keep in mind even if note fully articulated. As Adrian Parr notes, "the distinction is an important ingredient of [Deleuze's] differential ontology" (Parr, "Differentiation/Differenciation," in *The Deleuze Dictionary*, ed. Adrian Parr (New York: Columbia University Press, 2005). "Differentiation" is the structure of the virtual, which (for both Deleuze and Bergson) must not be simply "undifferentiated" (like Aristotle's matter). It is a mathematical term capturing the sense of an open system (rather than a Whole, see below) that is at once virtually articulated but not closed and fixed, and able to contain incompatible possibilities. "Differenciation" is the process of actualization of the virtual (but not the "realization" of a possible or pre-existing plan). Hence why Sauvagnargues emphasizes *différenciation* even if we usually read in Deleuze "differentiation" in this 1960 course. The distinction is made explicit in *Différence et répétition* (Paris: Presses Universitaires de France, 1968), 267–74 [*Difference and Repetition*, trans. Paul Patton (New York: Colombia University Press, 1993), 206–12]. I have corrected the English translation below to show where Deleuze begins working with this distinction.

5 TN: Sauvagnargues, "Deleuze avec Bergson," 153.

6 TN: The full version of these lecture notes was published in French as Gilles Deleuze, "Cours sur le chapitre III de *L'évolution créatrice* de Bergson," ed. Anne Sauvagnargues, in *Annales bergsonniennes*, vol. 2, *Bergson, Deleuze, la phénoménologie*, ed. Frédéric Worms (Paris: Presses Universitaires de France, 2004), 166–88. The English translation reproduced with permission here originally appeared as: Gilles Deleuze, "Lecture Course on Chapter Three of Bergson's 'Creative Evolution," trans. Bryn Loban, *SubStance* 36, no. 3 (2007): 72–90. This selection reproduced here is from pages 166–75/72–80.

7 TN: Note that I have altered the English translation of *l'intelligence* from "intelligence" to "the intellect," for the same reasons outlined above in the Translator's Introduction.

8 TN: Note that the English translation had lost Deleuze's intentional capitalization in these course notes. I have corrected the capitalization throughout (especially for Matter and the Intellect).

9 TN: *L'évolution créatrice*, 193.

BIBLIOGRAPHIES

WORKS BY HENRI BERGSON[a]

Essai sur les données immédiates de la conscience [1889]. Critical edition by Arnaud Bouaniche. Series editor Frédéric Worms. Paris: Presses Universitaires de France, 2007. [*Time and Free Will: An Essay on the Immediate Data of Consciousness*. Translated by F. L. Pogson. Montana: Kessinger Publishing Company, 1910.]

Matière et mémoire [1896]. Critical edition by Camille Riquier. Series editor Frédéric Worms. Paris: Presses Universitaires de France, 2008. [*Matter and Memory*. Translated by Nancy Margaret Paul and W. Scott Palmer. New York: Macmillan, 1911; New York: Zone Books, 1991.]

Le rire: Essai sur la signification du comique [1900]. Critical edition by Guillaume Sibertin-Blanc. Series editor Frédéric Worms. Paris: Presses Universitaires de France, 2007. [*Laughter: An Essay on the Meaning of the Comic* [1911]. Translated by Cloudsley Brereton and Fred Rothwell. Los Angeles: Green Integer, 1999.]

L'évolution créatrice [1907]. Critical edition by Arnaud François. Series editor Frédéric Worms. Paris: Presses Universitaires de France, 2013. [*Creative Evolution*. Translated by Arthur Mitchell. New York: Henry Holt and Company, 1911; *Creative Evolution*. Translated by Donald A. Landes. Foreword by Elizabeth Grosz. London: Routledge, 2023].

a Bergson's works appear here in chronological order of original publication.

L'énergie spirituelle [1919]. Critical edition by Élie During, Arnaud François, Stéphane Madelrieux, Camille Riquier, Guillaume Sibertin-Blanc, Ghislain Waterlot. Series editor Frédéric Worms. Paris: Presses Universitaires de France, 2009. [*Mind-Energy: Lectures and Essays.* Translated by H. Wildon Carr. New York: Henry Holt, 1920.]

Durée et Simultanéité [1922]. Critical edition by Élie During. Series editor Frédéric Worms. Paris: Presses Universitaires de France, 2009. [*Duration and Simultaneity.* Translated by Leon Jacobson with an introduction by Herbert Dingle. Indianapolis: Bobbs-Merrill, 1965; *Duration and Simultaneity: Bergson and the Einsteinian Universe.* Translated by Leon Jacobson. Translation of supplementary material by Mark Lewis and Robin Durie. Edited and selected by Robin Durie. 2nd edition. Manchester: Clinamen Press, 1999.]

Les deux sources de la morale et de la religion [1932]. Critical edition by Frédéric Keck and Ghislain Waterlot. Series editor Frédéric Worms. Paris: Presses Universitaires de France, 2013. [*The Two Sources of Morality and Religion.* Translated by R. Ashley Audra and Cloudesley Brereton, with the assistance of W. Horsfall Carter. London: Macmillan, 1935.]

La pensée et le mouvant [1934]. Critical edition by Arnaud Bouaniche, Arnaud François, Frédéric Fruteau de Laclos, Stéphane Madelrieux, Claire Marin, Ghislain Waterlot. Series editor Frédéric Worms. Paris: Presses Universitaires de France, 2013. [*The Creative Mind.* Translated by Mabelle L. Andison. New York: Philosophical Library, 1946].

Œuvres [1959]. Edited by André Robinet, with an introduction by Henri Gouhier. 6th edition. Paris: Presses Universitaires de France, 2001.

Mélanges. Edited and annotated by André Robinet, with the collaboration of Rose-Marie Mossé-Bastide, Martine Robinet, and Michel Gauthier. Paris: Presses Universitaires de France, 1972.

Hamelin, Octave, and Henri Bergson. *Fichte: Deux cours inédits.* Edited by Philippe Soulez and Fernand Turlot. Strasbourg: Presses Universitaires de Strasbourg, 1989.

Cours. Volume I, *Leçons de psychologie et de métaphysique.* Edited by Henri Hude. Paris: Presses Universitaires de France, 1990.

Cours. Volume II, *Leçons d'esthétique & Leçons de morale, psychologie et métaphysique.* Edited by Henri Hude. Paris: Presses Universitaires de France, 1992.

Cours. Volume III, *Leçons d'histoire de la philosophie moderne: Théories de l'âme.* Edited by Henri Hude. Paris: Presses Universitaires de France, 1995.

Cours. Volume IV, *Cours sur la philosophie grecque.* Edited by Henri Hude. Paris: Presses Universitaires de France, 2000.

Correspondances. Edited by André Robinet. Paris: Presses Universitaires de France, 2002.

"Cours de Bergson sur le *De rerum originatione radicali* de Leibniz [1898]." Edited by Matthias Vollet. In *Annales bergsoniennes III: Bergson et la science,* edited by Frédéric Worms, 35–52. Paris: Presses Universitaires de France, 2007.

Écrits philosophiques. Paris: Presses Universitaires de France, 2011.

Histoire de l'idée de temps: Cours au Collège de France, 1902–1903. Edited by Camille Riquier. Series editor Frédéric Worms. Paris: Presses Universitaires de France, 2016.

L'évolution du problème de la liberté: Cours au Collège de France, 1904–1905. Edited by Arnaud François. Series editor Frédéric Worms. Paris: Presses Universitaires de France, 2017.

Histoire des théories de la mémoire: Cours au Collège de France, 1903–1904. Edited by Arnaud François. Series editor Frédéric Worms. Paris: Presses Universitaires de France, 2018.

L'idée de temps: Cours au Collège de France, 1901–1902. Edited Gabriel Meyer-Bisch. Series editor Frédéric Worms. Paris: Presses Universitaires de France, 2019.

BIBLIOGRAPHY OF WORKS CITED BY BERGSON IN *L'ÉVOLUTION CRÉATRICE*

Aristotle. *On the Heavens.* Translated by J. L. Stocks. In *The Complete Works of Aristotle,* edited by Jonathan Barnes, 447–511. The revised Oxford translation. Vol. 1. Bolligen Series LXXI.2. Princeton: Princeton University Press, 1995.

Aristotle. *On the Soul.* Translated by J. A. Smith. In *The Complete Works of Aristotle,* edited by Jonathan Barnes, 641–92. The revised Oxford translation. Vol. 1. Bolligen Series LXXI.2. Princeton: Princeton University Press, 1995.

Aristotle. *Physics.* Translated by R. P. Hardie and R. K. Gaye. In *The Complete Works of Aristotle,* edited by Jonathan Barnes, 315–446. The revised Oxford translation. Vol. 1. Bolligen Series LXXI.2. Princeton: Princeton University Press, 1995.

Baldwin, James Mark. *Development and Evolution, including Psychophysical Evolution, Evolution by Orthoplasy, and the Theory of Genetic Modes.* New York and London: Macmillan, 1902.

Bastian, H. Charlton. *Le cerveau: Organe de la pensée chez l'homme et chez les animaux.* Vol. 1, *Les animaux.* Paris: G. Baillière et cie, 1882. [*The Brain as an Organ of Mind.* London: C. Kegan Paul & Co, 1880.]

Bateson, William. *Materials for the Study of Variation, Treated with Especial Regard to Discontinuity in the* Origin of Species. London: Macmillan and Co., 1894.

Bergson, Henri. *Essai sur les données immédiates de la conscience* [1889]. Critical edition by Arnaud Bouaniche. Series editor Frédéric Worms. Paris: Presses Universitaires de France, 2007. [*Time and Free Will: An Essay on the Immediate Data of Consciousness.* Translated by F. L. Pogson. Montana: Kessinger Publishing Company, 1910.]

Bergson, Henri. "L'idée de néant." *Revue philosophique de la France et de l'étranger* 62, no. 11 (1906): 449–66.

Bergson, Henri. "Introduction à la métaphysique." *Revue de métaphysique et de morale* 11, no. 1 (January 1903): 1–25. [This article later appeared with the same title in *La pensée et le mouvant* [1934], 177–227.] Critical edition by Arnaud Bouaniche, Arnaud François, Frédéric Fruteau de Laclos, Stéphane Madelrieux, Claire Marin, Ghislain Waterlot. Series editor Frédéric Worms. Paris: Presses Universitaires de France, 2013. [*The Creative Mind*, 133–69. Translated by Mabelle L. Andison. New York: Philosophical Library, 1946].

Bergson, Henri. *Matière et mémoire* [1896]. Critical edition by Camille Riquier. Series editor Frédéric Worms. Paris: Presses Universitaires de France, 2008. [*Matter and Memory.* Translated by Nancy Margaret Paul and W. Scott Palmer. New York: Macmillan, 1911; New York: Zone Books, 1991.]

Bergson, Henri. "Le paralogisme psycho-physiologique." *Revue de métaphysique* 12, no. 6 (November 1904): 895–908. [This article was reprinted as: "Le cerveau et la pensée: Une illusion philosophique." In *L'énergie spirituelle* [1919], 191–210. Critical edition by Arnaud François, Camille Riquier, Stéphane Madelrieux, Ghislain Waterlot, Guillaume Sibertin-Blanc, and Élie During. Series edited by Frédéric Worms. Paris: Presses Universitaires de France, 2009. It appears in English as: "Brain and Thought: A Philosophical Illusion." In *Mind-Energy: Lectures and Essays*, 231–55. Translated by H. Wildon Carr. New York: Henry Holt, 1920.]

Berthold, Gottfried. *Studien über Protoplasmamechanik.* Leipzig: Verlag von Arthur Felix, 1886.

Bethe, Albrecht. *Dürfen wir den Ameisen und Bienen psychische Qualitäten zuschreiben?* Special reprint from the *Archiv für die gesammte Physiologie* 70. Bonn: Verlag von Emil Strauss, 1898.

Blaringhem, Louis. "La notion d'espèce et la théorie de la mutation, d'après les travaux de Hugo de Vries." *L'année psychologique* 12 (1905): 95–112.

Boltzmann, Ludwig. *Vorlesungen über Gastheorie*. Vol. 2, *Theorie van der Waals'; Case mit zusammengesetzten Molekülen; Gasdissociation; Schlussbemerkungen*. Leipzig: Verlag von Johann Ambrosius Barth, 1898.

Bouvier, Eugène-Louis. "La nidification des abeilles à l'air libre." *Comptes rendus hebdomadaires des séances de l'Académie des sciences* 142, séance du 7 mai (1906): 1015–20.

Brandt, Alexander. "Ueber borstenartige Gebilde bei einem Hai und eine mutmaßliche Homologie der Haare und Zähne." *Biologisches Centralblatt* 18, no. 7 (1898): 257–70.

Brown-Séquard, Charles-Édouard. "Hérédité d'une affection due à une cause accidentelle: Faits et arguments contre les explications et les critiques de Weismann." *Archives de physiologie normale et pathologique*, 5th series, 4 (1892): 686–88.

Brown-Séquard, Charles-Édouard. "Nouvelles recherches sur l'épilepsie due à certaines lésions de la moelle épinière et des nerfs rachidiens." *Archives de physiologie normale et pathologique* 2 (1869): 211–20, 422–38, 496–503.

Busquet, Paul. *Les êtres vivants: Organisation–évolution*. Paris: Georges Carré et C. Naud, 1899.

Bütschli, Otto. *Untersuchungen über mikroskopische Schäume und das Protoplasma: Versuche und Beobachtungen zur Lösung der Frage nach den physikalischen Bedingungen der Lebenserscheinungen*. Leipzig: Wilhelm Engelmann, 1892. [*Investigations on Microscopic Foams and on Protoplasm: Experiments and Observations Directed Towards a Solution of the Question of the Physical Conditions of the Phenomena of Life*. Translated by E. A. Minchin. London: Adam and Charles Black, 1894.]

Buttel-Reepen, Hugo von. "Die phylogenetische Entstehung des Bienenstaates, sowie Mitteilungen zur Biologie der solitären und sozialen Apiden." *Biologisches Centralblatt* 23, no. 1 (1903): 4–31; no. 3 (1903): 89–108; no. 4 (1903): 129–54; no. 5 (1903): 183–95.

Calkins, Gary Nathan. "Studies on the Life-History of Protozoa." *Archiv für Entwickelungsmechanik der Organismen* 15, no. 1 (1902): 139–86.

Charrin, Albert. "L'hérédité en pathologie." *Revue générale des sciences pures et appliquées* 7, no. 1 (1896): 1–7.

Charrin, Albert, and Gabriel Delamare. "Hérédité cellulaire." *Comptes rendus hebdomadaires des séances de l'Académie des sciences* 133, séance du 1er juillet (1901): 69–71.

Charrin, Albert, Gabriel Delamare, and Gustave Moussu. "Transmission expérimentale aux descendants des lésions développées chez les ascendants." *Comptes rendus hebdomadaires des séances de l'Académie des sciences* 135, séance du 21 juillet (1902): 189–91.

Cope, Edward Drinker. *The Origin of the Fittest: Essays on Evolution.* New York: D. Appleton and Company, 1887.

Cope, Edward Drinker. *The Primary Factors of Organic Evolution.* Chicago: Open Court Publishing, 1896.

Cuénot, Lucien. "La nouvelle théorie transformiste: Jäger, Galton, Nussbaum et Weismann." *Revue générale des sciences pures et appliquées* 5, no. 3 (1894): 74–79.

Cuvier, Georges. "Sur un nouveau rapprochement à établir entre les classes qui composent le Règne animal." *Annales du Muséum d'histoire naturelle* 19 (1812): 73–84.

Darwin, Charles. *De la fécondation des Orchidées par les Insectes et des bons résultats du croisement.* Translated by Louis Rérolle. 2nd edition. Paris: C. Reinwald, 1891. [*The Various Contrivances by Which Orchids are Fertilised by Insects.* 2nd revised edition. London: John Murray, 1877.]

Darwin, Charles. *Les mouvements et les habitudes des plantes grimpantes.* Translated by Richard Gordon. 2nd edition. Paris: Reinwald, 1890. [*The Movements and Habits of Climbing Plants.* 2nd edition. London: John Murray, 1875.]

Darwin, Charles. *L'origine des espèces au moyen de la sélection naturelle ou la lutte pour l'existence dans la natuxre.* 6th edition. Translated by Edmond Barbier. Paris: C. Reinwald, 1887. [*On the Origins of Species by Means of Natural Selection, Or the Preservation of Favoured Races in the Struggle for Life* [1859]. 6th edition, with revisions and corrections. London: John Murray, 1872.]

Dastre, Albert. *La Vie et la Mort.* Paris: Ernest Flammarion, 1903. [*Life and Death.* Translated by W. J. Greenstreet. London: Walter Scott, 1911.]

Delage, Yves. "La conception polyzoïque des Êtres." *Revue scientifique* (*Revue rose*), 4th series, 5, no. 21 (May 23, 1896): 641–53.

Delage, Yves. *L'hérédité et les grands problèmes de la biologie générale.* 2nd edition. Paris: Schleicher Frères, 1903.

[Descartes, René. *Principles of Philosophy.* In *The Philosophical Writings of Descartes.* Vol. 1. Translated by John Cottingham, Robert Stoothoff, and Dugald Murdoch. Cambridge: Cambridge University Press, 1985.]

De Vries, Hugo. *Die Mutationstheorie: Versuche und Beobachtungen über die Entstehung von Arten im Pflanzenreich.* Vol. 1, *Die Entstehung der Arten durch Mutation.* Leipzig: Verlag von Veit, 1901; Vol. 2, *Elementare Bastardlehre.* Leipzig: Verlag von Veit, 1903. [*The Mutation Theory: Experiments and Observations on the Origin of Species in the Vegetable Kingdom.* Translated by J. B. Farmer and A. D. Darbishire. Vol. 1, *The Origin of Species by Mutation.* Chicago: The Open Court Publishing Company, 1909; Vol. 2,

The Origin of Varieties by Mutation. Chicago: The Open Court Publishing Company, 1910.]

De Vries, Hugo. *Species and Varieties: Their Origin by Mutation*. Edited by Daniel Trembly MacDougal. Chicago: The Open Court Publishing Company, 1905.

Driesch, Hans. *Die Lokalisation morphogenetischer Vorgänge: Ein Beweis vitalistischen Geschehens*. Leipzig: Wilhelm Engelmann, 1899. [A shortened and revised version appears in: *The Science and Philosophy of the Organism*. Vol. 1, *The Gifford Lectures Delivered Before the University of Aberdeen in the Year 1907*. London: Adam and Charles Black, 1908.]

Driesch, Hans. *Naturbegriffe und Natururteile: Analytische Untersuchungen zur reinen und empirischen Naturwissenschaft*. Leipzig: Wilhelm Engelmann, 1904. [An extensively revised version appears here: *The Science and Philosophy of the Organism*. Vol. 2, *The Gifford Lectures Delivered Before the University of Aberdeen in the Year 1908*. London: Adam and Charles Black, 1908.]

Driesch, Hans. *Die organischen Regulationen: Vorbereitungen zu einer Theorie des Lebens*. Leipzig: Wilhelm Engelmann, 1901. [A shortened and revised version appears here: *The Science and Philosophy of the Organism*. Vol. 1, *The Gifford Lectures Delivered Before the University of Aberdeen in the Year 1907*. London: Adam and Charles Black, 1908.]

Driesch, Hans. *Der Vitalismus als Geschichte und als Lehre*. Leipzig: Johann Ambrosius Barth, 1905. [A revised and rewritten version for an English-speaking audience appears as: *The History and Theory of Vitalism*. Translated by C. K. Ogden. London: Macmillan and Co., 1914.]

Du Bois-Reymond, Emil. *Über die Grenzen des Naturerkennens: Die sieben Welträthsel: Zwei Vorträge*. 5th edition. Leipzig: Verlag von Veit, 1882. ["The Limits of Our Knowledge of Nature." Translated by J. Fitzgerald. *Popular Science Monthly* 5 (May 1874): 17–32.]

Duhem, Pierre. *L'évolution de la mécanique*. Paris: Librairie Scientifique A. Hermann, 1905. [Pierre-Marie-Maurice Duhem, *The Evolution of Mechanics*. Translated by Michael Cole. Alphen aan den Rijn: Sijthoff and Noordhoff, 1980.]

Dunan, Charles. "Le problème de la vie." *Revue philosophique de la France et de l'étranger* 33, no. 1 (January 1892): 1–35; no. 2 (February 1892): 136–63; no. 5 (May 1892): 519–48.

Eimer, Theodor. *Die Entstehung der Arten auf Grund von Vererben erworbener Eigenschaften nach den Gesetzen organischen Wachsens*. Vol. 1, *Ein Beitrag zur einheitlichen Auffassung der Lebewelt*. Jena, Germany: Fischer, 1888. [*Organic Evolution as the Result of the Inheritance of Acquired Characteristics*

According to the Laws of Growth. Translated by J. T. Cunningham. London: Macmillan and Co., 1890.]

Eimer, Theodor. *Die Entstehung der Arten auf Grund von Vererben erworbener Eigenschaften nach den Gesetzen organischen Wachsens*. Vol. 2, *Orthogenesis der Schmetterlinge: Ein Beweis bestimmt gerichteter Entwickelung und Ohnmacht der natürlichen Zuchtwahl bei der Artbildung, zugleich eine Erwiderung an August Weismann*. Edited by C. Fickert. Leipzig: Engelmann, 1897. [A partial translation appears in English: *On Orthogenesis and the Impotence of Natural Selection in Species-Formation*. Translated by Thomas J. McCormack. Chicago: The Open Court, 1898.]

Évellin, François. *Infini et quantité: Étude sur le concept de l'infini en philosophie et dans les sciences*. Paris: Librairie Germer Baillière et cie., 1880.

Évellin, François. Letter to Ribot, April 22, 1881. *Revue philosophique de la France et de l'étranger* 11, no. 5 (May 1881): 564–68.

Fabre, Jean-Henri. *Nouveaux souvenirs entomologiques*. Paris: Delagrave, 1882.

Fabre, Jean-Henri. *Souvenirs entomologiques: Études sur l'instinct et les mœurs des insectes*. Première série [1879]. 3rd edition. Paris: Delagrave, 1894.

Fabre, Jean-Henri. *Souvenirs entomologiques: Études sur l'instinct et les moeurs des insectes* [1886]. Troisième série. 2nd edition. Paris: Delagrave, 1890.

Faraday, Michael. "A Speculation Touching Electric Conduction and the Nature of Matter [Letter to the Editor]." *The London, Edinburgh, and Dublin Philosophical Magazine and Journal of Science*, series 3, 24, no. 157 (January–June 1844): 136–44.

Fischel, Alfred. "Ueber die Regeneration der Linse." *Anatomischer Anzeiger: Centralblatt für die gesamte wissenschaftliche Anatomie. Amtliches Organ der anatomischen Gesellschaft* 14 (1898): 373–80.

Forel, Auguste. "Un aperçu de psychologie comparée." *L'année psychologique* 2 (1895): 18–44.

Foster, Michael. "Physiology: Part I.–General View." In *Encylopædia Britannica: A Dictionary of Arts, Sciences and General Literature*. Vol. 19, *PHY-PRO*, 8–23. 9th edition. Edinburgh: Adam and Charles Black, 1885.

Gaudry, Albert. *Essai de paléontologie philosophique: Ouvrage faisant suite aux* Enchaînements du monde animal dans les temps géologiques. Paris: Masson, 1896.

Giard, Alfred. *Controverses transformistes*. Paris: C. Naud, 1904.

Guérin, Paul. *Les connaissances actuelles sur la fécondation chez les Phanérogames*. Paris: A. Joanin, 1904.

Hartog, Marcus. "Sur les phénomènes de reproduction." *L'année biologique: Comptes rendus annuels des travaux de biologie générale* I (1895): 699–709.

Houssay, Frédéric. *La forme et la vie: Essai de la méthode mécanique en zoologie*. Paris: Schleicher Frères, 1900.

[Huxley, Thomas Henry. "*The Natural History of Creation*–By Dr. Ernest Haeckel [Review of *Natürliche Schöpfungsgeschichte*]." *The Academy* 1 (October 9, 1869): 13–14.]

Janet, Paul. *Les causes finales*. Paris: Germer Baillière, 1876. [*Final Causes*. Translated by William Affleck. Preface by Robert Flint. Edinburgh: T & T Clark, 1878.]

Jennings, Herbert S. *Contributions to the Study of the Behavior of Lower Organisms*. Washington: The Carnegie Institution of Washington, 1904.

Kant, Immanuel. *Critique of Pure Reason*. Translated and edited by Paul Guyer and Allen W. Wood. Cambridge: Cambridge University Press, 1998.

Lacombe, Paul. *De l'histoire considérée comme science*. Paris: Hachette, 1894.

Lalande, André. *La dissolution opposée à l'évolution dans les sciences physiques et morales*. Paris: Félix Alcan, 1899.

Laplace, Pierre-Simon de. *Théorie analytique des probabilités* [1812]. In *Œuvres complètes de Laplace*. Vol. 7, v–cliii. Paris: Gauthier-Villars, 1886. [*A Philosophical Essay on Probabilities*. Translated by Frederick William Truscott and Frederick Lincoln Emory. New York: John Wiley & Sons, 1902.]

Le Dantec, Félix. *L'individualité et l'erreur individualiste*. Preface by Alfred Giard. Paris: Félix Alcan, 1898.

Le Dantec, Félix. *Théorie nouvelle de la vie*. Paris: Félix Alcan, 1896.

Le Roy, Édouard. "Un positivisme nouveau." *Revue de métaphysique et de morale* 9, no. 2 (March–April 1901): 138–53.

Le Roy, Édouard. "Science et philosophie." *Revue de métaphysique et de morale* 7, no. 4 (July–August 1899): 375–425; no. 5 (September–October 1899): 503–62; no. 6 (November–December 1899): 708–31; 8, no. 1 (January–February 1900): 37–72.

Le Roy, Édouard. "Sur quelques objections adressées à la nouvelle philosophie." *Revue de métaphysique et de morale* 9, no. 3 (May–June 1901): 292–327; no. 4 (July–August 1901): 407–32.

Le Roy, Édouard, and Georges Vincent. "Sur l'idée de nombre." *Revue de métaphysique et de morale* 4, no. 5 (November 1896): 738–55.

Le Roy, Édouard, and Georges Vincent. "Sur la méthode mathématique." *Revue de métaphysique et de morale* 2, no. 5 (September–October 1894): 505–30; no. 6 (November–December 1894): 676–708.

Linden, Maria von. "L'assimilation de l'acide carbonique par les chrysalides de Lépidoptères." *Comptes rendus hebdomadaires des séances et mémoires de la Société de biologie* 59, no. 2, séance du 23 décembre (1905): 692–94.

Manacéïne, Marie de. "Quelques observations expérimentales sur l'influence de l'insomnie absolue." *Archives italiennes de biologie* 21 (1894): 322–25.

Marin, F. *"Sur l'origine des espèces."* Revue scientifique (*Revue rose*), 4th series, 16, no. 19 (November 9, 1901): 577–88.

Maupas, Émile. "Contribution à l'étude morphologique et anatomique des Infusoires ciliés." *Archives de zoologie expérimentale et générale,* 2nd series, 1 (1883): 427–664.

Metchnikov, Élie. *Études sur la nature humaine: Essai de philosophie optimiste.* Paris: Masson, 1903. [*The Nature of Man: Studies in Optimistic Philosophy.* English translation edited by P. Chalmers Mitchell. New York and London: G. P. Putnam's Sons, 1903.]

Metchnikov, Élie. "Revue de quelques travaux sur la dégénérescence sénile." *L'Année biologique: Comptes rendus annuels des travaux de biologie générale* 3 (1897): 249–66.

Minot, Charles Sedgwick. "On Certain Phenomena of Growing Old." In *Proceedings of the American Association for the Advancement of Science, 39th meeting, August 1890,* 271–89. Salem: The Permanent Secretary, 1891.

Möbius, Martin. *Beiträge zur Lehre von der Fortpflanzung der Gewächse.* Jena: Verlag von Gustav Fischer, 1897.

Morat, Jean-Pierre, and E. Dufourt. "Consommation du sucre par les muscles: Origine probable du glycogène musculaire." *Archives de physiologie normale et pathologique,* 5th series, 4 (1892): 327–36.

Morat, Jean-Pierre, and E. Dufourt. "Sur la consommation du glycogène des muscles pendant l'activité de ces organes." *Archives de physiologie normale et pathologique,* 5th series, 4 (1892): 457–64.

Morgan, Thomas Hunt. *Evolution and Adaptation.* New York and London: Macmillan, 1903.

Peckham, George Williams, and Elisabeth G. Peckham. *Wasps: Social and Solitary.* Introduction by John Burroughs. Illustrations by James H. Emerton. Westminster: Archibald Constable, 1905.

Perrier, Edmond. *Les colonies animales et la formation des organisms.* 2nd edition. Paris: Masson, 1898.

Plato. *Phaedrus.* Translated by Alexander Nehamas and Paul Woodruff. In *Complete Works,* edited by John. M. Cooper, with D. S. Hutchinson, 506–56. Indianapolis: Hackett Publishing Company, 1997.

Plato. *Timaeus.* Translated by Donald J. Zeyl. In *Complete Works,* edited by John. M. Cooper, with D. S. Hutchinson, 1224–91. Indianapolis: Hackett Publishing Company, 1997.

[Plotinus. *The Enneads.* Edited by Lloyd P. Gerson. Translated by George Boys-Stones, John M. Dillon, Lloyd P. Gerson, R. A. H. King, Andrew Smith, and James Wilberding. Cambridge: Cambridge University Press, 2018.]

Podwyssozki, W. "Compte-rendu: *Les Modifications du système nerveux central et des organes internes dans un cas de mort d'un Homme par suite d'inanition pendant 35 jours* par L. Tarakevich et S. Stchasny." *L'Année biologique: Comptes rendus annuels des travaux de biologie générale* 4 (1898): 338.

Podwyssozki, W. "Compte-rendu: *Sur la structure des colonies bactériennes* par S. Serkovski." *L'année biologique: Comptes rendus annuels des travaux de biologie générale* 4 (1898): 317.

Quinton, René. *L'eau de mer, milieu organique: Constance du milieu marin originel, comme milieu vital des cellules, à travers la série animale.* Paris: Masson, 1904.

Reinke, Johannes von. *Einleitung in die theoretische Biologie.* Berlin: Gebrüder Paetel, 1901.

Reinke, Johannes von. *Philosophie der Botanik.* Leipzig: Verlag von Johann Ambrosius Barth, 1905.

Reinke, Johannes von. *Die Welt als That: Umrisse einer Weltansicht auf Naturwissenschaftlicher Grundlage.* Berlin: Gebrüder Paetel, 1899.

Rhumbler, Ludwig. "Versuch einer mechanischen Erklärung der indirekten Zell und Kerntheilung." *Archiv für Entwickelungsmechanik der Organismen* 3, no. 4 (1896): 527–623.

[Romanes, George John. *Mental Evolution in Animals, with a Posthumous Essay on Instinct by Charles Darwin.* London: Kegan Paul, Trench, & Co., 1883.]

Roule, Louis. *L'embryologie générale.* Paris: C. Reinwald, 1893.

Salensky, W. "Heteroblastie." In *Proceedings of the Fourth International Congress of Zoology (Cambridge, 22–27 August, 1898),* edited by Adam Sedgwick, 111–18. London: C. J. Clay and Sons, 1899.

Samter, Max, and Richard Heymons. "Die Variationen bei *Artemia salina* Leach. und ihre Abhängigkeit von äußeren Einflüssen." *Abhandlungen der königlich preussischen Akademie der Wissenschaften,"* Abhandlungen nicht zur Akademie gehöriger Gelehrter: Physikalische Abhandlungen," Anhang 2 (1902): 1–62.

Saporta, Gaston de, and Antoine-Fortuné Marion. *L'évolution du règne vegetal: Les cryptogames.* Paris: Germer Baillière, 1881.

Scott, William Berryman. "On Variations and Mutations." *The American Journal of Science* 48, no. 287 (November 1894): 355–74.

Séailles, Gabriel. *Essai sur le génie dans l'art.* Paris: Germer Baillière, 1883.

Shaler, Nathaniel Southgate. *The Individual: A Study of Life and Death.* New York: D. Appleton and Company, 1900.

Shaler, Nathaniel Southgate. *The Interpretation of Nature* [1893]. Boston and New York: Houghton, Mifflin and Company, 1899.

Sigwart, Christoph. *Logik.* Vol. 1, *Die Lehre vom Urtheil, vom Begriff und vom Schluss.* 2nd edition, revised and augmented. Fribourg-en-Brisgau: Mohr,

1889. [*Logic*. Vol. 1, *The Judgment, Concept, and Inference*. 2nd edition, revised and enlarged. Translated by Helen Dendy. London: Swan Sonnenshein/New York: Macmillan, 1895.]

[Spencer, Herbert. *First Principles*. 2nd edition. London: Williams and Norgate, 1867.]

Vignon, Paul. "Recherches de cytologie générale sur les épithéliums: L'appareil pariétal, protecteur ou moteur: Le rôle de la coordination biologique." *Archives de zoologie expérimentale et générale*, 3rd series, 9 (1901): 371–715.

Voisin, Jules. *L'épilepsie*. Paris: Félix Alcan, 1897.

Voisin, Jules, and Albert Peron. "Recherches sur la toxicité urinaire chez les épileptiques." *Archives de neurologie: Revue des maladies nerveuses et mentales* 24 (1892): 178–202; 25 (1893): 65–72.

Weismann, August. *Aufsätze über Vererbung und verwandte biologische Fragen*. Jena: Verlag Gustav Fischer, 1892. [*Essays upon Heredity and Kindred Biological Problems*. Edited by Edward B. Poulton, Selmar Schönland, and Arthur E. Shipley. Oxford: Clarendon Press, 1889.]

Weismann, August. *Vorträge über Deszendenztheorie*. Vol. 2. Jena: Fischer, 1902. [*The Evolution Theory*. Translated by J. Arthur Thomson and Margaret R. Thomson. London: Edward Arnold, 1904.]

Wilson, Edmund Beecher. *The Cell in Development and Inheritance* [1896]. New York and London: Macmillan, 1897.

Wolff, Gustav. "Entwicklungsphysiologische Studien I: Die Regeneration der Urodelenlinse." *Archiv für Entwicklungsmechanik der Organismen* 1 (1895): 380–90.

ADDITIONAL WORKS CITED IN EDITORIAL ENDNOTES AND TRANSLATOR'S NOTES

Ansell-Pearson, Keith. *Thinking Beyond the Human Condition*. London: Bloomsbury, 2018.

Aristotle. *Metaphysics*. Translated by W. D. Ross. In *The Complete Works of Aristotle*, edited by Jonathan Barnes, 1552–728. The revised Oxford translation. Vol. 2. Bolligen Series LXXI.2. Princeton: Princeton University Press, 1995.

Aristotle. *Nichomachean Ethics*. Translated by W. D. Ross and revised by J. O. Urmson. In *The Complete Works of Aristotle*, edited by Jonathan Barnes, 1552–1728. The revised Oxford translation. Vol. 2. Bolligen Series LXXI.2. Princeton: Princeton University Press, 1995.

Balan, Bernard. "L'œil de la coquille Saint Jacques—Bergson et les faits scientifiques." *Raison présente*, no. 119 (1996): 87–106.

Belot, Gustave. "Une théorie nouvelle de la liberté." *Revue philosophique de la France et de l'étranger* 30 (1890): 361–92.

Bergson, Henri. "Aristotle's Concept of Place." Translated by J. K. Ryan. *Studies in Philosophy and the History of Philosophy* 5 (1970): 20–72.

Bergson, Henri. "Bergson à A. O. Lovejoy [Letter, May 10, 1911]." In *Écrits philosophiques*, 396–97. Paris: Presses Universitaires de France, 2011.

Bergson, Henri. *Cours*. Volume III, *Leçons d'histoire de la philosophie moderne: Théories de l'âme*. Edited by Henri Hude. Paris: Presses Universitaires de France, 1995.

Bergson, Henri. "Cours de Bergson sur la *De rerum originatione radicali de Leibniz* [1898]." Edited by Matthias Vollet. In *Annales bergsoniennes*. Vol. 3, *Bergson et la science*, edited by Frédéric Worms, 25–52. Paris: Presses Universitaires de France, 2007.

Bergson, Henri. *Les deux sources de la morale et de la religion* [1932]. Critical edition by Frédéric Keck and Ghislain Waterlot. Series edited by Frédéric Worms. Paris: Presses Universitaires de France, 2013. [*The Two Sources of Morality and Religion*. Translated by R. Ashley Audra and Cloudesley Brereton, with W. Horsfall Carter. London: Macmillan, 1935.]

Bergson, Henri. *Durée et Simultanéité* [1922]. Critical edition by Élie During. Series editor Frédéric Worms. Paris: Presses Universitaires de France, 2009. [*Duration and Simultaneity*. Translated by Leon Jacobson with an introduction by Herbert Dingle. Indianapolis: Bobbs-Merrill, 1965; *Duration and Simultaneity: Bergson and the Einsteinian Universe*. Translated by Leon Jacobson. Translation of supplementary material by Mark Lewis and Robin Durie. Edited and selected by Robin Durie. 2nd edition. Manchester: Clinamen Press, 1999.]

Bergson, Henri. *L'énergie spirituelle* [1919]. Critical edition by Élie During, Arnaud François, Stéphane Madelrieux, Camille Riquier, Guillaume Sibertin-Blanc, Ghislain Waterlot. Series editor Frédéric Worms. Paris: Presses Universitaires de France, 2009. [*Mind-Energy: Lectures and Essays*. Translated by H. Wildon Carr. New York: Henry Holt, 1920.]

Bergson, Henri. *Essai sur les données immédiates de la conscience* [1889]. Critical edition by Arnaud Bouaniche. Series editor Frédéric Worms. Paris: Presses Universitaires de France, 2007. [*Time and Free Will: An Essay on the Immediate Data of Consciousness*. Translated by F. L. Pogson. Montana: Kessinger Publishing Company, 1910.]

Bergson, Henri. *L'évolution du problème de la liberté: Cours au Collège de France, 1904–1905*. Edited by Arnaud François. Series editor Frédéric Worms. Paris: Presses Universitaires de France, 2017.

Bergson, Henri. *Histoire de l'idée de temps: Cours au Collège de France, 1902–1903.* Edited by Camille Riquier. Series editor Frédéric Worms. Paris: Presses Universitaires de France, 2016.

Bergson, Henri. *Histoire des théories de la mémoire: Cours au Collège de France, 1903–1904.* Edited by Arnaud François. Series editor Frédéric Worms. Paris: Presses Universitaires de France, 2018.

Bergson, Henri. *L'idée de temps: Cours au Collège de France, 1901–1902.* Edited Gabriel Meyer-Bisch. Series editor Frédéric Worms. Paris: Presses Universitaires de France, 2019.

Bergson, Henri. *Matière et mémoire* [1896]. Critical edition by Camille Riquier. Series editor Frédéric Worms. Paris: Presses Universitaires de France, 2008. [*Matter and Memory.* Translated by Nancy Margaret Paul and W. Scott Palmer. New York: Macmillan, 1911; New York: Zone Books, 1991.]

Bergson, Henri. *Mélanges.* Edited and annotated by André Robinet, with the collaboration of Rose-Marie Mossé-Bastide, Martine Robinet, and Michel Gauthier. Paris: Presses Universitaires de France, 1972.

Bergson, Henri. *Œuvres* [1959]. Edited and annotated by André Robinet. Introduction by Henri Gouhier. 6th edition. Paris: Presses Universitaires de France, 2001.

Bergson, Henri. *La pensée et le mouvant* [1934]. Critical edition by Arnaud Bouaniche, Arnaud François, Frédéric Fruteau de Laclos, Stéphane Madelrieux, Claire Marin, and Ghislain Waterlot. Series edited by Frédéric Worms. Paris: Presses Universitaires de France, 2013. [*The Creative Mind.* Translated by Mabelle L. Andison. New York: Philosophical Library, 1946.]

Berthelot, René. *Un romantisme utilitaire: Étude sur le movement pragmatiste.* Vol. 2, *Le pragmatisme chez Bergson.* Paris: Félix Alcan, 1913.

Brochard, Victor. Review of *Infini et quantité* by François Évellin. *Revue philosophique de la France et de l'étranger* 11, no. 4 (April 1881): 421–31.

Caro, Elme-Marie. *Le matérialisme et la science.* Paris: Hachette, 1867.

Clarke, Samuel, and Gottfried Wilhelm Leibniz. *The Leibniz-Clarke Correspondence: Together with Extracts from Newton's* Principia *and* Opticks. Edited by H. G. Alexander. Manchester: Manchester University Press, 1956.

Comte, August. *Premiers cours de philosophie positive: Préliminaires généraux et philosophie mathématique.* Paris: Presses Universitaires de France, 2007.

Cope, Edward Drinker. "The Foundations of Theism." *The Monist* 3, no. 4 (1893): 623–39.

Couturat, Louis. "Contre le nominalisme de M. Le Roy." *Revue de métaphysique et de morale* 7, no. 1 (January–February, 1900): 87–93

Delbœuf, Joseph. *La matière brute et la matière vivante: Études sur l'origine de la vie et de la mort*. Paris: Félix Alcan, 1887.

Deleuze, Gilles. *Bergsonism*. Translated by Hugh Tomlinson and Barbara Habberjam. New York: Zone Books, 1991.

Descartes, René. *Discourse on Method*. In *The Philosophical Writings of Descartes*. Vol. 1. Translated by John Cottingham, Robert Stoothoff, and Dugald Murdoch. Cambridge: Cambridge University Press, 1985.

Descartes, René. *Meditations on First Philosophy*. In *The Philosophical Writings of Descartes*. Vol. 2. Translated by John Cottingham, Robert Stoothoff, and Dugald Murdoch. Cambridge: Cambridge University Press, 1984.

Descartes, René. "To Mersenne, 27 February 1637." In *The Philosophical Writings of Descartes*. Vol. 3. Translated by John Cottingham, Robert Stoothoff, Dugald Murdoch, and Anthony Kenny. Cambridge: Cambridge University Press, 1991.

Dorfmeister, Georg. "Ueber den Einfluß der Temperatur bei der Erzeugung von Schmetterlingsvarietäten." *Mittheilungen des naturwissenschaftlichen Vereines für Steinmark* (1879): 3–8.

Fichte, J. G. *The Science of Knowledge*. Edited and translated by Peter Heath and John Lachs. Cambridge: Cambridge University Press, 1982.

Goddard, Jean-Christophe. "Bergson: Une lecture néo-platonicienne de Fichte." *Les Études philosophiques* 4, no. 59, "Bergson et l'idéalisme allemand" (October–December 2001): 465–77.

Goddard, Jean-Christophe. "Introduction à Fichte." In *La destination de l'homme*, by Johann Gottlieb Fichte, 9–13. Translated by Jean-Christophe Goddard. Paris: Flammarion, 1995.

Grosz, Elizabeth. *Becoming undone: Darwinian Reflections on Life, Politics, and Art*. Durham and London: Duke University Press, 2011.

Grosz, Elizabeth. "Bergson, Deleuze and the Becoming of Unbecoming." *parallax* 11, no. 2 (2005): 4–13.

Grosz, Elizabeth. *Time Travels: Feminism, Nature, Power*. Durham and London: Duke University Press, 2005.

Gueroult, Martial. "Bergson en face des philosophes." *Les Études bergsoniennes*. Vol. 5, *Bergson et l'histoire de la philosophie* (1960): 9–35.

Hamelin, Octave and Henri Bergson. *Fichte: Deux cours inédits*. Edited by Fernand Turlot and Philippe Soulez. Strasbourg: Presses Universitaires de Strasbourg, 1989.

Herring, Emily. "Henri Bergson, celebrity." *aeon*, May 6, 2019, https://aeon.co/essays/henri-bergson-the-philosopher-damned-for-his-female-fans

Huxley, Thomas Henry. *Evidence as to Man's Place in Nature*. London: Williams and Norgate, 1863.

Jacobi, Friedrich Heinrich. *Concerning the Doctrine of Spinoza in Letters to Herr Moses Mendelssohn* [1785]. In *The Main Philosophical Writings and the Novel Allwill*. Translated by George di Giovanni, 173–253. Montreal: McGill-Queens University Press, 1994.

Jankélévitch, Vladimir. *Henri Bergson* [1931/1959]. Edited by Alexandre Lefebvre and Nils F. Schott. Translated by Nils F. Schott. Durham and London: Duke University Press, 2015.

Juvenal. *The Sixteen Satires*. Translated by Peter Green. London: Penguin Books, 1998.

Kant, Immanuel. *Correspondence*. Translated and edited by Arnulf Zweig. Cambridge: Cambridge University Press, 1999.

Kant, Immanel. *Critique of Pure Reason*. Translated and edited by Paul Guyer and Allen W. Wood. Cambridge: Cambridge University Press, 1998.

Landes, Donald A. "Personality and Character: From a Privileged Image of *Durée* to the Core of a New Metaphysics." In *The Bergsonian Mind*, edited by Mark Sinclair and Yaron Wolf, 99–112. Routledge: London, 2021.

Le Dantec, Félix. "Pourquoi l'on devient vieux." *Revue philosophique de la France et de l'étranger* 43, no. 4 (avril 1897): 337–58; no. 5 (May 1897): 469–80.

Lévy-Bruhl, Lucien. Review of H. Bergson, *Essai sur les données immédiates de la conscience* (Alcan, 1889). *Revue philosophique de la France et de l'étranger* 29, no. 5 (1890): 519–38.

Locke, John. *An Essay Concerning Human Understanding*. Edited by Kenneth P. Winkler. Indianapolis: Hackett Publishing Company, 1996.

Miquel, Paul-Antoine. "De l'immanence de l'élan vital à l'émergence de la vie." *Annales bergsoniennes*. Vol. 3, *Bergson et la science*, edited by Frédéric Worms, 217–35. Paris: Presses Universitaires de France, 2007.

Perthes, Boucher de. *De la mâchoire humaine de Moulin-Quignon: Nouvelle découvertes en 1863 et 1864*. Paris: Jung-Teuttel, [1864].

Plato. *The Sophist*. Translated by Nicholas P. White. In *Complete Works*, edited by John. M. Cooper, with D. S. Hutchinson, 235–93. Indianapolis: Hackett Publishing Company, 1997.

Plotinus. *Quatrième Ennéade*. Translated by Émile Bréhier. Paris: Les Belles Lettres, 1927.

Ravaisson, Félix. *Of Habit / De l'habitude*. Translated by Clare Carlisle and Mark Sinclair. London: Continuum, 2008.

Ravaisson, Félix. *La philosophie en France au XIXe siècle*. Recueil de rapports sur le progrès des lettres et des sciences en France. Paris: Imprimeries Impériale, 1867.

Roux, Wilhelm. *Der Kampf der Theile im Organismus: Ein Beitrag zur Vervollständigung der mechanischen Zweckmässigkeitslehre*. Leipzig: Wilhelm Engelmann, 1881.

Ruyer, Raymond. "Bergson et le Sphex ammophile." *Revue de métaphysique et de morale* 64, no. 2 (1959): 163–79. ["Bergson and the Ammophila Sphex." Translated by Tano S. Posteraro, in "Instinct, Consciousness, Life: Ruyer contra Bergson," by Tano S. Posteraro and Jon Roffe. *Angelaki* 24, no. 5 (2019): 134–47. See the **Correspondence, Reception, and Commentaries** section.]

Schopenhauer, Arthur. *The World as Will and Representation*. Translated by E. F. J. Payne. Vol. 2. New York: Dover, 1966.

Séailles, Gabriel. *Léonard de Vinci, l'artiste & le savant: Essai de biographie philosophique*. Paris: Perrin, 1892.

Soulez, Philippe, and Frédéric Worms. *Bergson: Biographie*. Paris: Presses Universitaires de France, 2002.

Spencer, Herbert. *The Principles of Psychology*. London: Longman, Brown, Green, and Longmans, 1855.

Spinoza, Baruch. *Ethics, Treatise on the Emendation of the Intellect, and Selected Letters*. Translated by Samuel Shirley. Edited and introduced by Seymour Feldman. Indianapolis and London: Hackett Publishing Company, 1992.

Tannery, Paul. Letter to Ribot, April 10, 1881. *Revue philosophique de la France et de l'étranger* 11, no. 5 (May 1881): 561–63.

Weismann, August. *Die Bedeutung der sexuellen Fortpflanzung für die Selektions-Theorie*. Jena: Gustav Fischer, 1886.

Weismann, August. *The Germ-Plasm: A Theory of Heredity*. Translated by W. Newton Parker and Harriet Rönnfeldt. New York: Charles Scribner's Sons, 1893.

Weismann, August. "Ueber den Saison-Dimorphismus der Schmetterlinge." *Studien zur Descendez-Theorie*. Vol. 1, *Ueber den Saison-Dimorphismus der Schmetterlinge*. Leipzig: Wilhelm Engelmann, 1875. [*Studies in the Theory of Descent*. Translated and edited by Raphael Meldola, with a preface by Charles Darwin. London: Sampson Low, Marston, Searle, & Rivington, 1882.]

Worms, Frédéric. "La conception bergsonienne du temps." *Philosophie* 54, *Henri Bergson* (1997): 73–91.

ADDITIONAL WORKS CITED IN CORRESPONDENCE, RECEPTION, AND COMMENTARIES

Badiou, Alain. *Deleuze: The Clamor of Being*. Translated by Louise Burchill. Minneapolis: University of Minnesota Press, 2000.

Balthasar, Hans Urs von. "La philosophie de la vie chez Bergson et chez les Allemands modernes." In *Henri Bergson: Essais et témoignages inédits*, edited by Albert Béguin and Pierre Thévenaz, 364–70. Neuchâtel: La Baconnière, 1941.

Bassler, O. Bradley. "Motion and Mind in the Balance: The Transformation of Leibniz's Early Philosophy." *Studia Leibnitiana* 34, no. 2 (2002): 221–31.

Béguin, Albert, and Pierre Thévenaz, eds. *Henri Bergson: Essais et témoignages inédits*. Neuchâtel: La Baconnière, 1941.

Bergson, Henri. "À propos de 'L'évolution de l'intelligence géométrique'." *Revue de métaphysique et de morale* 16, no. 1 (1908): 28–33.

Bergson, Henri. "Cours sur Plotin." In *Cours*. Vol. IV, *Cours sur la philosophie grecque*, 17–78. Edited by Henri Hude. Paris: Presses Universitaires de France, 2000.

Bergson, Henri. *Écrits philosophiques*. Edited by Frédéric Worms. Paris: Presses Universitaires de France, 2011.

Bergson, Henri. *Ewolucja twórcza* [*Creative Evolution*]. Translated by Florian Znaniecki. Warsaw: Gebethner and Wolf, 1913.

Bergson, Henri. "Introduction: Vérité et réalité." In *Le pragmatisme*, 1–16. Translated by E. Le Brun. Paris: Ernest Flammarion, 1911.

Bergson, Henri. Letter to the Editor (Émile Borel), *Revue du mois* 4 (September 10, 1907): 351–54. Reprinted in *Mélanges*, 731–35.

Bergson, Henri. Letter to James, June 27, 1907. In "William James et M. Henri Bergson: Letters (1902–1910)." *Revue des deux mondes* (October 1933): 808–9.

Bergson, Henri. "Séance du 2 mai 1901: La parallélisme psycho-physique et la métaphysique positive." *Bulletin de la Société française de philosophie* 1 (June 1901): 33–71.

Bianco, Giuseppe. "Présentation du 'Commentaire' de Georges Canguilhem." In *Annales bergsoniennes*. Vol. III, *Bergson et la science*, edited by Frédéric Worms, 99–111. Paris: Presses Universitaires de France, 2007.

Borel, Émile. "L'évolution de l'intelligence géométrique." *Revue de métaphysique et de morale* 15, no. 6 (1907): 747–54.

Borel, Émile. *Leçons sur les séries divergentes*. Paris: Gauthier-Villars, 1901.

Borel, Émile. Letter to the Editor (Xavier Léon), *Revue de la métaphysique et de morale* 16, no. 2 (March 1908): 244–45.

Borel, Émile. "La logique et l'intuition en mathématiques." *Revue de métaphysique et de morale* 15, no. 3 (May 1907): 273–83.

Canguilhem, Georges. "Commentaire au troisième chapitre de *L'évolution créatrice*." *Bulletin de la Faculté des lettres de Strasbourg* 21, no. 5–6 (March–April 1943): 126–43; no. 7 (May 1943): 199–214. This also appears in *Annales bergsoniennes. Vol. III, Bergson et la science*, edited by Frédéric Worms, 113–60. Paris: Presses Universitaires de France, 2007.

Canguilhem, Georges. *Résistance, philosophie, biologique et histoire des sciences, 1940–1965. Œuvres complètes*, IV. Edited by Camille Limoges. Paris: Vrin, 2015.

Carr, H. Wildon. "Bergson's Theory of Knowledge." *Proceedings of the Aristotelian Society, 1908–1909*, New Series, 9 (1908–1909): 41–60.

Clairaut, Alexis-Claude. *Éléments de géométrie* [1741]. New edition. Edited by Honoré Regodt. Paris: Jules Delalain, 1853.

Delattre, Floris. "Samuel Butler et le bergsonisme, avec deux lettres inédits," *Revue anglo-américaine* 13, no. 5 (June 1936): 385–405.

Deleuze, Gilles. *Bergsonism*. Translated by Hugh Tomlinson and Barbara Habberjam. New York: Zone, 1988.

Deleuze, Gilles. "Cours sur le chapitre III de *L'évolution créatrice* de Bergson." Edited by Anne Sauvagnargues. In *Annales bergsoniennes II. Bergson, Deleuze, la phénoménologie*, edited by Frédéric Worms, 166–88. Paris: Presses Universitaires de France, 2004. The English translation is: Gilles Deleuze, "Lecture Course on Chapter Three of Bergson's 'Creative Evolution.'" Translated by Bryn Loban. *SubStance* 36, no. 3 (2007): 72–90.

Deleuze, Gilles. *Différence et répétition*. Paris: Presses Universitaires de France, 1968. [*Difference and Repetition*. Translated by Paul Patton. New York: Colombia University Press, 1993.]

Grivet, Jules (summary). "La théorie de la personne d'après Henri Bergson." *Études par des Pères de la Compagnie de Jésus* 129, no. 16 (November 1911): 449–85

Høffding, Harald. *La philosophie de Bergson: Exposé et critique*. Translated by Jacques de Coussanges. Paris: Félix Alcan, 1916.

James, William. *The Letters of William James*. Edited by his son Henry James, 290–94. Boston: The Atlantic Monthly Press, 1920.

James, William. *Pragmatism: A New Name for Some Old Ways of Thinking; Popular Lectures on Philosophy*. New York: Longmans, Green, and Co., 1907.

Jaspers, Karl. *Schelling: Größe und Verhängnis*. Munich: Piper, 1955.

Kant, Immanuel. *Lectures on Metaphysics*. Translated and edited by Karl Ameriks and Steve Naragon. Cambridge: Cambridge University Press, 1997.

Kant, Immanuel. *Theoretical Philosophy, 1755–1770*. Translated and edited by David Walford, with Ralf Meerbote. Cambridge: Cambridge University Press, 1992.

Lachelier, Jules. *Fondement et l'induction* [1871]. Edited by Thierry Leterre. Paris: Agora, 1993.

La Harpe, Jean de. "Souvenirs personnels d'un entretien avec Bergson." In *Henri Bergson: Essais et témoignages inédits*, edited by Albert Béguin and Pierre Thévenaz, 357–64. Neuchâtel: La Baconnière, 1941.

Landes, Donald A. "Merleau-Ponty: A Bergsonian in the Making." In *The Oxford Handbook of Modern French Philosophy*, edited by Dan Whistler and Mark Sinclair. Oxford: Oxford University Press, forthcoming.

Landes, Donald A. *Merleau-Ponty and the Paradoxes of Expression*. New York: Bloomsbury, 2013.

Le Dantec, Félix. "La biologie de M. Bergson." *Revue du mois*, vol. 4 (August 10, 1907): 230–41.

Le Dantec, Félix. *Le conflit: Entretiens philosophiques*. Paris: Armand Colin, 1901.

Le Dantec, Félix. *Éléments de philosophie biologique*. Paris: Félix Alcan, 1907.

Le Dantec, Félix. *Les influences ancestrales*. Paris: Ernest Flammarion, 1904.

Le Dantec, Félix. *Les lois naturelles: Réflexions d'un biologiste sur les sciences*. Paris: Félix Alcan, 1904.

Le Dantec, Félix. *La lutte universelle* (Paris: Ernest Flammarion, 1906).

Le Dantec, Félix. "L'ordre des sciences." *Revue philosophique de France et de l'étranger* 64 (July 1907): 1–21; (August 1907): 248–71.

Le Dantec, Félix. *Traité de biologie*. Paris: Félix Alcan, 1903.

Le Roy, Édouard. *Une philosophie nouvelle: Henri Bergson*. Paris: Félix Alcan, 1912.

Leibniz, Gottfried Wilhelm. *Discourse on Metaphysics* and *The Monadology*. Translated by George R. Montgomery. Amherst, NY: Prometheus Books, 1992.

Leibniz, Gottfried Wilhelm. *Theodicy: Essays on the Goodness of God, the Freedom of Man, and the Origin of Evil*. Edited by Austin Farrer. Translated by E. M. Huggard. London: Routledge and Kegan Paul, 1951.

Malraux, André. *La metamorphose des dieux*. Paris: Gallimard, 1957.

Marrati, Paola, and Todd Meyers. "Foreword: Life, as Such." In *Knowledge of Life*, by Georges Canguilhem, vii–xii. Translated by Stefanos Geroulanos and Daniela Ginsburg. New York: Fordham University Press, 2008.

Méray, Charles. *Nouveaux éléments de géométrie*. Paris: F. Savy, 1874.

Merleau-Ponty, Maurice. "Bergson in the Making." In *Signs*, translated by Richard C. McCleary, 182–91. Evanston: Northwestern University Press, 1964.

Merleau-Ponty, Maurice. *In Praise of Philosophy* [1953]. Translated by John Wild and James M. Edie. Evanston: Northwestern University Press, 1963.

Merleau-Ponty, Maurice. *Nature: Course Notes from the Collège de France*. Translated by Robert Vallier. Evanston: Northwestern University Press, 2003.

Merleau-Ponty, Maurice. *La Nature: Notes Cours du Collège de France*. Edited by Dominique Séglard. Paris: Seuil, 1995.

Merleau-Ponty, Maurice. *Phenomenology of Perception*. Translated by Donald A. Landes. London: Routledge, 2011.

Merleau-Ponty, Maurice. "La philosophie de l'existence [1959]." In *Parcours deux: 1951–1961*, 247–66. Paris: Verdier, 2000.

Merleau-Ponty, Maurice. *The Visible and the Invisible*. Translated by Alphonso Lingis. Evanston: Northwestern University Press, 1968.

Parr, Adrian. "Differentiation/Differenciation." In *The Deleuze Dictionary*, edited by Adrian Parr, 75–76. New York: Columbia University Press, 2005.

Posteraro, Tano S., and Jon Roffe. "Instinct, Consciousness, Life: Ruyer Contra Bergson." *Angelaki* 24, no. 5 (October 2019): 124–47.

Rageot, Gaston. Review of *L'évolution créatrice* by Henri Bergson. *Revue philosophique de la France et de l'étranger* 64 (1907): 73–85.

Ravaisson, Félix. *La philosophie en France au XIXe siècle* [1867]. 3rd edition. Paris: Librairie Hachette, 1889.

Rideau, Émile. *Les rapports de la matière et de l'esprit dans le bergsonisme*. Paris: Félix Alcan, 1932.

Russell, Bertrand. *The Analysis of Mind*. New York: Macmillan, 1921.

Ruyer, Raymond. "Bergson et le Sphex ammophile." *Revue de métaphysique et de morale* 64, no. 2 (1959): 163–79.

Sauvagnargues, Anne. "Cours sur le chapitre III de *L'évolution créatrice* de Bergson." Edited by Anne Sauvagnargues. In *Annales bergsoniennes*. Vol. II, *Bergson, Deleuze, la phénoménologie*, edited by Frédéric Worms, 166–88. Paris: Presses Universitaires de France, 2004.

Schelling, F. W. J. *System of Transcendental Idealism* (1800). Translated by Peter Heath. Introduction by M. Vater. Charlottesville: University Press of Virginia, 1978.

Solomon, J. Review of *Creative Evolution*. *Mind* 20, no. 79 (July 1911): 432–33.

Thibaudet, Albert. *Trente ans de vie française*. Vol. III, *Le bergsonisme*. 2 vols. Paris: Gallimard, 1923.

Tinbergen, Niko. *L'étude de l'instinct*. Paris: Payout, 1953. [*Social Behaviour in Animals, with Special Reference to Vertebrates* (London: Methuen, 1962)].

Tonquédec, Joseph de. "La clef des 'Deux Sources'." *Études par des Pères de la Compagnie de Jésus*, 213 (December 1932): 516–43; 667–83.

Tonquédec, Joseph de. "Comment interpréter l'ordre du monde?" *Études par des Pères de la Compagnie de Jésus* 114 (March 1908): 577–97.

Tonquédec, Joseph de. "M. Bergson est-il moniste?" *Études par des Pères de la Compagnie de Jésus* 130, no. 1 (February 1912): 506–16.

Tonquédec, Joseph de. *Sur la philosophie bergsonienne*. Paris: Beauchesne, 1936.

Valéry, Paul. *Charmes*. Commentary by Alain. Paris: Gallimard, 1929.

Vallier, Robert. Translator's Introduction to *Nature: Course Notes from the Collège de France*, by Maurice Merleau-Ponty, xiii–xx. Evanston: Northwestern University Press, 2003.

Wahl, Jean. *Vers le concret: Études d'histoire de la philosophie contemporaine*. Paris: Vrin, 1932.

NOTABLE REVIEWS OF AND CONTEMPORARY RESPONSES TO BERGSON'S *L'ÉVOLUTION CRÉATRICE*

Balfour, A. J. "Creative Evolution and Philosophic Doubt." *Hibbert Journal* 10, no. 1 (1911): 1–23.

Bergson's Huxley Lecture, delivered at the University of Birmingham on May 29, 1911, appears after Balfour's review: Bergson, Henri. "Life and Consciousness," *Hibbert Journal* 10, no. 1 (1911): 24–44. Balfour's article generated a series of responses: Lodge, Oliver, "Balfour and Bergson," *The Hibbert Journal* 10, no. 2 (1911): 290–307; Wolf, A. "Mr. Balfour on Teleology and Creative Evolution." *Hibbert Journal* 10, no. 2 (1911): 469–72. "Mr. Balfour's Objection to Bergson's Philosophy." *Current Literature* 51, no. 6 (December 1911): 659–61; "Mr. Balfour and M. Bergson." *Spectator* 1907, no. 4347 (September 21, 1911): 633–34.

Bode, B. H. Review: *Creative Evolution* by Henri Bergson and A. Mitchell. *The American Journal of Psychology* 23, no. 2 (1912): 333–35.

Borel, Émile. "L'évolution de l'intelligence géométrique." *Revue de métaphysique et de morale* 15, no. 6 (November–December 1907): 747–54. [See the **Correspondence, Reception, and Commentaries** section.]

Carr, H. Wildon. "II. Bergson's Theory of Knowledge." *Proceedings of the Aristotelian Society, 1908–1909*, New series, 9 (1908–1909): 41–60.

Cockerell, T. D. A. "The New Voice of Philosophy? *Dial* 51, no. 607 (October 1911): 253–55.

Husband, Mary Gilliland. *L'Évolution créatrice* par Henri Bergson. *International Journal of Ethics* 22, no. 4 (1912): 462–67.

Jaspers, Karl. Review of *L'évolution créatrice* (*Die schöpferische Entwicklung*, 1912) by Henri Bergson. *Zeitschrift für die gesamte Neurologie und Psychologie* 6 (1913): 885–86.

Jones, H. Gordon. "Bergson et l'évolution: *L'évolution créatrice*." *Revue positiviste internationale* 8, no. 5 (1913): 311–20.

Le Dantec, Félix. "La biologie de M. Bergson." *Revue du mois* 4, no. 8 (August 1907): 230–41. [Reprinted as: "Remarques sur la biologie de M. Bergson." In *Science et conscience, philosophie du XXe siècle* by Félix

Le Dantec, 217–39. Paris: Flammarion, 1908. Bergson wrote a response that appeared in the September 1907 issue of *Revue du Mois* as well as in Le Dantec's *Science et conscience, philosophie du XXe siècle*. For both Le Dantec's review and Bergson's response, see the **Correspondence, Reception, and Commentaries** section.]

Loveday, T. *L'évolution créatrice* by Henri Bergson. *Mind* 17, no. 67 (1908): 402–8.

Lovejoy, Arthur Oncken. "The Metaphysician of the Life-Force." *Nation* 89, no. 2309 (September 30, 1909): 298–301.

Mitchell, Arthur. Review of *L'évolution créatrice* by Henri Bergson. *Journal of Philosophy, Psychology, and Scientific Methods* 5, no. 22 (1908): 603–12.[a]

Pillon, François. Review of *L'évolution créatrice* by Henri Bergson. *Année philosophique 1907* 18 (1908): 182–84.

Quick, Oliver. "Bergson's 'Creative Evolution' and the Individual." *Mind* 22, no. 86 (1913): 217–30.

Rageot, Gaston. Review of *L'évolution créatrice* by Henri Bergson. *Revue philosophique de la France et de l'étranger* 64 (1907): 73–85.

Scott, J. W. "The Pessimism of *Creative Evolution*." *Mind* 22, no. 87 (July 1913): 344–60.

Sorel, Georges. Review of *L'évolution créatrice* by Henri Bergson. *Mouvement socialiste* 12, no. 191 (1907): 257–82; no. 193 (1907): 478–94; 13, no. 194 (1908): 34–52; no. 196 (1908): 184–94; no. 197 (1908): 276–94.

Thomson, John Arthur. "Biological Philosophy." *Nature* 87, no. 2189 (October 12, 1911): 475–77.

Tonquédec, Joseph de. "Comment interpréter l'ordre du monde?" *Études par des Pères de la Compagnie de Jésus* 114 (1908): 577–97. [See the **Correspondence, Reception, and Commentaries** section.]

Tonquédec, Joseph de. "M. Bergson est-il moniste?" *Études par des Pères de la compagnie de Jésus* 130, no. 1 (1912): 506–16. [See the **Correspondence, Reception, and Commentaries** section.]

Weber, Louis. Review of *L'évolution créatrice* by Henri Bergson. *Revue de métaphysique et de morale* 15, no. 5 (1907): 620–70.

a Arthur Mitchell was the first translator of *L'évolution créatrice* into English: *Creative Evolution*, trans. Arthur Mitchell (New York: Henry Holt and Co., 1911).

INDEX

As with the Editorial Notes above, the majority of page numbers in this Index refer to the French pagination of the standard edition of *L'évolution créatrice*, which are provided in the margins of this translation. I have, however, also included references to the Translator's Introduction, the *Correspondence, Reception, and Commentaries* section, and the Editorial Endnotes. To distinguish between references to Bergson's text and references to these other parts of the current edition, these latter page numbers are italicized and always appear as a group at the end of an entry or subentry.

For example, consider an entry that includes the following page numbers or ranges:

x–xi, 14, 41nC, 334–36, *xl, 394, 470n56–58, 582–83n4*

- Here the first four numbers or ranges would refer to the French pagination of *L'évolution créatrice*, included in the margins of Bergson's text above. Note, the number "41nC" refers to French page 41, footnote C in the current volume.
- The last four numbers or ranges would refer to the regular pagination of the other parts of this current edition. The numbers "*xl*" and "*394*" would refer to the pagination of the Translator's Introduction and the *Correspondence, Reception, and Commentaries* section, respectively. The range "*470n56–58*" indicates multiple consecutive endnotes (56–58) on page 470 of the current volume. The range "*582–83n4*" indicates that the relevant part of endnote 4 spans two pages.